Probability
and Stochastics Series

FRONTIERS IN QUEUEING

Models and Applications In Science and Engineering

PROBABILITY AND STOCHASTICS SERIES

Edited by Richard Durrett and Mark Pinsky

Probability
and Stochastics Series

FRONTIERS IN QUEUEING

Models and Applications In Science and Engineering

Edited by
Jewgeni H. Dshalalow

CRC Press
Boca Raton   New York   London   Tokyo

Acquiring Editor: Tim Pletscher
Assistant Managing Editor: Paul Gottehrer
Marketing Manager: Susie Carlisle
Direct Marketing Manager: Becky McEldowney
Cover Design: Dawn Boyd
Prepress: Kevin Luong

Library of Congress Cataloging-in-Publication Data

Frontiers in queueing : models and applications in science and engineering / edited by Jewgeni H. Dshalalow.
    p. cm. -- (Probability and stochastics)
  Includes bibliographical references (p. -) and index.
  ISBN 0-8493-8076-6 (alk. paper)
  1. --Queueing theory. I. Dshalalow, Jewgeni H. II. Series: Probability and stochastics series.
  T57.9.F76   1997
  519.8′2--dc21   96-49563
                                                                                      CIP

No claim to original U.S. Government works
International Standard Book Number 0-8493-8076-6
Library of Congress Card Number 96-49563
Printed in the United States of America 1 2 3 4 5 6 7 8 9 0
Printed on acid-free paper

Preface

Frontiers in Queueing, like its companion volume, *Advances in Queueing*, is written by a team of leading experts in their respective areas of research. Using a team to write a monograph within a prescribed framework is quite different from the more usual approach of editing a volume of disparate topics.

Although *Frontiers in Queueing* is a natural follow-up to the more theoretical *Advances in Queueing*, it can be read independently from the first volume. It includes a plethora of tools for the more practical reader, as well as a variety of queueing models and networks.

The book opens with a chapter by Ryszard Syski, one of the principal contributors to the contemporary theory of queueing as it is applied to telecommunications. It continues with such topics as retrial, multichannel and state dependent queues; polling, fluid and fractal models; and mobile communication and loss networks. Some topics are updates of popular surveys, and some appear for the first time. The last part of the book consists of techniques and statistical inference in queues, which completes the body of methods initiated in the first volume. Almost every chapter concludes with a list of open problems and research projects, which readers may find helpful in shaping their own research goals.

I deeply appreciate the generous efforts of all of the contributors to this volume. Their cooperation established a close working relationship that made this project a pleasure. My work would have been much more difficult without the help of my editorial assistant, Mr. Gary Russell, and impossible without the typing power and patience of Ms. Donn Miller-Harnish. My profound thanks are also due to the publisher and staff, especially Ms. Nora Konopka for her extraordinary efficiency and constant assistance, Tim Pletscher for his guidance, and Paul Gottehrer for his very thorough text proofreading and useful remarks. And last but not least, I would like to extend my gratitude to Mrs. Betsey Maupin and Professor Jay Yellen for many helpful hints while working on the final draft of the manuscript.

— Jewgeni H. Dshalalow
   Melbourne, Florida

Contributors

François Baccelli, INRIA, 2004 r-te des Lucioles, BP-93, 06902, Sophia-Antipolis, France

U. Narayan Bhat, Dean ad interim, Office of the Dean, Dedman College, Southern Methodist University, Dallas, TX 75275-0235, USA

Christos G. Cassandras, Department of Electrical and Computer Engineering, University of Massachusetts at Amherst, Amherst, MA 01003, USA

Hans Daduna, Institut für Mathematische Stochastik, Universität Hamburg, Bundesstraße 55, D-20146 Hamburg, Germany

Daryl J. Daley, Statistic Analysis Program, Centre of Mathematics and its Applications, School of Mathematical Sciences, Australian National University, Canberra ACT 0200, AUSTRALIA

Jewgeni H. Dshalalow, Department of Applied Mathematics, Florida Institute of Technology, 150 W. University Blvd., Melbourne, FL 32901, USA

Ashok Erramilli, Bellcore, NVC 2X-157, 331 Newman Spring Rd., Red Bank, NJ 07701, USA

Peter W. Glynn, Department of Operations Research, Stanford University, Stanford, CA 94305-4022, USA

Professor Yu-Chi Ho, Division of Applied Sciences, Pierce Hall, Harvard University, Cambridge, MA 02138, USA

David Jagerman, NEC USA, Inc., 4 Independence Way, Princeton, NJ 08540, USA

Hisashi Kobayashi, Department of Electrical Engineeering, Engineering Quadrangle, Princeton University, Princeton, NJ 08544, USA

Vladimir Kriman, Faculty of Industrial Engineering and Management, Technion Institute of Technology, Haifa 32000, Israel

Vidyadhar G. Kulkarni, Department of Operations Research, University of North Carolina at Chapel Hill, CB 3180, Chapel Hill, NC 27599, USA

H.M. Liang, Department of Mathematics, National Central University, Chung-Li, Taiwan

Brian L. Mark, Department of Electrical Engineeering, Engineering Quadrangle, Princeton University, Princeton, NJ 08544, USA

Benjamin Melamed, Rutgers University, RUTCOR, P.O. Box 5062, New Brunswick, NJ 08903, USA

Gregory K. Miller, Department of Statistical Science, Southern Methodist University, Dallas, TX 75275, USA

Onuttom Narayan, Department of Physics, University of California at Santa Cruz, Santa Cruz, CA 95064, USA

S. Subba Rao, Department of Information Systems and Operations Management, College of Business Administration, The University of Toledo, Toledo, OH 43606, USA

Reuven Y. Rubinstein, Faculty of Industrial Engineering and Management, Technion Institute of Technology, Haifa 32000, Israel

Ryszard Syski, Department of Mathematics, University of Maryland, College Park, MD 20742, USA

Hideaki Takagi, Institute of Socio-Economic Planning, University of Tsukuba, 1-1-1 Tennoudai, Tsukuba, Ibaraki 305, Japan

Henk C. Tijms, Department of Econometrics, Vrije University, De Boelelaan 1105, Amsterdam 1081 Hv., Holland

Marcelo Torres, Department of Operations Research, Stanford University Stanford, CA 94305-4022, USA

Walter Willinger, Bellcore, MRE 2P-372, 445 South St., Morristown, NJ 07690 -6438, USA

Serguei Zuyev, INRIA, 2004 r-te des Lucioles, BP-93, 06902, Sophia-Antipolis, France

Contents

Part I

Progress of Classical Queueing Models

Chapter 1
A personal view of queueing theory

Ryszard Syski

ABSTRACT This is a personal and highly subjective account of the development of queueing theory, reflecting the author's interest (or lack of it) in selected areas, and does not intend to be a complete representation of the field.

Two stages of development are distinguished. The first one, called Rise and Fall, concerns ideas and methods of early queueing theory. The second, called Revival and Separation, describes the recent situation of splitting queueing theory into two, namely the abstract and the practical.

CONTENTS

1.1 INTRODUCTION

Queueing theory (Q.T.), as we know it today, is over 50 years old. Although the pioneer work goes back to early years of this century, the rapid growth of Q.T. began in the early 1950s, after a slow start in the 1940s.

In the development of Q.T. it is possible to distinguish **two** stages of different length and different character. The first stage is the time of intensive activity and represents the rise and fall of Q.T. The second stage, which we witness now, represents revival and separation. It splits Q.T. into two main types according to abstract or practical interest. One can say that there are at present, two distinct and separate Q.T.'s.

The present paper is a personal and highly subjective account of the development of Q.T. It reflects the author's interest (or lack of it) in selected areas and does not intend to be a complete representation of the field. The author offers his apology to persons who may feel slighted in this account, together with the assurance that no slight has been intentional.

1.2 RISE AND FALL OF Q.T.

The first stage of development may be subdivided into three main periods (like geological epochs). The first is the period of pioneers. The names of A.K. Erlang (whose papers appeared during the period 1909-1920), E.C. Molina (1908), W.H.

Grinsted (1907), and T. Engset (1918) come immediately to mind. Many eminent writers began their work during the period 1920-1939. C. Palm (1907-1951) provided a solid theoretical background for modern theories which led to what are now Palm probabilities and regenerative processes; his *opus magnum* was recently (in 1988) translated into English [53].

Felix Pollaczek (1892-1982), whose papers appeared in the 1930s, was the first to obtain the complete solution for the GI/G/s queue. His single server theory (which he regarded later as *trivial*) was ready around 1957. Similarly, the general theory for multiserver systems was announced in 1953 and appeared in book form in 1961 [54]. There is no need to convince the reader that the work of Pollaczek exerted tremendous impact on the development of Q.T. On the occasion of his 100th birthday, a special issue of *AEÜ* (*Archiv für Elektronik und Übertragungstechnik*) honoring Pollaczek was published [62], where exposition of his work may be found in papers by workers who knew him personally. Pollaczek's method, based on multidimensional contour integrals, was at first received with polite admiration, but later its usefulness and power were established by P. Le Gall, J.W. Cohen and J.H.A. de Smit. This in turn led to a method based on boundary theory, developed by J.W. Cohen and O.J. Boxma in their book [15] published in 1983. Cohen made extensive use of this method in a series of papers devoted to random walks; see [14].

Other important writers from this period are C.D. Crommelin (delay systems, 1932-1934), E. Vaulot (first paper in 1924), A.Y. Khintchine (M/G/1 queue, 1932), L. Kosten (supplementary variables, 1937), and R. Wilkinson (teletraffic, 1931).

The application of stochastic processes marks the beginning of the second period in development of Q.T. which may be called a period of growth, lasting approximately from the 1940s to the late 1960s. The year 1939 may be regarded as the turning point because it was then that W. Feller introduced a birth-and-death process which was soon recognized as fundamental in Q.T. As pointed out by J.W. Cohen [12], the birth-and-death process was already anticipated in earlier work (in 1928) by T.C. Fry [30], himself a contributor to Q.T.

The fact that Q.T. was recognized as a branch of the theory of stochastic processes resulted in its rapid development. Conversely, previous work done in the earlier period, essentially restricted to problems in telephony, is now regarded as a precursor to the modern theory of stochastic processes. Thus, in this period, many mathematicians and workers in operations research became interested in queueing problems from the point of view of theory and its numerous applications to various real-life situations. In other words, Q.T. was established as a *bona fide* object of investigations.

During the 1940s, work on congestion problems in teletraffic continued, whereas work on stochastic processes was growing almost independently. Papers by C. Palm, F. Pollaczek, E. Vaulot, and L. Kosten continued to appear, but at the same time, papers were published on applications of stochastic processes to physics and biology by N. Arley, M.S. Bartlett, W. Feller, D.G. Kendall, J.E. Moyal, and others.

It should be mentioned that a note by E. Borel [5] on the waiting problem appeared in 1942, and in 1948 a study of a loss system by R. Fortet [28] appeared. Also, C. Jacobaeus wrote his early papers in the years 1945-1947 in which he laid the foundation for the study of link systems in telephone systems. A monograph on the work of A.K. Erlang, with a collection of his papers, was published by E.

Brockmeyer, H. L. Halstrom and A. Jensen in 1948, see [8] and [27].

It was in the early 1950s that Q.T. began to grow rapidly as a separate discipline. Its applications to investigations of the flow of traffic in systems involving humans, animals, and machines became the object of intensive study, although its relevance to teletraffic was never abandoned. Thus, Q.T. combined good mathematics with engineering and social sciences.

The first appearance in 1950 of the classic probability text by W. Feller was widely responsible for *probabilistic literacy* among general researchers. In his fundamental study of the M/G/1 queue published in 1951, D.G. Kendall [40] introduced a convenient notation to classify queueing systems, which is still in use with some extensions. However, more important is the fact that it was this paper by Kendall that stimulated the widespread applications of the regeneration points technique to Q.T.

In 1952, D.V. Lindley [47] obtained his integral equation for the waiting time distribution in GI/G/1 queues, and in 1953 W.L. Smith [64] investigated its solution using analysis in the complex plane (Rouche's theorem!). These papers initiated extensive work in subsequent years on integral equations of the Wiener-Hopf type in Q.T.

Thus, in 1955 J. Kiefer and J. Wolfowitz [43] extended Lindley's result to the GI/G/s waiting system, using an approach related to a random walk; their analysis led to an integral equation over a region in *s*-dimensional space, and they were able to prove the existence of the limiting distribution of the waiting time for a customer.

It is of interest to mention that in 1960 there appeared the English translation of the monograph *Mathematical Methods in the Theory of Queueing* by A.Y. Khintchine [42] (the Russian original appeared in 1955). Khintchine, the well-known contributor to the famous Wiener-Khintchine theorem and himself an eminent queueing theorist, was in the habit of writing short monographs in which he improved on results of other authors. In this case, he presented mathematical foundations of Q.T., based on his extension of works by Palm and Erlang; in particular, he gave the rigorous treatment of Palm functions. Although the book was lucidly written, it was not up-to-date when it appeared; see review [66].

During the 1950s, a new generation of young mathematicians became interested in queueing problems. For decades to come they produced the ever increasing flow of papers which continues to the present day. It is impossible to give proper justice to their work in a short essay like this one. It suffices to say that through their efforts and their strong devotion, they built Q.T. as it is known today.

In this period of intensive activity, new methods were introduced, solutions to problems of interest obtained, and many open problems suggested. A closer look at Q.T. led to the discovery of general principles and provided foundations in later years for the formulation of laws of general validity.

In the Netherlands, J.W. Cohen, working initially for a telephone company, wrote his first papers on teletraffic problems but switched subsequently to Q.T., exerting considerable influence on its development. His numerous papers range from theoretical problems to the most practical ones. In this period, he considered in particular, generalized Engset formulas and introduced "derived Markov chains." Later, in 1976, his monograph, *On Regenerative Processes in Queueing Theory*, appeared [10]. His famous treatise on *The Single Server Queue* appeared in 1969, and its second expanded edition was published in 1982 [11]. In this book, Cohen extended Pollaczek's method and showed its applicability, as already mentioned. As a testimonial to Cohen's lifelong contribution to Q.T., the *Liber Amicorum* book [6] was published on the occasion of his retirement in 1988.

Isolated in Hungary, L. Takács worked independently, introducing the concept of *virtual waiting time* and the *Takács integral equation*. Upon his arrival to the West in 1958, he became a prolific writer on Q.T., later becoming interested in combinatorial methods (the ballot theorem) and graph theory. His book, *Combinatorial Methods in the Theory of Stochastic Processes* [78], was published in 1967 and treats, in particular, problems in Q.T.; see also [77].

Honoring Takács for his contributions, the book *Studies in Applied Probability* [31] was published in 1994 on the occasion of his 70th birthday, containing papers by leading workers in Q.T. Another testimonial to Takács was published in 1994 as *A Special Jubilee Issue* in *JAMSA* [23], which also contains papers by leading workers in Q.T.

In the United States, J. Keilson, with keen interest in a variety of queueing problems and with profound physical intuition, combined in his papers good mathematics with significant practical applications. In 1965, his book, *Green's Function Methods in Probability Theory* [37], appeared, marking the first application of potential theoretic approach to stochastic processes, independent of works in modern probabilistic potential theory. This book introduced the concept of *compensation*, suggested by the boundary value problems in the classical potential theory, whose use led to fruitful applications (like an alternative to the Wiener-Hopf method). Keilson also investigated properties of distributions, especially stochastic monotonicity. His book, *Markov Chain Models - Rarity and Exponentiality* [38], appeared later in 1979. A book honoring Keilson for his life's work will appear shortly.

In France, P. Le Gall began his studies of Pollaczek's method, showing its advantages. His work concerns mainly teletraffic problems. His book, *Les Systemes Avec au sans Attente et les Processes Stochastiques* [46], was published in 1962; it contains a detailed exposition of evaluating Pollaczek contour integrals in practical problems.

Also in the United States, V.E. Beněs published, in 1963, his monograph on *General Stochastic Processes in the Theory of Queues* [3]. In this delightful short book, Beněs proposed a general approach free from explicit assumptions about the stochastic nature of the offered load to a single server queue (in particular, input and service processes need not be independent and need not be of the renewal type). It is remarkable that his method enjoys great revival nowadays because of recent applications to the "fluid approach" in traffic models. For further comments, see Section 3 below. In his other book [4], Beněs considered networks of queues and subsequently turned his attention to optimization problems.

In far away Australia, J. Gani investigated storage problems and biological applications and N.U. Prabhu began his work on queueing models, which later led to his studies of Wiener-Hopf theory. Prabhu's book, *Queues and Inventories* [55], appeared in 1965, and was followed in 1980 by its sequel, *Stochastic Storage Processes* [56]. Recently, Prabhu became interested in the statistical aspects of stochastic processes. In recognition of their work, both Gani and Prabhu were honored by books containing papers by workers in probability and statistics.

The author wrote his first paper on *congestion theory* in 1953, which foresaw his uncurable obsession with Markovian queue. His book, *Introduction to Congestion Theory in Telephone Systems* [73], appeared in 1960, and its second edition (augmented by a chapter on Markovian queues) was published several years later in 1986.

It is impossible to list here all researchers in the rapidly growing field of Q.T. Nevertheless, one should mention at least briefly some achievements of the late 1950s and early 1960s. The output theorem of P.J. Burke aroused great interest. J.F.C. Kingman, in a series of papers starting in 1961, began investigations of heavy traffic effects in queues, a topic taken up later by other researchers. His book, *On the Algebra of Queues* [44], published in 1966, introduced an important way of looking at queues. Special mention should be made of the short paper by J.D.C. Little [48], published in 1961, which initiated extensive research on what was to be known in later years as *Little's formula*.

Other researchers who contributed extensively to Q.T. include A.B. Clarke, B.W. Connolly, P. Finch, F.G. Foster, D.P. Gaver, A. Jensen, P.M. Morse, P. Naor, E. Reich, J. Riordan, J. Th. Runnenburg, T.L. Saaty, S. Stidham, D.M.G. Wishart, and many others. N.I. Bech, F. Capello, S. Ekberg, A. Elldin, A. Lotze and many others worked in teletraffic.

In due course, the time for books on topics in Q.T. arrived. After short notices in books by T.C. Fry and by W. Feller, already mentioned, there appeared extensive summaries or separate chapters in books on operations research and probability, e.g., T.L. Saaty in 1957, P.M. Morse in 1958, A.T. Bharucha-Reid in 1960, J.H.B. Kemperman in 1961.

Early books devoted entirely to Q.T. include *Elements of Queueing Theory with Application* by T.L. Saaty [62] and *Queues* by D.R. Cox and W.L. Smith [19], both appearing in 1961. As already mentioned, *Introduction to Congestion Theory in Telephone Systems* by R. Syski appeared in 1960, the book by Le Gall in 1962, and in the same year, *Stochastic Service Systems* by J. Riordan [57]. The monograph on *General Stochastic Processes* by V.E. Beneš appeared in 1963. L. Takács published his *Introduction to the Theory of Queues* in 1962 and in 1967 his monograph on *Combinatorial Methods* appeared. The book *Queues and Inventories* by N.U. Prahbu appeared in 1965, and the book *Introduction to Queueing Theory* by B.V. Gnedenko and I.N. Kovalenko was published in 1968; its second edition appeared in 1989 [34]. As already noted, the treatise on *The Single Server Queue* by J.W. Cohen appeared in 1969 and the monograph on *Mathematical Methods in the Theory of Queueing* by A.Y. Khintchine in 1960.

As Q.T. evolved, the need for personal contacts among researchers in the field became apparent. Queueing theoreticians met, usually during various professional meetings. In particular, the International Teletraffic Congress (ITC), initiated by A. Jensen, met for the first time in Copenhagen in 1955, the city of A.K. Erlang. Since then, ITC has met with great regularity in various cities of the world. Recently, the 13th Congress was again held in Copenhagen in 1991, and the last one in the south of France in 1994; the 15th Congress will meet in Washington, D.C. in 1997.

Devoted to problems in telecommunications, ITC held regular sessions on Q.T. Anticipating slightly, at the 10th ITC in Montreal in 1983, R. Syski organized a special session on *Recent Advances in Queueing Theory*, with J. Keilson, J.W. Cohen, and P. Kühn among the speakers. At the ITC-13 Congress in Copenhagen in 1991, J.W. Cohen (with the help of H.C. Tijms and R. Syski) organized a workshop on *Stochastic Modeling* with most papers devoted to Q.T. ITC also organizes special seminars devoted to basic problems in telecommunications at which papers on Q.T. are frequently presented (e.g., in Moscow 1984).

An important meeting of queueing theoreticians organized by W.L. Smith in

Chapel Hill in 1964 provided the unique opportunity for personal contacts and review of progress in Q.T. Proceedings of this symposium, published in 1965 under the title *Congestion Theory*, constitute both the authoritative book on Q.T. and a pleasant souvenir for participants [65].

Another meeting of importance took place in Lisbon in 1965, and its proceedings, edited by R. Cruon as *Queueing Theory* [20], contained among other papers, an attempt to standardize terminology in Q.T. Perhaps it may be mentioned that at both these meetings, Pollaczek's work was one of the principal topics of discussion. Another topic that aroused interest was the need for some kind of regular meetings among researchers in Q.T.; these meetings materialized during the next decade.

During the 1960s, intensive research in Q.T. prevailed and the number of papers grew exponentially. This was a period of great activity; many new complicated systems were examined and many new techniques were developed. Everyone was interested in queueing models; students wrote doctoral theses and social scientists and philosophers began to look at queues as a legitimate object of their investigations.

The third period began in the early 1970s. This was a period of further extensive development and consolidation of achievements; it was also a period of maturity. Unfortunately, this impressive growth led towards the end of the period to decline.

In its development in this period, Q.T. proceeded gradually from individual systems toward general consideration of types of stochastic processes and formulation of laws and methods applicable to sufficiently large classes of systems.

A successful formulation of a general theory proposed earlier by V.E. Beneš has already been mentioned. L. Takács stressed the role of fluctuation theory, J.W. Cohen achieved new results using regenerative processes, and J. Keilson and his numerous collaborators continued their work on various aspects of Q.T. In their joint paper [39] in 1973 on compensation measures, Keilson and Syski examined this concept for Markov chains. Subsequently, in papers appearing in 1977-1979, R. Syski extended these ideas to the perturbation of Markov chains in continuous time; see [75]. Th. Runnenburg pointed out the advantages of the method of collective marks, N.U. Prabhu examined the Wiener-Hopf technique, and P. Le Gall discussed the importance of stochastic integrals.

M. Neuts and his collaborators, in their many papers, investigated practically every conceivable problem in Q.T. Neuts introduced a rather extensive class of "phase-type distributions" [49], which found many applications; their connection with the perturbation model of Keilson was examined later in 1982 by R. Syski [69]. Following his interest in computational problems, Neuts developed several algorithms suitable for Q.T. [50].

W. Whitt studied weak convergence and its connection with point processes, general principles (PASTA property, Little's formula), and new methods of numerical calculations, especially in G/G/s systems. L. Kleinrock was interested in communication systems and their performance analysis. Important contributions to Q.T. obtained by D.J. Daley, C.R. Heathcote, and C.C. Heyde should be mentioned here. Numerous investigations in this period were devoted to specific topics of interest such as various forms of duality, Little's formula, conservation laws, limit theorems, and others.

During this period many new books on Q.T. were published. Among them the comprehensive text *Queueing Theory* by R.B. Cooper appeared in 1972, and its second extended edition appeared in 1981 [17]. An important contributor to Q.T. himself, Cooper published later in 1990 a survey of Q.T. in *Handbook of Operations Research and Management Science* [18]. In 1975, B.W. Connolly wrote *Lecture Notes on Queueing Systems*, and in 1974, *Fundamentals of Queueing Theory* by D. Gross and C.M. Harris appeared. A two-volume treatise on *Queueing Systems* by L. Kleinrock appeared in 1975-1976 [45]. In 1973, L. Kosten published his earlier study on *Stochastic Theory of Service Systems*. Monographs on *Regenerative Processes in Queueing Theory* by J.W. Cohen appearing in 1976, and on *Markov Chain Models* by J. Keilson appearing in 1976, have been already mentioned.

It is of interest to add that books on probability and stochastic processes have sections devoted to the mathematical analysis of Q.T.; e.g., P. Billingsley, *Probability and Measure* (first edition 1979, second 1985), a two-volume text entitled *A Course in Stochastic Processes* by S. Karlin and H.M. Taylor, first published in 1975 and again in 1981, and *Introduction to Stochastic Processes* by E. Çinlar, which appeared in 1975.

As books provided mainly consolidation and systematization of results, progress in Q.T. was visible in the amount of papers appearing during this period. It is impossible to list all individual writers in this personal account. The bibliography prepared by A. Doig in 1957 lists over 600 items, whereas the bibliographical list prepared in 1966 by T.L. Saaty contains several hundred more. A very complete list of references on Q.T. was given later in 1981 in the book by R.B. Cooper [17]; see also, his survey [18]. A comprehensive bibliography was published by H. Takagi [79] in 1991 and its recent update in 1995 by J.H. Dshalalow [24].

On the initiative of J. Keilson, the first Conference on Stochastic Processes and their Applications (SPATA) was held in Rochester in 1971. Its aim was to have all applied probabilists meet together to discuss the future of the field. The result was the successful series of conferences held practically every year, the latest (23rd) being held in 1995 in Singapore. Q.T. formed an important part of these conferences, but in later years they have become dominated by abstract probability.

At Rochester, the idea of a new journal was conceived, also called *SPATA*. The first issue appeared in 1973, with J.W. Cohen, J. Keilson, N.U. Prabhu and R. Syski being its "founding fathers." This new journal joined the well-respected "Gani journals," *Journal of Applied Probability* and *Advances in Applied Probability* (established already in the late 1960s), which for years to come published many papers on Q.T. Subsequently, the SPATA journal and conferences were taken over by the newly formed Bernoulli Society. During the 1980s, several other journals devoted exclusively to Q.T. were established.

On the other extreme, special systems with specified structural properties received considerable attention in the literature, especially in operations research studies. Special queue disciplines, priorities, scheduling, time sharing, buffer systems, variable number of servers, overflow, and output problems are just some illustrative examples. Also, numerical computations and simulation methods, control and optimizations of queues, and statistical analysis were among subjects of great interest.

Works of M. Yadin, M. Rubinovitch, B. Avi-Ithzak, I. Adiri, M. Hofri and E.G. Coffman, L. Kleinrock, M. Neuts, D.P. Gaver, G.F. Newell, F.P. Kelly, A. Descloux, P.J. Schweitzer, M.F. Ramalhoto, P. Kubat, R.L. Disney, R.V. Evans, and numerous others should be mentioned here.

Numerous conferences on Q.T., or containing special sessions devoted to it, were held during this period. SPATA and ITC held their regular meetings, and of special interest were sessions organized by TIMS, ORSA, SIAM, and IMS. Papers, books, and new journals made it rather difficult to keep up with the progress in the field.

Such intensive growth initially had its advantages but led eventually to stagnation.

Thus, after the initial influx of wealth of new ideas and concepts, new techniques and methods, and new areas of applications, there came a time of inevitable saturation. As the field grew too widespread and the papers thinner in new ideas, many researchers began to lose their interest and turned their attention instead to stochastic processes and other fields like biology and health statistics, telecommunications, and social sciences.

There was even a time when probability journals declined to accept papers on Q.T.; this was in particular true in the case of the influx of papers whose only novelty was a minor variation in classical models.

It was soon realized that in order to survive, there was a need to inquire deeper into the underlying stochastic aspects of queueing processes, as well as a need to analyze structurally more complex and practically important queueing systems. This trend has been already apparent in some of the works described above, but it evolved later with intensity under the influence of progress in the theory of stochastic processes and progress in computers. Thus, further development of Q.T. proceeded in two separate but closely related directions. But this is the theme of the next stage.

1.3 REVIVAL AND SEPARATION

The second stage of development in Q.T. is characterized on one hand by the tendency toward abstraction in queueing problems and on the other hand, by preoccupation with analysis of complex practical systems.

Applications of stochastic processes in Q.T. fall into two categories (which overlap to a great extent), namely:

(a)   general theories concerning a class of systems or a class of problems;

(b)   specific problems concerning an individual system.

The distinction is by no means sharp; it only represents two extremes of the broad spectrum of problems in Q.T. New approaches brought by developments in point processes, Markov processes, regenerative processes, and martingales were responsible for the fact that Q.T. enjoyed its revival. This trend already surfaced in the previous stage, but became dominant in the 1980s and still continues today.

Progress in computer systems opened new areas of investigations, and necessity for new methods became apparent. Unfortunately, young, eager computer scientists developed their own brand of Q.T., unaware that basic classical problems were solved decades ago.

Theoretical developments in stochastic processes as well as in statistical methods (inference on stochastic processes) led to important structural theorems of great interest to theoreticians, but these were frequently discarded by practitioners as "academic exercises."

On the other hand, concrete technological developments present challenging new mathematical problems that, if to be solved at all, require new powerful tools. As these methods are too advanced and complicated, practitioners satisfy themselves often with rather crude but simpler techniques (sufficient as the first approximation). Hence, the proverbial gap between theory and practice grows wider and wider at an increasing pace. Moreover, each of the theoretical disciplines developed rapidly for its own sake, and recent advances are accessible only to a few devoted specialists. This in turn makes it difficult for non-specialists and users to get acquainted with new methods and techniques; see [72].

As a consequence, there are today two Q.T.'s: one abstract, interested in very general formulations, a paradise for mathematicians; the other concrete, interested in practical systems, which occur in applications. Although both of these Q.T.'s have common roots, they grow apart from each other because of different interests. This is now clearly visible in publications and at conferences. Fortunately, there is a gray area where both of these groups overlap, but as usual, its representatives often receive mixed reception.

To present the account of the influence of all these developments on Q.T. is too ambitious a task, and would require a book. Only a few selected examples, mostly from category (*a*) mentioned above, must suffice.

Tendency towards abstraction was already present in previous periods, but it intensified with progress in stochastic processes. As it turns out, Q.T. provided splendid illustratives for the most esoteric mathematical concepts.

One of the early applications of martingales in Q.T. was the paper by D.P. Kennedy, which appeared in 1976 in SPATA [41]. The systematic use of martingales in queueing problems was explored in the book by P. Brémaud [7] published in 1981. During 1981-1983, W.A. Rosenkrantz [61] published several papers on martingales associated with queueing processes; l.c. also the book by P. Hall and C.C. Heyde [35] on applications of martingales, which appeared in 1980. At present, the use of martingales is widespread; see in particular, papers and books by S. Asmussen, F. Baccelli, A. Makowski, N.U. Prabhu, and W. Whitt.

As an another example of abstraction, one should mention a talk on the *Construction of Stationary Queues* presented by J. Neveu at the SPATA Conference [51] in Ithaca in 1983, in which deep properties of point processes were used.

Point processes found natural applications to Q.T., and literature in this area is voluminous. It suffices to mention here the treatise by P. Franken, D. König, U. Arndt, and V. Schmidt on *Queues and Point Processes* [29], published in 1982; also the book by Brémaud quoted above, the book by T. Rolski on *Stationary Random Processes Associated with Point Processes* [60], which appeared in 1981, and the book *Applied Probability and Queues* [1] by S. Asmussen, published in 1987.

An excellent exposition of modern probabilistic methods in Q.T. can be found in the book *Elements of Queueing Theory* [2] by F. Baccelli and P. Brémaud, which was published in 1994 (this is the expanded version of their monograph from 1987). The basic tool in this book is the theory of Palm probabilities (Palm martingale calculus) applied to queueing systems with ergodic inputs. This approach, together with the theory of stochastic recurrences, provides a natural framework for the study of conservation laws and stochastic comparison of queues.

After very abstract development (in particular, by French probabilists), the regenerative processes, which actually evolved from regeneration points analysis, provided tools for the study of general queueing processes as well as for better understanding of interesting special systems. Here, work by E. Çinlar, H. Kaspi, and others should be mentioned; see also, the short monograph by J.W. Cohen listed above. It may be added that use of Palm measures in abstract probability can be found in the book by R.K. Getoor [33].

As already mentioned earlier, queueing problems can be treated as boundary value problems. This point of view was presented by J.W. Cohen and O.J. Boxma in their monograph [15]; also see papers by G. Fayolle and by J.P.C. Blanc. Evolved from interpretation of Pollaczek's integral equations, this method provides solution to many concrete problems.

In his recent book, *Analysis of Random Walks* [14] published in 1992, J.W. Cohen used this method to study problems of absorption in the boundary of a set in n-dimensional random walks. The basis of his analysis is the *hitting point identity* (a generalization of Wald's one-dimensional identity), which relates the generating function of the joint distribution of the entrance time to the boundary (from a point) and the components of the hitting point. The solution is then obtained by finding zeros of an appropriate kernel (like in Wiener-Hopf factorization). In particular, this method yields new results in the study of workload processes.

Potential theory enters to Q.T. through a connection with Markov processes. Indeed, probabilistic potential theory provided far-reaching unification of treatment and opened new areas of research. Thus, the Dirichlet problem and Poisson equation form a basis for investigation of passage times, which include in particular, busy periods and waiting times. This point of view was expressed by R. Syski in his papers on Markovian queues; see also his book on *Passage Times* [75].

A mentioned earlier, development of the "fluid approach" to queueing models created a revival of interest in the theory proposed by E.V. Beněs in his book [3]. L. Decreusefond and A.S. Ustunel examined connection of the Beněs equation and stochastic calculus in [21]; R. Syski described Beněs' approach in terms of point processes and compared it with Brémaud's martingale approach [76]. See also the comments on technical applications below.

Investigations of general properties of classes of queueing systems continue today and exhibit strength, vigor, and novelty. Researchers in applied probability put new life in Q.T. Their works use advanced mathematical methods and possess heavy analytic flavor. Even a quick look at current publications in journals and in congressional proceedings provides convincing testimony.

Again, it is impossible to list all these recent investigations in this short essay, so only a few examples are stated here. "Old Guard," now of retirement age and still going strong, may serve as the illustration of this new trend. Progress in the field is continued by a new generation of researchers in stochastic processes, applied probability, operations research, and statistics.

For example, in his monograph [80], which is based on his thesis and was published in 1981, van Doorn examined the role of stochastic monotonicity in birth-death processes following J. Keilson. O.J. Boxma analyzed polling stations in a 1991 paper, and authored numerous papers on various aspects of queueing systems. J.H. Dshalalow investigated (in 1994) first passage problems in stochastic processes, including queueing [22]. Such problems involve times of the first entrance to a set (crossing a fixed level) and are of importance in theory and in applications (e.g., waiting times, busy periods).

Areas that received attention recently include: networks of queues (L. Kleinrock, F.P. Kelly, E.V. Beněs, R. Disney), conservation laws (O.J. Boxma, R. Schassberger, W. Whitt, R.W. Wolff), Markov chains (R.L. Tweedie, E. Seneta, V.V. Kalashnikov), queueing models (R.B. Cooper, D.J. Daley, B.T. Doshi, T.J. Ott, M. Rubinovitch), stochastic processes (S. Asmussen, W. Whitt, D. König, J.H. Dshalalow), optimization/computation (M. Hofri, P.J. Schweitzer, H.C.

Tijms, U. Yechiali), and statistics (C.C. Heyde, D.L. Iglehart, N.U. Prabhu, S. Resnik).

This short list is, of course, highly incomplete and subjective, and the reader should consult survey books and papers for a more balanced picture. In particular, the survey by R.B. Cooper [18], already mentioned, may be helpful. For a current state of affairs in Q.T. see the recently published book *Advances in Queueing*, edited by J.H. Dshalalow [25]. His paper [24] also gives the informative survey of mathematical methods in early Q.T.

New journals devoted to applied probability and stochastic processes appeared in the last decade and frequently publish papers in Q.T. These include *Stochastic Models*, with M.F. Neuts as the editor, and the *Journal of Applied Mathematics and Stochastic Analysis*, edited by J.H. Dshalalow. Even *The Annals of Applied Probability* does not avoid Q.T. papers.

Various conferences on applied probability, meeting frequently around the globe, have sessions devoted to Q.T.

Turning now to another kind of Q.T., namely that dealing with specific problems concerning an individual system, considerable technological progress in computers and telecommunication networks opened new problems that required new methods of solution. Classical Q.T. is often not sufficient for these problems and more use of advances in stochastic processes needs to be made. The principal object of interest is characterization of the performance of a particular system on hand, and not existence theorems or the like.

Structural complexity of an investigated system frequently forced researchers to look for approximate numerical answers. It is this search for concrete solutions to practical problems which created the split in Q.T.

Numerous conferences, meetings, and workshops were devoted to these practical problems. ORSA, TIMS, and the newly formed INFORMS became almost exclusively platforms for papers describing technical details of some practical system. A recent INFORMS conference in Boca Raton (the third in the the series) was a very successful meeting in this respect. Even in ITC during recent years, the number of theoretical papers has declined. As already mentioned, SPATA conferences went in the opposite direction to become highly theoretical.

Among a few problems which are of interest to both Q.T.'s is the analysis of the ATM (Asynchronous Transfer Mode) system. The difficulty here is the "burstiness" of traffic. Several mathematical descriptions have been proposed, but no final word has yet been said. J.W. Roberts [58] used the fluid approach, based on Beneš' equation, to describe the system; see also a recent paper by L.C.G. Rogers [59]. I. Norros proposed the use of self-similar processes [52]. W. Burakowski et al. presented their work in a recent paper [9].

An international group of mathematicians and engineers interested in modern telecommunication systems, and ATM in particular, publishes their work in reports known as COST reports; see [9], [52].

At the thirteenth ITC in 1991, there was a special workshop on ATM that had several papers on queueing. Proceedings of this workshop and the workshop on stochastic modeling were issued jointly under the title *Queueing, Performance, and Control in ATM*, edited by J.W. Cohen and C.D. Pack [16]. The splitting into two Q.T.'s was clearly visible here.

1.4 FUTURE

It is risky to make predictions. However, as this author is going to retire soon, he is not afraid to offer his view on future developments.

As claimed in this essay, there are at present two Q.T.'s, divided according to their interest. One is highly abstract and the other highly practical. It seems that this split will continue to grow wider and wider.

Progress in the theory of stochastic processes (especially point, regenerative, and stationary processes) will influence new approaches to Q.T. This may be in the form of new methods, new interpretations, and the development of new theories with wide applicability. Researchers in abstract probability usually do not have Q.T. in mind; different talents are required to find applicability of their results. The history of Q.T. provides examples of this fortunate relationship. Other examples, not discussed in this essay, are diffusion approximation, the large deviations technique, and random fields. One may hope that the near future will bring applications of superprocesses, the object of current research in stochastic processes.

Progress in technical developments of systems involving various forms of traffic created the need for mathematical analysis of performance of individual systems. This brings new problems which require new tools, and the search for these tools is of great practical importance. This is clearly visible not only in teletraffic theory (ATM systems, LAN traffic, networks), but also in other disciplines where Q.T. methods are used (biological and health studies, computers). As already mentioned, simulation and numerical analysis are frequently the only way to obtain approximate results.

It is therefore hoped that the gap between these two Q.T.'s may eventually be diminished. Idealistically, this could be achieved when theoreticians learn about practical problems and practitioners learn about theory. In present times of great specialization, this is highly unrealistic. Nevertheless, one could try to work in this direction, at least with our students in universities, by stressing the importance of theory *and* applications. Otherwise, researchers in Q.T. 1 and Q.T. 2 could not find a common language.

As an example of open problems in the gray area where both Q.T.'s meet, the following are suggested:

i)   description of traffic in the ATM system,

ii)   flow of traffic through network of queues,

iii)   description of a queueing system in terms of multidimensional regenerative processes.

There is still room for a Ph.D. thesis!

ACKNOWLEDGEMENT

This is an updated version of an invited talk at GTE Laboratories, Waltham, MA, in January 1992. The author wishes to thank Dr. J. Keilson for his invitation (which produced the first version of this paper) and for his hospitality.

Thanks are also due to Dr. R.B. Cooper for his comments on the first version of this paper, and for references [18] and [79].

BIBLIOGRAPHY

The present account was based on opinions expressed by the author in his papers and books [66]-[76], where also detailed references can be found. Other references listed here are to works discussed in the text. For the complete list of references, see bibliographies complied by R.B. Cooper [17], [18], J.H. Dshalalow [24], and H. Takagi [79].

[1]   Asmussen, S., *Applied Probability and Queues*, J. Wiley, New York 1987.

[2]   Baccelli, F. and Brémaud, P., *Elements in Queueing Theory (Palm-Martingale Calculus and Stochastic Recurrences)*, Springer, New York 1994.

[3]   Beneš, V.E., *General Stochastic Processes in the Theory of Queues*, Addison-Wesley, Reading, MA 1963.

[4]   Beneš, V.E., *Mathematical Theory of Connecting Networks and Telephone Traffic*, Academic Press, New York 1965.

[5]   Borel, E., Sur l'emploi du theoreme de Bernoulli pour faciliter le calcul d'une infinite de coefficients.   Application au probleme de l'attente a un guichet, *Comptes Rendus Acad. Sci. Paris* **214** (1942), 452-456.

[6]   Boxma, O.J. and Syski, R., (eds.), *Queueing Theory and its Applications, Liber Amicorum for J.W. Cohen*, North Holland, Amsterdam 1988.

[7]   Brémaud, P., *Point Processes and Queues (Martingale Dynamics)*, Springer, New York 1981.

[8]   Brockmeyer, E., Halstron, H.L. and Jensen, A., *The Life and Works of A.K. Erlang*, Trans. Danish Acad. Techn. Sci. **2**, Copenhagen 1948.

[9]   Burakowski, W., Syski, R., Szablowski, P. and Bak, A., Functionals of Poisson processes in ATM traffic modeling, *Doc. COST* **242**(93) (1993), 12 pages.

[10]   Cohen, J.W., *On Regenerative Processes in Queueing Theory*, Springer, New York 1976.

[11]   Cohen, J.E., *The Single Server Queue*, 2nd revised ed., North Holland, Amsterdam 1982.

[12]   Cohen, J.W., Who introduced the birth-and-death technique?, In: *Proc. 11th ITC* (ed. by M. Akiyama), North Holland (1985), 36-39.

[13]   Cohen, J.W., for collected works of J.W. Cohen, see [6], 1986.

[14]   Cohen, J.W., *Analysis of Random Walks*, IOS Press, Amsterdam 1992.

[15]   Cohen, J.W. and Boxma, O.J., *Boundary Value Problems in Queueing System Analysis*, North Holland, Amsterdam 1983.

[16]   Cohen, J.W. and Pack, C.D., (eds.), *Queueing, Performance and Control in ATM*, North Holland, Amsterdam 1991.

[17]   Cooper, R.B., *Introduction to Queueing Theory*, 2nd edition, North Holland, Amsterdam 1981.

[18]   Cooper, R.B., Queueing theory, In: *Handbooks in Operations Research and Management Science: Stochastic Models*, (ed. by D.P. Heyman and M.J. Sobel), **2** (1990), 469-518.

[19]   Cox, D.R. and W.L. Smith, *Queues*, Methuen, London 1961.

[20]   Cruon, R., (ed.), *Queueing Theory*, English Univ. Press, London 1967.

[21]   Decreusefond, L. and Ustunel, A.S., The Beneš equation and stochastic calculus of variations, *SPATA* **57** (1995), 273-284.

[22]   Dshalalow, J.H., On termination time processes, *Studies in Applied Probability (Papers in honor of L. Takács)*, J. Appl. Prob. **31A** (1994), 325-336.

[23]   Dshalalow, J.H., (ed.), Special Jubilee issue in honor of Lajos Takács, *J. Appl. Math. and Stoch. Analysis* 7:3 (1994).

[24]   Dshalalow, J.H., An anthology of classical queueing methods, In: *Advances in Queueing*, (ed. by J.H. Dshalalow), CRC Press (1995).

[25]   Dshalalow, J.H., (ed.), *Advances in Queueing*, CRC Press, Boca Raton, FL 1995.

[26]   Dshalalow, J.H. and Syski, R., Lajos Takács and his work, In: *Special Jubilee issue in honor of L. Takács, J. Appl. Math. and Stoch. Analysis* 7:3 (1994), 215-237.

[27]   Erlang, A.K., for collected works of A.K. Erlang, see [8].

[28]   Fortet, R., Sur la probabilitie de perte d'un appelle telephonique, *C.R. Acad. Sci. Paris* **226** (1948), 1502-1504.

[29]   Franken, P., König, D., Arndt, U. and Schmidt, V., *Queues and Point Processes*, J. Wiley, Chichester 1982.

[30]   Fry, Th.C., *Probability and its Engineering Uses*, van Nostrand, New York 1928.

[31]   Galambos, J. and Gani, J., (eds.), *Studies in Applied Probability (Papers in Honor of L. Takács), J. Appl. Prob.* **31A** (1994).

[32]   Gani, J., (ed.), *The Craft of Probabilistic Modeling (A Collection of Personal Accounts)*, Springer Verlag, New York 1986.

[33]   Getoor, R.K., *Excessive Measures*, Birkhäuser, Boston 1990.

[34]   Gnedenko, B.V. and Kovalenko, I.N., *Introduction to Queueing Theory*, 2nd edition, Birkhäuser, Boston 1989.

[35]   Hall, P. and Heyde, C.C., *Martingale Limit Theory and its Applications*, Academic Press, New York 1980.

[36]   Kaspi, H. and Rubinvitch, M., Regenerative sets and their applications to Markov storage systems, In: *Queueing Theory and its Applications, Liber Amicorum for J.W. Cohen*, (ed. by O.J. Boxma and R. Syski), North Holland, Amsterdam (1988), 413-427.

[37]   Keilson, J., *Green's Functions Methods in Probability Theory*, Griffin, London 1965.

[38]   Keilson, J., *Markov Chain Models - Rarity and Exponentiality*, Springer, New York 1979.

[39]   Keilson, J. and Syski, R., Compensation measures in the theory of Markov chains, *SPATA* **2** (1974), 59-72.

[40]   Kendall, D.G., Some problems in the theory of queues, *J. Roy. Stat. Soc.* **13**:B (1951), 151-185.

[41]   Kennedy, D.P., Some martingales related to cumulative sum tests and single-server queues, *SPATA* **4** (1976), 261-270.

[42]   Khintchine, A.Y., *Mathematical Methods in the Theory of Queueing*, Griffin, London 1960.

[43]   Kiefer, J. and Wolfowitz, J., On the theory of queues with many servers, *Trans. Amer. Math. Soc.* **78** (1955), 1-18.

[44]   Kingman, J.F.C., *On the Algebra of Queues*, Methuen, London 1966.

[45]   Kleinrock, L., *Queueing Systems* (I, II), J. Wiley, New York 1975-76.

[46]   Le Gall, P., *Les Systemes avec ou sans Attente et les Processus Stochastiques*, Dunod, Paris 1962.

[47]   Lindley, D.V., The theory of queues with a single server, *Proc. Camb. Phil. Soc.* **48** (1952), 277-289.

[48]   Little, J.D.C., A proof for queueing formula $L = \lambda W$, *Oper. Res.* **9** (1961),

383-387.

[49]   Neuts, M.F., Probability distributions of phase type, In: *Liber Amicorum for Prof. H. Florin*, Univ. Louvain (1975), 173-205.

[50]   Neuts, M.F., *Matrix-Geometric Solutions in Stochastic Models*, The Johns Hopkins Univ. Press, Baltimore 1981.

[51]   Neveu, J., Construction of stationary queues, Invited paper, *12th SPATA Conference*, abstract *SPATA* **17** (1984), 11.

[52]   Norros, I., Studies on a model for connectionless traffic, based on fractional Brownian motion, *COST* **242TD**(92), (1992), 12 pages.

[53]   Palm, C., *Intensity Variations in Telephone Traffic*, North-Holland, Amsterdam 1988 (translation of the original paper from 1943).

[54]   Pollaczek, F., for collected works of F. Pollaczek, see [63].

[55]   Prabhu, N.U., *Queues and Inventories*, Wiley, New York 1965.

[56]   Prabhu, N.U., *Stochastic Storage Processes*, Springer, New York 1980.

[57]   Riordan, J., *Stochastic Service Systems*, Wiley, New York 1962.

[58]   Roberts, J.W., (ed.), *Performance Evaluation and Design of Multiservice Networks*, COST **224**, final report CEC (1992).

[59]   Rogers, L.C.G., Fluid models in queueing theory and Wiener-Hopf factorization of Markov chains, *Ann. Appl. Prob.* **4** (1994), 390-413.

[60]   Rolski, T., *Stationary Random Processes Associated with Point Processes*, Springer-Verlag, New York 1981.

[61]   Rosenkrantz, W.A., Some martingales associated with queueing and storage processes, *Z. Wahrs. Verv. Gebiete* **58** (1981), 205-222.

[62]   Saaty, T.L., *Elements of Queueing Theory with Applications*, McGraw-Hill, New York 1961.

[63]   Schreiber, F., Cohen, J.W. and Syski, R., (eds.), Special Issue on Teletraffic Theory and Engineering in Memory of Felix Pollaczek, *AEÜ*, **47** (1993), 273-466.

[64]   Smith, W.L., On the distribution of queueing times, *Proc. Camb. Phil. Soc.* **49** (1953), 449-461.

[65]   Smith, W.L. and Wilkinson, W.E., (eds.), *Congestion Theory*, Univ. North Carolina Press, Chapel Hill 1965.

[66]   Syski, R., Review: A.Y. Khinchine, *Mathematical Methods in the Theory of Queueing*, Incorporated Statistician, London (1961), 57-61.

[67]   Syski, R., Markovian queue, In: *Congestion Theory*, (ed. by W.L. Smith and W.E. Wilkinson), Univ. of North Carolina Press, Chapel Hill (1965), 170-227.

[68]   Syski, R., Queueing theory symposium - Introduction and summary, *Proc. XX TIMS Meeting*, (ed. by E. Shlifer), Jerusalem Academic Press **2** (1975), 507-508.

[69]   Syski, R., Phase-type distribution and perturbation model, *Appl. Mathematicae* **17** (1982), 377-399.

[70]   Syski, R., Markovian queues in teletraffic theory, In: *Proc. 3rd Int. ITC Seminar on Fundamentals of Teletraffic Theory*, Inst. Probl. Info. Trans., Moscow (1984), 430-440.

[71]   Syski, R., Multiserver queues, In: *Encyclopedia of Statistical Sciences*, (ed. by S. Kotz, N.L. Johnson and C.B. Read), J. Wiley **5** (1985), 727-732.

[72]   Syski, R., Markovian models - An essay, In: *The Craft of Probabilistic Modeling (A Collection of Personal Accounts)*, (ed. by J. Gani), Springer Verlag, New York (1986), 109-125.

[73] Syski, R., *Introduction to Congestion Theory in Telephone Systems*, 2nd edition, North Holland, Amsterdam 1986.

[74] Syski, R., Markov processes and teletraffic, In: *Queueing Theory and its Applications, Liber Amicorum for J.W. Cohen*, (ed. by O.J. Boxma and R. Syski), North Holland, Amsterdam (1988), 95-126.

[75] Syski, R., *Passage Times for Markov Chains*, IOS Press, Amsterdam 1992.

[76] Syski, R., Comments on a single server queue, *J. Appl. Math. and Stoch. Analysis* 8:4 (1995), 331-340.

[77] Takács, L., for collected works of L. Takács, see [23] and [31].

[78] Takács, L., *Combinatorial Methods in the Theory of Stochastic Processes*, Wiley, New York 1967.

[79] Takagi, H., *Queueing Analysis*, North Holland, Amsterdam 1991.

[80] van Doorn, E., *Stochastic Monotonicity and Queueing Applications of Birth-Death Processes*, Springer, New York 1981.

Chapter 2
Retrial queues revisited

V.G. Kulkarni and H.M. Liang

ABSTRACT A retrial system consists of a primary service facility and an orbit. Customers can arrive at the service facility either from outside the system or from the orbit. If an arriving customer is blocked from entering the service facility, he/she may join the orbit and conduct a retrial later. Otherwise, he/she enters the service facility and gets served.

This paper collects together existing material regarding seven main aspects of retrial queueing systems: (*i*) Stability, (*ii*) Queue Lengths, (*iii*) Waiting Times and Number of Retrials, (*iv*) Asymptotic Analysis, (*v*) Busy Periods and Server Idle Times, (*vi*) Monotonicity, and (*vii*) Control. In each of these topics, we present the existing results and then discuss open problems.

CONTENTS

2.1 INTRODUCTION

A retrial queueing system consists of a service facility and an *orbit*, as shown schematically in Figure 1. Customers arrive from outside at rate λ, and demand i.i.d. service with mean τ and variance σ^2. An arriving customer enters the service facility if the facility is not full. Otherwise he/she enters the orbit with probability α_0 and attempts service after a random amount of time, called *retrial time*. (He/she leaves the system with probability $1 - \alpha_0$.) Unless otherwise mentioned, the orbit capacity is assumed to be infinite. The successive retrial times are i.i.d. with mean τ_r and variance σ_r^2. Let α_n be the probability that a customer returns to the orbit after the nth unsuccessful retrial (and leaves the system without service with probability $1 - \alpha_n$). If $\alpha_n = 1$, $n \geq 0$, the customers are called *persistent*; otherwise they are called *impatient*.

We designate the retrial queues by using the nomenclature of the usual queueing systems for the service facility. Thus an M/G/2/7 retrial queue has a service facility with two servers and capacity seven, the external arrival process is Poisson

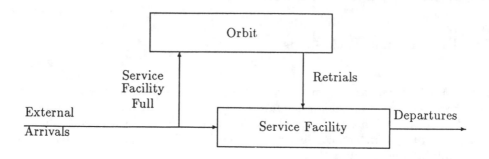

Figure 1. Schematic diagram of a retrial queue.

and the service times have general distribution. Unless otherwise mentioned, the retrial times are assumed to be i.i.d. exponential random variables with rate $\theta = 1/\tau_r$ and the customers are assumed to be persistent. For the impatient customers, the following three cases have received the most attention:

1. $\alpha_0 < 1, \alpha_n = \alpha = 1$ for $n \geq 1$.
2. $\alpha_0 = 1, \alpha_n = \alpha < 1$ for $n \geq 1$.
3. $\alpha_n = \alpha < 1$ for $n \geq 0$.

Let $N(t)$ be the number of customers in the orbit and $C(t)$ be the number of customers in the service facility at time t. We are interested in the bivariate process $\{(C(t), N(t)), t \geq 0\}$, called the CN *process*. If the external arrival process is a renewal process, it is easy to see that the CN process is a regenerative process, which regenerates whenever a customer enters an empty system.

There is a large literature devoted to retrial queues. It is mainly motivated by applications to telephone and telecommunications systems. There are two excellent survey papers on retrial queues: Yang and Templeton [75] and Falin [31]. The latter paper, especially, contains detailed references to Russian literature on retrial queues. The current paper by no means claims to contain the complete bibliography on retrial queues. Rather, it is a collection of results that the authors find interesting and worthy of further research. Naturally, there is a certain degree of subjectivity involved in the selection and any omissions are not to be taken as casting doubts on the usefulness or relevance of the omitted material.

We concentrate on the following main aspects of the study of retrial queues: (i) Stability, (ii) Queue Lengths, (iii) Waiting Times and Number of Retrials, (iv) Asymptotic Analysis, (v) Busy Periods and Server Idle Times, (vi) Monotonicity, and (vii) Control.

In Section 2.2 we study stability issues. Recently (see Meyn and Tweedie [63]), these issues have been discussed extensively, with examples of queueing systems that can be unstable even if traffic intensity is less than one. Retrial queues also provide examples of such a phenomenon. (See Liang and Kulkarni [61].) Stability conditions are extremely difficult to establish for general retrial queues.

Section 2.3 is devoted to the results on queue lengths, more precisely, the number of customers in the system. This is the most extensively studied aspect of the retrial queues as is apparent from the references. In section four we describe the results on waiting times. The waiting time is closely related to the number of retrials a customer conducts before he/she enters service and hence we include those results in this section. In Section 2.5 we study the asymptotic behavior of steady-state queue lengths and waiting times as the retrial rate approaches zero or infinity, or as the traffic intensity approaches zero or one.

Section 2.6 covers results about busy periods and server idle times. The relation between these two quantities is far more complicated in retrial queues than in the standard queueing models. Section 2.7 is devoted to the monotonicity of retrial queues. This involves the study of the CN process as a function of input parameters, e.g., the retrial times, the service times, arrival rates, etc. Finally, in Section 2.8, we discuss control problems in retrial queues. There is not much literature on this topic, but we think it is an area that deserves more attention.

In each section we present the results for each type of retrial queue that has been studied. Most of the time we present the main results. When a result involves too much notation we omit it and just cite a reference to save space. At the end of each section we give a brief discussion of what we perceive to be open and noteworthy problems in the related areas. We hope this organization will be beneficial to the reader.

2.2 STABILITY

In this section we present the currently available results on the stability of the CN process. Let $\rho = \lambda \tau$. We list the results for several special classes of the retrial queueing systems.

M/M/1/1. In this case, the CN process is a continuous time Markov chain (CTMC) and the following results are known:

$$\rho < 1 \Rightarrow CN \text{ process is positive recurrent,} \tag{2.1}$$

$$\rho = 1, \theta\tau \geq 1 \Rightarrow CN \text{ process is null recurrent,} \tag{2.2}$$

$$\rho = 1, \theta\tau < 1 \Rightarrow CN \text{ process is transient,} \tag{2.3}$$

$$\rho > 1 \Rightarrow CN \text{ process is transient.} \tag{2.4}$$

See Falin [24].

M/G/1/1. Let N_n be the number of customers in the orbit at the nth departure epoch. Due to the exponential retrial times, $\{N_n, n \geq 0\}$ is a discrete time Markov chain. By using Pakes' lemma [65], one can show that the DTMC is positive recurrent if

$$\rho < 1. \tag{2.5}$$

The above condition implies the stability of the CN process. If $\rho > 1$ the DTMC is transient, and hence the CN process is not stable. There are no results for the case $\rho = 1$. See Deul [10], Falin [24], Greenberg [38], Greenberg and Wolff [39], and Wolff [72].

M/G/1/1 with impatient customers. In the case of $\alpha_n = \alpha = 1$ for all $n \geq 1$, the stability condition is shown to be

$$\rho\alpha_0 < 1. \tag{2.6}$$

See Falin [28].

In the case of $\alpha < 1$, the system is stable as long as the mean retrial time is finite. (See Fayolle and Brun [36].) This follows because the system can be majorized by an M/G/∞ queue. A similar argument can be used to extend the result to G/G/1/1 retrial queues with impatient customers.

G/G/1/1. Liang and Kulkarni [61] consider the case of a renewal arrival process and general i.i.d. service times and i.i.d. retrial times. By using constant interarrival times and retrial times, they show that $\rho < 1$ is not a sufficient condition for stability of the CN process. However, they show that $\rho < 1$ is indeed a sufficient condition for the stability of the system with persistent customers if the interarrival times and retrial times are finite mixtures of Erlang random variables and the service times are general. Since the family of mixtures of Erlang distributions is dense in the continuous distributions (see Tijms [70]), their result covers most continuous cases. They prove this result by using Foster's criterion [37] for an embedded DTMC in a majorizing system.

M/M/s/s. Deul [10], Falin [22, 24], and Hanschke [43] show that the necessary and sufficient stability condition is

$$\lambda \tau < s. \tag{2.7}$$

They use the embedded Markov chain approach and Foster's criterion [37] to establish the above inequality.

Open problems. Liang and Kulkarni [61] assume that the service facility is a single server queue with no waiting room. It remains to extend their analysis to more general service facilities and infinite mixtures of Erlangs as the interarrival time and retrial time distributions. The results on the case with impatient customers need to be extended to the case of general retrial times. The interesting special case $\rho = 1$ also merits further research.

2.3 QUEUE LENGTHS

In this section we collect the results about the steady state distributions of the CN process assuming they exist. Let C and N be the number of customers in the service facility and in the orbit in steady state, respectively. Let

$$P_{in} = P(C = i, N = n).$$

As before, we list the known results according to the retrial queueing models.

M/M/1/1. The following results are from Jonin and Sedol [45]. See also Choo and Conolly [7].

$$P_{00} = (1 - \rho)^{1 + \frac{\lambda}{\theta}}, \tag{2.8}$$

$$P_{0n} = \frac{\rho^n}{n! \theta^n} (1 - \rho)^{1 + \frac{\lambda}{\theta}} \prod_{i=0}^{n-1} (\lambda + i\theta), \quad n \geq 1, \tag{2.9}$$

$$P_{1n} = \frac{\rho^{n+1}}{n! \theta^n} (1 - \rho)^{1 + \frac{\lambda}{\theta}} \prod_{i=0}^{n} (\lambda + i\theta), \quad n \geq 0. \tag{2.10}$$

The results on the transient distribution of $(C(t), N(t))$ are available in Falin [13, 27].

M/G/1/1. We state the results for the generating functions of the steady state distributions of the CN process. Let $\widetilde{B}(s)$ be the Laplace-Stieltjes transform (LST) of the service time distribution and define $k(z) = \widetilde{B}(\lambda - \lambda z)$. The main results are:

$$P_0(z) = \sum_{n=0}^{\infty} z^n P_{0n} = (1-\rho)\exp\left\{\frac{\lambda}{\theta}\int_1^z \frac{1-k(u)}{k(u)-u}du\right\}, \tag{2.11}$$

$$P_1(z) = \sum_{n=0}^{\infty} z^n P_{1n} = \frac{\lambda(1-z)}{k(z)-z}P_0(z). \tag{2.12}$$

In particular, the generating function of the marginal distribution of the queue length is

$$P(z) = \sum_{n=0}^{\infty} z^n P(N=n) = \frac{1-z}{k(z)-z}P_0(z). \tag{2.13}$$

From the generating functions, one can derive the mean and variance of N as given below (here $s^{(n)}$ is the nth moment of the service time):

$$E(N) = \frac{\lambda^2}{1-\rho}\left(\frac{\tau}{\theta}+\frac{s^{(2)}}{2}\right), \tag{2.14}$$

$$Var(N) = \left(\frac{1}{1-\rho}\right)\left(\frac{\lambda^3 s^{(3)}}{3}+\frac{\lambda^3 s^{(2)}}{2\theta(1-\rho)}+\frac{\lambda^4 (s^{(2)})^2}{4(1-\rho)}+\frac{\lambda\rho}{\theta}+\frac{\lambda^2 s^{(2)}}{2}\right). \tag{2.15}$$

The marginal distribution of the number of customers in service is given by

$$P(C=1) = \rho, \quad P(C=0) = 1-\rho. \tag{2.16}$$

So, the mean number of customers in the system is $E(C+N) = \rho + E(N)$.

These results are obtained by Keilson et al. [49] by using elapsed service times as a supplementary variable. The same results are obtained by various authors by various methods. See Alexandrov [1], Falin [11], and Jonin and Brezgunova [46].

In the transient state, the system can be described by the process $(C(t), N(t), I(t), \zeta(t))$ where $I(t)$ is the number of departures in the time interval $[0, t]$, and $\zeta(t)$ is the elapsed service time at time t if there is a customer in service. Falin [21, 27] studies the distributions

$$P_{0ni}(t) = P\{C(t)=0, N(t)=n, I(t)=i\},$$

$$P_{1ni}(t,x) = \frac{d}{dx}P\{C(t)=1, N(t)=n, I(t)=i, \zeta(t)<x\}.$$

He derives the LST of the partial generating functions of these distributions and also the LST of the mean queue length $N(t)$ and the blocking probability $P_1(t) = P(C(t)=1)$ assuming the initial state is empty.

M/G/1/1 with general retrial times. Yang et al. [73] and Liang [57] independently study the case when the retrial times are not exponential. They assume the retrial times are finite mixtures of Erlangs and develop a numerical algorithm to obtain the limiting distribution of the CN process.

M/G/1/1 with impatient customers. The retrial queue with impatient customers was first formulated by Cohen [8], where he studied case 3. Recently, Yang and Templeton [74] obtained some exact system measures for this case. They derived the probability $\pi(\gamma)$ that any arbitrary customer (retrial customer) leaves the system without service in terms of the server utilization $\rho_s = P(C=1)$ as follows:

$$\pi = 1 - \frac{\rho_s}{\rho}, \tag{2.17}$$

$$\gamma = 1 - \frac{\rho_s - \rho(1-\rho_s)}{\alpha\rho_s\rho}. \tag{2.18}$$

The moment generating functions of N and $C + N$ were also derived, and the means were obtained as follows:

$$E(N) = \left(1 - \frac{\rho_s}{\rho}\right)\left(\frac{\alpha\lambda}{(1 - \alpha)\theta}\right), \tag{2.19}$$

$$E(C + N) = \rho_s + \left(1 - \frac{\rho_s}{\rho}\right)\left(\frac{\alpha\lambda}{(1 - \alpha)\theta}\right). \tag{2.20}$$

They also proposed an algorithm to calculate the server utilization ρ_s.

The model of the case where $\alpha_0 < 1$ and $\alpha_n = 1$ for all $n \geq 1$ is similar to the case of persistent customers. Hence one can obtain the generating functions of marginal distributions $P_0(z)$, $P_1(z)$, and $P(z)$ as follows:

$$P_0(z) = \frac{1 - \rho\alpha_0}{1 + \rho(1 - \alpha_0)}\exp\left\{\frac{\lambda}{\theta}\int_1^z \frac{1 - k(u)}{k(u) - u}du\right\}, \tag{2.21}$$

$$P_1(z) = \lambda\frac{1 - z}{k(z) - z}P_0(z), \tag{2.22}$$

where $k(z) = \widetilde{B}(\lambda\alpha_0(1 - z))$. The queue length distribution has the generating function given by

$$P(z) = \frac{1 - k(z) + \alpha_0(k(z) - z)}{\alpha_0(k(z) - z)}P_0(z), \tag{2.23}$$

$$E(N) = \frac{\lambda^2\alpha_0}{1 - \rho\alpha_0}\left(\frac{\tau}{\theta} + \frac{s^2}{2(1 + \rho(1 - \alpha_0))}\right). \tag{2.24}$$

The marginal distribution of number of customers in service is

$$P_0 = \frac{1 - \rho\alpha_0}{1 + \rho(1 - \alpha_0)}, \tag{2.25}$$

$$P_1 = \frac{\rho}{1 + \rho(1 - \alpha_0)}. \tag{2.26}$$

See Lubacz and Roberts [62].

No exact solution is known for the case $\alpha_n = \alpha_1 < 1$ for $n \geq 1$. An approximation is given by Greenberg [38]. For the special case when the service times are exponential, Falin [14] derives the partial generating functions $P_0(z)$, $P_1(z)$, and P_1 and $E(N)$. See also Jonin and Sedol [48], Fayolle and Brun [36], and Bretschneider [2].

M/M/2/2. Jonin and Sedol [47] find the stationary distrubtion of the CN process as follows:

$$P_{00} = \left\{\sum_{j=0}^{\infty}\frac{\lambda^j}{\theta^j j!}\prod_{k=0}^{j-1}\frac{(\lambda + k\theta)^2 + k\theta}{2 + 3\lambda + 2\theta + 2k\theta}\right.$$
$$\left.\cdot\left[1 + \lambda + j\theta + \frac{(1 + \lambda + (j + 1)\theta)((\lambda + j\theta)^2 + j\theta)}{2 + 3\lambda + 2(j + 1)\theta}\right]\right\}^{-1}, \tag{2.27}$$

$$P_{0j} = \frac{\lambda^j}{\theta^j j!}\prod_{k=0}^{j-1}\frac{(\lambda + k\theta)^2 + k\theta}{2 + 3\lambda + 2\theta + 2k\theta}P_{00}, \tag{2.28}$$

$$P_{ij} = (\lambda + j\theta)\frac{\lambda^j}{\theta^j j!}\prod_{k=0}^{j-1}\frac{(\lambda + k\theta)^2 + k\theta}{2 + 3\lambda + 2\theta + 2k\theta}P_{00}, \tag{2.29}$$

$$P_{2j} = (1 + \lambda + (j+1)\theta) \frac{\lambda^j}{\theta^j j!} \prod_{k=0}^{j} \frac{(\lambda + k\theta)^2 + k\theta}{2 + 3\lambda + 2\theta + 2k\theta} P_{00}. \qquad (2.30)$$

Hanschke [43] shows that the generating function can be described by hypergeometric functions. Keilson et al. [49] propose a recursive algorithm to compute the steady state distribution.

M/M/s/s. Due to the complexity of the analysis, no explicit formulas are obtained for any main probability characteristics of retrial queues with more than two servers. However, several approximations and numerical methods have been proposed. Wilkinson [71] proposes an approximation method in which he truncates the orbit capacity to a finite integer that reduces to a system in a finite dimensional state space. He then derives a finite set of linear equations which yield the steady state distributions. The truncated system converges to the original system as the capacity increases to infinity. The convergence problems are studied by Cohen [8], Falin [23], and Greenberg [38].

Other approximation methods are also proposed. Both methods proposed by Falin [19] and by Greenberg [39] involve an adjustment of the retrial rates; while Neuts and Rao [64] use matrix-geometric approach. Greenberg [39] also obtains upper bounds for the performance measures.

M/G/1/B. Ridout [66] derives the steady state equations for the probabilities P_{in}. However, there are no closed form solutions. Hence, Ridout suggests an iterative procedure to approximate the P_{in}'s starting from analytical results when $B = 1$. The system performance measures, such as mean number of retrial customers, mean queue length, mean waiting time, etc., are expressed in terms of the P_{in}'s.

M/G/1/B with impatient customers and finite orbit. In the model studied by Hashida and Kawashima [44], the orbit capacity O is finite and $\alpha_n = \alpha$, $n \geq 0$. They derive equations for the probabilities P_{in}. Although the state space is finite, the calculation time grows rapidly when B and O become large. An efficient procedure is developed to calculate the exact values from which the system performance measures follow.

M/G/s/B with impatient customers. Stepanov and Tsitovich [69] study the case $\alpha_n = 1$, $n \geq 0$, but with rate γ a customer may leave the system without service while waiting in the buffer. They derive the necessary condition for stability

$$\lambda < s\tau + (B - s)\gamma \qquad (2.31)$$

and study the system characteristics under heavy traffic.

$M^X/G/1/1$. Assume that customers arrive at the system according to a compound Poisson process. If the service facility is occupied, all arrivals enter the orbit. Otherwise, one of the arrivals enters the service and the others (if any) join the orbit. The results with batch arrivals are similar to those for the M/G/1/1 retrial queue.

The generating function of the queue length distribution is similar to that in the M/G/1/1 retrial queue. Let $g^{(k)}$ be the kth moment of the batch size, with $g = g^{(1)}$. The first moments are given by

$$E(C + N) = \rho + \frac{1}{2(1-\rho)} \left\{ \lambda^2 s^{(2)} g^{(2)} + \lambda s(g^{(2)} - g) + \frac{2\lambda\rho}{\theta} + \frac{2\lambda(g-1)}{\theta} \right\}, \qquad (2.32)$$

$$E(N) = E(C + N) - \rho, \qquad (2.33)$$

$$E(W) = \frac{E(N)}{\lambda g}, \qquad (2.34)$$

$$E(R) = \theta E(W). \qquad (2.35)$$

See Falin [11] and Yang and Templeton [75] for different approaches. Falin [15] also studies the busy periods and transient distributions.

$M^X/G/1/1$ with impatient customers. This model is studied by Inamori et al. [5] under the assumption that the retrial times are constant. The authors develop an approximation algorithm.

M/G/1/1 with multiclass customers. Here the arrivals to a retrial queue belong to n distinct types. Type i customers arrive according to a Poisson process with rate λ_i; they demand i.i.d. service times with mean τ_i, and have retrial rate θ_i.

Hanschke [41] first considers the Poisson case when $\theta_i = \theta$ for all i. Kulkarni [53] studies the case with two types of customers. For each type i, he derives explicit formulas for the mean number of customers, the mean waiting time, and the mean number of retrials of type i customers. Later, Falin [18] generalizes the results to n types of customers. Furthermore, Kulkarni [52] studies an interesting game theoretic problem arising out of this model. Both Kulkarni [54] and Falin [30] consider the models with multiclass batch arrivals.

Miscellaneous. Kulkarni and Choi [55] study a retrial queue where the server is subject to failures and repairs. They consider several preemptive policies and derive expressions for the generating functions and the expected number in the system by using the theory of Markov regenerative processes. In a later paper, Choi and Kulkarni [3] study a retrial queue with customer feedback. See also Hanschke [41]. Choi et al. [6] study an application of retrial queues to telecommunication protocols. Choi and Park [4] consider a standard M/G/1 retrial queue with Bernoulli feedback. See also Choi et al. [5]. Keilson and Servi [50] consider a retrial queue with Markov-modulated arrivals. Falin [33] studies a retrial queue with priorities.

Open problems. As it follows from this section, queue lengths processes are the most extensively studied characteristics in the area of retrial queues. The most important open problem would be to establish results (analytical, numerical, and approximations) in the case of general retrial times. For example, the model of Erlangian retrial times can be analyzed by a multidimensional continuous time Markov chain. However, no analytical solutions are available. Another open problem is to find an analytical expression for the probability to find the server busy in an M/G/1/1 retrial queue with impatient customers. A third area of work is the case of a general service facility consisting of a queueing network. There is a single node where the external input to the network takes place. These externally arriving customers join the orbit if the network is in an "undesirable" state (e.g., if the node capacity is full, if the network contains more than a certain number of customers, etc.). Thus the area of loss networks can be enhanced with the results from retrial queues.

2.4 WAITING TIMES AND THE NUMBER OF RETRIALS

The waiting time of an arbitrary customer is the time between the moment he/she enters the system and the moment he/she enters service. Clearly the waiting time is a sum of a random number of retrial times and hence it is important to study the number of retrial attempts by a customer. Let W be the waiting time and R be the number of retrial attempts of a customer in the steady state. For a general retrial queue with i.i.d. retrial times, these two random variables are related as follows:

$$E(e^{-W}) = E((\widetilde{T}(s))^R),\tag{2.36}$$

where $\widetilde{T}(s)$ is the LST of a retrial time. The above equation yields

$$E(W) = E(R)\tau_r,\tag{2.37}$$

$$E(W^2) = E(R^2)\tau_r^2 + E(R)\sigma_r^2.\tag{2.38}$$

We will briefly describe the results on waiting time W, followed by those on R.

M/G/1/1. Falin and Fricker [34] derive the LST of W and the generating function of R. The expressions are complicated and hence are not reproduced here. From this, or by using Little's formula, we get

$$E(W) = \frac{\lambda}{1-\rho}\left(\frac{s^{(2)}}{2} + \frac{\tau}{\theta}\right).\tag{2.39}$$

Further results are derived for the case of exponential service times as described below.

M/M/1/1. Let μ be the service rate when the service times are exponential. Equation (2.39) becomes

$$E(W) = \frac{\rho}{1-\rho}\left(\frac{1}{\mu} + \frac{1}{\theta}\right).\tag{2.40}$$

The following expression for variance is derived by Falin [12]:

$$Var(W) = \frac{\rho}{(1-\rho)^2(2-\rho)}\left(\frac{4-2\rho+\rho^2}{\mu^2} + \frac{4}{\mu\theta} + \frac{4-4\rho+\rho^2}{\theta^2}\right).\tag{2.41}$$

See also Hanschke [42].

The problem of obtaining the distribution of W is extremely complicated. Falin and Fricker [34] give an expression for the LST of W. Later Grishechkin [40] extends the results. Falin [26] also obtains the generating function of R. Other work on waiting time for this model can be found in Choo and Conolly [7], Falin [12], Kulkarni [51], and Conolly [9].

Open problems. As can be seen from this section, the waiting time aspect is not as well studied as the queue length in retrial queues. In many of the models described in the previous sections, there are no known results for the waiting time and number of retrials.

2.5 ASYMPTOTIC ANALYSIS

Asymptotic analysis is the study of the retrial queueing systems as the traffic intensity ρ goes to zero or one, or as the retrial rate θ goes to zero or infinity.

M/G/1/1. As $\theta \to \infty$, the system converges to the corresponding classical M/G/1 queue. (See Falin [20].) As $\theta \to 0$, N is asymptotically Gaussian with

$$E(N) = \frac{\lambda\rho}{(1-\rho)\theta},\tag{2.42}$$

$$Var(N) = \frac{(\lambda^3 s^{(2)} + 2\lambda\rho - 2\lambda\rho^2)}{2(1-\rho)^2\theta}. \tag{2.43}$$

Falin [13, 25] also studies some interesting qualitative properties of the distribution. He shows that in heavy traffic, i.e., for $\rho = \lambda\tau \to 1$, the distribution of the scaled queue $(1-\rho)N$ converges weakly to the gamma distribution with

$$E(N) = \frac{\theta}{\tau} + \frac{s^{(2)}}{2\tau^2}, \tag{2.44}$$

and

$$Var(N) = \frac{s^{(2)}}{2\tau^2}\left(\frac{\theta}{\tau} + \frac{s^{(2)}}{2\tau^2}\right). \tag{2.45}$$

Another interesting result states that when the service time distribution is NBUE (new better than used in expectation), the number of customers in the system is stochastically smaller than a negative binomial random variable with distribution

$$\frac{\rho^n}{n!\theta^n}(1-\rho)^{1+\frac{\lambda}{\theta}}\prod_{i=0}^{n}(\lambda + i\theta) \quad n \geq 0. \tag{2.46}$$

See Falin [27] for more details.

Falin [26] also studies the asymptotic characteristics of the waiting time. He shows that

$$\lim_{\rho \to 1} E\exp(-s(1-\rho)W) = \int_0^\infty \left(1+\frac{su}{\mu}\right)^{-1-\frac{\mu}{\theta}}e^{-u}du, \tag{2.47}$$

$$\lim_{\theta \to 0} E\exp(-s\theta W) = 1 - \rho + \rho\frac{1-\rho}{1-\rho+s}. \tag{2.48}$$

M/M/s/s. As the retrial rate tends to infinity, the steady state distribution converges to that of the classical M/M/s queue (see Falin [20]). On the other hand, when the retrial rate decreases to zero, the limiting system can be viewed as an Erlang loss model with an increased arrival rate (see Cohen [8]). He also shows that the number of customers in service C and a scaled number of retrial customers N are asymptotically independent, where C is Erlang and N is Gaussian. Under heavy traffic, that is when $\lambda\tau \to s$, the following results are obtained by Stepanov [67]. Let $\epsilon = 1 - \frac{\lambda}{s}$; then,

$$E(C) = 1 - s\epsilon + o(\epsilon), \tag{2.49}$$

$$E(N) = \frac{1+\theta}{\theta\epsilon} - 2\frac{1+\theta}{\theta} + o(1). \tag{2.50}$$

Falin and Suharev [35] also show that ϵN converges weakly to the gamma distribution with mean and variance both equal to $\frac{1+\theta}{\theta}$.

M/M/s/s with impatient customers. In the case of the high retrial rate, if $\alpha_n = \alpha = 1$, $n \geq 1$, the system converges to the classical M/M/s queue with balking (see Falin [22]). If $\alpha_n = \alpha < 1$, $n \geq 1$, the limiting system is the Erlang loss system M/M/s/0, (see Falin [29]).

In heavy traffic in the case when $\alpha_n = \alpha$, $n \geq 1$, Stepanov [67] shows that

$$E(C) = 1 - \frac{(1-\alpha)s}{(\alpha_0 + 1 - \alpha)\lambda} + o\left(\frac{1}{\lambda}\right), \tag{2.51}$$

$$E(N) = \frac{\lambda\alpha_0}{\theta(1-\alpha_0)} - \frac{\alpha s}{\theta(1-\alpha)(\alpha_0 + 1 - \alpha)} + o(1). \tag{2.52}$$

Also, N is asymptotically Gaussian with mean and variance both equal to $\frac{\lambda\alpha_0}{\theta(1-\alpha)}$. See Falin and Suharev [35].

Miscellaneous. Falin [16] considers the case with special service time. The service time consists of two stages of service. In addition to being impatient in the orbit, customers may leave the system or go to the orbit after the first stage of service. Again the asymptotic system characteristics are obtained in light traffic. Other references on multiservers with impatient customers can be found in Jonin and Sedol [47, 48], Le Gall [56], and Falin [16].

Open problems. An interesting open problem is: can one devise an appropriate asymptotic region in which the queue length process, appropriately normalized, approaches a diffusion process? It stands to reason that Ornstein-Uhlenbeck processes might appear in such limiting regions, since the rate of departure from the orbit is proportional to the number of customers in the orbit.

2.6 BUSY PERIODS AND SERVER IDLE TIMES

The busy period is defined to be the time interval from the time of arrival to an empty system to the time of the first service completion leaving no customers in the system.

M/M/1/1. The busy period problem for this system is first studied by Choo and Conolly [7]. They derive equations for the density function of the busy period and develop an algorithm to obtain numerical solutions. An error in their analysis is discussed in Kulkarni [51]. See also Conolly [9].

M/G/1/1. Falin [13] studies the busy periods for the M/G/1/1 retrial queue. (See also Aleksandrov [1].) He derives the joint LST of the length of the busy period L and the number of customers M served in a busy period. The expression is quite involved and hence is not reproduced here. The expressions for $E(L)$ and $E(M)$ for the case $\rho < 1$ are as follows:

$$E(M) = \frac{1}{1-\rho}\exp\left\{\frac{\lambda}{\theta}\int_0^1 \frac{1-k(u)}{k(u)-u}du\right\}, \tag{2.53}$$

$$E(L) = \frac{E(M)-1}{\lambda}.$$

Falin [15] generalizes this result for the k-busy period, i.e., the time between an arrival to the system when the server is idle and with $k-1$ customers in the orbit until the system is exhausted.

Another question of interest is the server idle time. In the classical G/G/1 queue, the server idle time is equal to the system idle time, which is related to the busy period in a busy cycle. However, in a retrial queue the server can be idle even when there are customers in the system (orbit). Hence, one cannot derive the server idle time from the information on busy periods.

M/M/1/1. Choo and Conolly [7] study the server idle time I for the M/M/1/1 retrial queue. They derive the kth moment of server idle time in the steady state in terms of the distribution of number of customers at a departure epoch. For example,

$$E(I) = \frac{1-\rho}{\lambda}. \tag{2.54}$$

Yang and Templeton [75] derive the LST of I as follows:

$$E(e^{-sI}) = \frac{1}{\lambda+s}\left[\lambda + s\int_0^1 x^{\frac{s+\lambda}{\theta}}dQ(x)\right], \tag{2.55}$$

where $Q(s)$ is the moment generating function of the limiting distribution of the number of customers at an arbitrary departure epoch.

Open problems. There is clearly a host of unfinished work in extending these results to other retrial queueing systems mentioned in Section 2.3.

2.7 MONOTONICITY

Monotonicity can be thought of as a "descriptive" approach to the analysis of a queueing system. Instead of studying a performance measure in a quantitative fashion, this approach attempts to reveal a relationship between performance measures and the parameters of the system. This approach is first used for the retrial queues by Liang and Kulkarni [6] in displaying a relation between the retrial times and the queue length in a single server retrial system. They study a general case with a general arrival process and general service times. There is a finite waiting room in the primary queue. The retrial time distribution is a finite mixture of Erlangs, for which they define a new dominance relation called K-dominance. (Intuitively, the K-*dominance* condition says that the longer a retrial customer stays in the orbit, the less likely it is that he/she will attempt service. This is somewhat analogous to the decreasing failure rate concept.) They show that longer retrial times with respect to the K-dominance result in more customers in the system and in the orbit in the stochastic sense. The proof is based on sample path arguments. Falin [32] studies stochastic inequalities for the number in an M/G/1 queue using explicit formulas.

In a working paper, Liang [58] studies the monotonicity relation between the service time and the queue length in the system for a G/G/1/C retrial system with general retrial times. As one might expect, longer service times in the stochastic sense cause more customers in the system, provided the retrial times have decreasing failure rates.

Open problems. Since this is a relatively recent approach to the problem, there are quite a few topics of interest that remain to be investigated. For example, one might want to know how the number of retrial attempts is affected by the service time distribution or by the retrial time distribution. Another topic is the extension of the available results to multiple servers retrial systems.

2.8 OPTIMAL CONTROL

A simple control problem arises if we assume that there is a control mechanism that determines whether a customer requesting service goes to the orbit or to the queue. Liang [57, 59] studies the case of an M/M/1/∞ retrial queue with exponential retrial times. There is a waiting cost rate function associated with the number of customers in the queue and in the orbit. The cost for a customer waiting in the queue is higher than that for a customer waiting in the orbit. The goal is to minimize the long run expected total discounted cost.

Liang [57] shows that an optimal policy can be characterized by a switching curve as follows: Suppose the state of the system is described by (i, j), where i is the number of customers in the primary queue, and j is the number of customers in the orbit. For every j, there is a critical number $i^*(j)$ such that if an arriving customer finds the system in state (i, j), the optimal decision is to route the customer to the primary queue if $i \leq i^*(j)$. Furthermore, $i^*(j)$ is a *decreasing* function of j! She conjectures that there is also a critical number $j^*(i)$ such that the optimal decision is to route the customer to the primary queue in state (i, j) if $j \leq j^*(i)$. Furthermore, $j^*(i)$ is a *decreasing* function of i! Liang studies a special case with a finite state space and with a linear cost function. The numerical results support the conjecture.

Open problems. Besides further investigation of the switching curve, one might be interested in the study of control of service rate, which includes the control of the number of servers in the multiple servers case. Another interesting control problem is one with a limited amount of information: Suppose each customer requesting service can only observe the number in the primary queue, but not the number of customers in the orbit. Based on this information and the past observations, he/she must decide whether to join the primary queue now or conduct a retrial later. This problem can be formulated as a sequential game theory problem, since there is no central decision maker. Each customer tries to minimize his/her own costs. What will be an optimal strategy for each customer?

2.9 CONCLUSIONS

In this paper we attempted to bring together some of the existing results on retrial queues. We did not touch upon several other aspects, such as output processes, retrial queueing networks, etc. This does not imply that they are not important. We hope that we have convinced the reader that retrial queueing systems are interesting with applications and that some of the readers will be inspired enough to tackle a few of the open problems.

BIBLIOGRAPHY

[1] Aleksandrov, A.M., A queueing system with repeated orders, *Eng. Cybernetics.* **12**:3 (1974), 1-4.

[2] Bretschneider, G., Repeated calls with limited repetition probability, In: *Proc. 6th Int. Teletraffic Congress*, Munich (1970), 431/1-434/5.

[3] Choi, B.D. and Kulkarni, V.G., Feedback retrial queueing systems, In: *Queueing and Related Models* (ed. by U.N. Bhat and I.V. Basawa), Oxford University Press, New York, 93-105.

[4] Choi, B.D. and Park, K.K., The M/G/1 retrial queue with Bernoulli schedule, *Queueing Sys.* **7** (1990), 209-218.

[5] Choi, B.D., Park, K.Y., and Pearce, C.E.M., An M/M/1 retrial queue with control policy and general retrial times, *Queueing Sys.* **3** (1993), 257-292.

[6] Choi, B.D., Shin, Y.W., and Ahn, W.C., Retrial queues with collision arising from unslotted CMSA/CD protocol, *Queueing Sys.* **4** (1992), 335-356.

[7] Choo, Q.H. and Conolly, B.W., New results in the theory of repeated orders queueing systems, *J. Appl. Probab.* **16** (1979), 631-640.

[8] Cohen, J.W., Basic problems of telephone traffic theory and the influence of repeated calls, *Philips Telecom. Rev.* **18** (1957), 49-100.

[9] Conolly, B.W., Letter to the editor, *J. Appl. Probab.* **19** (1982), 904-905.

[10] Deul, N., Stationary conditions for multiserver queueing systems with repeated calls, *Elektronische Informationsverarbeitung und Kybernetik* **10-**

12:16 (1980), 607-613.

[11] Falin, G.I., Aggregate arrival of customers in a one-line system with repeated calls, *Ukrain. Math. J.* **28** (1976), 561-565.

[12] Falin, G.I., Waiting time in a single-channel queueing system with repeated calls, *Vestnik Moscow Univ. Ser. 15 Comput. Math. Cybernet* **4** (1977), 83-87.

[13] Falin, G.I., A single-line system with secondary orders, *Eng. Cybernet.* **17**:2 (1979), 76-83.

[14] Falin, G.I., An M/G/1 system with repeated calls in heavy traffic, *Vestnik Moscow Univ. Ser. 1 Math. Mech.* **6** (1980), 48-50.

[15] Falin, G.I., Functioning under nonsteady conditions of a single-channel system with group arrival of requests and repeated calls, *Ukrain. Math. J.* **33** (1981), 429-432.

[16] Falin, G.I., Investigation of weakly loaded switching systems with repeated calls, *Eng. Cybernet.* **19**:3 (1981), 69-73.

[17] Falin, G.I., Calculation of probabilistic characteristics of a multi-channel queue with repeated calls, *Vestnik Moskow Univ. Ser. 15, Vychisl. Math. Cybernet* **1** (1983), 35-41.

[18] Falin, G.I., The influence of inhomogeneity of the composition of subscribers on the functioning of telephone systems with repeated calls, *Eng. Cybernet.* **21**:6 (1983), 21-25.

[19] Falin, G.I., On the accuracy of a numerical method of calculation of characteristics of systems with repeated calls, *Elektrosvyaz* **8** (1983), 35-36.

[20] Falin, G.I., Asymptotic investigation of fully available switching system with high repetition intensity of blocked calls, *Moscow Univ. Math. Bull.* **39**:6 (1984), 72-77.

[21] Falin, G.I., Continuous approximation for a single server system with an arbitrary service time under repeated calls, *Eng. Cybernet.* **22**:2 (1984), 66-71.

[22] Falin, G.I., On sufficient conditions for ergodicity of multichannel queueing systems with repeated calls, *Adv. Appl. Probab.* **16** (1984), 447-448.

[23] Falin, G.I., Comparability to migration processes, *Probab. Theory Appl.* **2** (1986), 392-396.

[24] Falin, G.I., On ergodicity of multichannel queueing systems with repeated calls, *Sov. J. Comput. Syst. Sci.* **25**:1 (1986), 60-65.

[25] Falin, G.I., On heavily loaded systems with repeated calls, *Sov. J. Comput. Syst. Sci.* **24**:4 (1986), 124-128.

[26] Falin, G.I., On the waiting time process in a single-line queue with repeated calls, *J. Appl. Probab.* **23**:1 (1986), 185-192.

[27] Falin, G.I., Single-line repeated orders queueing systems, *Math. Operationsforschung und Statistik, Optimization* **5** (1986), 649-667.

[28] Falin, G.I., Estimations of error in approximation of countable Markov chains associated with models of repeated calls, *Vestnik Moscow Univ. Ser. 1 Math. Mech.* **2** (1987), 12-15.

[29] Falin, G.I., Multichannel queueing systems with repeated calls under high intensity of repetition, *J. Inform. Processing Cybernet.* **1** (1987), 37-47.

[30] Falin, G.I., On a multiclass batch arrival retrial queue (letter to the editor), *Adv. Appl. Probab.* **20**:2 (1988), 483-497.

[31] Falin, G.I., A survey of retrial queues, *Queueing Sys.* **7** (1990), 127-168.

[32] Falin, G.I., Stochastic inequalities for M/G/1 queue, *OR Letters* (1994), (to appear).

[33] Falin, G.I., Artalejo, J.R., and Martin, M., On the single server retrial queue with priority customers, *Queueing Sys.* **14** (1993), 439-455.

[34] Falin, G.I. and Fricker, C., On the virtual waiting time in an M/G/1 retrial queue, *J. Appl. Probab.* **28**:2 (1991), 446-460.

[35] Falin, G.I. and Suharev, Y.I., Singular perturbed equations and asymptotic investigation of stationary characteristics of retrial queues, *Vestnik Moscow Univ. Ser. 1, Math. Mech.* **5** (1988), 7-10.

[36] Fayolle, G. and Brun, M.A., On a system with impatience and repeated calls, In: *Queueing Theory and its Applications*, CWI Monographs **7**, North-Holland, Amsterdam-New York (1988), 283-303.

[37] Foster, F.G., On the stochastic matrices associated with certain queueing processes, *Ann. Math. Stat.* **24** (1953), 355-360.

[38] Greenberg, B.S., M/G/1 queueing systems with returning customers, *J. Appl. Probab.* **26**:1 (1989), 152-163.

[39] Greenberg, B.S. and Wolff, R.W., An upper bound on the performance of queues with returning customers, *J. Appl. Probab.* **24**:2 (1987), 466-475.

[40] Grishechkin, S.A., Multiclass batch arrival retrial queues analyzed as branching processes with immigration, *Queueing Sys.* **11**:4 (1992), 395-418.

[41] Hanschke, T., The M/G/1/1 queue with repeated attempts and different types of feedback effects, *OR Spektrum* **7** (1985), 209-215.

[42] Hanschke, T., A computational procedure for the variance of the waiting time in the M/M/1/1 queue with repeated calls, In: *Operations Research Proceedings* 1985, Springer-Verlag, Berlin 1986.

[43] Hanschke, T., Explicit formulas for the characteristics of the M/M/2/2 queue with repeated attempts, *J. Appl. Probab.* **24** (1987), 486-494.

[44] Hashida, O. and Kawashima, K., Buffer behavior with repeated calls, *Electronics and Commun. in Japan* **62-B** (1979), 3-27.

[45] Inamori, H., Sawai, M., Endo, T., and Tanabe, K., An automatically repeated call model in NTT public facsimile, In: *Proc. 11th Int. Teletraffic Congress* (1985), 1017-1023.

[46] Jonin, G.L. and Brezgunova, N.M., One-line system with repeated calls in the case of gamma-distribution occupation time, *Latvian Math. Yearbook* **11** (1972), 65-71.

[47] Jonin, G.L. and Sedol, J.J., Investigation of telephone systems with repeated calls, *Latvian Math. Yearbook* **7** (1970), 71-83.

[48] Jonin, G.L. and Sedol, J.J., Telephone systems with repeated calls, In: *Proc. 6th Int. Teletraffic Congress* Munich (1970), 435/1-435/5.

[49] Keilson, J., Cozzolino, J., and Young, H., A service system with unfilled requests repeated, *Oper. Res.* **16** (1968), 1126-1137.

[50] Keilson, J. and Servi, L.D., The matrix M/M/∞ system: retrial models and Markov modulated sources, *Adv. Appl. Probab.* **25**:2 (1993), 453-471.

[51] Kulkarni, V.G., Letter to the editor, *J. Appl. Probab.* **19** (1982), 901-904.

[52] Kulkarni, V.G., A game theoretic model for two types of customers competing for service, *Oper. Res. Lett.* **2** (1983), 119-122.

[53] Kulkarni, V.G., On queueing systems with retrials, *J. Appl. Probab.* **20** (1983), 380-389.

[54] Kulkarni, V.G., Expected waiting times in a multiclass batch arrival retrial queue, *J. Appl. Probab.* **23**:1 (1986), 144-154.

[55] Kulkarni, V.G. and Choi, B.D., Retrial queues with server subject to breakdowns and repairs, *Queueing Sys.* **7** (1990), 191-208.

[56] Le Gall, P., The repeated call model and the queue with impatience, In: *Proc. Third Int. Seminar on Teletraffic ·Theory* (1984), 290-293.

[57] Liang, H.M., *Retrial Queues*, Ph.D. Thesis, Univ. of North Carolina at Chapel Hill 1991.

[58] Liang, H.M., Monotonicity properties of single-server retrial queues with respect to service time, working paper (1994).

[59] Liang, H.M., Optimal control of single server retrial queues, working paper (1994).

[60] Liang, H.M. and Kulkarni, V.G., Monotonicity properties of single-server retrial queue, *Stochastic Models* **3**:3 (1993), 373-400.

[61] Liang, H.M. and Kulkarni, V.G., Stability condition for a single server retrial queue, *Adv. Appl. Probab.* **25** (1993), 690-701.

[62] Lubacz, J. and Roberts, J., A new approach to the single server repeat attempts system with balking, In: *Proc. 3rd Int. Seminar on Teletraffic Theory* Moscow (1984), 290-293.

[63] Meyn, S.P. and Tweedie, R.L., *Markov Chains and Stochastic Stability*, Springer-Verlag, New York 1993.

[64] Neuts, M.F. and Rao, B.M., Numerical investigation of a multiserver retrial model, *Queueing Sys.* **7**:2 (1990), 169-189.

[65] Pakes, A.G., Some conditions for ergodicity and recurrence of Markov chains, *Oper. Res.* **17** (1969), 1048-1061.

[66] Ridout, G.E., A study of retrial queueing systems with buffers, Master's Thesis, Univ. of Toronto, Toronto 1984.

[67] Stepanov, S.N., Asymptotic formulae and estimations for probabilistic characteristics of full-available group with absolutely persistent subscribers, *Probl. of Control and Info. Theory* **12**:5 (1983), 361.

[68] Stepanov, S.N., *Numerical Methods for Calculation for Systems with Repeated Calls*, Nauka, Moscow 1983.

[69] Stepanov, S.N. and Tsitovich, I.I., The model of a full-available group with repeated calls and waiting positions in the case of extreme load, *Prob. of Control and Info. Theory* **14**:1 (1985), 25.

[70] Tijms, H.J., *Stochastic Modeling and Analysis: A Computational Approach*, John Wiley, New York 1985.

[71] Wilkinson, R.I., Theories for toll traffic engineering in the USA, *Bell Syst. Techn. J.* **35**:2 (1956), 421-507.

[72] Wolff, R.W., *Stochastic Modeling and the Theory of Queues*, Prentice-Hall, Englewood Cliffs 1989.

[73] Yang, T., Posner, M.J.M., and Templeton, J.G.C., An approximation method for the M/G/1 retrial queue with general retrial times, working paper (1990).

[74] Yang, T., Posner, M.J.M., and Templeton, J.G.C., The M/G/1 retrial queue with nonpersistent customers, *Queueing Sys.* **7** (1990), 209-218.

[75] Yang, T. and Templeton, J.G.C., A survey on retrial queues, *Queueing Systems* **2** (1987), 203-233.

Chapter 3

Some results for the mean waiting-time and workload in GI/GI/k queues[1]

Daryl J. Daley

ABSTRACT   This chapter studies stationary GI/GI/k queueing systems with first-come first-served discipline and generic interarrival and service times T and S, respectively, both with first two moments finite and (relative) traffic intensity $\rho \equiv ES/(kET) < 1$. The components of the stationary Kiefer-Wolfowitz vector,

$$W \equiv (W_1, \ldots, W_k) =_d W_{(n)} \text{ (all } n\text{)}, \quad W_{(n)} \equiv (W_{n1}, \ldots, W_{nk}),$$

of workloads form an associated family of random variables. The first moments are bounded below as in

$$\sum_{i=1}^{k} (EW_i - \rho EW_1) \geq (ES^2 - ESET)_+ /2ET,$$

the bound being tight in D/GI/k systems for which $P\{S = jET$ for some $j = 0, 1, \ldots\} = 1$. If $ES^2 < kESET = (ES)^2/\rho$, then there is a system for which the mean waiting-time $EW_1 = 0$. By considering the limit as $\epsilon \to 0$ of a sequence of systems with two-point service time distributions specified by

$$P\{S = ES + \sigma/\epsilon\} = \epsilon^2 P\{S = ES - \sigma\epsilon\} = \epsilon^2/(1 + \epsilon^2),$$

where $\sigma^2 = \text{var}\, S$, an asymptotic decomposition result is established for the sum $EZ \equiv \sum_{i=1}^{k} (EW_i - \rho EW_1)$ amongst systems with given first two moments for S and T; it is new even for the single-server case. From this and further detailed asymptotic results it follows that when $\rho < 1 - k^{-1}$, amongst systems with given first two moments finite for S and T, there is always a sequence of systems for which $\inf EW_1 = 0$. Heuristic calculations indicate the nature of all the EW_i when $1 - k^{-1} < \rho < 1$. Evidence concerning conjectured upper bounds on EW_1 is reviewed.

CONTENTS

---

[1]This work was done in part while visiting the Center for Stochastic Processes, Statistics Department, University of North Carolina at Chapel Hill, supported by US AFOSR F49620-82-C0009.

3.1 INTRODUCTION

The properties of stationary stable k-server queueing systems GI/GI/k developed in this paper are mostly analogues of results for the simpler single-server system GI/GI/1. The general thrust of them concerns inequalities for the mean waiting time, but this quantity is harder to study in the k-server system because simple expressions for it also involve the mean workload at arrival epochs: this happens to coincide with the mean waiting time when $k = 1$ but not otherwise. Another contrasting result is that the mean waiting time is hardly affected by very occasional, very large service times unless the system is within one server of being overloaded.

We consider those first-come first-served (FCFS) k-server queueing systems with renewal arrival process, independent identically distributed (i.i.d.) service times (i.e., GI/GI/k systems in Kendall's notation) for which the generic interarrival and service time random variables (r.v.s) T and S, respectively, have distributions in the class Q_2 defined by prescribing the first two moments ES, ES^2, ET, ET^2, all finite, and such that the *relative traffic intensity* ρ, defined by

$$\rho \equiv \frac{ES}{kET}, \quad \text{satisfies } 0 < \rho < 1 \tag{3.1}$$

(strictly, then, $Q_2 = Q_2(ET, ES, ET^2, ES^2)$). Under this last condition, Kiefer & Wolfowitz [20] showed that there exists a unique stationary distribution for the random vector

$$W_{(n)} \equiv (W_{n1}, \ldots, W_{nk}) \quad (W_{n1} \leq W_{n2} \leq \ldots \leq W_{nk})$$

of occupation times or *workloads* of the various servers at an arrival epoch. Such random vectors satisfy

$$W_{(n+1)} = \Re((W_{n1} + S_n - T_n)_+, (W_{n2} - T_n)_+, \ldots, (W_{nk} - T_n)_+), \tag{3.2}$$

where S_n and T_n, distributed like S and T, are mutually independent and independent of $W_{(n)}$, and the operator $\Re$ orders its k arguments as the ascending components of a k-vector: equivalently, the components of $W_{(n+1)}$ and $W_{(n)}$ are related by

$$W_{n+1,i} = \begin{cases} (\min[W_{n,i+1}, \max(W_{n1} + S_n, W_{ni})] - T_n)_+ & (i = 1, \ldots, k-1), \\ (\max(W_{n1} + S_n, W_{nk}) - T_n)_+ & (i = k). \end{cases} \tag{3.3}$$

The same sort of argument as underlies (3.3) yields, for example,

$$W_{n+2,k} + W_{n+2,k-1}$$

$$= (W_{n+1,k} - T_{n+1})_+ + (\max(W_{n+1,k-1}, W_{n+1,1} + S_{n+1}) - T_{n+1})_+$$

$$= \sum_{i=0}^{1} (\max(W_{n+i,k-i}, W_{n+i,1} + S_{n+i}) - T_{[n+i,n+1]})_+,$$

where for integers $i \leq j$, $T_{[i,j]} = T_i + \ldots + T_j$; more generally, for positive integers $s \leq k$,

$$\sum_{i=0}^{s-1} W_{n+s,k-i}$$

$$= \sum_{i=0}^{s-1} (\max(W_{n+i,k-i}, W_{n+i,1} + S_{n+i}) - T_{[n+i,n+s-1]})_+. \tag{3.4}$$

For brevity, write $W = (W_1, \ldots, W_k)$ for a random vector with the stationary distribution for $\{W_{(n)}\}$, and

$$L = W_1 + \ldots + W_k \tag{3.5}$$

for a generic stationary total workload r.v. $L_n \equiv W_{n1} + \ldots + W_{nk}$. W_1 is the generic stationary waiting time r.v.

In a single-server system, the total workload and waiting time r.v.s coincide, whereas in the k-server system they are clearly distinct except for the trivial case that $W_{n1} = W_{nk} = 0$ a.s. This distinction affects the k-server analogue of the Pollaczek-Khinchin formula, for example, in an M/GI/k system, we have (e.g., Brumelle [5])

$$ETEL - ESEW_1 = \tfrac{1}{2}ES^2, \tag{3.6}$$

while in D/G$_{\mathbb{Z}_+ (ET)}$/k systems (where G$_{\mathbb{Z}_+ (ET)}$ denotes any discrete distribution for S concentrated on the lattice $\{0, ET, 2ET, \ldots\}$), we shall show that

$$ETEL - ESEW_1 = \tfrac{1}{2}(ES^2 - ESET)_+. \tag{3.7}$$

This equation is just the discrete analogue of (3.6), and the fact that it is an *equation* rather than an inequality enables us to deduce, just as Ott [32] showed for GI/GI/1 that

$$\inf_{\mathbb{Q}_2}(ET - ES)EW = \tfrac{1}{2}(ES^2 - ESET)_+, \tag{3.8}$$

that in GI/GI/k we have

$$\inf_{\mathbb{Q}_2}(ETEL - ESEW_1) = \tfrac{1}{2}(ES^2 - ESET)_+. \tag{3.9}$$

As another analogue of (3.8) concerning EW_1, we establish that

$$\text{when } \rho < 1 - k^{-1}, \quad \inf_{\mathbb{Q}_2} EW_1 = 0. \tag{3.10}$$

We have not been able to establish what holds when $1 - k^{-1} \leq \rho < 1$: on the basis of heuristic calculations we conjecture that the inequality

$$2k^2\rho(1 - \rho)ETEW_1 \geq [1 - k(1 - \rho)][ES^2 - (2k - 1)ESET + k(k - 1)(ET)^2]_+$$

$$= [1 - k(1 - \rho)](\text{var}S - (kET - ES)[ES - (k - 1)ET])_+ \tag{3.11}$$

holds, and is tight.

It is convenient for our discussion to introduce the generic r.v. (cf. the left-hand sides of (3.6) and (3.7))

$$Z \equiv (ET)L - (ES)W_1. \tag{3.12}$$

This paper has no claims to survey the breadth of work on particular many-server systems. Ovuworie [33] did such for the literature in the latter 1970s. Kimura [22] includes over 60 references concerning approximations for particular many-server systems. See also Stoyan [37].

3.2 RELATIONS CONCERNING FIRST MOMENTS

The basic equations from which we work are relations (3.3) and the fact, as follows from (3.2), that

$$W =_d \Re((W_1 + S - T)_+, (W_2 - T)_+, \ldots, (W_k - T)_+), \qquad (3.13)$$

where S and T are independent of $(W_1, \ldots, W_k)$ on the right-hand side. In the single-server case we have just

$$W =_d (W + S - T)_+. \qquad (3.14)$$

Kingman [23] initiated a fruitful method of studying EW in the single-server case by taking expectations in (3.14), leading via first moments to

$$E(T - S - W)_+ = ET - ES = (1 - \rho)ET, \qquad (3.15)$$

and via second moments (i.e., square and take expectations) to

$$2ET(1 - \rho)EW = E(T - S)^2 - E(T - S - W)_+^2 \qquad (3.16)$$
$$= \text{var}(T - S) - \text{var}(T - S - W)_+.$$

The k-server analogues of (3.15) and (3.16), as for example in Kingman [25], are derived by exploiting the symmetry of the sum and sum of squares of the components of a k-vector. Define

$$R_n = (T_n - S_n - W_{n1})_+ + (T_n - W_{n2})_+ + \ldots + (T_n - W_{nk})_+;$$

then, equating the expectations of L_n and L_{n+1} yields

$$ER_n = kET - ES = k(1 - \rho)ET, \qquad (3.17)$$

while equating the sums of squares of the elements of $W_{(n)}$ and $W_{(n+1)}$ yields (cf. Z at (3.12))

$$2EZ = 2(ETEL - ESEW_1) = 2(kET - ES)EW_1 + 2ET\sum_{i=2}^{k}(EW_i - EW_1)$$
$$= E(S - T)^2 + (k-1)ET^2 - \sum_{i=1}^{k} E(T - \delta_{i1}S - W_i)_+^2, \qquad (3.18)$$

where $\delta_{i1} = 1$ for $i = 1$, and $\delta_{i1} = 0$ otherwise. Suzuki & Yoshida [39] gave an alternative representation of (3.18) starting from

$$L_{n+1} = \sum_{i=1}^{k} (W_{ni} + \delta_{i1}S_n - T_n)_+ = L_n + S_n - kT_n + R_n;$$

rewrite it in the form

$$L_{n+1} - R_n = L_n + S_n - kT_n$$

so that, after taking second moments and eliminating the common variances,

$$-2\,\text{cov}(R_n, L_{n+1}) + \text{var}\,R_n = \text{var}(S_n - kT_n)$$
$$= -2\,\text{cov}(R_n, (L_{n+1} - kW_{n+1,1}) + kW_{n+1,1}) + \text{var}\,R_n.$$

Since $R_n W_{n+1,1} = 0$ a.s.,

$$-2\,\text{cov}(R_n, kW_{n+1,1}) = 2ER_n EkW_{n+1,1} = 2k^2(1 - \rho)ETEW_1,$$

and thus, as given by Suzuki & Yoshida [39],

$$2k^2(1 - \rho)ETEW_1 = \text{var}(S - kT) - \text{var}\,R_n + 2\,\text{cov}(R_n, L_{n+1} - kW_{n+1,1}). \qquad (3.19)$$

Further relations that depend on the use of symmetric functions of $(W_1, \ldots, W_k)$ are given in Whitt [41].

Example 3.1. (D/D/k) The functions $W_i(ET, ES)$ that occur in the completely deterministic system D/D/k are directly computable: elementary algebra gives, for $0 < \rho < 1$ as we are assuming,

$$\mathrm{E}W_i(ET, ES) = {}_{\mathrm{a.s.}} W_i(ET, ES) = {}_{\mathrm{a.s.}} (ES - (k + 1 - i)ET)_+$$
$$= (\rho k - (k + 1 - i))_+ ET, \tag{3.20}$$

which is positive or zero according to whether $i >$ or $\leq 1 + k(1 - \rho)$.

Example 3.2. (GI/D/k) In the k-server queue with constant service times $S = {}_{\mathrm{a.s.}} ES < kET$, the component r.v.s $\widetilde{W}_i$ of the stationary workload vector $\widetilde{W}$ satisfy

$$\widetilde{W}_i = {}_d \begin{cases} (\widetilde{W}_{i+1} - T)_+ & (i = 1, \ldots, k - 1), \\ (\widetilde{W}_1 + ES - T)_+ & (i = k). \end{cases} \tag{3.21}$$

Thus,

$$\widetilde{W}_1 = {}_d (\widetilde{W}_1 + ES - T_{[1,k]})_+, \tag{3.22}$$

i.e., $\widetilde{W}_1$ behaves like the stationary waiting time r.v. in a particular single-server system $\mathrm{GI}^{*k}/\mathrm{D}/1$, where GI^{*k} denotes the k-fold convolution power of the inter-arrival time distribution of GI/D/k.

In particular, the stationary workload vector in M/D/k can be studied via the stationary waiting-time r.v. of $\mathrm{E}_k/\mathrm{D}/1$.

Example 3.3. In a stationary GI/D/k system with $k \geq 3$, $\{\mathrm{E}W_i : i = 1, \ldots, k\}$ is a monotone convex increasing sequence.

Proof. Monotonicity is a consequence of the definition of the Kiefer-Wolfowitz vector W of stationary workloads. For nonnegative real numbers x, y and z, $x + (x - y - z)_+ \geq (x - y)_+ + (x - z)_+$. Now for $1 < j < k$,

$$\widetilde{W}_j = {}_d (\widetilde{W}_{j+1} - T_1)_+ = {}_d (\widetilde{W}_{j+1} - T_2)_+,$$
$$\widetilde{W}_{j-1} = {}_d (\widetilde{W}_j - T_2)_+ = {}_d (\widetilde{W}_{j+1} - T_1 - T_2)_+, \tag{3.23}$$

so

$$\mathrm{E}\widetilde{W}_{j+1} + \mathrm{E}\widetilde{W}_{j-1} = \mathrm{E}\widetilde{W}_{j+1} + \mathrm{E}(\widetilde{W}_{j+1} - T_1 - T_2)_+$$
$$\geq \mathrm{E}(\widetilde{W}_{j+1} - T_1)_+ + \mathrm{E}(\widetilde{W}_{j+1} - T_2)_+ = 2\mathrm{E}\widetilde{W}_j.$$

Remark 3.4. The same argument shows that the mean of the stationary workload vector in a k-server system with the cyclic allocation discipline (e.g., Wolff [44]) is convex in its components. The convexity property does hold for stable GI/M/k systems (see Example 3.5), and as a light-traffic limit of GI/GI/k systems (this follows easily from Daley and Rolski [13]). Whether or not it holds in general for GI/GI/k we do not know; see the discussion following Example 3.5.

Example 3.5. (Convexity of $\{\mathrm{E}W_i\}$ in GI/M/k) The stable k-server queueing system GI/M/k is amenable to study via the embedded Markov chain $\{X_n\}$ consisting of the number of customers already in the system at customer arrival epochs (Kendall [19]). Write $p_j = \mathrm{P}\{X_n = j\}$ for the stationary distribution of this chain, and $P_j = p_0 + \ldots + p_j$ for its cumulative distribution. Let $E(\lambda)$ denote a r.v. with the negative exponential distribution on $(0, \infty)$ with mean $1/\lambda$, and write E_r ($r = 1, 2, \ldots$) for r.v.s i.i.d. like $E(k\mu)$, where $1/\mu$ is the mean service time. Then, using the properties of exponential r.v.s, the components W_i of the generic stationary workload vector W have the representation,

$$W_1 = {}_d \begin{cases} 0 & \text{with prob. } P_{k-1}, \\ E_1 + \ldots + E_{j+1-k} & \text{with prob. } p_j \ (j = k, k+1, \ldots), \end{cases} \tag{3.24}$$

and, for $i = 2, \ldots, k$,

$$W_i =_d \begin{cases} 0 & \text{with prob. } P_{k-i}, \\ E(j\mu) + \ldots + E([k-i+1]\mu) & \text{with prob. } p_j \ (j = k-i+1, \ldots, k-1), \\ E_1 + \ldots + E_{j+1-k} + E([k-1]\mu) \\ \qquad + \ldots + E([k-i+1]\mu) & \text{with prob. } p_j \ (j = k, k+1, \ldots), \end{cases}$$

$$=_d \begin{cases} 0 & \text{with prob. } P_{k-i}, \\ E([k-i+1]\mu) & \text{with prob. } p_{k-i+1}, \\ W_{i-1} \mid \{W_{i-1} > 0\} + E([k-i+1]\mu) & \text{with prob. } 1 - P_{k-i+1}, \end{cases} \quad (3.25)$$

where the two random variables in the last representation are mutually independent. Thus, for $1 < i < k$, $E(W_i - W_{i-1}) = (1 - P_{k-i})/[(k-i+1)\mu]$, and hence

$$E(W_{i+1} - 2W_i + W_{i-1}) = \frac{1 - P_{k-i-1}}{(k-i)\mu} - \frac{1 - P_{k-i}}{(k-i+1)\mu}$$
$$= (1 - P_{k-i})\left(\frac{1}{(k-i)\mu} - \frac{1}{(k-i+1)\mu}\right) + \frac{p_{k-i}}{(k-i)\mu},$$

which, being positive, proves the convexity property.

Why might the same property hold true in GI/GI/k? The Kiefer-Wolfowitz workload vector satisfies a recursion relation that entails the incremental operation $W_{n1} + S_n$, the reordering operator $\mathfrak{R}$ which takes this incremented component and 'slots it into place' amongst the other components $\{W_{ni} : i = 2, \ldots, k\}$, and then decrements all components of the vector equally, subject to the action of the nonnegative truncation operator $(\cdot)_+$. On average, the difference between adjacent components that are 'appreciable' is thus an interarrival time, as is plain in Example 3.3. The 'slotting' operation tends mostly to locate $W_{n1} + S_n$ as the basis of the largest component of $W_{(n+1)}$, so the differences between adjacent workload components at the $(n+1)^{th}$ arrival epoch are, roughly speaking, those of the components with index one larger than at the previous arrival epoch. Specifically, for $i = 2, \ldots, k-1$, $\Delta_{n+1,i-1} \equiv W_{n+1,i} - W_{n+1,i-1}$ equals

$$\begin{cases} (W_{n,i+1} - T_n)_+ - (W_{ni} - T_n)_+ \leq \Delta_{ni}, & \text{if } W_{n1} + S_n \geq W_{n,i+1}, \\ (W_{n1} + S_n - T_n)_+ - (W_{ni} - T_n)_+ < \Delta_{ni}, & \text{if } W_{ni} \leq W_{n1} + S_n < W_{n,i+1}, \\ (W_{ni} - T_n)_+ - (W_{n1} + S_n - T_n)_+ < \Delta_{n,i-1}, & \text{if } W_{n,i-1} < W_{n1} + S_n \leq W_{ni}, \\ (W_{ni} - T_n)_+ - (W_{n,i-1} - T_n)_+ \leq \Delta_{n,i-1}, & \text{if } W_{n1} + S_n \leq W_{n,i-1}, \end{cases}$$

$$(3.26)$$

so for some quantities ϑ_{nj} satisfying $0 \leq \vartheta_{nj} \leq 1$ $(j = 1, 2)$ and determined by $W_{n1}, W_{n,i-1}, W_{ni}, W_{n,i+1}, S_n,$ and T_n,

$$\Delta_{n+1,i-1} = \begin{cases} \vartheta_{n1}\Delta_{ni}, & \text{if } W_{n1} + S_n \geq W_{ni}, \\ \vartheta_{n2}\Delta_{n,i-1}, & \text{if } W_{n1} + S_n \leq W_{ni}. \end{cases} \quad (3.27)$$

Using indicator variables like $I_{\{W_{n1} + S_n < W_{ni}\}}$, we can write

$$E(\Delta_{n+1,i-1}) = E[\vartheta_{n1}\Delta_{ni}I_{\{W_{n1}+S_n \geq W_{ni}\}} + \vartheta_{n2}\Delta_{n,i-1}I_{\{W_{n1}+S_n < W_{ni}\}}].$$

Now we can appeal to stationarity and write

$$E(\Delta_{i-1}) + E[(1-\vartheta_{n1})\Delta_{ni}I_{\{W_{n1}+S_n \geq W_{ni}\}}]$$
$$= E[\Delta_i - E[(\Delta_{ni} - \theta_{n2}\Delta_{n,i-1})I_{\{W_{n1}+S_n < W_{ni}\}}],$$

so

$$E\Delta_i - E\delta_{i-1} = E[(1-\theta_{n1})\Delta_{ni}I_{\{W_{n1}+S_n > W_{ni}\}}]$$
$$+ E[(\Delta_{ni} - \vartheta_{n2}\Delta_{n,i-1})I_{\{W_{n1}+S_n < W_{ni}\}}]. \tag{3.28}$$

If the conjecture is false, then for some i the left-hand side is negative. For this to be so, since the first term on the right-hand side is never negative, the other term there must be negative, which means that, on the set where $\Delta_{n1} + \ldots + \Delta_{n,i-1} > S_n$, the contribution to the expectation of Δ_{ni} is smaller than the contribution to the expectation of $\vartheta_{n2}\Delta_{n,i-1}$. Intuitively, this seems most unlikely.

Note that

$$EZ = (1-\rho)ETEL + \rho ET\sum_{i=1}^{k-1} iE\Delta_{k-i} = ET\sum_{i=1}^{k-1} iE\Delta_{k-i} + (kET - ES)EW_1.$$

Convexity would imply from these relations, respectively, that

$$EZ \leq (1-\rho)ETEL + \tfrac{1}{2}ES(EW_k - EW_1) \leq \tfrac{1}{2}kET(EW_k - 2\rho EW_1),$$
$$EZ \geq ET[\tfrac{1}{2}k(k-1)E\Delta_1] + (kET - ES)EW_1. \tag{3.29}$$

3.3 CONTINUITY RESULTS FOR FAMILIES OF GI/GI/k SYSTEMS

We use the following continuity properties of GI/GI/k systems.

Continuity Theorem 3.6. (e.g., Borovkov [4], §28, Theorem 3) *Let sequences of GI/GI/k systems be determined by generic r.v.s $\{(T^{(\nu)}, S^{(\nu)}): \nu = 1, 2, \ldots\}$ that converge weakly to limits $(T^{(\infty)}, S^{(\infty)})$, such that the first moments converge, $ET^{(\nu)} \to ET^{(\infty)}$, $ES^{(\nu)} \to ES^{(\infty)}$, with $ES^{(\infty)} < kET^{(\infty)}$. Then*

(i)   the corresponding stationary workload vectors $\{W^{(\nu)}\}$ converge weakly, and

(ii)   the limit $W^{(\infty)}$ corresponds to the system determined by the limit pair $(T^{(\infty)}, S^{(\infty)})$.

Such weak convergence implies via Fatou's lemma that

$$\liminf_{\nu \to \infty} EW_i^{(\nu)} \geq EW_i^{(\infty)} \quad (i = 1, \ldots, k). \tag{3.30}$$

Section 3.5 below gives sufficient conditions for equality to hold in (3.30) for $i = 1, \ldots, k-1$ for certain sequences $\{S^{(\nu)}\}$, and indicates what may hold concerning $EW_i^{(\infty)}$ when the inequality at (3.30) is strict.

Given a GI/GI/k system with $\rho < 1$, consider a related system with generic service time r.v. $S^{(\epsilon)}$ say, expressible in terms of a parameter β and an indicator r.v. $I^{(\epsilon)}$, which is independent of S, as in

$$S^{(\epsilon)} = S(1 - I^{(\epsilon)}\beta\epsilon + (1 - I^{(\epsilon)})(\beta/\epsilon))$$
$$= \begin{cases} S(1-\beta\epsilon) & \text{with probability } 1/(1+\epsilon^2), \\ S(1+\beta/\epsilon) & \text{with probability } \epsilon^2/(1+\epsilon^2), \end{cases} \tag{3.31}$$

where $P\{I^{(\epsilon)} = 0\} = 1 - Pr\{I^{(\epsilon)} = 1\} = 1/(1+\epsilon^2)$; in (31), $0 < \beta < \infty$ and

$0 < \epsilon < 1/\beta$, so that for all such ϵ, $S^{(\epsilon)} \geq 0$ a.s., $ES^{(\epsilon)} = ES > 0$, and $E(S^{(\epsilon)})^2 = (1 + \beta^2)ES^2$. Denote by

$$W(T, S^{(\epsilon)}) \equiv (W_1(T, S^{(\epsilon)}), \ldots, W_k(T, S^{(\epsilon)})) \equiv (W_1^{(\epsilon)}, \ldots, W_k^{(\epsilon)}) \equiv W^{(\epsilon)} \qquad (3.32)$$

the stationary workload vector corresponding to $(T, S^{(\epsilon)})$. Sequences $\{T^{(\epsilon)}\}$ and $\{W(T^{(\epsilon)}, S)\}$ are defined similarly. Our first proposition is an immediate consequence of the Continuity Theorem.

Proposition 3.7. *For (T, S) in $\mathcal{Q}_2$, the k-vectors $W(T, S^{(\epsilon)})$ and $W(T^{(\epsilon)}, S)$ converge weakly as $\epsilon \to 0$ to the k-vector $W(T, S)$.*

We turn now to the main subject of this paper, the first moment properties of these sequences. Here the similarity of the two sequences evident in Proposition 3.7 disappears, because if (3.34) were to hold with lim in place of lim inf, an inconsistency with (3.35) would result.

Proposition 3.8. *Let a $GI/GI/k$ system with $(T, S) \in \mathcal{Q}_2$ and $0 < \rho < 1$ be given.*

(a)
$$\lim_{\epsilon \to 0} EW_i(T^{(\epsilon)}, S) = EW_i(T, S) \quad (i = 1, \ldots, k). \qquad (3.33)$$

(b)
$$\liminf_{\epsilon \to 0} EW_i(T, S^{(\epsilon)}) \geq EW_i(T, S) \quad (i = 1, \ldots, k), \qquad (3.34)$$

while for the random variable $Z^{(\epsilon)} \equiv Z(T, S^{(\epsilon)})$, $\lim_{\epsilon \to 0} EZ^{(\epsilon)}$ exists and satisfies

$$\lim_{\epsilon \to 0} EZ^{(\epsilon)} = EZ(T, S) + \tfrac{1}{2}\beta^2 ES^2. \qquad (3.35)$$

Proof. The inequality at (3.34) is a consequence of Fatou's Lemma as at (3.30).

To prove (3.33), observe that for a system with $\{T_n\}$ replaced by $\{\check{T}_n^{(\delta)}\} \equiv \{(1 - \delta)T_n\}$ with $0 < \delta < 1 - \rho$, the stationary workload vector $\check{W}^{(\delta)}$ exists and satisfies $\check{W}^{(\delta)} \geq_d W$ componentwise (this follows from observations in Kiefer & Wolfowitz [20]). Further, the weak convergence of $\check{W}^{(\delta)}$ to W is monotone in $\delta \to 0$ so either of the dominated and monotone convergence theorems implies that $\lim E\check{W}^{(\delta)} = EW_i$ $(i = 1, \ldots, k)$. Now consider $W_i(T^{(\epsilon)}, S)$. For every ϵ, $W_i(\check{T}^{(\beta\epsilon)}, S) \geq_d W_i(T^{(\epsilon)}, S)$ componentwise, so the monotone convergence of $E\check{W}^{(\beta\epsilon)}$ and the weak convergence of $W_i(T^{(\epsilon)}, S)$ and $W_i(\check{T}^{(\beta\epsilon)}, S)$ to the same limit proves (3.33).

To prove (3.35), start by noting that from (3.18) we have

$$2EZ^{(\epsilon)} = E(S^{(\epsilon)} - T)^2 + (k-1)ET^2 - \sum_{i=1}^{k} E(T - \delta_{i1}S^{(\epsilon)} - W_i^{(\epsilon)})_+^2$$

$$= \mathrm{var}S + \beta^2 ES^2 + E(T - ES)^2 + (k-1)ET^2 - \sum_{i=1}^{k} EA_2(\delta_{i1}S^{(\epsilon)} + W_i^{(\epsilon)}) \qquad (3.36)$$

where the function $A_2(x) \equiv E(T - x)_+^2$ is a continuous bounded function of x in $0 \leq x < \infty$. $S^{(\epsilon)}$ here is independent of each component of $W^{(\epsilon)}$ so, using Proposition 7, $\delta_{i1}S^{(\epsilon)} + W_i^{(\epsilon)}$ converges weakly to $\delta_{i1}S + W_i$ as $\epsilon \to 0$. Then by bounded convergence,

$$EA_2(\delta_{i1}S^{(\epsilon)} + W_i^{(\epsilon)}) \to EA_2(\delta_{i1}S + W_i) = E(T - \delta_{i1}S - W_i)_+^2.$$

Consequently, $2EZ^{(\epsilon)}$ converges to a limit as $\epsilon \to 0$, and by inspection, $\lim_{\epsilon \to 0} 2EZ^{(\epsilon)} - \beta^2 ES^2$ equals

$$\mathrm{E}(T-S)^2 + (k-1)\mathrm{E}T^2 - \sum_{i=1}^{k} \mathrm{E}(T - \delta_{i1}S - W_i)_+^2 = 2\mathrm{E}Z(T,S).$$

Strictly, the limit of $\mathrm{E}Z^{(\epsilon)}$ depends on β, and to reflect this we could write $Z(T, S^{(\epsilon)}; \beta)$. Write $c^2 = (\mathrm{var}S)/(\mathrm{E}S)^2$ for the coefficient of variation of S; we then have the identities

$$\lim_{\epsilon \to 0} \mathrm{E}Z(T, S^{(\epsilon)}; \beta) - \mathrm{E}Z(T, S; 0) = \lim_{\epsilon \to 0} \mathrm{E}Z(\mathrm{E}T, S^{(\epsilon)}; \beta) - \mathrm{E}Z(\mathrm{E}T, S; 0)$$

$$= \lim_{\epsilon \to 0} \mathrm{E}Z(\mathrm{E}T, (\mathrm{E}S)^{(\epsilon)}; \beta(1 + c^2)^{1/2}) - \mathrm{E}Z(\mathrm{E}T, \mathrm{E}S; 0). \qquad \square \tag{3.37}$$

This relation indicates that we can describe the effect of rarely occurring large service times on the mean workload, as reflected in $\mathrm{E}Z$, via just the two moments of S and T.

The class of distributions $\mathbb{Q}_2$ includes the two-point distributions defined by

$$S = D^{(\epsilon)} = \begin{cases} \mathrm{E}S - \sigma\epsilon & \text{with probability } 1/(1+\epsilon^2), \\ \mathrm{E}S + \sigma/\epsilon & \text{with probability } \epsilon^2/(1+\epsilon^2), \end{cases} \tag{3.38}$$

where $0 < \epsilon \leq \mathrm{E}S/\sigma$, $\mathrm{E}S/\sigma$, $\mathrm{E}S$ and $\sigma^2 \equiv \mathrm{var}\,S$ being prescribed. Setting $\beta = \sigma\mathrm{E}S$ in (3.35) in the special case $k = 1$ yields the following particular result.

Corollary 3.9. *In the family of stationary* $GI/D^{(\epsilon)}/1$ *queues,*

$$(\mathrm{E}T - \mathrm{E}S)\lim_{\epsilon \to 0} \mathrm{E}W(T, \mathrm{E}S^{(\epsilon)}) = (\mathrm{E}T - \mathrm{E}S)\mathrm{E}W(T, \mathrm{E}S) + \tfrac{1}{2}\mathrm{var}\,D^{(\epsilon)}. \tag{3.39}$$

This result shows in particular that the upper bound (Kingman [23]; Daley [8])

$$\mathrm{E}W(T, S) \leq \frac{\rho(2-\rho)\mathrm{var}\,T + \mathrm{var}\,S}{2(1-\rho)\mathrm{E}T} \tag{3.40}$$

is tight in a D/GI/1 system. This tightness result was demonstrated by explicit calculation in an example in Trengove [40] (also, see Daley, Kreinin & Trengove [14]); the method used above is more elegant and, as shown, is applicable to GI/GI/k.

Proposition 3.10. (Comparison of k- and $(k-1)$-server systems) *Let there be given a GI/GI/k system with* $\mathrm{E}S < (k-1)\mathrm{E}T$ *(equivalently,* $\rho < 1 - k^{-1}$*), with stationary workload vector* W*. The GI/GI/$(k-1)$ system with the same generic interarrival and service time r.v.s* T *and* S *has a stationary* $(k-1)$*-dimensional workload vector* W' *which dominates in distribution the first* $k-1$ *components of* W*. When* (T, S) *is in* $\mathbb{Q}_2$*, all components of* W' *have a finite first moment.*

Proof. The condition $\rho < 1 - k^{-1}$ for the given GI/GI/k system implies that the $(k-1)$-server system in question is stable, i.e., the stationary r.v. W' is well-defined. Loynes' construction as used in studying GI/GI/k systems (e.g., Borovkov [4], §27, or Sigman [35], §6.2) then shows that W' is componentwise larger in distribution than the first $k-1$ components of the stationary k-vector W as asserted. $\qquad \square$

In our proof of Proposition 3.15 below we use a further elaboration of this comparison result. For a given stationary realization $W_{(n)}$, construct a realization $\{W'_{(n)}: n \geq 0\}$ of a GI/GI/$(k-1)$ system by setting $W'_{(0)}$ equal to the first $k-1$ components of $W_{(0)}$ so that $W'_{(0)} \leq_d W'$. Then $W'_{(n)} \leq_d W'$ for all $n \geq 0$, and $W'_{(n)}$ converges weakly to W' by the Continuity Theorem. Further, for (T, S) in $\mathbb{Q}_2$, $\mathrm{E}W'$ has all its components finite, so by the dominated convergence theorem $\mathrm{E}W'_{(n)}$ converges to $\mathrm{E}W'$.

3.4 SOME BOUNDS INVOLVING EZ IN GI/GI/k SYSTEMS

Proposition 3.11. *In a GI/GI/k system,*

$$EZ = ETEL - ESEW_1 = ET \sum_{i=1}^{k} E(W_i - \rho W_1) \geq \tfrac{1}{2}(ES^2 - ESET)_+ . \quad (3.41)$$

Equality holds in any D/GI/k system for which $ES^2/ES \geq ET$ *if and only if the support of the r.v.* S *is contained in* $\mathbb{Z}_+(ET)$, *i.e.,*

$$P\{S = jET \text{ for some } j = 0,1,\ldots\} = 1. \quad (3.42)$$

Proof. Equality at (3.41) comes from the definitions at (3.34) and (3.1). For the inequality, express the right-hand side of (3.18) in the form

$$ES^2 - ESET + \sum_{i=1}^{k} (E[T(T - \delta_{i1}S)] - E[(T - \delta_{i1}S - W_i)(T - \delta_{i1}S - W_i)_+])$$
$$= ES^2 - ESET + \sum_{i=1}^{k} E(T[T - \delta_{i1}S - (T - \delta_{i1}S - W_i)_+])$$
$$+ \sum_{i=1}^{k} E[(W_i + \delta_{i1}S)(T - \delta_{i1}S - W_i)_+].$$

The last term is nonnegative by inspection, while in each expectation in the other sum, the second factor is monotonic nonincreasing in each of the three independent r.v.s T, $-\delta_{i1}S$, and W_i which, being independent, form an associated family, so the ith term in the sum is bounded below by the product of expectations

$$ET[ET - \delta_{i1}ES - E(T - \delta_{i1}S - W_i)_+],$$

and the sum itself is therefore bounded below by

$$ET[kET - ES - \sum_{i=1}^{k} E(T - \delta_{i1}S - W_i)_+] = ET[kET - ES - ER_n] = 0,$$

using (3.17). This proves the inequality at (3.41).

In a D/GI/k system with S as at (3.42), i.e., a $D/G_{\mathbb{Z}_+(ET)}/k$ system, all r.v.s W_i are a.s. nonnegative integer multiples of ET also because, by (3.3), the possible points of support of the set of r.v.s $\{W_i\}$ coincide with those of

$$\{(S_1 + \ldots + S_j - \ell ET)_+ : 1 \leq j \leq \ell = 1,2,\ldots\}. \quad (3.43)$$

Write

$$p_0 = P\{S = 0\}, \quad \pi_i = P\{W_i = 0\} \quad (i = 1,\ldots k).$$

Then from (3.17), each term $(T - \delta_{i1}S - W_i)_+ = 0$ or ET, so

$$k(1 - \rho)ET = \sum_{i=1}^{k} E(T - \delta_{i1}S - W_i)_+ = (p_0\pi_1 + \pi_2 + \ldots + \pi_k)ET. \quad (3.44)$$

Similarly, in (3.18),

$$\sum_{i=1}^{k} E(T - \delta_{i1}S - W_i)^2_+ = (p_0\pi_1 + \pi_2 + \ldots + \pi_k)(ET)^2 = k(1 - \rho)(ET)^2. \quad (3.45)$$

Thus, for any $D/G_{\mathbb{Z}_+(ET)}/k$ system,

$$2(ETEL - ESEW_1) = ES^2 - ESET + ET(kET - ES) - k(1 - \rho)(ET)^2$$

$$= \mathrm{E}S^2 - \mathrm{E}S\mathrm{E}T = \mathrm{E}[S(S - \mathrm{E}T)] = \sum_{j=2}^{\infty} j(j-1)(\mathrm{E}T)^2 \mathrm{P}\{S = j\mathrm{E}T\},$$

i.e., equality holds at (3.41) as asserted.

For the converse, observe that the terms $\mathrm{E}[(W_i + \delta_{i1}S)(T - \delta_{i1}S - W_i)_+]$ vanish if and only if one or the other of the factors vanishes a.s., i.e., for each $i = 1, \ldots, k$,

$$1 = \mathrm{P}\{W_i + \delta_{i1}S = 0\} + \mathrm{P}\{\mathrm{E}T \le \delta_{i1}S + W_i\}$$
$$= 1 - \mathrm{P}\{0 < W_i + \delta_{i1}S < \mathrm{E}T\}. \tag{3.46}$$

For stationarity with $\rho < 1$ we must have, from $\mathrm{P}\{W_1 = 0\} > 0$ and (3.3) with $i = 1$,

$$\mathrm{P}\{W_1 + S < \mathrm{E}T\} + \mathrm{P}\{W_2 < \mathrm{E}T\} > 0$$

(cf. (3.44)), so recalling that the points of support of W_i must coincide with those of the sums at (3.43) we conclude that this set must be $\{j\mathrm{E}T : j = 0, 1, \ldots\}$ or a subset, and that this set must therefore also contain the support of S.

Example 3.12. (Extremal GI/GI/k system with $\mathrm{P}\{W_1 = 0\} = 1$) Since W_1 is a nonnegative random variable, $\mathrm{E}W_1 = 0$ if and only if $W_1 =$ a.s. 0. Now by an argument similar to that used above, $W_1 =_{\text{a.s.}} 0$ if and only if $S \le_{\text{a.s.}} kT$ (for, if not, starting from $W_{(0)} =_{\text{a.s.}} 0$, we can construct $W_{(n)}$ with $W_{n1} > 0$, hence we cannot have $W_1 =_{\text{a.s.}} 0$). Consider the particular $D/G_{\mathbb{Z}_+(\mathrm{E}, T)}/k$ system in which $\mathrm{P}\{S = k\mathrm{E}T\} = \rho = 1 - \mathrm{P}\{S = 0\}$. Then $W_1 =_{\text{a.s.}} 0$ by inspection, and $\mathrm{E}S^2 = \rho k^2(\mathrm{E}T)^2 = k\mathrm{E}T\mathrm{E}S = (\mathrm{E}S)^2/\rho$. This is the largest second moment for S compatible with $S \le_{\text{a.s.}} kT$.

Now if $\operatorname{var} T > 0$ and $\mathrm{E}S^2 < k\mathrm{E}S\mathrm{E}T$, then it can be verified (via some tedious algebra) that we can find positive $h < \mathrm{E}T$ such that a lattice with span h supports S and T which are in $\mathbb{Q}_2$ (i.e., have the prescribed first two moments), with S supported by $\{0, h, \ldots, kh\}$ and T by $\{h, 2h, \ldots\}$. For such a system, $S \le_{\text{a.s.}} kT$ so $W_1 =_{\text{a.s.}}$ as before.

Proposition 3.13. (Crude upper bounds on $\mathrm{E}W_i$.)

$$\mathrm{E}W_i \le \begin{cases} \dfrac{\mathrm{E}Z/\mathrm{E}T}{k(1-\rho)} & (i \le 1 + k\rho), \\[3mm] \dfrac{\mathrm{E}Z/\mathrm{E}T}{[k-(i-1)](1-\rho)} & (1 + k\rho < i \le k). \end{cases} \tag{3.47}$$

Proof. The components of W are ordered as $W_1 \le \ldots \le W_k$ by definition. Use this with the definition of Z at (3.12) to write

$$[k - (i-1)]\mathrm{E}W_i \le \sum_{j=i}^{k} \mathrm{E}W_j = (\mathrm{E}Z\mathrm{E}T) + k\rho\mathrm{E}W_1 - \sum_{j=1}^{i-1} \mathrm{E}W_j. \tag{3.48}$$

This yields the case $i = 1$ of (3.47). For general i the right-hand side of (3.48) is bounded above by

$$(\mathrm{E}Z/\mathrm{E}T) + [k\rho - (i-1)]\mathrm{E}W_1 \le (\mathrm{E}Z/\mathrm{E}T)(1 + [k\rho - (i-1)]_+(k - k\rho)),$$

from which the rest of (3.47) now follows. $\square$

Remark 3.14. We use the inequality (3.47) to yield with (3.36) the inequality

$$(1-\rho)(k+1-i)\mathrm{E}W_i^{\prime(\epsilon)} \le \frac{\mathrm{E}Z^{\prime(\epsilon)}}{\mathrm{E}T} \le \frac{\beta^2 \mathrm{E}S^2 + \mathrm{E}(T-S)^2 + (k-1)\mathrm{E}T^2}{2\mathrm{E}T}, \tag{3.49}$$

which is a bound that is independent of ϵ.

3.5   LOWER BOUNDS AND LIMITS FOR EW_1 IN CERTAIN GI/GI/k SYSTEMS

The next result, when taken in conjunction with (3.35) and (3.30), leads ultimately to the statement at (3.10).

Proposition 3.15. *When* $\rho < 1 - k^{-1}$,

$$\lim_{\epsilon \to 0} EW_i(T, D^{(\epsilon)}) = \begin{cases} EW_i(T, ES) & (i = 1, \ldots, k-1), \\ EW_k(T, ES) + \frac{1}{2}\beta^2 ET & (i = k). \end{cases} \tag{3.50}$$

Remark 3.16. In outline, the proof that follows proceeds by showing that in 'rare' periods when a customer with an unusually large service time is present in the system, one server becomes preoccupied with such a 'rare' customer while the rest of the system behaves like a stable GI/GI/$(k-1)$ system. Since such customers are rare, the perturbation to EW_i is negligible for $i \leq k-1$, while their rate of occurrence just balances out the additional workload to give the extra term when $i = k$ (evaluating this term is the essence of Lemma 3.20 in the Appendix).

We prove the asymptotic result at (3.50) for the particular family of service times $\{D^{(\epsilon)}\}$, but we believe that a similar result holds true more generally for $\{S^{(\epsilon)}\}$ (see Subsection 3.5.2). Accordingly, those parts of the proof that would apply in that context of greater generality are so written.

Proof of Proposition 3.15. Recall that doubly infinite sequences $\{W_{(n)}^{(\epsilon)}\}$, $\{S_n^{(\epsilon)}\}$, and $\{T_n\}$ exist satisfying (3.3) with each $W_{(n)}^{(\epsilon)}$ having the stationary distribution. For given $\beta > 0$ and ϵ in $0 < \epsilon < 1/\beta$, regard the r.v.s $S_n^{(\epsilon)}$ as being specified by two independent sequences $\{S_n\}$ and indicator r.v.s $\{I_n^{(\epsilon)}\}$ as at (3.31). For any integers n and $\alpha > 0$, define the sequence of disjoint events A_{nj} for $j = 0, \ldots, \alpha$ by

$$A_{nj} = \{I_{n-j}^{(\epsilon)} = 1, I_{n-r}^{(\epsilon)} = 0 \ (r = 0, \ldots, j-1)\}, \tag{3.51}$$

and set

$$A_n = A_{n0} \cup \cdots \cup A_{n\alpha} = \{\textstyle\sum_{j=0}^{\alpha} I_{n-j}^{(\epsilon)} \geq 1\}. \tag{3.52}$$

Note that

$$P(A_n^c) = 1 - \frac{\epsilon^2}{1+\epsilon^2} \sum_{j=0}^{\alpha} \frac{1}{(1+\epsilon^2)^j} = (1+\epsilon^2)^{-\alpha-1} \approx \exp(-\epsilon^2(\alpha+1)), \tag{3.53}$$

which converges to 1 when $\epsilon \to 0$ and $\alpha \to \infty$ as in Condition A, which we assume throughout.

Condition A. $\epsilon \to 0$ and $\alpha \to \infty$ in such a way that $\epsilon\alpha \to \infty$ and $\epsilon^2\alpha \to 0$.

The $\alpha + 2$ events $A_{n0}, \ldots, A_{n\alpha}, A_n^c$ constitute a partition of the state space of the process $\{W_n^{(\epsilon)}\}$, so we can write

$$W_{n+1,k}^{(\epsilon)} \geq (\max(W_{nk}^{(\epsilon)}, W_{n1}^{(\epsilon)} + S_n^{(\epsilon)}) - T_n)_+ I(A_n^c)$$

$$+ \sum_{j=0}^{\alpha} (W_{n-j,1}^{(\epsilon)} + S_{n-j}^{(\epsilon)} - T_{[n-j,n]})_+ I(A_{nj})$$

$$\equiv (\max(W_{nk}^{(\epsilon)}, W_{n1}^{(\epsilon)} + S_n^{(\epsilon)}) - T_n)_+ I(A_n^c) + \sum_{j=0}^{\alpha} J_{3j}^{(\epsilon)} \text{ say,} \tag{3.54}$$

where $T_{[\cdot,\cdot]}$ is as before at (3.4). When $\epsilon \to 0$, $(\max(W_{nk}^{(\epsilon)}, W_{n1}^{(\epsilon)} + S_n(1-\beta\epsilon)) - $

$T_n)_+ I(A_n^c)$ converges weakly, by (3.53) and the Continuity Theorem, to $(\max(W_{nk}, W_{n1} + S_n) - T_n)_+ = W_{n+1,k}$, so by Fatou's lemma,

$$\liminf_{\epsilon \to 0} \mathrm{E}[(\max(W_{nk}^{(\epsilon)}, W_{n1}^{(\epsilon)} + S_n^{(\epsilon)}) - T_n)_+ I(A_n^c)]$$

$$\geq \mathrm{E}W_{n+1,k}(T, S) = \mathrm{E}W_k(T, S). \tag{3.55}$$

Each term $J_{3j}^{(\epsilon)} = (W_{n-j,1}^{(\epsilon)} + S_{n-j}(1 + \beta/\epsilon) - T_{[n-j,n]})_+ I(A_{nj})$, i.e., each $J_{3j}^{(\epsilon)}$ is the product of two independent r.v.s. Then, using Condition A in conjunction with Lemma 3.20 in the limit step below,

$$\sum_{j=0}^{\alpha} \mathrm{E}(J_{3j}^{(\epsilon)}) = \epsilon^2 \sum_{j=0}^{\alpha} (1+\epsilon^2)^{-j-1} \mathrm{E}(W_{n-j,1}^{(\epsilon)} + S_{n-j}(1+\beta/\epsilon) - T_{[n-j,n]})_+$$

$$\geq \epsilon^2 \sum_{j=0}^{\alpha} (1+\epsilon^2)^{-j} \mathrm{E}(\beta S/\epsilon - T_{[0,j]})_+$$

$$\to \tfrac{1}{2} \beta^2 \mathrm{E}S^2 / \mathrm{E}T \quad (\epsilon \to 0). \tag{3.56}$$

Thus, referring to (3.54) and combining (3.55) and (3.56),

$$\liminf_{\epsilon \to 0} \mathrm{E}W_k(T, S^{(\epsilon)}) \geq \mathrm{E}W_k(T, S) + \tfrac{1}{2}\beta^2 \mathrm{E}S^2 / \mathrm{E}T. \tag{3.57}$$

Refer again to (3.54) and write

$$W_{n+1,k}^{(\epsilon)} \geq (W_{n1}^{(\epsilon)} + S_n^{(\epsilon)} - T_n)_+ - (W_{n1}^{(\epsilon)} + S_n^{(\epsilon)} - T_n)_+ I(A_n)$$

$$+ \sum_{j=0}^{\alpha} (W_{n-j,1}^{(\epsilon)} + S_{n-j}(1+\beta/\epsilon) - T_{[n-j,n]})_+ I(A_{nj}) \tag{3.58}$$

$$\equiv J_1^{(\epsilon)} - J_2^{(\epsilon)} + \sum_{j=0}^{\alpha} J_{3j}^{(\epsilon)} \quad \text{say.} \tag{3.59}$$

We show that when Condition A holds,

$$\mathrm{E}(J_2^{(\epsilon)}) = \mathrm{E}\left(\sum_{j=0}^{\alpha} (W_{n1}^{(\epsilon)} + S_n^{(\epsilon)} - T_n)_+ I(A_{nj})\right) \to 0 \quad (\epsilon \to 0). \tag{3.60}$$

Consider on each A_{nj} a system defined on the same arrival epochs but modified at the arrival of the $(n-j)^{th}$ customer and all subsequent arrivals to evolve as a GI/GI/k' system, where $k' = k - 1$; specifically, at these epochs the service times $\{S_r^{(\epsilon)}: r \geq n - j\}$ are i.i.d. like the original sequence $\{S_n^{(\epsilon)}\}$, and the first k' components of $\{W_{(r)}^{(\epsilon)}\}$ (for the same r) are replaced by those of the k'-vectors $\{W'^{(\epsilon)}_{(r)}\}$ which are the workload vectors of a GI/GI$^{(\epsilon)}$/k' system for which $W'^{(\epsilon)}_{n-j,i} = W_{n-j,i}^{(\epsilon)}$ $(i = 1,\ldots,k')$ and subsequent $W'^{(\epsilon)}_{(r)}$ are determined by the recurrence relations at (3.2) (equivalently, (3.3)), with k replaced by k'. By Proposition 3.10 and the construction noted below it, on A_{nj}, $W_i'^{(\epsilon)} \geq_d W'^{(\epsilon)}_{ri}$ for $r \geq n - j$, and by Proposition 3.13 and (3.49) there is a finite constant $C > \mathrm{E}W_i'^{(\epsilon)}$ with C independent of ϵ. Consequently, since A_{nj} is independent of $W_{ni}'^{(\epsilon)}$,

$$\mathrm{E}[(W_{n1}^{(\epsilon)} + S_n^{(\epsilon)} - T_n)_+ I(A_{nj})] \leq \mathrm{E}[(W_1'^{(\epsilon)} + S^{(\epsilon)} - T)_+ I(A_{nj})]$$

$$= \mathrm{E}(W_1'^{(\epsilon)} + S^{(\epsilon)} - T)_+ \epsilon^2 (1+\epsilon^2)^{-j-1} \leq C\epsilon^2/(1+\epsilon^2)^{j+1}. \tag{3.61}$$

Combining this with other terms, $\mathrm{E}(J_2^{(\epsilon)}) \leq C[1 - (1+\epsilon^2)^{-\alpha-1}] = o(1)$ under Condition A, so the limit at (3.60) holds.

Recall equation (3.4). It relates the sum of the s largest workloads at the $(n+s)^{th}$ customer arrival, for any fixed positive integer $s < k$, to the waiting times at the s previous arrival epochs, a subset of the s largest workloads at those epochs, and the times between customer arrivals. Since $1 = I(A_n^c) + I(A_n) =$

$I(A_n^c) + I(A_{n0}) + \ldots + I(A_{n\alpha})$, we can separate the term $i = 0$ in the sum below and write, much as at (3.54) and (3.58),

$$
\begin{aligned}
\sum_{i=0}^{s-1} W_{n+s,k-i}^{(\epsilon)} \geq & \sum_{i=1}^{s-1} (\max(W_{n+i,k-i}^{(\epsilon)}, W_{n+i,1}^{(\epsilon)} + S_{n+i}^{(\epsilon)}) \\
& - T_{[n+i,n+s-1]}) + (I(A_n^c) + I(A_n)) \\
& + (\max(W_{nk}^{(\epsilon)}, W_{n1}^{(\epsilon)} + S_n^{(\epsilon)}) - T_{[n,n+s-1]}) + I(A_n^c) \\
& + \sum_{j=0}^{\alpha} (W_{n-j,1}^{(\epsilon)} + S_{n-j}^{(\epsilon)} - T_{[n-j,n+s-1]}) + I(A_{nj}) \qquad (3.62)
\end{aligned}
$$

$$
\begin{aligned}
\geq & \sum_{i=1}^{s-1} (W_{n+i,1}^{(\epsilon)} + S_{n+i}^{(\epsilon)} - T_{[n+i,n+s-1]}) + (I(A_n^c) + I(A_n)) \\
& + (W_{n1}^{(\epsilon)} + S_n^{(\epsilon)} - T_{[n,n+s-1]}) + I(A_n^c) \\
& + \sum_{j=0}^{\alpha} (W_{n-s,1}^{(\epsilon)} + S_{n-j}(1+\beta/\epsilon) - T_{[n-j,n+s-1]}) + I(A_{nj}) \qquad (3.63)
\end{aligned}
$$

$$
\begin{aligned}
\geq & \sum_{i=0}^{s-1} (W_{n+i,1}^{(\epsilon)} + S_{n+i}^{(\epsilon)} - T_{[n+i,n+s-1]}) + \\
& - (W_{n1}^{(\epsilon)} + S_n^{(\epsilon)} - T_{[n,n+s-1]}) + I(A_n) \\
& + \sum_{j=0}^{\alpha} (W_{n-j,1}^{(\epsilon)} + S_{n-j}(1+\beta/\epsilon) - T_{[n-j,n+s-1]}) + I(A_{nj}). \qquad (3.64)
\end{aligned}
$$

Take expectations and use the stationarity of $W_{(n)}$; then we have, more generally than (3.54),

$$
\begin{aligned}
\sum_{i=1}^{s} \mathrm{E} W_{k+1-i}^{(\epsilon)} \geq & \sum_{i=1}^{s} \mathrm{E}(W_1^{(\epsilon)} + S^{(\epsilon)} - T_{[1,s+1-i]}) + \\
& - \mathrm{E}[(W_{n1}^{(\epsilon)} + S_n^{(\epsilon)} - T_{[n,n+s-1]}) + I(A_n)] \\
+ \sum_{j=0}^{\alpha} \epsilon^2 (1+\epsilon^2)^{-j-1} \mathrm{E}(W_{n-j,1}^{(\epsilon)} & + S_{n-j}(1+\beta/\epsilon) - T_{[n-j,n+s-1]}) + \cdot
\end{aligned}
\qquad (3.65)
$$

On the right-hand side here, the first two terms equal

$$
s\mathrm{E}W_1^{(\epsilon)} + s\mathrm{E}S^{(\epsilon)} - \tfrac{1}{2}s(s+1)\mathrm{E}T + \sum_{i=1}^{s} \mathrm{E}(T_{[1,s+1-i]} - (W_1^{(\epsilon)} + S^{(\epsilon)}))_+; \qquad (3.66)
$$

also, under Condition A, the second term is $o(1)$ by the same argument as used above in establishing (3.60), and, appealing to Lemma 20, the last term equals $(\tfrac{1}{2}\beta^2 \mathrm{E}S^2/\mathrm{E}T)[1 - o(1)]$. Substitute for the left-hand side of (3.65) from below (3.35) and collect terms for the right-hand side. This gives, after multiplication by $2\mathrm{E}T$,

$$
\begin{aligned}
\beta^2 \mathrm{E}S^2 + \mathrm{E}(T-S)^2 + (k-1)\mathrm{E}T^2 - \sum_{i=1}^{k} \mathrm{E}(T - \delta_{i1}S^{(\epsilon)} - W_i^{(\epsilon)})_+^2 \\
- 2\mathrm{E}T \sum_{i=1}^{k-s} \mathrm{E}W_i^{(\epsilon)} + 2\mathrm{E}S\mathrm{E}W_1^{(\epsilon)} \\
\geq 2s\mathrm{E}T \; \mathrm{E}W_1^{(\epsilon)} + 2s\mathrm{E}T\mathrm{E}S^{(\epsilon)} - s(s+1)(\mathrm{E}T)^2
\end{aligned}
$$

$$+ 2ET \sum_{i=1}^{s} \mathrm{E}(T_{[1, s+1-i]} - (W_1^{(\epsilon)} + S^{(\epsilon)}))_+$$
$$+ o(1) + \beta^2 \mathrm{E}S^2[1 - o(1)].$$

Rearranging terms gives, under Condition A,

$$2[(s+1)ET - ES]\mathrm{E}W_1^{(\epsilon)} + 2ET \sum_{i=2}^{k-s} \mathrm{E}W_i^{(\epsilon)}$$

$$\leq \beta^2 \mathrm{E}S^2 o(1) + o(1) + \mathrm{E}(T-S)^2 + (k-1)ET^2 - \sum_{i=1}^{k} \mathrm{E}(T - \delta_{i1} S^{(\epsilon)} - W_i^{(\epsilon)})_+^2$$

$$- 2sETES + s(s+1)(ET)^2 - 2ET \sum_{i=1}^{s} \mathrm{E}(T_{[1, s+1-i]} - S^{(\epsilon)} - W_1^{(\epsilon)})_+ . \tag{3.67}$$

The right-hand side here converges as $\epsilon \to 0$, using the same argument as for Proposition 8, to

$$\mathrm{E}(T-S)^2 + (k-1)ET^2 - \sum_{i=1}^{k} \mathrm{E}(T - \delta_{i1} S - W_i)_+^2$$

$$- 2sETES + s(s+1)(ET)^2 - 2ET \sum_{i=1}^{s} \mathrm{E}(T_{[1, s+1-i]} - S - W_1)_+ \tag{3.68}$$

$$= 2ET \sum_{i=1}^{k} \mathrm{E}W_i - 2ESEW_1 + 2sETEW_1 - 2ET \sum_{i=1}^{s} \mathrm{E}(W_1 + S - T_{[1, s+1-i]})_+$$

$$= 2[(s+1)ET - ES]\mathrm{E}W_1 + 2ET \sum_{i=2}^{k} \mathrm{E}W_i$$

$$- 2ET \sum_{i=1}^{s} \mathrm{E}(W_1 + S - T_{[1, s+1-i]})_+ . \tag{3.69}$$

Specialize to $GI/D^{(\epsilon)}/k$ and $GI/D/k$ (see Example 3.2), recalling that $\mathrm{E}\widetilde{W}_1 = ES - T_{[1, k+1-i]})_+$. Then the second sum in (3.69) equals

$$\sum_{i=1}^{s} \mathrm{E}\widetilde{W}_{k-s+i} = \sum_{i=k-s+1}^{k} \mathrm{E}\widetilde{W}_i,$$

so (3.69) itself equals

$$2[(s+1)ET - ES]\mathrm{E}\widetilde{W}_1 + 2ET \sum_{i=2}^{k-s} \mathrm{E}\widetilde{W}_i.$$

Put $s = k-1$. The last term on the right-hand side vanishes, the coefficient of $\mathrm{E}\widetilde{W}_1$ is positive, and therefore, from (3.67) and the positive of $(s+1)ET - ES$ for $s = k-1$,

$$\limsup_{\epsilon \to 0} \mathrm{E}\widetilde{W}_1^{(\epsilon)} \leq \mathrm{E}\widetilde{W}_1.$$

This with (3.30) establishes the case $i = 1$ of (3.50).

From (3.35) we have

$$\lim_{\epsilon \to 0} \left[ET \sum_{i=1}^{k} (\mathrm{E}W_i^{(\epsilon)} - \mathrm{E}W_i) - ES(\mathrm{E}W_1^{(\epsilon)} - \mathrm{E}W_1) \right] = \tfrac{1}{2}\beta^2 \mathrm{E}S^2, \tag{3.70}$$

so from the case $i = 1$ of (3.50), we must have, when $\rho < 1 - k^{-1}$, that

$$\lim_{\epsilon \to 0} \sum_{i=1}^{k} (\mathrm{E}W_i^{(\epsilon)} - \mathrm{E}W_i) = \frac{\tfrac{1}{2}\beta^2 \mathrm{E}S^2}{ET} \leq \sum_{i=1}^{k} \liminf_{\epsilon \to 0} (\mathrm{E}W_i^{(\epsilon)} - \mathrm{E}W_i) \text{ by (3.60) and (3.34),}$$

$$\leq \liminf_{\epsilon \to 0} \sum_{i=1}^{k} (\mathrm{E}W_i^{(\epsilon)} - \mathrm{E}W_i). \tag{3.71}$$

Since all these terms are nonnegative, equality must hold throughout (3.71), whence it follows that $\lim_{\epsilon \to 0} EW_i^{(\epsilon)}$ exists for all $i = 2, \ldots, k$ and that (3.50) holds. The proof is complete.   □

Combining Example 3.2 with Proposition 3.15 yields the corollary below, of which the particular case $i = 1$ gives $\lim_{\epsilon \to 0} EW_1(D/D^{(\epsilon)}/k) = 0$ for $\rho < 1 - k^{-1}$, asserted earlier at (3.10).

Corollary 3.17. *When* $\rho < 1 - k^{-1}$,

$$\lim_{\epsilon \to 0} EW_i(D/D^{(\epsilon)}/k) = \begin{cases} (ES - (k+1-i)ET)_+ (i - 1 - k(1-\rho))_+ ET & (i < k), \\ (k\rho - 1)_+ ET + \tfrac{1}{2}(\mathrm{var}S)/ET & (i = k). \end{cases} \tag{3.72}$$

3.5.1 Heuristics for the case $1 - k^{-1} < \rho < 1$

To see what may happen in $GI/D^{(\epsilon)}/k$ when $1 - k^{-1} < \rho < 1$, consider the vector $W_{(n+j)}^{(\epsilon)}$ on A_{n0}. For $1 \le j <$ some critical value N' say, $W_{n+j,k}^{(\epsilon)}$ decreases by T_j between successive arrivals while for $i = 1, \ldots, k-1$, $W_{n+j,i}^{(\epsilon)}$ *on average* increases by $ES/(k-1) - ET$ (actually, these $W_{n+j,i}^{(\epsilon)}$ stay approximately constant for blocks of $k - 1$ successive arrival epochs, and then increase by an amount distributed like $S - T_{[1, k-1]}$; Example 3.18 at the end of this subsection may help). The critical value N' is determined by the epoch where this total increase in $W_{n+j,k-1}^{(\epsilon)}$ balances with the result of the decrease in $W_{n+j,k}^{(\epsilon)}$ from $W_{n+1,k}^{(\epsilon)} = W_{n1}^{(\epsilon)} + ES(1 + \beta/\epsilon) - T_n$, i.e., the net change in $W_{n+j,k}^{(\epsilon)}$. Retaining only the first order term,

$$\frac{\beta S}{\epsilon} - N'ET \approx N'\left(\frac{ES}{k-1} - NT\right), \tag{3.73}$$

i.e., $N' \approx [(k-1)/ES]\beta S/\epsilon$. Thereafter, all components of $W_{(n+j)}^{(\epsilon)}$ decrease by $ET - ES/k$ on average, for further

$$N'' \equiv \frac{N'[ES/(k-1) - ET]}{ET - ES/k} = \frac{N'[1 - k(1-\rho)]}{(k-1)\rho(1-\rho)} \tag{3.74}$$

arrivals. Then the net effect of any (rarely occurring) large service demand from the arrivals at $n, n-1, \ldots$ in increasing $EW_{n+1,k}^{(\epsilon)}$ is approximately equal to

$$\frac{\epsilon^2}{1+\epsilon^2} E\left[N'\left(\frac{\beta S}{\epsilon} + \frac{\beta S}{\epsilon} - N'ET\right) + N''\left(\frac{\beta S}{\epsilon} - N'ET\right)\right] + o(1),$$

i.e.,

$$EW_k(T, S^{(\epsilon)}) - EW_k(T, S) + o(1) \tag{3.75}$$

$$= \tfrac{1}{2}\epsilon^2 E\left[N'\left(\frac{2\beta S}{\epsilon} - \frac{k-1}{ES} \cdot \frac{\beta S}{\epsilon} \cdot ET + \frac{1 - k(1-\rho)}{(k-1)(1-\rho)} \cdot \frac{\beta S}{\epsilon}\left(1 - \frac{(k-1)ET}{ES}\right)\right)\right]$$

$$= \tfrac{1}{2}\beta^3 ES^2 \left[\frac{k-1}{ES}\left(2 - \frac{k-1}{k\rho} + \frac{1 - k(1-\rho)}{(k-1)(1-\rho)} \cdot \frac{1 - k(1-\rho)}{k\rho}\right)\right]$$

$$= \frac{\tfrac{1}{2}\beta^2 ES^2}{k^2\rho(1-\rho)ET}\left(\rho + (1-\rho)(k-1)^2\right). \tag{3.76}$$

The net effect from the same source in increasing $EW_{n+1,i}^{(\epsilon)}$ $(i = 1, \ldots, k-1)$ is approximately equal to

$$\epsilon^2 E\left[\frac{N' + N''}{2}\left(\frac{\beta S}{\epsilon} - N'ET\right)\right] + o(1),$$

i.e., for $i = 1, \ldots, k-1$,

$$EW_i(T, S^{(\epsilon)}) - EW_i(T, S) + o(1)$$

$$= \tfrac{1}{2}\epsilon^2 E\left[N'\left(1 + \frac{1 - k(1-\rho)}{(k-1)(1-\rho)}\right)\frac{\beta S}{\epsilon}\left(1 - \frac{(k-1)ET}{ES}\right)\right] \tag{3.77}$$

$$= \frac{\tfrac{1}{2}\beta^2 ES^2}{k^2(1-\rho)ET} \cdot \frac{1 - k(1-\rho)}{1 - (1-\rho)}. \tag{3.78}$$

Equations (3.76) and (3.78) would imply that

$$E(Z^{(\epsilon)} - Z) + o(1)$$

$$= \frac{\tfrac{1}{2}\beta^2 ES^2}{k^2\rho(1-\rho)ET}\left(ET[\rho + (1-\rho)(k-1)^2] + [(k-1)ET - ES][1 - k(1-\rho)]\right)$$

$$= \tfrac{1}{2}\beta^2 ES^2,$$

as is consistent with (3.35) in Proposition 3.8.

In the particular case of a $GI/D^{(\epsilon)}/k$ system, $\beta^2 ES^2 = \text{var} S^{(\epsilon)}$, so we then anticipate that, for $i = 1$,

$$EW_1(GI/D^{(\epsilon)}/k) - EW_1(GI/D/k) \approx \frac{k^{-2}\text{var} S^{(\epsilon)}}{2(1-\rho)ET} \cdot \frac{1 - k(1-\rho)}{1 - (1-\rho)}. \tag{3.79}$$

We refer to this expression in Section 3.6.

Example 3.18. (Numerical case of $D/D^{(\epsilon)}/3$) Consider a 3-server system with constant interarrivals a unit time apart, service times $= 2.7$, and just one extraordinary service time $= 7$. Then the following sequence of workload vectors shows the stationary regime and the transitory effects of this extraordinary service time. The stationary workload vector resumes at the 20th arrival after the extraordinary one. In terms of N' and N'' at (3.73) and (3.74), if we impute $N' = 4$, then (3.74) gives $N'' \approx 140/9 = 15.6$ and $N' + N'' \approx 20$.

W_{n3} ...	1.7	1.7	6	5	4	3	2.4	3.1	2.7	2.1	2.8
W_{n2} ...	.7	.7	.7	1.7	1.7	2.4	2	1.4	2.1	1.7	1.1
W_{n1} ...	0	0	0	0	.7	.7	1.4	1	.4	1.1	.7

2.4	1.8	2.5	2.1	1.7	2.2	1.8	1.7	1.9	1.7	1.7	1.7 ...
1.8	1.4	.8	1.5	1.1	.7	1.2	.8	.7	.9	.7	.7 ...
.1	.8	.4	0	.5	.1	0	.2	0	0	0	...

3.5.2 Conjectured limits concerning $EW_i(GI/GI/k)$

The results of the heuristics just given can be combined with Proposition 3.15 as follows.

Conjecture 3.19. *In a stable GI/GI/k system with moments in* $\mathbb{Q}_2$,

$$\lim_{\epsilon \to 0} EW_i(T, S^{(\epsilon)}) - EW_i(T, S)$$

$$
= \begin{cases}
\dfrac{[1 - k(1 - \rho)]_+}{1 - (1 - \rho)} \cdot \dfrac{k^{-2}\beta^2 ES^2}{2(1 - \rho)ET} & (i = 1, \ldots, k - 1), \\[2ex]
\dfrac{\frac{1}{2}\beta^2 ES^2}{ET} \times \begin{cases} 1 & (\rho < 1 - k^{-1}), \\ f_k(\rho) & (1 - k^{-1} < \rho < 1), \end{cases} & (i = k),
\end{cases}
\tag{3.80}
$$

where $f_k(\rho) = \dfrac{\rho + (-\rho)(k - 1)^2}{k^2 \rho(1 - \rho)}$.

This conjecture extends the result of Proposition 3.15 from deterministic service times to general service times with finite second moment. So far as we can tell, the crucial part of the proof of Proposition 3.15 showing that equality holds in (3.50) for $i = 1$, depends on its inequalities (used before taking lim sup after (3.69)) having the strict cyclic allocation of customers that occurs with constant service times. Yet, the principles embodied in the heuristics of Section 3.5.1 would appear just as applicable to $0 < \rho < 1 - k^{-1}$, so there seems to be no reason why the result at (3.50) should not extend to GI/GI/k systems in $\mathbb{Q}_2$.

3.6 UPPER BOUNDS ON THE MEAN WAITING TIME EW_1

It is convenient in this section to use notation such as

$$W_i(T, S; k) \tag{3.81}$$

involving the generic interarrival time T and service time S of a GI/GI/k system. Our aim here is to review evidence that supports the conjecture that

$$EW_1(T, S; k) \leq EW(T, S/k; 1). \tag{3.82}$$

The truth of (3.82) would imply that of the weaker inequality

$$EW_1(T, S; k) \leq \frac{\rho(2 - \rho)\mathrm{var}T + k^{-2}\mathrm{var}S}{2(1 - \rho)ET}, \tag{3.83}$$

and the still weaker inequality

$$EW_1(T, S; k) \leq \frac{\mathrm{var}T + k^{-2}\mathrm{var}S}{2(1 - \rho)ET}. \tag{3.84}$$

It is known, as a result without any restrictive conditions (Wolff [44], [45]; Gittins [18]), that

$$EW_1(T, S; k) \leq EW(T_1 + \ldots + T_k, S; 1) \tag{3.85}$$

where $T_1, \ldots, T_k$ are independent r.v.s with $T_i =_d T$ $(i = 1, \ldots, k)$. Using bounds for single-server systems, (3.85) yields

$$EW_1(T, S; k) \leq \frac{\rho(2 - \rho)\mathrm{var}T + k^{-1}\mathrm{var}S}{2(1 - \rho)ET}. \tag{3.86}$$

The essence of the conjectured strengthening (3.82) and (3.83) of (3.85) and (3.86), respectively is in the change of the multiplier k^{-1} at (3.86) to k^{-2}. It is proper to note here that the discussion in this section has essentially no overlap with the comparative comments in Daley, Kreinin, and Trengove [14] on the possibilities of extending methods used to derive bounds on EW_1 in GI/GI/1 systems to GI/GI/k systems. That paper refers to a 1984 version of the present paper.

Let $\text{Exp}(\mu)$ denote a generic r.v. with negative exponential distribution on $(0,\infty)$ and mean μ^{-1}, so that when $S =_d \text{Exp}(\mu)$ we have a GI/M/$\cdot$ system in Kendall's notation. Makino [30] used an embedded Markov chain technique to show that the sequence

$$\{EW_1(T, k\,\text{Exp}(\mu); k) : k = 1, 2, \ldots\} \tag{3.87}$$

decreases monotonically in k; (3.82) then follows for GI/M/$\cdot$ systems, a result also derived by Brumelle [5]. Mori [31] suggested that (3.87) may hold in the more general sense that whenever (as we are assuming throughout) we have a queueing system with relative traffic intensity $\rho < 1$,

$$\{EW_1(T, kS; k) : k = 1, 2, \ldots\} \tag{3.88}$$

is a decreasing sequence. Certainly, for constant service times, we have

$$W_1(T, ES; 1) \geq_d W_1(T, kES; k) \geq_d W_1(T, kk'ES; kk') \tag{3.89}$$

for positive integers k and k', and hence (3.82); the proof of (3.89) follows directly from the representation

$$W_1(T, kES; k) =_d W(T_1 + \ldots + T_k, kES; 1) =_d \sup_{n \geq 0} \left\{ \sum_{j=1}^{nk} (ES - T_j) \right\}. \tag{3.90}$$

Now $EW(T, S; 1) = \sum_{n=1}^{\infty} n^{-1} E(S_1 + \ldots + S_n - T_1 - \ldots - T_n)_+$, and for i.i.d. r.v.s $X, X_1, X_2, \ldots$

$$a_n \equiv \frac{E(X_1 + \ldots + X_n)_+}{n} \geq \frac{E(X_1 + \ldots + X_{n+1})_+}{n+1} = a_{n+1}, \tag{3.91}$$

whence an alternative derivation of (3.82) for GI/D/k systems is feasible, but we have not been able to verify (3.88) for GI/D/k systems. It would be sufficient to have $\{a_n\}$ a *convex* decreasing sequence, but the convexity need not hold, as for example when $P\{X = \pm 1\} = 1$. The stronger conjecture that $W_1(T, kS; k) \geq_d W_1(T, (k+1)S; k+1)$ is readily disproved by taking $T = 0$ with high probability and T large with low probability, even though this monotonicity property does hold for GI/M/k systems. In passing, we observe that Mori [31] claimed to have proved the monotonicity of (3.88) (see below his equation (3.73)). However, perusal of his proof shows that his argument proceeds essentially via the claim, which is referred to a sample path argument, that

$$W_1(T, kS; k) \geq_d W_1(T, (k+1)S; k+1);$$

this is proved, correctly, for the case $k = 1$, but need not be true in general.

In terms of general distributions for S but a restricted range of ρ, Susuki & Yoshida [39] established (3.84) when $\rho < k^{-1}$, and Köllerström [26] verified Kingman's [24] conjecture for heavy traffic conditions, i.e., for suitable sequences of distributions for S and T subject to $\rho \to 1 - 0$, and hence $\rho(2 - \rho) \to 1$, that $(1 - \rho)W_1$ converges weakly to an exponential r.v. with mean $\frac{1}{2}(\text{var}\,T + k^{-2}\text{var}\,S)/ET$.

On the basis of (3.19) and bearing in mind these observations, Mori [31] noted that if at equation (3.19), $2\text{cov}(R_n, \ldots) - \text{var}\,R_n \leq 0$, then (3.84) holds. Independently of Mori's work, I made in Daley [8], the stronger conjecture that (3.83) holds, the reasoning being as follows. Write $v_{(n)} = (V_{n1}, \ldots, V_{nk}) = \Re(W_{n1} + S_n, W_{n2}, \ldots, W_{nk})$, so that at (3.19)

$$\mathrm{var}\, R_n = \mathrm{var}\left(\sum_i (T_n - V_{ni})_+\right)$$

$$= \sum_i \mathrm{var}(T_n - V_{ni})_+ + 2\sum_{i<j} \mathrm{cov}((T_n - V_{ni})_+, (T_n - V_{nj})_+).$$

I showed in Daley [8] that for any nonnegative x, y and r.v. X with $\mathrm{E}X_+^2 < \infty$,

$$\mathrm{E}[(X-x)_+(X-y)_+] \geq [\mathrm{E}X_+^2/(\mathrm{E}X_+)^2]\mathrm{E}(X-x)_+\mathrm{E}(X-y)_+,$$

first condition on $V_{(n)}$ and then use the fact that the components of $V_{(n)}$ are associated to deduce that

$$\mathrm{cov}((T_n - V_{ni})_+, (T_n - V_{nj})_+) \geq (\mathrm{var}T/(\mathrm{E}T)^2)\mathrm{E}(T_n - V_{ni})_+\mathrm{E}(X-y)_+;$$

and thus

$$\mathrm{var}R_n \geq (\mathrm{var}\, T/(\mathrm{E}T)^2)(\mathrm{E}R_n)^2 = k^2(1-\rho)^2 \mathrm{var}T. \tag{3.92}$$

Equivalently,

$$\mathrm{var}(kT_n - S_n) - \mathrm{var}\, R_n \leq k^2\rho(2-\rho)\mathrm{var}\, T + \mathrm{var}\, S. \tag{3.93}$$

Next, in the covariance term in (3.19),

$$L_{n+1} - kW_{n+1,1} = \sum_{i=2}^{k} ((V_{ni} - T_n)_+ - (V_{n1} - T_n)_+)$$

is an increasing function of $-T_n$, V_{n1}, and each $V_{ni} - V_{n1}$ $(i = 2, \ldots, k)$, whereas

$$R_n = \sum_{i=1}^{k} (T_n - V_{ni})_+ = (T_n - V_{n1})_+ + \sum_{i=2}^{k} (T_n - V_{n1} - (V_{ni} - V_{n1}))_+$$

is a decreasing function of these $k+1$ arguments. Then if the family of r.v.s

$$\{V_{n1}, V_{n2} - V_{n1}, \ldots, V_{nk} - V_{n1}, -T_n\} \tag{3.94}$$

could be shown to be associated, the covariance term in (3.19) could not be positive, and thus (3.83) would follow. While it is true that $\{V_{n1}, V_{n2}, \ldots, V_{nk}, -T_n\}$ is an associated family (cf., e.g., Baccelli and Bremaud [2] for a recent textbook account of W being associated), we have not been able to demonstrate this property for the family at (3.94). For example, in the simplest non-trivial case, namely, of a GI/GI/2 system, we have

$$V_{n+1,2} - V_{n+1,1} = |(V_{n1} - T_n)_+ + S_{n+1} - (V_{n2} - T_n)_+|$$

$$= |(V_{n1} - T_n)_+ + S_{n+1} - (V_{n2} - V_{n1} + V_{n1} - T_n)_+|,$$

and since the right-hand side is not monotonic in $V_{n2} - V_{n1}$, we cannot use monotonicity in an induction argument to demonstrate that $\{V_{n1}, V_{n2} - V_{n1}\}$ is an associated family.

Referring again to the derivation of (3.19), we can write

$$\tfrac{1}{2}\left[\mathrm{var}(S - kT) - \mathrm{var}R_n\right] = -\mathrm{cov}(R_n, L_{n+1})$$

$$= -\mathrm{cov}\left(\sum_{i=1}^{k} (V_{ni} - T_n)_+, \sum_{j=1}^{k} (T_n - V_{nj})_+\right)$$

$$= -\sum_{i=2}^{k} \sum_{j=1}^{i-1} \mathrm{E}[(V_{ni} - T_n)_+(T_n - V_{nj})_+]$$

$$+ \sum_{i=1}^{k} \sum_{j=1}^{k} \mathrm{E}(V_{ni} - T_n)_+ \, \mathrm{E}(T_n - V_{nj})_+ \tag{3.95}$$

using the identity $(V_{ni} - T_n)_+ (T_n - V_{nj})_+ =_{\text{a.s.}} 0$ for $j \geq i$ because $V_{n1} \leq \ldots \leq V_{nk}$,

$$= - \sum_{i=2}^{k} \sum_{j=1}^{i-1} \mathrm{cov}((V_{ni} - T_n)_+, (T_n - V_{nj})_+) + \sum_{i=1}^{k} \sum_{j=i}^{k} \mathrm{E}W_i \mathrm{E}(T - V_j)_+$$

$$\geq \sum_{i=1}^{k} \sum_{j=i}^{k} \mathrm{E}W_i \mathrm{E}(T - V_j)_+, \tag{3.96}$$

the inequality here coming from association. Combining all this yields

$$2k(1 - \rho)\mathrm{E}T\mathrm{E}W_1 + \sum_{i=2}^{k} \sum_{j=i}^{k} \mathrm{E}W_i \mathrm{E}(T - V_j)_+$$
$$\leq \mathrm{var}(S - kT) - \mathrm{var}R_n \leq k^2 \rho(2 - \rho)\mathrm{var}T + \mathrm{var}S. \tag{3.97}$$

This appears to be too weak to deduce (3.83).

The heuristic argument leading to the relations at (3.76) and (3.78) serves to suggest why the bound at (3.86), deduced from (3.85), may be capable of improvement to (3.83). In Section 3.5, the system $GI/D^{(\epsilon)}/k$ with the FCFS discipline mostly operates in a cyclic fashion. Occasionally, a customer with a very large service time perturbs this behavior so that $\{W_{n1}\}$ then reflects, for a while, the transient behavior of a $GI/D^{(\epsilon)}/(k-1)$ system, and the effect if any of the perturbation on $\mathrm{E}W_1$ occurs via a pair of (cyclical) sharing arrangements, the first being over $k-1$ servers, as the W_{ni} for $1 \leq i \leq k-1$ show an increasing trend with n, and the other over all k servers, as all W_{ni} have a decreasing trend. A precise interpretation of the conjecture at (3.82) is that equal sharing of *all* demands for service by all k servers may not be as efficient (as measured by $\mathrm{E}W_1$) as isolating the large service times and sharing the smaller service times by an FCFS discipline that is close to the cyclic discipline for such smaller demands. Observe that the relation at (3.79), if true, is consistent with the truth of (3.83).

If the conjecture (3.83) is not true generally, then it would appear likely that it should be false in some $GI/B/k$ system (here, B denotes a two-point distribution with support 0 and $\mathrm{E}S^2/\mathrm{E}S$), as all known examples of $GI/GI/1$ yielding the extremal mean waiting times $\inf \mathrm{E}W_1$ and $\sup \mathrm{E}W_1$ over the family of generic S with specified $\mathrm{E}S$ and $\mathrm{E}S^2$ are either $GI/B/1$ or limits of $GI/D^{(\epsilon)}/1$ systems.

To examine the $GI/B/k$ possibility, regard the service time r.v. of a $GI/B/k$ system as the product $(\mathrm{E}S^2/\mathrm{E}S)I_\beta$ of an indicator r.v. I_β, with $\mathrm{P}\{I_\beta = 1\} = \beta \equiv (\mathrm{E}S)^2/\mathrm{E}S^2$, and the constant $\mathrm{E}S^2/\mathrm{E}S = \mathrm{E}S/\beta$. Then, as shown in the unpublished work of Daley & Mackisack,

$$W_1(GI/B/k) \equiv W_1(T, (\mathrm{E}S^2/\mathrm{E}S)I_\beta; k)$$
$$=_d W_1(T_1 + \ldots + T_{\nu(\beta)}, (\mathrm{E}S^2/\mathrm{E}S); k) \tag{3.98}$$

where $T_1, T_2, \ldots$ are i.i.d. like T and $\nu(\beta)$ is independent of $\{T_n\}$ with $\mathrm{P}\{\nu(\beta) = j\} = \beta(1-\beta)^{j-1}$ $(j = 1, 2, \ldots)$. Now $W_1(T, \mathrm{E}S; k) =_d W(T_1 + \ldots + T_k, \mathrm{E}S; 1)$ for $GI/D/k$ systems (recall Example 3.2), so

$$W_1(GI/B/k) =_d W(T_1 + \ldots + T_{\nu_1(\beta)} + \ldots + T_{\nu_k(\beta)}, \mathrm{E}S^2/\mathrm{E}S; 1)$$
$$\equiv W(T^*, S^*; 1) \text{ say}, \tag{3.99}$$

where $\nu_1(\beta),\ldots,\nu_k(\beta)$ are i.i.d. like $\nu(\beta)$. Recall that $E\nu(\beta) = 1/\beta$ and $\mathrm{var}\,\nu(\beta) = (1-\beta)/\beta^2$. Then, in such a single-server system,

$$ET^* = kE\nu/(\beta)ET = kET/\beta, \quad \mathrm{var}T^* = kE\nu(\beta)\mathrm{var}T + (k\mathrm{var}\nu(\beta))(ET)^2,$$

$$(3.100)$$

$$ES^* = ES^2/ES = ES/\beta, \quad \mathrm{var}S^* = 0, \text{ and } \rho^* = [ES/\beta]/[kET/\beta] = \rho.$$

Consequently, applying the case $k = 1$ of (3.83) (known to be true from Daley [8]),

$$EW_1(\mathrm{GI/B}/k) = EW(T^*, S^*; 1)$$

$$\leq \frac{\rho^*(2-\rho^*)\mathrm{var}T^* + \mathrm{var}\,S^*}{2(1-\rho^*)ET^*} = \frac{\rho(2-\rho)[(k/\beta)\mathrm{var}T + (k(1-\beta)/\beta^2)(ET)^2]}{2(1-\rho)(k/\beta)ET}$$

so

$$2(1-\rho)ETEW_1(\mathrm{GI/B}/k) - \rho(2-\rho)\mathrm{var}T \leq \rho(2-\rho)(\beta^{-1}-1)(ET)^2$$

$$= \rho(2-\rho)(\mathrm{var}S)k^{-2}/\rho^2$$

$$= (2\rho^{-1}-1)k^{-2}\mathrm{var}S. \qquad (3.101)$$

Thus, while not consistent with (3.83) for general k, it is plausible that the discrepant factor $2\rho^{-1} - 1$ may be attributed entirely to the use of (3.83) in the case $k = 1$, a suggestion that is even more believable when it is noted that for small β, $T_1 + \ldots + T_{\nu(\beta)}$ has approximately the exponential distribution and for $E_k/\mathrm{GI}/1$ systems, the factor $\rho(2-\rho)$ in (3.83) can be replaced by ρ^2.

As a final remark, but one which we have not been able to exploit, we observe that when an r.v. Y is the random sum of a geometric number of r.v.s, such as $Y = T_1 + \ldots + T_{\nu(\beta)}$ above, then it has a weak new-worse-than-used (NWU) property, namely, $P\{Y > x + y\} \geq (1-\beta)P\{Y > x\}P\{Y > y\}$, as well as having its mean residual life bounded above,

$$E(Y - x \mid Y > x) \leq [\beta^{-1} + (\mathrm{var}T)/(ET)^2]ET.$$

The weak NWU property yields $E(Y - x \mid Y > x) \geq (\beta^{-1} - 1)ET = EY - ET$, and hence an alternative upper bound on $EW_1(\mathrm{GI/B}/k)$ can be derived using (3.98), (3.90), and (3.17) (cf., e.g., Stoyan, [37], §5.5). However, it does not improve on the bound at (3.101).

APPENDIX

Lemma 3.20. *For a non-negative r.v. Y independent of the i.i.d. nonnegative r.v.s $X, X_1, X_2, \ldots$, for which $EX^2 < \infty$ and $EY^2 < \infty$, and for integers $\alpha(\epsilon)$ such that $\epsilon\alpha(\epsilon) \to \infty$ and $\epsilon^2\alpha(\epsilon) \to 0$ as $\epsilon \to 0$,*

$$\lim_{\epsilon \to 0} \epsilon^2 \sum_{j=1}^{\alpha(\epsilon)} (1+\epsilon^2)^{-j} E(Y/\epsilon - (X_1 + \ldots + X_j))_+ = \frac{EY^2}{2EX}. \qquad (3.102)$$

Proof. Let F^{j*} denote the j-fold convolution of the d.f. F of X, and write $H = \sum_{j=1}^{\infty} F^{j*}$ for the zero-deleted renewal function generated by F. Recall from Lorden [27] that

$$\lambda x - 1 \leq H(x) \leq \lambda x + \lambda^2 \mathrm{var}X \quad (\text{all } x > 0) \qquad (3.103)$$

(cf., also Carlsson & Nerman [6]; Daley [9]). Observe that the expression at (3.102) is bounded above by

$$\epsilon^2 \sum_{j=1}^{\alpha(\epsilon)} \int_0^{Y/\epsilon} F^{j*}(u)du = \epsilon^2 \int_0^{Y/\epsilon} H(u)du - \epsilon^2 \int_0^{Y/\epsilon} \sum_{j=\alpha(\epsilon)+1}^{\infty} F^{j*}(u)du, \quad (3.104)$$

and bounded below by the product of this upper bound with $(1+\epsilon^2)^{-\alpha(\epsilon)} \approx \exp(-\epsilon^2\alpha(\epsilon)) \to 1$, $(\epsilon \to 0)$. Consequently, it is enough to demonstrate that the expression at (3.104) converges to the limit at (3.102). From (3.103) it follows that the first term on the right-hand side of (3.104) converges to $\frac{1}{2}\lambda EY^2$ as $\epsilon \to 0$, so we have only to show that the other term converges to zero.

Since

$$\sum_{j=\alpha(\epsilon)+1}^{\infty} F^{j*}(u) = \int_0^u H(u-v)dF^{\alpha(\epsilon)*}(u) \le H(u)F^{\alpha(\epsilon)*}(u),$$

the modulus of this other term is bounded by $E(P\{X_1 + \ldots + X_{\alpha\epsilon} \le Y/\epsilon\} [\frac{1}{2}\lambda Y^2 +$ (const.)$\epsilon Y])$. The probability term here $\approx P\{\epsilon\alpha(\epsilon)EX < Y\}$ by the central limit theorem, so the dominated convergence theorem then implies that when $\epsilon\alpha(\epsilon) \to \infty$ as $\epsilon \to 0$, the bound $\to 0$ as $\epsilon \to 0$. The lemma is proved.   □

It is clear from the proof that the result remains true when the sum at (3.102) starts from $j = s$ for some fixed integer s.

The lemma can also be established without assuming $EX^2 < \infty$, but since we do not need it, we do not give the proof.

ACKNOWLEDGEMENTS

I thank Teunis Ott for showing me his paper in manuscript. This work was begun during a visit to the Statistics Department of the University of North Carolina at Chapel Hill following conversation with Ward Whitt at Bell Laboratories. I also thank Tomasz Rolski for his interest, and the referees of a much older version for their most helpful comments.

BIBLIOGRAPHY

[1]   Asmussen, S., *Applied Probability and Queues*, Wiley, Chichester 1987.

[2]   Baccelli, F. and Brémaud, P., *Elements of Queueing Theory: Palm-Martingale Calculus and Stochastic Recurrences*, Springer-Verlag, Berlin 1994.

[3]   Barlow, R.E. and Proschan, F., *Statistical Theory of Reliability and Life Testing*, Holt, Rinehart and Winston, New York 1975.

[4]   Borovkov, A.A., *Stochastic Processes in Queueing Theory*, Springer-Verlag, New York 1976 (translated from Russian edition, 1972).

[5]   Brumelle, S.L., Bounds on the wait in a GI/M/k queue, *Mgt. Sci.* **19** (1973), 773-777.

[6]   Carlsson, H. and Nerman, O., An alternative proof of Lorden's renewal inequality, *Adv. Appl. Prob.* **18** (1986), 1015-1016.

[7]   Daley, D.J., Stochastically monotone Markov chains, *Z. Wahrs.* **10** (1968), 305-317.

[8]   Daley, D.J., Inequalities for the moments of tails of random variables with a queueing application, *Z. Wahrs.* **41** (1977), 139-143.

[9]   Daley, D.J., Tight bounds for the renewal function of a random walk, *Ann. Prob.* **8** (1980), 615-621.

[10] Daley, D.J., Certain optimality properties of the first-come first-served discipline for G/G/s queues, *Stoch. Proc. Appl.* **25** (1987), 301-308.

[11] Daley, D.J. and Rolski, T., A light traffic approximation for a single-server queue, *Math. Oper. Res.* **9** (1984), 624-628.

[12] Daley, D.J. and Rolski, T., Light traffic approximations in queues, *Math. Oper. Res.* **16** (1991), 57-71.

[13] Daley, D.J. and Rolski, T., Light traffic approximations in many-server queues, *Adv. Appl. Prob.* **24** (1992), 202-218.

[14] Daley, D.J., Kreinin, A. Ya. and Trengove, C.D., Inequalities concerning the waiting-time in single-server queues: A survey. In: *Queueing and Related Models* (ed. by U.N. Bhat and I.V. Basawa), Oxford University Press, New York (1992), 177-223.

[15] Daley, D.J. and Vere-Jones, D., *An Introduction to the Theory of Point Processes*, Springer-Verlag, New York 1988.

[16] Foss, S.G., Approximation of multichannel queueing systems, *Sibirski Mat. Zh.* **21**:6 (1980), 132-140 (in Russian). *Sib. Math. J.* **21** (1980), 851-857 (trans.).

[17] Franken, P., König, D., Arndt, U. and Schmidt, V., *Queues and Point Processes*, Akademie-Verlag, Berlin and Wiley, Chichester 1982.

[18] Gittins, J.C., A comparison of service disciplines for GI/G/m queues, *Math. Oper. Stat. Ser. Optim.* **9** (1978), 255-260.

[19] Kendall, D.G., Stochastic processes occurring in the theory of queues and their analysis by means of the imbedded Markov chain, *Ann. Math. Stat.* **24** (1953), 338-354.

[20] Kiefer, J. and Wolfowitz, J., On the theory of queues with many servers, *Trans. Amer. Math. Soc.* **78** (1955), 1-18.

[21] Kiefer, J. and Wolfowitz, J., On the characteristics of the general queueing process, with applications to random walk, *Ann. Math. Stat.* **27** (1956), 147-161.

[22] Kimura, T., Approximations for multi-server queues: system interpolations, *Queueing Sys.* **17** (1994), 347-382.

[23] Kingman, J.F.C., Some inequalities for the GI/G/1 queue, *Biometrika* **49** (1962), 315-324.

[24] Kingman, J.F.C., The heavy traffic approximation in the theory of queues, In: *Proc. Symp. Congestion Theory* (ed. by W.L. Smith and W.E. Wilkinson), Univ. North Carolina Press, Chapel Hill, NC (1965), 137-159.

[25] Kingman, J.F.C., Inequalities in the theory of queues, *J. Roy. Stat. Soc. Ser. B* **32** (1970), 102-110.

[26] Köllerström, J., Heavy traffic theory for queues with several servers, I, *J. Appl. Prob.* **11** (1974), 544-552.

[27] Lorden, G., On excess over the boundary, *Ann. Math. Stat.* **41** (1970), 520-527.

[28] Loulou, R., Two sample-path inequalities for G/G/k queues, *INFOR* **21** (1983), 136-144.

[29] Loynes, R.M., The stability of a queue with non-independent interarrival and service times, *Proc. Cambridge Philos. Soc.* **58** (1962), 497-520.

[30] Makino, T., Investigation of the mean waiting time for queueing system with many servers, *Ann. Inst. Stat. Math.* **21** (1968), 357-366.

[31] Mori, M., Some bounds for queues, *J. Oper. Res. Soc. Japan* **18** (1975), 152-181.

[32] Ott, T.J., Simple inequalities for the D/G/1 queue, *Oper. Res.* **35** (1987), 589-597.

[33] Ovuworie, G.C., Multi-channel queues: a survey and bibliography, *Internat. Stat. Rev.* **48** (1980), 49-71.

[34] Rolski, T., Remarks on Palm probabilities of point processes which are not simple, manuscript (1989).

[35] Sigman, K., *Stationary Marked Point Processes: An Intuitive Approach*, Chapman and Hall, New York 1995.

[36] Slivnyak, I.M., Stationary streams of homogeneous random events, *Vest. Harkov. Gos. Univ. Ser. Mech.-Math.* **32** (1966), 73-116.

[37] Stoyan, D., *Comparison Methods for Queues and Other Stochastic Models*, John Wiley, Chichester 1983. (English translation and revision, Ed. D.J. Daley.)

[38] Stoyan, D. and Stoyan, H., Inequalities for the mean waiting time in single-line queueing systems, *Izv. Akad. Nauk SSSR Techn. Kibernet.* **1974**:6 (1974), 104-106.

[39] Suzuki, T. and Yoshida, Y., Inequalities for many-server queue and other queues, *J. Oper. Res. Soc. Japan* **21** (1970), 29-42.

[40] Trengove, C.D., Bounds for the mean waiting time in queues, M.Sc. thesis, Univ. of Melbourne (1978), unpublished.

[41] Whitt, W., The effect of variability in the GI/G/s queue, *J. Appl. Prob.* **17** (1980), 1062-1071. Correction, *J. Appl. Prob.* **21** (1984), 445-446.

[42] Whitt, W., Comparison conjectures about the M/G/s queue, *Oper. Res. Letters* **2** (1983), 203-209.

[43] Whitt, W., An interpolation approximation for the mean workload in a GI/G/1 queue, *Oper. Res.* **37** (1989), 936-952.

[44] Wolff, R.W., An upper bound for multi-channel queues, *J. Appl. Prob.* **14** (1977), 884-888.

[45] Wolff, R.W., Upper bounds on work in system for multi-channel queues, *J. Appl. Prob.* **24** (1987), 547-551.

Chapter 4

Queueing systems with state dependent parameters

Jewgeni H. Dshalalow

ABSTRACT This chapter surveys queueing systems with processes and characteristics (service, waiting time, and idle and busy period disciplines) dependent on "internal sources" of the system, such as queue length process and unfinished work. Some particular state dependent models, such as single-server systems with (q, N)-policy, multi-channel systems, and machine repair problems, are discussed in detail.

CONTENTS

4.1 INTRODUCTION

Queues *with state dependent parameters* include systems whose input stream, servicing process, waiting time process, idle and busy periods and their disciplines, and other features (such as vacations) depend on "internal processes," i.e., processes generated by the system itself. (One should not confuse queues with state dependent parameters with those whose dependence is generated by an external source, such as with *Markov modulated input* or *semi-Markov input*.)

An abundance of studies of systems with state dependent parameters published by now makes the task of surveying or even listing all of them very challenging if not impossible. The literature on systems with state dependent parameters includes, among other things, such populous topics as queues with vacationing server, retrials (see Chapter 2 by Kulkarni and Liang in this book), priorities, and feedback, having been themselves subject to comprehensive surveys in the past. Consequently, we will exclude these categories from the discussion. The objective of our chapter (which is much less ambitious) is to classify the most significant work and survey some recent results from the rest in this area.

At first, we will list all major processes and characteristics that will be subject to interdependence in queueing systems:

1. *The queueing process.* This most frequently applies to the number of customers

in the systems including those being in service. Although in some papers the "queue" stands for the entire group of lined up customers, throughout this chapter we shall be using the term *queue* to mean the cumulative number of customers in the system.

2. *Input process* - ordinary or bulk.
3. *Service process* - ordinary or batch.
4. *Waiting room* or, equivalently, the *size of the buffer*. This is zero or finite or infinite or variable.
5. *Idle* and *busy period disciplines* which specify the rules of when to begin idle and busy periods, respectively.
6. *Waiting time process*, frequently regarded as the *"actual" waiting time process*. This is the period of time an arriving customer is to spend in the waiting room.
7. *Sojourn time* is customer's waiting time plus the time spent by the customer beyond the waiting room, most often in service.
8. *Unfinished work*, or *workload*, or *work backlog*, or *work*. This is the period of time needed for the servicing facility to process all customers present in the system at a particular observation point (t), including a customer (or customers) in service.
9. *Virtual waiting time* or *Takács process*. This is the period of time a fictious customer would have to wait had he arrived at a particular observation point (t), not including his service time. The *virtual waiting time* coincides with *workload* and makes sense for FIFO disciplines only.

The vast majority of queueing systems with state dependent parameters fall into the three main categories: queue length dependent parameters, workload dependent parameters, and interdependence of input and service. Based on the available literature, the three main categories of the state dependence naturally break into the following subcategories.

Queue length state dependence

1. Input, including interarrival times and sizes of arriving batches of customers. The input intensity can increase or decrease depending on the queue length. In particular, a long queue can "discourage" arriving customers to the degree that the input becomes sparse to allow the queue drop. An instance of a "rigid" discouragement is when customers do not line up at all if the queue length hits some specified level. (This describes queues with finite waiting room or loss systems.)
   a) The input process "continuously" depends on the queueing process. For instance, the input rate λ is a function of the queueing process $Q(t)$ at each moment of time t, i.e., $\lambda = \lambda(Q(t))$. We will call them *ic-dependent* systems.
   b) The input depends on the queueing process only at some specified epochs of time, such as the moments of service completions (*id-dependence*) or customers' arrivals (*ia-dependence*). Loss systems are a trivial example of the latter case.
2. Service, including durations of service, sizes of processing batches of customers, and the number of servers. In particular, it requires the service process to intensify when the queue raises and slow down (by choosing more economical means) when the queue drops.
   a) The service process continuously depends on the queueing process by having the service intensity μ vary with the queueing process $Q(t)$ as $\mu = \mu(Q(t))$.

b) The service process varies only upon some specific epochs, such as service completions and arrival instants.

3. Idle period disciplines.

a) Normally, any idle period begins when a completely processed customer (or batch of customers) leaves the system empty. Then, the server (or servers) rests until a customer enters the system. In various instances, especially when the server processes more than one customer at a time, an idle period will start whenever the queue drops below some specified level r. Such a rule is referred to as the *quorum discipline* (or *q-discipline*, or just *quorum*). r is called the *quorum level*. If $r = 1$, then such a queueing system is called *exhaustive*.

b) Unless specified, the server always rests and is ready to restore his activity as soon as a new arrival takes place. Alternatively, the server can be assigned to another task to work on during which time he is unavailable to those arrivals that may occur. Consequently, the server is said to *leave on vacation*. If after any such vacation the server returns to the system and finds a customer (or customers) he starts to process them immediately. Otherwise, he rests. This type of vacation is called a *single vacation*, as opposed to *multiple vacations*. In the latter case, a vacation is in fact a sequence of single vacation segments. Upon the end of each vacation segment, i.e., return of the server, he checks on the system, and if the system is still empty he leaves on the next vacation segment. There are other synonyms used for "vacations," such as *removable servers*, *continuously operating servers*, and *T-policy*.

c) K- or E-limited discipline requires the server to turn off or take vacations when K customers have been processed since the beginning of a busy period, or if the queue becomes exhausted.

d) Any combination of the two to three idle period disciplines, such as quorum with vacations and E-limited discipline with vacations.

4. Busy period disciplines.

a) In many real-world situations, after each idle period, the server needs a "warm-up" time to begin his first service, also called the *start-up*. In most cases, the presence of start-up periods is nothing but service dependence on the queue length. However, if service is batch and a busy period starts with the queue length less than the server capacity, the server can admit some customers that arrive during the start-up period.

b) Usually, service is restored as soon as any new arrival takes place. However, to reduce the number of switch-overs, and thereby start-up periods, it is prudent to have the queue accumulate to some N customers and then activate service. Such a rule is known as *N-policy* and the number N is called the *threshold level*.

5. Combinations of the above disciplines. For instance, the (q,N)-policy is often referred to as (v,N)-policy or *hysteretic* discipline. The word "hysteretic," originated in physics, applies well to this situation. Indeed, with $q \leq N$, the system turns off when the queue crosses q from above and turns on when it crosses N from below.

6. Other disciplines. This category includes such disciplines as *feedback* and *retrials*, which are themselves independent subjects in queueing. Of the named disciplines, retrial systems are a relatively recent and important area of research and this subject is updated by Kulkarni and Liang as Chapter 2 in this book.

Workload dependence

1. Input flow.

   a) Limited waiting time. Customers arriving to the system can be discouraged not only by a long line (which is an obvious hindrance), but also by the long time they may spend in the queue before they are accepted to service. There are a few different instances of customers' "impatience," among them with customer *awareness* and customer *unawareness*. In the former case, customers are able to estimate their wait and if it exceeds their patience they choose not to join the queue (although it is not always realistic to expect customers to estimate accurately their waiting time). In the second case, customers always join the queue and leave the system unprocessed as soon as their wait exceeds their patience.

   b) The input rate λ depends continuously on the unfinished work v_t, i.e., $\lambda = \lambda(v_t)$.

2. Service. Similarly, the service rate $\mu = \mu(v_t)$ continuously depends on the amount of unfinished work.

3. Busy period discipline. While being idle, the server would not activate his service unless the workload would accumulate to at least D units of time. This is referred to as the *D-policy*. (The D-policy is similar to the N-policy, but the latter is applied to the queueing process.)

4. Idle period discipline. This is similar to the q-policy that is applied to the queueing process. The server will initiate an idle period (or go on vacation) as soon as upon service completion the workload drops below level w. We will call it the *w-policy*.

Interdependence of input and service

This class of systems includes those where the interarrival and service times of every customer are not independent. Most frequently, their joint distributions are special bivariate exponential.

In the following sections we survey most recent pertinent work on queues with input and service processes dependent on internal sources of the system, such as queueing process, workload, and input and service processes themselves. The dependence can be continuous or it may take place upon some specific stopping times, such as arrival and departure epochs. Among other types of dependencies, the following are identified: the effect of queueing and workload processes on busy and idle period disciplines, such as quorum, N-, D-, and T-policies, and hysteresis. All classes of state dependent queues are also broken into single server, multiserver, and closed queues. The reference list is also categorized accordingly. Since models fall into several classes, for convenience some of them are referenced more than once. The reference list is preceded by the section of open problems and future research directions.

4.2 QUEUE LENGTH STATE DEPENDENCE. SINGLE SERVER SYSTEMS

4.2.1 Systems with queue length dependent input and service. Continuous dependence

In this section we first consider a system of $M/G/1$ type with continuous input dependence (*ic-dependence*) and service time dependence upon departure epochs (*sd-dependence*). Let $T = \{T_0, T_1, \ldots\}$ be the sequence of successive service completions and let $\{B_0, B_1, \ldots\}$ be a sequence of arbitrary probability distribution functions (PDF's) with finite means $\{b_0, b_1, \ldots\}$. Given that $Q_n = i$, the $(n+1)$th service will be distributed B_i and the input is a pure birth process with birth rates λ_j $(j \geq i)$. Such a system was studied by Schellhaas [27], van Hoorn [18], and Tijms and van Hoorn [30].

The transition probabilities, $a_{ij}(t)$, of the birth process on time interval $[0,t]$ satisfy the following system of forward Kolmogorov differential equations:

$$\left.\begin{array}{c} \frac{d}{dt}a_{ij}(t) = -\lambda_j a_{ij}(t) + \lambda_{j-1}a_{i,j-1}(t)I_{\{j > i\}}(i,j), \quad j \geq i \\[2mm] a_{ij}(0) = \delta_{ij}, \end{array}\right\} \tag{4.1}$$

where I_A is the indicator function of a set A.

Under the above assumptions about the input and service processes, the sequence (Q_n, T_n) is a Markov renewal process. Denote by $q(t) = (q_{ij}(t) = \mathbb{P}^i\{Q_1 = j,\, T_1 \leq t\})$ the semi-Markov kernel. Then,

$$q_{ij}(t) = \begin{cases} 1 - e^{-\lambda_0 t}, & i = 0, j = 1 \\[2mm] \int_0^t a_{i,j+1}(u)B_i(du), & i \geq 1, j \geq i-1 \\[2mm] 0, & otherwise, \end{cases} \tag{4.2}$$

where $a_{ij}(u)$ is a solution to the above system (4.1) with $\sum_{i \epsilon E} a_{ij}(t) = 1$.

If E is infinite the following conditions are sufficient for the ergodicity of the queueing process $Q(t)$ and the embedded Markov chain Q_n [27]:

E1.   There is a $u > 0$ such that $\sup\{B_i(u): i = 0,1,\ldots\} < 1$.

E2.   There is a $\lambda^* < \infty$ and an $m \in \mathbb{N}$ such that $\sup\{\lambda_i : i \geq m\} \leq \lambda^*$ and $\sup\{\lambda^* b_i : i \geq m\} < 1$.

Let $p = (p_i : i \in E)$ and $\pi = (\pi_i : i \in E)$ be stationary probabilities of the embedded Markov chain Q_n and the queueing process $Q(t)$, respectively. Given the above ergodicity conditions E1 and E2, vectors p and π are related to each other as follows:

$$\pi_1 = \frac{(p_1 - p_0)}{\lambda_1 D}, \quad \pi_k = \frac{p_k}{\lambda_k D}, \text{ for } k \neq 1 \tag{4.3}$$

with

$$D = \frac{p_0}{\lambda_0} + \sum_{k=1}^{|E|-1} p_k b_k, \tag{4.4}$$

where $|E|$ stands for the cardinal number of set E.

The probabilities π_k can be found from the following recursive relations:

$$\pi_k = \frac{1}{c_k}\left(d_{k1}\lambda_0\pi_0 + \sum_{i=1}^{k-1}d_{ki}\lambda_i p_i\right), \tag{4.5}$$

where

$$c_k = \int_0^\infty e^{-\lambda_k t}B_k(dt) \tag{4.6}$$

and

$$d_{ki} = \int_0^\infty a_{ik}(t)[1 - B_i(t)]dt, \tag{4.7}$$

$$i = 1,\ldots,|E|-1 \text{ if } E \text{ is finite, otherwise, } i = 1,\ldots.$$

From these formulas one can choose any $\pi_0 \in (0,1)$ and calculate the other π_k's from the above formulas and normalize them thereafter.

For special service time distributions, such as PH-distributions, the above algorithm is computationally simpler. For more details, see [27].

Knessl et al. [19] studied the "continuous" state dependence in the following $M/G/1$-type of system by using the supplementary variable techniques and singular perturbation methods. Let $Y(t)$ denote the service time elapsed from the beginning of service until time t. Then $(Q(t),Y(t))$ is a two-dimensional Markov process. Let

$$p_j(t,y) = \frac{\partial}{\partial y}\mathbb{P}\{Q(t) = j, Y(t) \le y\}, \; j = 0,1,\ldots, \; y \ge 0. \tag{4.8}$$

The arrival and service rates (λ_n and μ_n, respectively) depend on the queue length as follows. Let

$$\mathbb{P}\{Q(t+dt) - Q(t) = 1 \mid Y(t) = y, Q(t) = n\} = \lambda_n dt + o(dt) \tag{4.9}$$

and

$$\mathbb{P}\{Q(t+dt) - Q(t) = -1 \mid Y(t) = y, Q(t) = n\} = \mu_n(y)dt + o(dt), \tag{4.10}$$

where

$$\mu_n(y) = \frac{b_n(y)}{\int_y^\infty b_n(x)dx} \tag{4.11}$$

is the instantaneous departure rate associated with the service density $b_n(y)$. We call this type of service dependence *sc-dependence*. The densities $p_j(t,y)$ satisfy the forward Kolmogorov equations:

$$\frac{\partial}{\partial t}p_n(t,y) = -\frac{\partial}{\partial y}p_n(t,y) - [\lambda_n + \mu_n(y)]p_n(t,y) + \lambda_{n-1}p_{n-1}(t,y), \; n \ge 2, \tag{4.12}$$

$$p_n(t,0) = \int_0^\infty p_{n+1}(t,x)\mu_{n+1}(x)dx, \; n \ge 2, \tag{4.13}$$

$$\frac{\partial}{\partial t}p_1(t,y) = -\frac{\partial}{\partial y}p_1(t,y) - [\lambda_1 + \mu_1(y)]p_1(t,y), \tag{4.14}$$

$$p_1(t,0) = \lambda_0\pi_0(t) + \int_0^\infty p_2(t,x)\mu_2(x)dx, \tag{4.15}$$

$$\frac{\partial}{\partial t}\pi_0(t) = -\lambda_0 p_0(t) + \int_0^\infty p_1(t,x)\mu_1(x)dx, \tag{4.16}$$

where $\pi_j(t) = \mathbb{P}\{Q(t) = j\}$.

Without ic- and sc-dependence, the above equations are solved in terms of z- and Laplace transforms. However, the dependence of the rates on n does not permit the ease of transforms traditionally applied to systems with no state dependence. Approximation techniques are therefore used. The idea is to introduce the "scaled variables" ξ and η such that

$$\xi = \varepsilon n \text{ and } \eta = \frac{y}{\varepsilon}, \tag{4.17}$$

where ε is a small parameter. Due to (4.17), the state dependent rates (4.9-4.10) are defined as

$$\mathbb{P}\{Q(t+dt) - Q(t) = 1 \mid Y(t) = \varepsilon\eta, Q(t) = \tfrac{\xi}{\varepsilon}\} = \tfrac{1}{\varepsilon}\lambda(\xi)dt + o(dt), \tag{4.18}$$

$$\mathbb{P}\{Q(t+dt) - Q(t) = -1 \mid Y(t) = \varepsilon\eta, Q(t) = \tfrac{\xi}{\varepsilon}\} = \tfrac{1}{\varepsilon}\mu(\eta,\xi)dt + o(dt). \tag{4.19}$$

The probability density $p_n(t,y)$ will be transformed into

$$p_n(t,y)dy = P(t,\xi,\eta)d\eta, \tag{4.20}$$

or in the stationary case,

$$p_n(y)dy = P(\xi,\eta)d\eta. \tag{4.21}$$

Therefore, under the above transformations, system (4.12-4.16) in the steady state will be written in the form:

$$\tfrac{\partial}{\partial\eta}P(\eta,\xi) = \lambda(\xi-\varepsilon)P(\eta,\xi-\varepsilon) - [\lambda(\xi)+\mu(\eta,\xi)]P(\eta,\xi), \tag{4.22}$$

$$P(0,\xi) = \int_0^\infty P(x,\xi+\varepsilon)\mu(x,\xi+\varepsilon)dx, \tag{4.23}$$

$$\xi \geq 2\varepsilon \text{ (which corresponds to } n \geq 2\text{)},$$

$$\tfrac{\partial}{\partial\eta}P(\eta,\xi) = -[\lambda(\xi)+\mu(\eta,\xi)]P(\eta,\xi), \tag{4.24}$$

$$P(0,\varepsilon) = \frac{\lambda(0)}{\varepsilon}\pi_0 + \int_0^\infty P(x,2\varepsilon)\mu(x,2\varepsilon)dx, \tag{4.25}$$

$$\frac{\lambda(0)}{\varepsilon}\pi_0 = \int_0^\infty P(x,\varepsilon)\mu(x,\varepsilon)dx, \tag{4.26}$$

where π_n, $n = 0,1,\ldots$, is the stationary distribution of the queueing process. System (4.22-4.26) is augmented by the normalization condition

$$\pi_0 + \varepsilon\sum_{j\geq 1}\int_0^\infty P(\eta,j\varepsilon)d\eta = 1. \tag{4.27}$$

Specified conditions are imposed on the traffic intensity λ/μ to ensure the ergodicity of the queueing process. Those conditions are considered in several regions with respect to the scale variables. In one of the regions, the solution of system (4.22-4.26) is sought in the form

$$P(\eta,\xi) \sim \exp\left(-\tfrac{1}{\varepsilon}[\psi_0(\eta,\xi)+\varepsilon\psi_1(\eta,\xi)+\varepsilon^2\psi_2(\eta,\xi)+\ldots]\right), \tag{4.28}$$

where ψ_j are unknown functions of η and ξ. Then, P in the form (4.28) is substituted into system (4.22-4.26) and the coefficients of each power of ε are then equated to zero. The authors of [19] show that the first two terms of the asymptotic expansion (4.28) satisfy second order nonlinear differential equations. Once found, ψ_0 and ψ_1 can be substituted into first order linear differential equations in ψ_j, $j \geq 2$. Some special cases yield explicit solutions. For the rest of this discussion, the reader is referred to the original paper [19] and to Chapter 12, Applications of Singular Perturbation Methods, pp. 311-336, by Knessl and Tier in [271].

Queueing systems with finite waiting space or loss systems are those with the input dependent on the queue length and therefore belong to this section [7-9,20, 23,24]. Occasionally, they are interpreted as *systems with discouraged arrivals*. However, the traditional class of queues with discouragements includes birth and death models with variable input and service rates [15-17,22], or even more specifically, those with input rates inversely proportionally to the state of the system [26], and $M/D/1$ and $D/M/1$ models [22]. While the latter deals with the steady state mode, the birth and death type models mostly analyze the queueing process in the transient mode.

4.2.2 State dependence upon departure epochs

This class includes systems with dependence of service on the queue length, but only upon departure epochs. (Call it, for convenience, *sd-dependence*.) Perhaps it is also a more realistic scenario of service dependence in contrast with "continuous" dependence presented above. The vast majority of queueing systems with this type of dependence are of $M/G/1$ type, which we assume for the following discussion of basic systems.

Let $T = \{T_0 = 0, T_1, T_2, \ldots\}$ be the sequence of successive service completions and $\{B_0, B_1, \ldots\}$ a sequence of arbitrary PDF's with finite means $\{b_0, b_1, \ldots\}$. Given that $Q_n = Q(T_n+) = i$, the $(n+1)$th service will be distributed according to B_i. We assume that the input is compound Poisson with intensity λ, and we assume the common probability generating function (p.g.f.) $a(u)$ and mean α for sizes of arriving batches. The transition probability matrix for the embedded Markov chain Q_n of the basic system is in the form

$$
P = \begin{pmatrix}
p_{00} & p_{01} & p_{02} & \cdots \\
p_{10} & p_{11} & p_{12} & \cdots \\
0 & p_{21} & p_{22} & \cdots \\
0 & 0 & p_{32} & \cdots \\
\cdots & \cdots & \cdots & \cdots
\end{pmatrix}.
$$

The p.g.f.'s of the embedded process Q_n and the continuous time parameter queueing process in the steady state, $P(u)$ and $\pi(u)$, respectively, satisfy the following relation:

$$
\pi(u) = \frac{\alpha(1-u)}{1-a(u)}P(u). \tag{4.29}
$$

(See Dshalalow [45].) In particular, when the input is orderly it follows that $\pi(u) = P(u)$, previously obtained by Schäl [73].

However, with no further restrictions on the B_i's, a tractable analytic form for $P(u)$ is impossible. The assumption that $B_i = B$ for $i \geq M$ (a nonnegative integer) makes an ideal compromise between an analytically tractable solution and loss of generality. Indeed, the number M can be assumed to be arbitrarily large to the degree that the queue would hardly reach given the ergodicity condition $\rho = \alpha b \lambda < 1$. The transition probability matrix P is reduced to the form of a so-called *delta*-2 (Δ_2-) *matrix*, called so in Abolnikov and Dukhovny [253] (who studied ergodicity conditions for various Markov chains occurring in queueing). In this matrix, the ith row (where $i \geq M$) contains the entries $k_0, k_1, \ldots$, preceded by $i-1$ zeros. The p.g.f. $A_i(u)$ of the ith row is $u^{i-1}B^*(\lambda - \lambda\alpha(u))$, where B^* stands for the Laplace-Stieltjes transform (LST) of the PDF B. So, under this assumption, the p.g.f. $P(u)$ for the embedded process in the equilibrium can be expressed in a Kendall-like formula whose numerator would contain probabilities $p_0, \ldots, p_{M-1}$ satisfying a system of linear equations, or, alternatively, they can be found recursively from a finite system of linear equations; see Abolnikov and Dshalalow [1] for a more general quorum system.

Sd-dependent systems evolved from the sixties to the present time in an abundance of papers [32-39,44-55,60-63,68-71,73-79,83,84] and they frequently reduce to one or two "control levels." An sd-dependent system is said to have *control level L* and the service time is assigned to *mode L* if the service time distribution of a customer is B_L given that the queue length was L immediately after the completion of previous service. With one control level L a system has two service modes with PDF's $B_0 = \ldots = B_{L-1}$ and $B_L = B_{L+1} = \ldots$ and one switch. Suzuki and Ebe [78] optimized the value of control level L switching between a slow and speed service mode.

Sd-dependent systems can also be modified by allowing the input to depend on the queueing process by altering its rates upon departure epochs only, i.e., with *id-dependence*. While the assumption about continuous dependence of the input rate on the queue length (discussed in the previous section) is undoubtedly more realistic, it does not enable one to find a solution for the steady state distribution of the queueing process in a closed form. Assume that the input intensity alters only upon service completions and does not change over inter-departure periods (which are referred to as *service cycles*). Given $Q_n = i$, we also impose the same restriction $\lambda_i = \lambda$ and $\alpha_i(u) = \alpha(u)$, $i \geq M$, which does not deprive the solution for the p.g.f. $P(u)$ from its closed form under the scenario of sd-dependence. Examples of such systems will be demonstrated in the upcoming sections in combination with other state dependent characteristics.

The id-dependence was defined and studied in Dshalalow [261,262,267] in connection with input intensities, optimization of various types of objective functions, and conservation laws such as Little's formula. In terms of [267] the id-dependence was referred to as *modulation*. The general notion of modulation, according to [267], is based on variation of parameters of a stochastic process perturbed by another process continuously or upon a sequence of stopping times.

Another, less populous subcategory of state dependent queues that belongs to this class is one with *hysteretic* or *H-policy*. The basic hysteretic discipline modifies a one-control level sd-dependence system in the following way. Suppose L and M ($\geq L$) are two control levels. If the queue drops below level $L+1$ the system switches to service mode 1. If the queue raises above level M the system switches to service mode 2. If the queue length is between $L+1$ and M a mode assigned to the system from the preceding service will be preserved. An obvious advantage of a two-control level H-policy system compared to a one-level sd-dependence system is that the presence of a fixed control level may yield an unwanted abundance of switch-overs when the queue oscillates about that level, whereas the H-policy offers a significantly more flexible rule reducing the number of switch-overs to a minimum by placing the numbers L and M far apart.

A multilevel H-policy is predominantly employed in $M/M/1$ type queues [87,89,91]. The queueing process in $M/G/1$ H-policy (even bilevel) systems is much more difficult to analyze, since the embedded chain Q_n is not Markovian. Bahary and Kolesar [85] studied a multilevel hysteretic control in the $M/G/1$ queue, but their method did not lead to any closed analytical solution. A natural idea would be to investigate a relevant two-dimensional chain [88,90]. However, Dukhovny [86] in 1976 used an alternative approach by considering an auxiliary embedded Markov process defined as

$$\xi_n = \begin{cases} Q_n, & \text{if after } T_n \text{ the system will be in mode 1} \\ Q_n + M - L, & \text{otherwise.} \end{cases} \tag{4.30}$$

With this transformation of Q_n, ξ_n uniquely defines the value of Q_n and the service mode. Indeed, given $\xi_n \leq M$, it is easy to show that the service mode is 1 and $Q_n = \xi_n$. Otherwise, the service mode is 2 and $Q_n = \xi_n - M + L$. Dukhovny studied Q_n via the Markov chain ξ_n in its transient and stationary modes and obtained the corresponding p.g.f.'s in a tractable analytical form.

4.2.3 Quorum Systems

Most *quorum systems* apply to single server queues and they define the idle period discipline. In other words, a quorum yields a rule (*q-policy*) that specifies beginnings of idle periods. It is generated by the "inability" of the server to process less than r customers in one batch, where $r \geq 1$. Consequently, idle periods begin when the queue drops below quorum level r, i.e., crosses quorum level r "from above." To start a busy period, the system needs to accumulate to (the same number) r customers again. When reaching r "from below" we call r the *threshold level*. [In a more general situation, the threshold level can be different from the quorum level, say equals N ($\geq r$), and it is referred to as *N-policy*. The combination of the q- and N-policies is known as (*q,N*)-*policy* or *hysteretic control* of which the q- and N-policies are special cases with $N = r$ and $r = 1$, respectively.] Note that without batch service it is unreasonable to apply q-policy, as it would prolong idle periods and increase the number of switch-overs instead of balancing each other.

The first quorum system goes back to Takács [277] (pp. 81-112) in his 1962 monograph. This was a "rigid" version with quorum-threshold level r, the same as server capacity. In 1967, Neuts [127] studied a more flexible (r,R)-quorum system with quorum-threshold level r and server capacity R. (See also Neuts [128].) This means that the nth servicing batch has the size equal to $\max\{r, \min\{Q_n, R\}\}$. Leaving the server capacity R fixed, one can target r for possible optimization. Note that conventional batch service queues are nothing else but ($1,R$)-quorum systems, such as one studied by Chaudhry and Templeton [102]; see also monograph [254] by the same authors. The term *quorum* is mentioned for the first time in that monograph by Chaudhry and Templeton (p. 179).

A vivid example of an (r,R)-quorum system is a transportation queue. Usually, a bus or truck does not want to begin transportation unless its capacity is "challenged." Another example is due to G. Newell cited in Neuts' paper [127]. Consider a main road and minor roads in junction. A traffic light on the main road interrupts its traffic flow after a certain length of time if at least r cars have activated a tripplate on the minor road. Otherwise, the light stays green until r cars have arrived. The red cycle is timed so that at most R cars can merge during it. We count as successive service times the time required for the platoon to merge, plus the fixed length of the green cycle on the main road. The model represents the queue on the minor road under the assumption that cars arriving during the time that vehicles ahead are merging must wait for the next cycle. Quorum systems can also model various packet-switched communication systems (which exchange messages among users by transmitting sequences of packets comprising messages).

Of those single-server systems most quorum systems are of the $M/G/1$ type, in notation $M/G^{r,R}/1$, in which idle periods are Erlang distributed with variable phases. Given that upon completion of a service, a departing group left i customers behind and $i < r$, the idle period will consist of $r - i$ exponential phases corresponding to $r - i$ successive arrivals. Q-policy systems were studied by Chaudhry, Gupta, and Madill [103], Chaudhry, Madill, and Briére [104], Neuts [128,130], Jacob, Krishnamoorthy, and Madhusoodanan [119], Dshalalow and Russell [109], and Gold and Tran-Gia [115]. Easton and Chaudhry [113] and Kambo and Chaudhry [120] studied the $E_k/M^{r,R}/1$ system. (Neuts and Nadarajan [248] considered a multi-channel Markovian quorum system.) Neuts [274, p. 184] considered the $GI/PH^{r,R}/1$ system applying matrix-geometric techniques. A useful bibliography on quorum systems is contained in the monograph by Medhi [273]. Among other characteristics, queueing, waiting time, and busy period processes, both at departure epochs and with continuous time parameter, are studied.

The transition probability matrix of the embedded queueing process (over departure epochs) in a typical $M/G/1$ queue with quorum is of the form

$$
P = \begin{pmatrix}
p_{00} & p_{01} & p_{02} & \cdots \\
\cdots & \cdots & \cdots & \cdots \\
p_{r0} & p_{r1} & p_{r2} & \cdots \\
0 & p_{r+1,1} & p_{r+1,2} & \cdots \\
0 & 0 & p_{r+2,2} & \cdots \\
\cdots & \cdots & \cdots & \cdots
\end{pmatrix},
$$

with $p_{r+1,1} = p_{r+2,2} = p_{r+3,3} = \cdots = k_0,$ $p_{r+1,2} = p_{r+2,3} = \cdots = k_1,$ ....
Such matrices were studied by Abolnikov and Dukhovny [253] and called there Δ_{r+1}-*matrices*. The generating function of the limiting distribution of the queueing process at departure epochs generalizes Kendall's formula for an $M/G/1$ system and contains R probabilities $p_0,...,p_{R-1}$ which are determined from a system of R linear equations. An example of these techniques will be given in Subsections 4.2.6 and 4.2.7 in connection with more general systems.

A quorum system becomes more complex when the input is bulk. Given the natural extension of the busy period policy to start a busy period when the queue crosses level r for the first time, the idle period consists of randomly many phases at the beginning of a busy period (the *first passage time*) and the queue level (the *first excess level*) is no longer fixed and equals r. Models of this kind were studied by H.-S. Lee and Srinivasan [148], H.-S. Lee [149], H.W. Lee et al. [150,151], and in a sequence of papers by Abolnikov and Dshalalow [92-94], Abolnikov et al. [95-100], Dshalalow [105-108], Dshalalow and Tadj [111], Muh [155,156], and Tadj [132], all preceded by the work on the *first excess level theory* initiated in Abolnikov and Dshalalow [252] and further developed in Dshalalow [263-265]. (See the discussion in the following subsection.) H.W. Lee et al. [150,151] were using another method to find the first excess level which will be discussed in Subsection 4.2.8.

The q-policy is usually combined with other state dependent characteristics. Neuts [127] allowed the service time distributions to depend on the size of processing batches (which is basically a queue length dependence). Similar assumptions were made in [128]. Another important combination with quorum is the T- or vacation policy. (See, for instance, Dshalalow [106], Dshalalow and Yellen [112], and Jacob and Madhusoodanan [118].) Recall that the basic T-policy requires idle periods to last T units of time, where T is a random variable (r.v.). It is interpreted as a single vacation the server leaves on upon the end of the last busy period and is not interrupted regardless of the system status. The *multiple vacations policy* combined with quorum requires the server to go on vacations as many times as needed to accumulate at least r customers. In this case, regardless of the type of the input (i.e., bulk or ordinary), the first passage time (i.e., the epoch when the server begins a busy period) and the first excess level at this instant of time is mandated by the first excess level theory applied by Dshalalow [106] and

Dshalalow and Yellen [112] to bulk input systems with multiple vacations. The q-policy is combined with the N-policy in work by Muh [155,156]. While the q-policy specifies the beginning of an idle period, the N-policy defines when to terminate it. A combination of the two allows a greater flexibility of a system, especially if their parameters are targeted for optimization.

One of the generalizations of the quorum policy is to allow the quorum level to be an r.v., instead of a fixed number. This system was treated in Dshalalow [105,107,108] and is the subject of Subsections 4.2.4-4.2.6.

4.2.4 Systems with state dependent input and random quorum (or queue length dependent service batch). Introduction

In this class of models, the input stream varies only upon instants of service completions. Although this type of modulation is more restrictive compared to that treated in Subsection 4.2.1, it is possible, in return, to employ more features for the input, service, and idle/busy period disciplines and still get explicit formulas rather than algorithms. The demonstration of these techniques will be based on one particular but typical model studied in Dshalalow [108].

To make this section self-contained, we will outline all major notation for the model:

$Q(t)$ - queueing process (with right continuous paths)
$\tau = \{\tau_1, \tau_2, \ldots\}$ - arrival instants
$T = \{T_0 = 0, T_1, \ldots\}$ - service completions
$\sigma = \{\sigma_1, \sigma_2, \ldots\}$ - service lengths
$Q_n = Q(T_n)$
$X = \{X_1, X_2, \ldots\}$ - sizes of arriving groups
$C(A) = \sum_{j \geq 0} I_A(T_j)$ gives the number of service completions

on time set A (i.e., C is the counting process associated with the point process T), where I_A is the indicator function of a set A.

Let $C_t = C([0,t])$. Then, $\widetilde{Q}(t) = Q(T_{C_t})$ is the queue length upon

a service completion prior to t, i.e., $\widetilde{Q}$ is the minimal embedded semi-Markov process.

$\mathbb{P}^i$ denotes the probability measure induced by the initial value i of the queueing process $Q(0) = Q_0 = i$. For the associated functionals will use the notation $\mathbb{E}^i[\cdot]$.

The model we are going to present is of $M/G/1$ type with state dependent bulk input, state dependent batch service, and state dependent random server capacity as follows.

Input. Customers arrive in batches at instants τ in accordance with the state dependent batch arrival process. This means that both the intensity of the arrival point process τ and the size of arriving batches X depend on queue lengths at instants T. More specifically, given $Q_0 = i$, the input in interval $[0, T_1)$ is compound Poisson with intensity λ_i and with batches of size $X_r^{(i)}, r = 1, 2, \ldots$, associated with the p.g.f.

$$a_i(z) = \mathbb{E}^i[z^{X_1^{(i)}}] = \mathbb{E}[z^{X_1^{(i)}} \mid Q_0 = i]. \tag{4.31}$$

Service process. All customers are processed by a single server in batches of random size. Upon completion of the $(n-1)$th service, the server generates the nth capacity c_n ($\leq r$, where r is some positive integer ≥ 1) which is assigned to the size of the nth group the server is going to process. Correspondingly, it takes a batch of the size c_n from the queue if available and processes it for a random time σ_n. Should the queue length be insufficient for that capacity the server will rest until the queue replenishes to c_n. Since customers arrive in batches, it would be unrealistic to expect the queue to fit c_n exactly. Therefore, the server will wait until the queue reaches or exceeds level c_n for the first time (the *first passage time*), after which the server will immediately resume its work and complete the nth service after σ_n units of time. [Note that the server may need to begin every busy period with a start-up time. This will automatically be implemented into a more general service dependence policy.] Consequently, the period $[T_{n-1}, T_n)$, which we call the nth *service cycle*, occasionally includes an idle period (which ends at the very moment the queue "crosses" level c_n) and the nth service period, which is of duration σ_n.

Now, given $Q_0 = i$, the batch size c_1 (or service capacity) generated for the first service is distributed as

$$\mathbb{P}^i\{c_1 = s\} = \mathbb{P}\{c_1 = s \mid Q_0 = i\} = \begin{cases} g_s^{(i)}, & s = 1, \ldots, r \\ 0, & otherwise. \end{cases} \qquad (4.32)$$

Let $g_i(z) = \mathbb{E}^i[z^{r-c_1}]$. The PDF of the first service time will be given by $B_i(x) = \mathbb{P}^i\{\sigma_1 \leq x\}$ with the finite mean b_i and the LST $\beta(\theta)$ defined for $Re(\theta) \geq 0$. Given Q_{n-1}, the service time in the interval $[T_{n-1}, T_n)$ is independent of the input stream. Any effect of the processes acting within interval $[T_{n-1}, T_n)$ on the processes on the upcoming service cycles is reduced to the value of the queueing process Q_n only.

Idle and busy period disciplines. Given $Q_0 = i < c_1$, the server begins to rest until the queue crosses level c_1 for the first time at the first passage time τ_ν. Let $X_1^{(i)}, X_2^{(i)}, \ldots$ denote the sequence of arriving batches of customers during the first service cycle and let

$$\nu = \nu_1 = \inf\{k = 0, 1, \ldots : A_k = X_0^{(i)}(= Q_0) + X_1^{(i)} + X_2^{(i)} + \ldots + X_k^{(i)} \geq c_1\}$$

be the *termination index*. The sequence, $A_0, A_1, \ldots$, is a discrete-valued delayed renewal process. Since c_1 is crossed at one of the moments $\tau_1, \tau_2, \ldots$, the first passage time will take place at τ_ν. The queue length at this time, called the *first excess level*, will be

$$A_\nu = Q(\tau_\nu) = Q_0 + X_1^{(i)} + X_2^{(i)} + \ldots + X_k^{(i)}. \qquad (4.33)$$

If $Q_0 = i \geq c_0$ then $\nu = 0$ and the idle period does not begin. The first passage time τ_ν will be assigned to 0 and the first excess level will be Q_0. Otherwise, the first idle period ends at τ_ν and the first busy period starts at the same time τ_ν with the initial queue length A_ν.

4.2.5 First excess level analysis

The information on ν, τ_ν, and A_ν is pivotal to the investigation of the queueing, waiting time, and busy period processes. The analysis of these characteristics in terms of the functional $\mathbb{E}^i[\xi^\nu e^{-\theta \tau_\nu} u^{A_\nu} \mid c_1 = s] = \mathcal{F}_s^{(i)} = \mathcal{F}_s^{(i)}(\xi, \theta, u)$ was rendered in several papers [263-265] for more general processes. For convenience, we bring only the essential facts about $\mathfrak{F}_s^{(i)}$. Following formula (9.8), p. 251 of [265],

$$\mathfrak{F}_s^{(i)}(\xi,\theta,u) = u^i - u^i(1 - \xi h_i(\theta)a_i(u))\mathfrak{D}_x^{s-i-1}\left\{\frac{1}{1 - \xi h_i(\theta)a_i(ux)}\right\}, \quad (4.34)$$

where $h_i(\theta) = \lambda_i/(\lambda_i + \theta)$ and the operator $\mathfrak{D}$ is defined as

$$\mathfrak{D}_x^k(\cdot) = \begin{cases} \frac{1}{k!}\lim_{x\to 0}\frac{\partial^k}{\partial x^k}\frac{1}{1-x}(\cdot), & k \geq 0 \\ \\ 0, & k < 0. \end{cases} \quad (4.35)$$

The expression

$$G_s^{(i)} = G_s^{(i)}(\xi,\theta,u) = u^i\mathfrak{D}_x^{s-i-1}\left\{\frac{1}{1 - \xi h_i(\theta)a_i(ux)}\right\}, \quad (4.36)$$

called the *generator of the first excess level*, according to [252], satisfies the formula

$$G_s^{(i)} = \sum_{j\geq 0}\mathbb{E}^i[\xi^\nu e^{-\theta\tau_\nu}u^A\nu I_{\{0,\dots,s-1\}}(A_j)\,|\,c_1 = s] \quad (4.37)$$

and, for every fixed ξ and θ, is the $(s-1)$th degree polynomial.

We will also use the notation for the main part of the generator frequently used in various formulas throughout this chapter. Let

$$\Delta_s^{(i)}(u) = \mathfrak{D}_x^{s-i-1}\left\{\frac{1}{1 - a_i(ux)}\right\} \quad (4.38)$$

and

$$\Delta_s^{(i)} = \Delta_s^{(i)}(1). \quad (4.39)$$

Therefore, $\Delta_s^{(i)}(u)$ is an $(s-1-i)$th degree polynomial.

Now, consider the functionals of the following marginal distributions.

Termination index:

$$T_s^{(i)}(\xi) = \mathfrak{F}_s^{(i)}(\xi,0,1) = \mathbb{E}^i[\xi^\nu\,|\,c_1 = s] = 1 - (1-\xi)\mathfrak{D}_x^{s-i-1}\left\{\frac{1}{1 - \xi a_i(x)}\right\}, \quad (4.40)$$

which is equivalent to

$$T_s^{(i)}(\xi) = 1 - (1-\xi)G_s^{(i)}(\xi,0,1). \quad (4.41)$$

The mean termination index is then

$$\overline{T}_s^{(i)} = \mathbb{E}^i[\nu\,|\,c_1 = s] = G_s^{(i)}(1,0,1) = \Delta_s^{(i)}. \quad (4.42)$$

First passage time:

$$P_s^{(i)}(\theta) = \mathfrak{F}_s^{(i)}(1,\theta,1) = \mathbb{E}^i[e^{-\theta\tau_\nu}\,|\,c_1 = s] = 1 - (1 - h_i(\theta))G_s^{(i)}(1,\theta,1)$$

$$= 1 - (1 - h_i(\theta))\mathfrak{D}_x^{s-i-1}\left\{\frac{1}{1 - h_i(\theta)a_i(x)}\right\}. \quad (4.43)$$

The mean first passage time is

$$\overline{P}_s^{(i)} = \mathbb{E}^i[\tau_\nu\,|\,c_1 = s] = \frac{1}{\lambda_i}\overline{T}_s^{(i)} = \frac{1}{\lambda_i}\Delta_s^{(i)}. \quad (4.44)$$

First excess level:

$$L_s^{(i)}(u) = \mathbb{E}^i[u^{A_\nu} \mid c_1 = s] = u^i - [1 - a_i(u)]G_s^{(i)}(1,0,u)$$

$$= u^i - u^i[1 - a_i(u)]\Delta_s^{(i)}(u). \tag{4.45}$$

Therefore, the mean value of the first excess level is

$$\overline{L}_s^{(i)} = \mathbb{E}^i[A_\nu \mid c_1 = s] = i + \alpha_i \Delta_s^{(i)}. \tag{4.46}$$

The expression

$$I_s^{(i)} = \alpha_i \Delta_s^{(i)} \tag{4.47}$$

is the mean value of the increment of the first excess level given $c_1 = s$. It is also the expected number of customers that arrive during an idle period given the server capacity $c_1 = s$.

4.2.6   Analysis of the systems with state dependent input and random quorum

<u>Embedded</u> <u>Queueing</u> <u>Process</u>. Let $M_i(A)$ give the number of customers that enter the system on time set $A \subseteq [T_0, \infty)$ given that $Q_0 = i$. Then, the transition of the queueing process from its state Q_0 (at time $T_0 = 0$) to Q_1 (at T_1) occurs in accordance with the relation

$$Q_1 = A_\nu - c_1 + M_{Q_0}([\tau_\nu, T_1)) \tag{4.48}$$

which can be extended for all transitions of the Markov chain Q_n using (4.48) as a recursive relation and modifying accordingly the indices.

The original queueing process $Q(t)$ with continuous time parameter is semi-regenerative relative to the point process T and the Markov renewal process $(Q,T) = (Q_n, T_n)$. As mentioned, $\widetilde{Q}(t)$ is the associated minimal semi-Markov process. The next section will discuss how to derive the main characteristic of $Q(t)$ based on those for the embedded process Q_n.

The following are the "consistency" assumptions imposed on state dependency of input and service. For $i > J \geq r - 1$, we set

$$\lambda_i = \lambda, \ a_i(u) = a(u), \ g_s^{(i)} = g_s \ (s = 1,\ldots,r), \ \alpha_i = a_i'(1) = \alpha, \ B_i = B, \ b_i = b. \tag{4.49}$$

All related functionals will drop their indices correspondingly. The system still carries almost the same generality by choosing J arbitrarily large at levels all real-world systems would hardly reach. The above constraint on $J \geq r - 1$ is a pure convenience, since we can always further suppress the state dependency by variation of indices.

Theorem 4.1. *The embedded process Q_n is ergodic if and only if*

$$\rho = \alpha b \lambda < \overline{g} = \mathbb{E}^i[c_1], \ i > J, \tag{4.50}$$

where $\mathbb{E}^i[c_1]$, $i > J$, is the mean server capacity.

Proposition 4.2.   *The p.g.f. $A_i(u) = \mathbb{E}^i[u^{Q_1}]$ of the ith row of the transition probability matrix P satisfies the following relations:*

$$A_i(u) = u^{i-r} K_i(u) \Gamma_i(u), \tag{4.51}$$

where

$$K_i(u) = \beta_i(\lambda_i - \lambda_i a_i(u)) \tag{4.52}$$

and

$$\Gamma_i(u) = g_i(u) - \sum_{s=i+1}^r g_s^{(i)} u^{r-s-i} [u^i - L_s^{(i)}(u)]. \qquad (4.53)$$

Theorem 4.3. (Dshalalow [49]) *Given the ergodicity condition* $\rho < \overline{g}$, *the p.g.f.* $P(u)$ *of the stationary probability distribution* $p = (p_0, p_1, \ldots)$ *of the embedded process* Q_n *satisfies the following Kendall's type formula:*

$$P(u) = \frac{\sum_{i=0}^J p_i u^i [K_i(u) \Gamma_i(u) - K(u) g(u)]}{u^r - K(u) g(u)}. \qquad (4.54)$$

Probabilities $p_0, \ldots, p_J$ *are the unique solution of the system of linear equations:*

$$\sum_{i=0}^J p_i \frac{d^k}{du^k} [A_i(u) - u^i] \Big|_{u=u_z} = 0, \; k = 0, \ldots, k_z - 1; \; z = 0, \ldots, Z, \qquad (4.55)$$

$$\sum_{i=0}^J p_i \Big(\rho_i - \rho + \sum_{s=i+1}^r g_s^{(i)} I_s^{(i)} \Big) + \rho = \overline{g} + \sum_{i=0}^J p_i (\overline{g}_i - \overline{g}), \qquad (4.56)$$

where u_z $(z = 0, \ldots, Z)$ *are the distinct roots of the function* $u \mapsto u^r - K(u) g(u)$ *in the compact region* $\{z \in \mathbb{C}: |z| \le 1\} \backslash \{(0,1)\}$ *so that the root* u_z *has the multiplicity* k_z *with* $\sum_{z=0}^{Z-1} k_z = r - 1$ *and* $u_Z = 0$ *is a root of multiplicity* $J - r + 1$.

Consequently, (4.55) provides J equations and (4.56) gives another equation derived from the condition $P(1) = 1$. There is a practical alternative to reduce the number of equations in (4.55) to only $r - 1$ by applying a method similar to one in Abolnikov and Dshalalow [33]. This would be worthwhile only when J is large compared to r since in this case equations (4.55) would lose their elegance.

Interestingly enough, equation (4.56) is a manifestation of the so-called "conservation law," which reconciles the server load with system capacity. The latter is defined as follows.

Let $\beta^{(i)} = \mathbb{E}^i[T_1]$ be the expected length of the service cycle given $Q_0 = i$ and let $\vec{\beta} = (\beta^{(i)}; i = 0, 1, \ldots)^{\mathrm{T}}$. [Then, the scalar product $p\vec{\beta}$ obviously gives the mean value of a service cycle in the stationary mode given equilibrium condition (4.50).] We will also use the vectors $\vec{\lambda} = (\lambda_i; i = 0, 1, \ldots)^{\mathrm{T}}$ and $\vec{\alpha} = (\alpha_i; i = 0, 1, \ldots)^{\mathrm{T}}$.

Let vector $\vec{\alpha} \circ \vec{\beta} \circ \vec{\lambda}$ denote the Hadamard product of the vectors $\vec{\alpha}$, $\vec{\beta}$, and $\vec{\lambda}$. Since the scalar product $p(\vec{\alpha} \circ \vec{\beta} \circ \vec{\lambda})$ carries the information about the service and idle periods, as well as the effect of the input, it is called the *system intensity*. The above conservation property of the system is elucidated by the following fact from Dshalalow [49].

Proposition 4.4. *The left-hand side of equation* (4.56) *equals the system intensity.*

Let

$$\mathcal{L} = \sum_{i=0}^\infty p_i \overline{g}_i = \sum_{i=0}^J p_i \overline{g}_i + \overline{g} \sum_{i=J+1}^\infty p_i. \qquad (4.57)$$

The value $\mathcal{L}$ is obviously the *mean (stationary) server load*. Now, in light of the last proposition, equation (4.56) implies that, in equilibrium, the system intensity and the mean server load $\mathcal{L}$ are equal. This result is obvious for the classical $M/G/1$ system whose system intensity is $\rho + p_0$ (where $\rho = b\lambda$) and equal to 1, the (mean) server load.

Finally, the mean service cycle satisfies the formula:

$$p\vec{\beta} = b + \sum_{i=0}^J p_i \Big\{ b_i - b + \sum_{s=i+1}^r \frac{1}{\lambda_i} g_s^{(i)} \overline{T}_s^{(i)} \Big\}. \qquad (4.58)$$

Consider a few special cases. With no input modulation and service control, formula (4.54) reduces to

$$P(u) = \frac{[a(u) - 1]K(u)\sum_{i=0}^{r-1}p_i\sum_{s=i+1}^{r}g_s u^{r-s}G_s^{(i)}(1,0,\theta)}{u^r - K(u)g(u)}. \qquad (4.59)$$

In another special case with a fixed server capacity we set $g_s^{(i)} = 1$ for $s = r$ and $g_s^{(i)} = 0$ for $s = 1,...,r-1$, to yield $g_i(u) = 1$ and consequently $\Gamma_i(u) = u^{-i}L_r^{(i)}(u)$ and $A_i(u) = u^{i-r}K_i(u)L_r^{(i)}(u)$. The latter reduces $P(u)$ in (4.59) to

$$P(u) = \frac{\sum_{i=0}^{J}p_i\{K_i(u)L_r^{(i)}(u) - u^i K(u)\}}{u^r - K(u)}. \qquad (4.60)$$

Formula (4.60) represents the embedded queueing process in the system with (rigid) fixed server capacity (known as an "*r*-quorum" system), more general than a system with ordinary input studied by Dshalalow and Russell [50], but more special than that in Abolnikov and Dshalalow [33]. In the latter it was assumed that the server capacity is R but the minimal batch to be processed is of size r ($\leq R$). Thus, formula (4.60) deals with a particular case in [33] for $r = R$. Finally, if the input is orderly the first excess level is equal to either r or to the queue length at the beginning of service and formula (4.60) further reduces to

$$P(u) = \frac{\sum_{i=0}^{J}p_i u^i\{K_i(u)u^{(r-i)^+} - K(u)\}}{u^r - K(u)}, \qquad (4.61)$$

where the operator $(\cdot)^+$ denotes $\sup\{(\cdot),0\}$.

Queueing Process with Continuous Time Parameter. We apply semi-regenerative techniques to the process $Q(t)$, essentially using the results for the embedded process Q_n. Let functional matrix $h(t) = (h_{ik}(t); i,k = 0,1,...)$ denote the semi-regenerative kernel, i.e.,

$$h_{ik}(t) = \mathbb{P}^i\{Q(t) = k, T_1 > t\}, \qquad (4.62)$$

and let

$$\delta_{in}(t) = \mathbb{P}^i\{M_i([0,t] = n\}. \qquad (4.63)$$

Lemma 4.5. *$h(t)$ satisfies the following formula:*

$$h_{ik}(t) = \sum_{s=i+1}^{min(k,r)}K_{ik}^{(s)}(t)g_s^{(i)}$$

$$+ \delta_{i,k-i}(t)\left\{[1 - B_i(t)]\sum_{s=1}^{min(i,r)}g_s^{(i)} + \sum_{s=k+1}^{r}g_s^{(i)}\right\}, \quad 0 \leq i \leq k, \qquad (4.64)$$

$$h_{ik}(t) = 0, \quad 0 \leq k < i, \qquad (4.65)$$

where

$$K_{ik}^{(s)}(t) = \sum_{m=s-i}^{k-i}\int_0^t \varphi_s^{(i)}(t-u,m+i)\delta_{i,k-i-m}(u)[1 - B_i(u)]du, \qquad (4.66)$$
$$0 \leq i < s \leq min(k,r),$$

and $\varphi_s^{(i)}(\cdot,\cdot)$ stands for the conditional joint probability density function of the first passage time and the first excess level given $c_1 = s$ and $Q_0 = i$.

Now, given equilibrium condition (4.50), the semi-regenerative kernel $h(t)$ is integrable over $\mathbb{R}_+$ with the notation

$$H = \int_0^\infty h(t)dt. \qquad (4.67)$$

Let $H_i(u)$ be the p.g.f. of the ith row of matrix H and let $H(u)$ denote the functional vector $(H_i(u); i = 0,1,...)^\mathrm{T}$. The main convergence theorem for semi-regenerative processes (*c.f.* Çinlar [255]) can be rephrased as follows [49]:

Given the ergodicity condition (4.50) for the embedded process Q_n, the stationary distribution $\pi = (\pi_0, \pi_1, \ldots)$ exists and its p.g.f. $\pi(u)$ satisfies the formula:

$$p\vec{\beta}\pi(u) = pH(u). \tag{4.68}$$

The last formula and Lemma 4.5 lead to the following.

Theorem 4.6. *Given the ergodicity condition $\rho < \overline{g}$ for the embedded process Q_n, the stationary distribution π of the queueing process $Q(t)$ exists, is independent of the initial state, and satisfies the formula:*

$$p\vec{\beta}\pi(u) = P(u)\gamma(u) + \sum_{i=0}^{J} p_i u^i [\gamma_i(u) - \gamma(u)]$$

$$+ \sum_{i=0}^{r-1} p_i K_i(u) \frac{1}{\lambda_i} \sum_{s=i+1}^{r} g_s^{(i)} G_s^{(i)}(1,0,u), \tag{4.69}$$

where

$$\gamma_i(u) = \frac{1}{\lambda_i} \frac{1 - K_i(u)}{1 - a_i(u)}, \tag{4.70}$$

and $P(u)$ satisfies formula (4.54).

If we drop the input modulation and service control, the first sum on the right-hand side of (4.69) will vanish and $\lambda \alpha p \vec{\beta}$ will be equal to $p(\vec{\alpha} \circ \vec{\beta} \circ \vec{\lambda})$, which, in turn, is by Proposition 4.4 the mean server load $\mathcal{L}$, to yield

$$\frac{\mathcal{L}}{\alpha}\pi(u) = \lambda\gamma(u)P(u) + K(u) \sum_{i=0}^{r-1} p_i \sum_{s=i+1}^{r} g_s^{(i)} G_s^{(i)}(1,0,u), \tag{4.71}$$

where $P(u)$ satisfies formula (4.59). Finally, the case of r-quorum discussed in connection with formula (4.60) will simplify the last equation to

$$\frac{\mathcal{L}}{\alpha}\pi(u) = \lambda\gamma(u)P(u) + K(u) \sum_{i=0}^{r-1} p_i G_r^{(i)}(1,0,u), \tag{4.72}$$

with the use of (4.60) for $P(u)$.

Idle and busy periods. By standard probability arguments, the LST of an idle period in the steady state is

$$\mathfrak{I}(\theta) = \frac{\sum_{i=0}^{r-1} p_i \sum_{s=i+1}^{r} g_s^{(i)} P_s^{(i)}(\theta)}{\sum_{i=0}^{r-1} p_i \sum_{s=i+1}^{r} g_s^{(i)}}. \tag{4.73}$$

Therefore, the expected value of the idle period is

$$\mathbb{E}[I_{per}] = \frac{\sum_{i=0}^{r-1} p_i \sum_{s=i+1}^{r} g_s^{(i)} \lambda_i \overline{T}_s^{(i)}}{\sum_{i=0}^{r-1} p_i \sum_{s=i+1}^{r} g_s^{(i)}}. \tag{4.74}$$

Let $\mathbb{E}[B_{per}]$ stand for the mean value of a busy period in the steady state and let π_{idle} be the probability to reach the server idle at any time in equilibrium. Then, π_{idle} satisfies either of the equivalent expressions:

$$\pi_{idle} = \sum_{i=0}^{r-1} \pi_i \sum_{s=i+1}^{r} g_s^{(i)} = \frac{\mathbb{E}[I_{per}]}{\mathbb{E}[I_{per}] + \mathbb{E}[B_{per}]}. \tag{4.75}$$

This yields the formula

$$\mathbb{E}[B_{per}] = \frac{\mathbb{E}[I_{per}](1 - \pi_{idle})}{\pi_{idle}} \tag{4.76}$$

for the mean value of a busy period.

4.2.7   N-policy systems and systems with hysteretic control

The N- (or *threshold*) policy is a rule specifying the beginning of a busy period. Namely, once the server enters an idle period, it specifies how many customers (N, the threshold level) should be accumulated in the queue before the server turns on. One of the main reasons for employing N-policy is to minimize server switch-overs, usually followed by start-ups. Of course, this prolongs idle periods thereby increasing the unfinished work or, equivalently, customers' sojourn time. The latter, as mentioned, is balanced out by the reduction of switch-overs, and the optimal value of N becomes a reasonable objective. Another common name for N-policy is *removable server(s)*, called so by Yadin and Naor [168] (who were the first to introduce N-policy systems in 1963) and their followers [165,169-172].

N-policy is most often combined with vacations. Once the system is exhausted, the server goes on a single vacation and upon his return he checks on the queue, and if the queue has accumulated to at least N customers he begins service; otherwise, he rests and starts servicing as soon as the queue reaches level N. In the case of multiple vacations, when the system is exhausted, the server initiates a sequence of vacation segments. This sequence (not any particular segment) is interrupted as soon as the system accumulates to at least N customers. Systems with N-policy and vacations were studied in papers [101,138,141,148,149,153,157,160,162,165].

A combination of q- and N-policies, the (q,N)-policy, is also referred to as *hysteresis* (or *(v,N)-policy*). Indeed, systems switch-overs are governed by the queue crossing levels r and N ($\geq r$) from above and below, respectively. Typical extensions are bulk arrivals and batch service. In the latter case, the server capacity is assumed to equal R ($\geq r$), and, consequently, the server takes a minimum of r customers at once, if available. Otherwise, it runs idle (or even takes vacations). The maximal servicing batch size cannot exceed R. Since all available systems in this category are of $M/G/1$ type we will denote them by $M^X/G_N^{r,R}/1$ or $M^X/G_N^{r,R},V/1$, where V stands for "vacations." The study of the most general class of models $M^X/G_N^{r,R}/1$ to date was rendered by Muh [155,156], who considered three systems with different relations of N and R and the assumption about the start-ups of busy periods. In one of the models, customers in excess of level N who arrived during the start-up time are admitted to a servicing batch at the beginning of a busy period to fit the capacity R. The use of the first excess level theory [252,263-265] enabled one to know the start of a busy period and the queue level at this epoch.

We will demonstrate the analysis of $M^X/G_N^{r,R}/1$ systems essentially similar to the models studied by Muh [155,156]. We will use the same notation as in the beginning of Subsection 4.2.4. In the addition, we formalize the following main processes and disciplines.

Input. Customers arrive in batches at instants τ in accordance with the state dependent batch arrival process. It means that both the intensity of the arrival point process τ and arriving batch sizes X depend on the queue length at instants T. More specifically, given $Q_0 = i$, the input in interval $[0,T_1)$ is compound Poisson with intensity λ_i and with batches $X_r^{(i)}, r = 1,2,\ldots$, associated with the p.g.f.

$$a_i(u) = \mathbb{E}^i[u^{X_1^{(i)}}]. \tag{4.77}$$

Service process. All customers are processed by a single server in batches of size at least r but not exceeding R. If the queue length Q_n upon the completion of the nth service (at T_n) is not less than r (≥ 1) the server takes $\min\{Q_n,R\}$ $(R \geq r)$ customers in one group and immediately starts processing this group and completes service at $T_n + \sigma_n$. Otherwise, the server rests until the queue crosses threshold level N for the first time, after which a busy period begins.

Idle and busy period disciplines obey hysteretic policy with control levels r and N $(\geq r)$. That means an idle period starts when the queue crosses level r from above and ends (i.e., a following busy period begins) when the queue crosses level N from below. Since the input is bulk, the queue will obviously be more likely to exceed, rather than reach, threshold level N. Occasionally, the server can augment his servicing batch with some customers arriving during every start-up period.

We will present three different systems studied by Muh in his work [155,156]:

1) $r \leq N \leq R$ is the "plain model," i.e., no customers having arrived during a start-up period are added to the first servicing batch. The batch size assigned to service is $\min\{Q(\tau_\nu),R\}$, which, between N and R, is specified at the instant τ_ν of the first passage of level N.

2) $r \leq R \leq N$. The batch size of the first group is R, since $Q(\tau_\nu) \geq R$.

3) $r \leq N \leq R$ with add-on customers, i.e., customers arriving during a start-up period are added to the first servicing batch, excluding those in excess of server capacity R. Consequently, the batch assigned to service is $\min\{Q(\tau_\nu) + W_{n+1}, R\}$, where W_{n+1} is the number of customers arrived during the start-up period past T_n and the following idle period.

Muh studied the embedded Markov chain Q_n under the above assumptions in the above models 1, 2, and 3. We will apply the results of first level excess analysis in 4.4.2 to the functional

$$\mathbb{E}^i[\xi^\nu e^{-\theta\tau_\nu}u^{A_\nu} \mid c_1 = N] = \mathcal{F}_N^{(i)} = \mathcal{F}_N^{(i)}(\xi,\theta,u)$$

with a slightly modified notation adapted to the needs of models 1-3. Namely, in the above formula, c_1 is no longer server capacity but the value of the threshold level N.

Let $M_i(A)$ give the number of customers that enter the system on time set $A \subseteq [T_0,\infty)$ given that $Q_0 = i$. Then the embedded Markov chain is subject to the following transitions.

Model 1. $r \leq N \leq R$ (plain model):

$$Q_1 = \begin{cases} (Q_0 - R)^+ + M_{Q_0}([0,T_1)), & Q_0 \geq r \\ (A_\nu - R)^+ + M_{Q_0}([\tau_\nu,T_1)), & Q_0 < r. \end{cases} \tag{4.78}$$

Model 2. $r \leq R \leq N$:

$$Q_1 = \begin{cases} (Q_0 - R)^+ + M_{Q_0}([0,T_1)), & Q_0 \geq r \\ A_\nu - R + M_{Q_0}([\tau_\nu,T_1)), & Q_0 < r. \end{cases} \tag{4.79}$$

Model 3. $r \leq N \leq R$ with add-on customers:

$$Q_1 = \begin{cases} (Q_0 - R)^+ + M_{Q_0}([0,T_1)), & Q_0 \geq r \\ (A_\nu + W_1 - R)^+ + M_{Q_0}([\tau_\nu + w_1,T_1)), & Q_0 < r, \end{cases} \tag{4.80}$$

where W_1 denotes the number of customers arrived during the start-up time following the service completion at $T_0 = 0$ and w_1 is the length of this start-up period.

To proceed with the models we need some preliminaries. Let $f(u) = \mathbb{E}[u^X]$ for some nonnegative integer-valued r.v. X. For a nonnegative integer p define the operator

$$\mathcal{H}^p f(u) = \mathbb{E}[u^{(X-p)^+}]. \tag{4.81}$$

Lemma 4.7. (Dshalalow and Yellen [112]) $\mathcal{H}^p$ *is a linear operator with fixed points at constant functions and with the analytic representation*

$$\mathcal{H}^p f(u) = u^{-p} f(u) + \mathcal{D}_x^{p-1}\{f(x) - u^{-p} f(ux)\}, \tag{4.82}$$

where $\mathcal{D}$ is defined in (4.35).

Now we impose some restrictions to state dependency quite similar to those in (4.49), but for analytic convenience the restrictions are even tighter. Namely, we assume that $\lambda_i = \lambda$, $B_i = B$, $a_i(u) = a(u)$, $i \geq R$.

Proposition 4.8. *The embedded process Q_n is ergodic if and only if*

$$\rho = \alpha b \lambda < R. \tag{4.83}$$

Now we will again make use of formulas (4.45,4.47) and other formulas on first excess level theory for $s = N$.

Theorem 4.9. *Given the ergodicity condition $\rho < R$, the p.d.f. $P(u)$ of the stationary distribution $p = (p_0, p_1, \ldots)$ satisfies the following formulas:*

Model 1

$$P(u) = \frac{\sum_{i=0}^{R-1} p_i \{u^R K_i(u) \mathcal{H}^R L_N^{(i)}(u) - u^i K(u)\}}{u^R - K(u)}. \tag{4.84}$$

Model 2

$$P(u) = \frac{\sum_{i=0}^{r-1} p_i \{K_i(u) L_N^{(i)}(u) - u^i K(u)\} + \sum_{i=r}^{R-1} p_i \{u^R K_i(u) - u^i K(u)\}}{u^R - K(u)}. \tag{4.85}$$

Model 3

$$P(u) = \frac{\sum_{i=0}^{r-1} p_i \{u^R K_i(u) \mathcal{H}^R (L_N^{(i)} J_i)(u) - u^i K(u)\} + \sum_{i=r}^{R-1} p_i \{u^R K_i(u) - u^i K(u)\}}{u^R - K(u)},$$

$$\tag{4.86}$$

where

$$J_i(u) = \gamma(\lambda_i - \lambda_i a_i(u)) \tag{4.87}$$

and $\gamma_i(\theta)$ is the LST of the p.d.f. of a start-up time given that $Q_0 = i$, $i = 0, \ldots, r-1$.

Since $J_i(u) = 1$ (which corresponds to a start-up time equal to zero *a.s.*), $L_N^{(i)}(u) = u^i$, and, consequently, $\mathcal{H}^R u^i = u^{(i-R)^+}$, for $i \geq r$, we can rewrite formula (4.86) for $P(u)$ in a more compact form without the second sum in the numerator:

Model 3

$$P(u) = \frac{\sum_{i=0}^{R-1} p_i \{u^R K_i(u) \mathcal{H}^R (L_N^{(i)} J_i)(u) - u^i K(u)\}}{u^R - K(u)}. \tag{4.88}$$

The following is concerned with the derivation of the unknown probabilities p_i, $i = 0, \ldots, R - 1$, and the conservation property.

Model 1. Denote

$$Q_1^{Res} = (A_\nu - R)^+ = (Q(\tau_\nu) - R)^+. \tag{4.89}$$

Obviously, Q_1^{Res} is the residual queue length at the beginning of the first service (initiating a busy period) less the batch size to be processed. Therefore, $A_\nu - Q_1^{Res}$ is this batch size. Taking into account that A_ν is the queue length at the beginning of the first service regardless of whether or not an idle period is present, i.e., with A_ν assigned to Q_0 in the latter case (see formula (4.33) for the definition of A_ν), the first batch size can be also be interpreted as the first server load with the notation

$$l_1 = A_\nu - Q_1^{Res}. \tag{4.90}$$

Evidently, given Q_n, the nth service load l_n can be derived from the first excess level analysis. From the last two formulas we find the mean server load $\mathcal{L}$ in equilibrium:

$$\mathcal{L} = \sum_{i \geq 0} p_i \mathbb{E}^i[l_1] = \sum_{i=0}^{r-1} p_i(\overline{L}_N^{(i)} - Q_{res}^{(i)}) + \sum_{i=r}^{R-1} p_i i + \sum_{i \geq R} p_i R, \tag{4.91}$$

where

$$Q_{res}^{(i)} = \mathbb{E}^i[Q_1^{Res}]. \tag{4.92}$$

Since

$$R + Q_{res}^{(i)} = R + \mathbb{E}^i[(A_\nu - R)^+] = \frac{d}{du}\Big(u^R \mathcal{H}^R L_N^{(i)}(u)\Big)\Big|_{u=1},$$

we will obtain by using Lemma 4.7 that

$$Q_{res}^{(i)} = \overline{L}_N^{(i)} - R + \mathfrak{D}_x^{R-1}\Big\{RL_N^{(i)}(x) - x\frac{d}{dx}L_N^{(i)}(x)\Big\}. \tag{4.93}$$

Using formula (4.47) for the mean increment $I_N^{(i)}$ of the first excess level, formula (4.91) for the mean server load $\mathcal{L}$ can be reduced to

$$\mathcal{L} = \sum_{i=0}^{r-1} p_i(I_N^{(i)} - Q_{res}^{(i)}) + \sum_{i=0}^{R-1} p_i(i - R) + R. \tag{4.94}$$

Now we will discuss the conservation property of the system. From formula (4.84) for $P(u)$ we obtain the conservation equation in the form

$$\sum_{i=0}^{r-1} p_i(-Q_{res}^{(i)}) + \sum_{i=0}^{R-1} p_i(i - R) + R = \rho + \sum_{i=0}^{R-1} p_i(\rho_i - \rho), \tag{4.95}$$

by equating $P(1) = 1$, where

$$\rho_i = \alpha_i b_i \lambda_i. \tag{4.96}$$

Recall that $\beta^{(i)} = \mathbb{E}^i[T_1]$ is the expected length of the service cycle given $Q_0 = i$ and that $\vec{\beta} = (\beta^{(i)}; i = 0, 1, \ldots)^{\mathrm{T}}$. (The scalar product $p\vec{\beta}$ gives the mean value of a service cycle in the stationary mode given the equilibrium condition $\rho < R$.) With $\vec{\lambda} = (\lambda_i; i = 0, 1, \ldots)^{\mathrm{T}}$ and $\vec{\alpha} = (\alpha_i; i = 0, 1, \ldots)^{\mathrm{T}}$, $\vec{\alpha} \circ \vec{\beta} \circ \vec{\lambda}$ is the Hadamard product of the vectors $\vec{\alpha}$, $\vec{\beta}$, and $\vec{\lambda}$. Also recall that the scalar product $p(\vec{\alpha} \circ \vec{\beta} \circ \vec{\lambda})$ is the *system intensity*. Since obviously

$$\beta^{(i)} = \begin{cases} b_i + \frac{1}{\lambda_i}\overline{T}_N^{(i)}, & i < r \\ b_i, & i \geq r, \end{cases} \tag{4.97}$$

we have that

$$\alpha_i \beta^{(i)} \lambda_i = \begin{cases} \rho_i + I_N^{(i)}, \ i < r \\ \rho_i, \ i \geq r, \end{cases} \tag{4.98}$$

and therefore the system intensity $p(\vec{\alpha} \circ \vec{\beta} \circ \vec{\lambda})$ is

$$p(\vec{\alpha} \circ \vec{\beta} \circ \vec{\lambda}) = \rho + \sum_{i=0}^{R-1} p_i(\rho_i - \rho) + \sum_{i=0}^{r-1} p_i I_N^{(i)}. \tag{4.99}$$

We see that the value $\sum_{i=0}^{r-1} p_i I_N^{(i)}$ needs to be added to the left- and the right-hand sides of the conservation equation (4.95) to obtain the mean server load $\mathcal{L}$ and the system intensity $p(\vec{\alpha} \circ \vec{\beta} \circ \vec{\lambda})$. Therefore, in equilibrium the mean server load $\mathcal{L}$ and the system intensity $p(\vec{\alpha} \circ \vec{\beta} \circ \vec{\lambda})$ coincide, thereby establishing the conservation law for model 1.

Conservation equation (4.95) is also useful as one of R equations to find the unknowns $p_0, \dots, p_{R-1}$. The other $R-1$ equations are given by the following theorem.

Theorem 4.10. *The unknown probabilities $p_0, \dots, p_{R-1}$ can be found from the following system of linear equations (augmented by conservation equation (4.95)):*

$$\sum_{i=0}^{R-1} p_i \frac{d^s}{du^s} \{ K_i(u) \mathcal{H}^R L_N^{(i)}(u) - u^i \} \Big|_{u = u_z} = 0, \ s = 0, \dots, k_z - 1; \ z = 1, \dots, Z, \tag{4.100}$$

where u_z $(z = 1, \dots, Z)$ are distinct roots of $u^R - K(u)$ in the closed unit disk centered at zero with the deleted point $(0,1)$ such that u_z has multiplicity k_z with $\sum_{z=1}^{Z} k_z = R - 1$.

Model 2. If l_1 stands again for the first server load, we have that

$$\mathbb{E}^i[l_1] = \begin{cases} i, \ i = r, \dots, R-1 \\ R, \ i = 0, \dots, r-1, R, R+1, \dots \end{cases}$$

and therefore the mean server load $\mathcal{L}$ in equilibrium is

$$\mathcal{L} = R + \sum_{i=r}^{R-1} p_i(i - R). \tag{4.101}$$

The formulas for the mean ith service cycle and the system intensity $p(\vec{\alpha} \circ \vec{\beta} \circ \vec{\lambda})$ are identical to (4.97-4.99) for model 1, and hence,

$$p(\vec{\alpha} \circ \vec{\beta} \circ \vec{\lambda}) = \rho + \sum_{i=0}^{R-1} p_i(\rho_i - \rho) + \sum_{i=0}^{r-1} p_i I_N^{(i)}. \tag{4.102}$$

Now, from (4.85) for $P(1) = 1$ we get the conservation equation in the form

$$\rho + \sum_{i=0}^{R-1} p_i(\rho_i - \rho) + \sum_{i=0}^{r-1} p_i I_N^{(i)} = R + \sum_{i=r}^{R-1} p_i(i - R), \tag{4.103}$$

of which the left-hand side is the system intensity and the right-hand side is the mean server load, i.e.,

$$p(\vec{\alpha} \circ \vec{\beta} \circ \vec{\lambda}) = \mathcal{L}. \tag{4.104}$$

The conservation equation in one of its forms will also be used as one of the R equations in $p_0, \dots, p_{R-1}$. The other $R-1$ equations are as follows.

Theorem 4.11. *The unknown probabilities $p_0, \dots, p_{R-1}$ can be found from the following system of linear equations (augmented by conservation equation (4.103)):*

$$\sum_{i=0}^{R-1} p_i \frac{d^s}{du^s} \{ K_i(u) \Gamma_N^{(i)}(u) - u^i \} \Big|_{u = u_z} = 0,$$

$$s = 0,\ldots,k_z - 1; \ z = 1,\ldots,Z, \tag{4.105}$$

where $u_z \ (z = 1,\ldots,Z)$ *are distinct roots of* $u^R - K(u)$ *in the closed unit disk centered at zero with the deleted point* $(0,1)$ *such that* u_z *has multiplicity* k_z *with* $\sum_{z=1}^{Z} k_z = R - 1$, *and* $\Gamma_N^{(i)}(u)$ *is defined as*

$$\Gamma_N^{(i)}(u) = \begin{cases} u^{-R} L_N^{(i)}(u), & i < r \\ 1, & i \leq r. \end{cases} \tag{4.106}$$

Model 3. The residual queue length at the beginning of the first service is

$$Q_1^{Res} = (A_\nu + W_1 - R)^+, \tag{4.107}$$

where W_1 is the number of customers that arrive during the first start-up period. Thus, $A_\nu + W_1 - Q_1^{Res}$ is the size of a batch to be processed during the first service period. With the notation

$$\overline{p}_i = \mathbb{E}^i[W_1] \ \text{and} \ Q_{res}^{(i)} = \mathbb{E}^i[Q_1^{Res}], \tag{4.108}$$

the mean value $\mathcal{L}_i$ of the first server load given $Q_0 = i$ is

$$\mathcal{L}_i = \begin{cases} \overline{L}_N^{(i)} + \overline{p}_i - Q_{res}^{(i)}, & i = 0,\ldots,r-1 \\ i, & i = r,\ldots,R-1 \\ R, & i \geq R. \end{cases} \tag{4.109}$$

Then, the mean server load $\mathcal{L} = \sum_{i \geq 0} p_i \mathcal{L}_i$ and, therefore, with

$$\overline{p}_i = \alpha_i \lambda_i w_i \ (w_i \ \text{is the mean start-up period given} \ Q_0 = i) \tag{4.110}$$

we have

$$\mathcal{L} = \sum_{i=0}^{r-1} p_i(I_N^{(i)} - Q_{res}^{(i)}) + \sum_{i=0}^{R-1} p_i(i - R) + R + \sum_{i=0}^{r-1} p_i \overline{p}_i. \tag{4.111}$$

The ith mean service cycle $\beta^{(i)} = \mathbb{E}^i[T_1]$ is

$$\beta^{(i)} = \begin{cases} \frac{1}{\lambda_i} \overline{T}_N^{(i)} + w_i + b_i, & i = 0,\ldots,r-1 \\ b_i, & i = r,r+1,\ldots, \end{cases} \tag{4.112}$$

which yields

$$\alpha_i \beta^{(i)} \lambda_i = \begin{cases} I_N^{(i)} + \overline{p}_i + \rho_i, & i = 0,\ldots,r-1 \\ \rho_i, & i = r,r+1,\ldots, \end{cases} \tag{4.113}$$

and therefore the system intensity is

$$p(\vec{\alpha} \circ \vec{\beta} \circ \vec{\lambda}) = \rho + \sum_{i=0}^{R-1} p_i(\rho_i - \rho) + \sum_{i=0}^{r-1} p_i I_N^{(i)} + \sum_{i=0}^{r-1} p_i \overline{p}_i. \tag{4.114}$$

The conservation equation is identical to (4.95) for model 1:

$$\sum_{i=0}^{r-1} p_i(-Q_{res}^{(i)}) + \sum_{i=0}^{R-1} p_i(i - R) + R = \rho + \sum_{i=0}^{R-1} p_i(\rho_i - \rho). \tag{4.115}$$

Hence, adding

$$\sum_{i=0}^{r-1} p_i I_N^{(i)} + \ \sum_{i=0}^{r-1} p_i \overline{p}_i$$

to the left- and right-hand side of (4.115), we obtain the mean server load and the system intensity, respectively, thereby stating formally that $\mathcal{L} = p(\vec{\alpha} \circ \vec{\beta} \circ \vec{\lambda})$. And finally we have the following.

Theorem 4.12. *The unknown probabilities $p_0, \ldots, p_{R-1}$ can be found from the following system of linear equations (augmented by conservation equation (4.115)):*

$$\sum_{i=0}^{R-1} p_i \frac{d^s}{du^s}\{K_i(u)\mathcal{H}^R(L_N^{(i)}J_i)(u) - u^i\}\Big|_{u = u_z} = 0, \ s = 0, \ldots, k_z - 1; \ z = 1, \ldots, Z,$$

(4.116)

where u_z $(z = 1, \ldots, Z)$ are distinct roots of $u^R - K(u)$ in the closed unit disk centered at zero with the deleted point $(0,1)$ such that u_z has the multiplicity k_z with $\sum_{z=1}^{Z} k_z = R - 1$.

4.2.8 Special cases and other queue length dependent models

Example 4.13. Consider a special case of the $M^X/G_N^{r,R}/1$, model 2, for $r = R = 1$, studied by H.W. Lee et al. [150]; in our notation, therefore, this is $M^X/G_N^{1,1}/1$. Although the authors did not use the first excess level theory, they established a quite elegant algorithm (perhaps effective for special cases only) to obtain the characteristics similar to the first excess level. The paper [150] deals both with the embedded and continuous time parameter processes by using supplementary variables techniques, assuming no queue length dependence of the input and service and assuming that the service time distribution is absolutely continuous.

H.W. Lee et al. [150] obtained the p.g.f. $\pi(u)$ for the queueing process with continuous time parameter in equilibrium:

$$\pi(u) = \mathbf{K}(u)\frac{\sum_{i=0}^{N-1} \delta_i u^i}{\sum_{i=0}^{N-1} \delta_i},$$

(4.117)

where $\mathbf{K}(u) = (1-\rho)K(u)\dfrac{u-1}{u-K(u)}$ is Kendall's p.g.f. of the queueing process with continuous time parameter in equilibrium for the $M^X/G/1$ system (for $N = 1$), and δ_i can be obtained recursively from

$$\delta_0 = 1, \ \delta_i = \sum_{s=1}^{i} a_s \delta_{i-s},$$

(4.118)

where a_s is the probability that an arriving batch is of size s.

For a comparison with (4.117), consider formula (4.85) for this special case in the scenario of model 2. Formally, (4.85) reduces to

$$P(u) = \frac{p_0\left[uK(u)\mathcal{H}^1 L_N^{(0)}(u) - K(u)\right]}{u - K(u)}.$$

(4.119)

From the conservation equation for model 2 we get

$$p_0 = \frac{1-\rho}{I_N^{(0)}}$$

(4.120)

with

$$I_N^{(0)} = \alpha \Delta_N^{(0)}.$$

(4.121)

Now, applying Lemma 4.7 we have

$$\mathcal{H}^1 L_N^{(0)}(u) = \tfrac{1}{u} L_N^{(0)}(u),$$

which reduces formula (4.119) to

$$P(u) = \mathbf{K}^+(u)\frac{\Delta_N^{(0)}(u)}{\Delta_N^{(0)}}, \tag{4.122}$$

where $\mathbf{K}^+(u)$ is Kendall's formula for the p.g.f. of the embedded queueing process for the $M^X/G/1$ system in the steady state:

$$\mathbf{K}^+(u) = \frac{K(u)[a(u)-1]}{u - K(u)}\frac{1-\rho}{\alpha}. \tag{4.123}$$

Based on the relationship between $\mathbf{K}(u)$ and $\mathbf{K}^+(u)$ it is quite plausible to assume that

$$\frac{\sum_{i=0}^{N-1}\delta_i u^i}{\sum_{i=0}^{N-1}\delta_i} = \frac{\Delta_N^{(0)}(u)}{\Delta_N^{(0)}}. \tag{4.124}$$

The results for all three models can be extended to the queueing processes with continuous time parameter by using semi-regenerative techniques, and they will also enable one to validate or reject conjecture (4.124).

Example 4.14. Another special case, this time of model 1, was treated in Abolnikov and Dshalalow [92] for $N = r$; thereby in notation $M^X/G_r^{r,R}/1$. This restriction does not simplify formula (4.84). However, the authors of [92] derived the p.g.f. $\pi(u)$ for the process with continuous time parameter:

$$p\vec{\beta}\pi(u) = \tfrac{1}{\lambda}\gamma(u)P(u) + \sum_{i=0}^{R-1}p_i u^i\tfrac{1}{\lambda_i}\{K_i(u)\Delta_r^{(i)}(u) + \gamma_i(u) - \gamma(u)\}, \tag{4.125}$$

where

$$\gamma_i(u) = \frac{1 - K_i(u)}{1 - a_i(u)} \tag{4.126}$$

and γ is like γ_i with index i dropped from all terms. The quantity $p\vec{\beta}$ is the mean service cycle in equilibrium and can be obtained from formula (4.97):

$$p\vec{\beta} = b + \sum_{i=0}^{R-1}p_i(b_i - b) + \sum_{i=0}^{r-1}p_i\frac{1}{\lambda_i}\Delta_r^{(i)}. \tag{4.127}$$

Remark 4.15. An interesting modification of hysteretic control systems applies to two servers and "hysteretic parameters" $r_1 \le r_2 \le N_1 \le N_2$. When the queue crosses levels r_2 and r_1 from above, the system turns off one of the two available servers and both servers, respectively. When the queue accumulates to N_1 and N_2 customers then one of the servers and both servers are launched, respectively. However, the analytical complexity deterred the authors Bell [246] and Rhee and Sivazlian [249] from studying systems more general than $M/M/2$.

4.3 WORKLOAD DEPENDENT SYSTEMS

4.3.1 Queues with limited waiting time (impatient customers)

This class of queues with waiting time dependent parameters includes two types of systems in which customers are discouraged by waiting times. The first one is where customers are unaware of the situation or, more precisely, not being able to estimate their waiting times, line up and drop the line unprocessed or even with service interruption when their waiting times exceed their patience. The second sub-class includes systems with customers aware of their waiting times and thus being able to choose whether or not to join the line. Both types deal with $GI/G/1$ queueing systems with generally distributed customers' impatience, in notation

$GI/G/1 + GI$.

Queues with limited waiting times were a popular subject from the sixties through the eighties, studied in the works of Barrer [175] in 1957, Daley [176] in 1964 and [177] in 1965, Gnedenko and Kovalenko [268] in 1966, Takács [186] in 1974, Baccelli et al. [174] in 1984, de Kok and Tijms [178] in 1985, and others. Most general models were studied by Daley [177] and Baccelli et al. [174]. Perhaps the most elegant result was presented in Takács [186].

Takács [186] considered an $M/G/1$ type system with state dependent service and a limited waiting time. In this system, no customer may stay in the system longer than some fixed time τ. [If a customer's waiting time is longer than τ he leaves the system unprocessed.] In addition, any service time is distributed according to B_0 when this initiates a busy period, and according to B otherwise. We denote such system by $M/G/1 + \tau$. Let v_t be the virtual waiting time (Takács process) or unfinished work and w_n the waiting time of the nth customer. Takács [186] established the following result.

Theorem 4.16. *The limiting distribution*

$$\lim_{n\to\infty} \mathbb{P}^i\{w_n \le x\} = \lim_{t\to\infty} \mathbb{P}^i\{v_t \le x\} = F(x)/F(\tau) \qquad (4.128)$$

exists and is independent of the initial value of ν_0. $F(x)$ is a continuous distribution function for $x \ge 0$, equal to 0, for $x < 0$ and $F(0) = 1$. The LST $\phi(\theta)$ of $F(x)$ is given by

$$\phi(\theta) = 1 + \frac{\lambda[1 - \beta_0(\theta)]}{\theta - \lambda[1 - \beta(\theta)]}, \ Re(\theta) > \lambda(1 - \omega), \qquad (4.129)$$

where ω is the smallest positive real zero of the equation

$$\omega = \beta(\lambda - \lambda\omega). \qquad (4.130)$$

Daley [177] and Baccelli et al. [174] considered $GI/G/1$ queues with impatient customers whose waiting or sojourn times are limited by generally distributed random variables. Among other things, the authors of [174] considered a class of models with impatient customers where an arriving customer is aware of when his waiting (or sojourn) time is "beyond his patience" and in this case he leaves the system without lining up.

Let w_n denote the system workload at time $\tau_n -$, i.e., at the instant of time just before the nth arrival. This would be the nth customer's waiting time if he chooses to wait (not counting his service time), also called the *offered waiting time*. Suppose g_n is the time of patience of the nth customer. Then the nth customer will join the queue only if $g_n > w_n$. It is assumed that $\{g_n, n = 1,...\}$ is a sequence of independent, identically and generally distributed r.v.'s independent of the arrival and service processes. Thus, for this model we will use the symbol $GI/G/1 + GI$.

We introduce the following notation.

> $A(x)$ - interarrival time PDF
> $B(x)$ - service time PDF
> $G(x)$ - patience time PDF (G need not converge to 1 when $x\to\infty$, but $G(0) = 0$)
> $W_n(x)$ - the PDF of w_n
> $\frac{1}{\lambda}$ - the mean interarrival time
> b - the mean service time
> $\rho = \lambda b$

P_n - the nth busy period
C_n - the nth busy cycle
l_n - the number of customers arrived at the system during the nth busy period
m_n - the number of customers served during the nth busy period
v_t - the *virtual offered waiting time*
$V(x) = \lim\limits_{t\to\infty} \mathbb{P}\{v_t \leq x\}$

$W(x) = \lim\limits_{n\to\infty} \mathbb{P}\{w_n \leq x\}$

v_∞ - the virtual offered waiting time in equilibrium with the PDF $V(x)$
w_∞ - the (actual) offered waiting time in equilibrium
$V^*(\theta)$ and $W^*(\theta)$ - LST's of the virtual and actual waiting times in the steady state, respectively
$\pi = \lim\limits_{n\to\infty} \mathbb{P}\{$the nth customer is rejected$\}$

σ_n - service time of the nth customer

$$\gamma(\theta) = \begin{cases} \rho, \ \theta = 0 \\ \lambda\mathbb{E}[\frac{1}{\theta}(1 - e^{-\theta\sigma_1})], \ \theta > 0. \end{cases}$$

The following are general characteristics and performance measures for the $GI/G/1 + GI$ queue. w_n is a Markov chain with state space $\mathbb{R}_+$. $W_n(x)$ satisfies the recursive integral equations

$$W_{n+1}(x) = \int_0^\infty dA(t)\Big\{W_n(t+x) - \int_{u=0}^{t+x}[1 - G(u)][1 - B(t + x - u)]dW_n(u)\Big\},$$
$$n \geq 0. \qquad (4.131)$$

Let $\overline{A} = \inf\{t\colon A(t) = 1\}$, $\overline{B} = \sup\{t\colon B(t) = 0\}$ and let $\overline{A} > \overline{B}$. The process w_n is ergodic provided the condition

$$\rho[1 - G(\infty)] < 1 \qquad (4.132)$$

is met. The following holds true:

$$V(0) = \frac{\mathbb{E}[C_1 - P_1]}{\mathbb{E}[C_1]}, \qquad (4.133)$$

$$(1 - \pi)\rho = 1 - V(0). \qquad (4.134)$$

$$\mathbb{E}[e^{-\theta v_\infty}] = \mathbb{P}\{v_\infty = 0\} + (1 - \pi)\gamma(\theta)\mathbb{E}[e^{-\theta w_\infty}]. \qquad (4.135)$$

The latter yields the Pollaczek-Khintchine type formula for the $GI/G/1 + GI$ system,

$$\mathbb{E}[v_\infty] = (1 - \pi)\rho\Big\{\mathbb{E}[w_\infty] + \frac{\lambda}{2}\frac{\mathbb{E}[\sigma_1^2]}{\mathbb{E}[\sigma_1]}\Big\}, \qquad (4.136)$$

generalizing the known result for $GI/G/1$.

Example 4.17. $M/G/1 + GI$ queue. This model formally generalizes a special case of Takács' $M/G/1 + \tau$ queue with no queue length state dependence. Now, preserving all of the above characteristics, assume that $A(x)$ is exponentially distributed with parameter λ. Let

$$V(t,x) = \mathbb{P}\{v_t \leq x\},$$

$$V^*(t,\theta) = \mathbb{E}[e^{-\theta v_t}],$$

$$V_G^*(t,\theta) = \int_{0-}^{\infty} e^{-\theta x}[1 - G(x)]d_x V(t,x),$$

$$V_G^*(\theta) = \int_{0-}^{\infty} e^{-\theta x}[1 - G(x)]dV(x),$$

$$\overline{\beta}(\theta) = \lambda\frac{1 - \beta(\theta)}{\theta}.$$

Then $V(t,x)$ satisfies the following integro-differential equation of Takács type:

$$\frac{1}{\theta}\frac{\partial V^*(t,\theta)}{\partial t} = V^*(t,\theta) - V(t,0) - \overline{\beta}(\theta)V_G^*(t,\theta). \tag{4.137}$$

The stationary mode exists given condition (4.132). $V(x) = W(x)$ and (4.137) reduces to

$$V^*(\theta) = V(0) + \overline{\beta}(\theta)V_G^*(\theta). \tag{4.138}$$

Given an exponentially distributed impatience (with parameter c), i.e., the system $M/G/1 + M$, (4.138) further reduces to

$$V^*(\theta) = V(0)\left[1 + \gamma(\theta)\sum_{n=1}^{\infty}\prod_{k=1}^{n}\gamma(\theta + kc)\right], \; Re(\theta) \geq 0, \tag{4.139}$$

where

$$V(0) = \left[1 + \rho\sum_{n=1}^{\infty}\prod_{k=1}^{n}\gamma(kc)\right]^{-1}. \tag{4.140}$$

4.3.2 Systems with workload dependent input and service rates

In this class of queues with state dependent parameters, both the input and service rates are "continuously" dependent on the amount v_t of unfinished work at time t. The idea of such dependence and singular perturbation techniques were developed by Knessl et al. in their work [189-192], which includes systems ranging from $M/M/1$ to $GI/G/1$ types.

In the case of the $M/G/1$ queueing system the authors of [189] assume that the input rate λ (initially time dependent, $\lambda = \lambda(t)$) depends on v_t, i.e., $\lambda = \lambda(v_t)$. The dependence of service on the workload is rendered as follows. If a customer arriving at time t needs S units of service time, then $\mathbb{P}\{S \leq x \mid v_t\} = B(x,v_t)$ with the probability density function $b(x,v_t) = (\partial/\partial x)B(x,v_t)$. Therefore, the probability density function $b(\cdot,\cdot)$ of service depends on workload v_t only upon arrival; v_t remains a Markov process as it is in the plain $M/G/1$ queueing system.

The authors use similar techniques as in Subsection 4.2.1 by introducing scale variables for parameters in the modified Takács integro-differential equation and seeking the limiting distribution for the virtual waiting time process in the form of asymptotic expansion.

Systems with workload dependent service rates were studied by Callahan [193], Cohen [194], Posner [199], and Tijms [200].

4.3.3 D-policy queues

The D-policy is very similar to the N-policy: both determine when to end an idle period and begin the following busy period. However, in the case of the D-policy, a busy period starts when the cumulative workload crosses level D. In this case, a (single) server should take a fixed number of customers in a batch, or else the workload would be hard to define. In all known systems analyzed the server capacity has always been 1. There have been just a few articles on D-policy, perhaps because of somewhat limited analytical tractability of outcoming results.

The first work on D-policy was rendered by Balachandran [201] in 1971 and Balachandran and Tijms [203] in 1973 applied to $M/G/1$ type systems. They targeted the optimal system control by balancing switching and holding costs. The penalty is exacted each time the system turns on (in which case a greater D is desirable) and for holding on at a constant rate per unit of unfinished work (which, on contrary, tends to minimize D), i.e., per unit time of customer's waiting time. The authors of [203] derived a formula for the infinite horizon average cost as a functional of D that enabled one to determine the best value of D. In addition, they compared these results with similar results obtained in the N-policy system, and based on special cases they conjectured that the D-policy is superior to the N-policy. The latter was confirmed by Boxma [204] in 1976.

Interestingly, if the holding cost is taken per unit of customer's sojourn time, the results of optimization differ significantly from those relative to the holding cost per customer's waiting time. In fact, Rubin and Zhang [207] in 1988 showed that, given this assumption in the $M/G/1$ system, the D-policy is no longer superior to the N-policy.

In all these instances, knowledge of the distribution of the waiting time process is crucial in obtaining average cost functionals and employing control, especially when the holding cost is nonlinear. Waiting time processes in N-policy $M/G/1$ systems were studied by Neuts [157] in 1986 and Shanthikumar [161] in 1981, and in D-policy $GI/G/1$ systems by Li and Niu [206] in 1992.

In [206], the authors define the distribution of the "waiting time process" w of a "randomly selected customer" as

$$\mathbb{P}\{w \le x\} = \lim_{n \to \infty} \frac{1}{n} \sum_{i=1}^{n} I_{[0,x]}(w_i), \tag{4.141}$$

where I_A is the indicator function of a set A and w_i is the waiting time of the ith customer. The interpretation of (4.141) is as follows. $\frac{1}{n}$ is the probability of selecting any one of the first n arriving customers and $I_{[0,x]}(w_i)$ is the conditional probability for the ith customer, once selected, to have a waiting time less than or equal to x. Therefore, as $n \to \infty$, (4.141) gives the waiting time distribution as seen by a randomly selected arriving customer. In fact, (4.141) gives the long-run average (or proportion) associated with the waiting time process. The authors of [206] established a rather complex expression for $\mathbb{P}\{w \le x\}$ in the D-policy $GI/G/1$ system which is more explicit for $M/G/1$ and $GI/M/1$ special cases, especially for the first moment of $\mathbb{P}\{w \le x\}$.

4.4 INTERDEPENDENCE OF INPUT AND SERVICE

This class of queues includes a correlation between input and service rates. A typical analysis of these models is to state a monotonicity property of the waiting time process in terms of its dependency on the correlation coefficient. In 1995, Chao [211] studied a system with the assumption that the interarrival time t_n between the $(n-1)$th and nth customers and the service time s_n of the nth customer have the bivariate exponential distribution:

$$\mathbb{P}^i\{t_n > t, s_n > s\} = e^{-\alpha t - \beta s - \gamma min\{s,t\}}, \ s > 0, \ t > 0. \tag{4.142}$$

The covariance between t_n and s_n is

$$Cov(t_n, s_n) = \frac{\gamma}{(\alpha + \beta + \gamma)(\alpha + \gamma)(\beta + \gamma)} \tag{4.143}$$

and the correlation coefficient is

$$\rho = \frac{\gamma}{\alpha + \beta + \gamma}. \tag{4.144}$$

Keeping arrival and service rates, $\lambda = \alpha + \gamma$ and $\mu = \beta + \gamma$, constant, let us vary the correlation coefficient ρ, which is equivalent to changing α, β, and γ simultaneously. In terms of λ and μ, the correlation coefficient can be rewritten as

$$\rho = \frac{\lambda + \mu}{\lambda + \mu - \gamma} - 1. \tag{4.145}$$

An r.v. X is said to be *less than or equal to* a random variable Y *in nondecreasing convex order* if $\mathbb{E}[f(X)] \leq f(Y)$ for any Borel monotone nondecreasing and convex function f. Let $w_n(\rho)$ be the waiting time of the nth customer with the correlation coefficient ρ. Chao [211] proved the following theorem.

Theorem 4.18. *The waiting time $w(\rho)$ of a customer in the steady state is a monotone nonincreasing function of ρ and nondecreasing in convex order. That is, for $\rho_1 \leq \rho_2$,*

$$\mathbb{E}[f\{w(\rho_1)\}] \geq \mathbb{E}[f\{w(\rho_2)\}]. \tag{4.146}$$

Specifically, for $f(x) = x$, $\mathbb{E}[w(\rho_1)] \geq \mathbb{E}[w(\rho_2)]$, i.e., the waiting time process is nonincreasing in the correlation coefficient.

Earlier, Mitchell and Paulson [216] and Conolly and Choo [212] in 1979, analyzed an $M/M/1$ type of system with the assumption that the interarrival time t_n between the $(n-1)$th and nth customer and the service time s_n of the nth customer have the bivariate exponential distribution

$$g(t,s) = \mathbb{P}^i\{t_n > t, s_n > s\} = (1 - \rho)\lambda\mu e^{-\lambda t - \mu s} I_0[2(\rho\lambda\mu st)^{1/2}], \tag{4.147}$$

with correlation coefficient $\rho \in [0,1)$, where $t \geq 0, s \geq 0$, and

$$I_0(z) = \sum_{n=0}^{\infty} \frac{(z/2)^{2n}}{(n!)^2}$$

is the modified Bessel function of the first kind and order zero.

Given g, the probability density function $a_n(w)$ of the waiting time w_n of the nth customer satisfies the following recursive relation:

$$a_{n+1}(w) = \int_0^\infty \int_w^\infty a_n(w)g(t,s)dtds + \int_0^\infty \int_0^w a_n(w+t-s)g(t,s)dtds. \tag{4.148}$$

The Laplace transform $\alpha(\theta)/\alpha(0)$ of the steady state density satisfies the functional equation

$$\alpha(\theta) = \frac{\mu[1 - \rho(\theta + \mu)]\alpha(h(\theta))}{(\theta + \mu(1-\rho))(\theta + \mu(1 - \frac{\lambda}{\mu}))}, \tag{4.149}$$

which has a solution in the form [216]

$$\alpha(\theta) = \left(\frac{\mu(1-\rho)}{\theta + \mu(1-\rho)}\right)\left(\frac{\theta + \mu}{\theta + \mu(1 - \frac{\lambda}{\mu})}\right) \prod_{j=1}^{\infty} \left(\frac{\mu(1-\rho)}{z_j + \mu(1-\rho)}\right)\left(\frac{z_j + \mu}{z_j + \mu(1 - \frac{\lambda}{\mu})}\right), \tag{4.150}$$

where

$$z_0 = \theta, \; z_j = h(z_{j-1}), \; j = 1,2,\ldots, \; h(x) = 1 - \frac{\mu\rho}{x+\mu}. \qquad (4.151)$$

The authors of [212] and [216] presented numerical values for the mean and variance of the stationary waiting time distribution. Numerical results indicate that with the given assumption on interarrival and service periods dependency, the mean waiting time is less than or equal to that of the usual $M/M/1$ system. Niu [217] in 1981 confirmed this conjecture analytically generalizing it even to $M/G/1$ systems.

4.5   QUEUE LENGTH DEPENDENT MULTI-CHANNEL QUEUES

In this class of $G/M/m$ type systems, the input stream varies upon instants of departures of customers from the source. The demonstration of these techniques will be based on two particular but typical models studied in Dshalalow [228,236]. We shall be using the following notation throughout the remainder of this chapter:

> $Q(t)$ - queueing process (with left continuous paths)
> $T = \{T_0 = 0, T_1, \ldots\}$ - arrival instants
> $Q = \{Q_n = Q(T_n-); n = 0,1,\ldots\}$ queueing process embedded in $Q(t)$ over T.
> Given $Q_n = i$, the length of the $(n+1)$th interarrival time $T_{n+1} - T_n$ is distributed in accordance with the PDF $A_i(x)$, $i = 0,1,\ldots$, chosen from the sequence $\vec{A} = (A_0, A_1, \ldots)$ of arbitrary PDF's with $A_i(0+) = 0$, $A_i(\infty) = 1$, $\mathbb{E}^i[T_1] = a_i < \infty$, $\vec{a} = (a_0, a_1, \ldots)^{\mathrm{T}}$. Within the interval $(T_n - T_{n+1}]$ the service durations in any of the m parallel channels are conditionally independent given $Q_n = i$ and exponentially distributed with parameter $\mu_i \in (\mu_0, \mu_1, \ldots)$.

In terms of Kendall's notation the system is $G_0, G_1, \ldots / M_0, M_1, \ldots / m / w$, where m is the number of channels and w is the size of the buffer or waiting room, which can range from 0 to ∞. Denote by E the state space of the queueing process $Q(t)$.

4.5.1 Queueing process with continuous time parameter

Denote by $C(A) = \sum_{j \geq 0} I_A(T_j)$ the number of arrivals on time set A (i.e., C is the counting process associated with the point process $\underset{\sim}{T}$), where I_A is the indicator function of a set A and let $C_t = C([0,t))$. Then $\widetilde{Q}(t) = Q(T_{C_t})$ is the queue length immediately before T_{C_t}-th arrival (prior to t), and $\widetilde{Q}$ is the embedded minimal semi-Markov process associated with the Markov renewal process (Q,T). We will also use the process $V(t) = T_{C_t+1} - t$. The three-dimensional process

$$\mathbb{Q}(t) = (\widetilde{Q}(t), Q(t), V(t)) \qquad (4.152)$$

is apparently weak Markov with state space $E \times E \times \mathbb{R}_+$. Assume that its Markov semi-group (P_t) is absolutely continuous and let

$$\pi_{jk}^{(i)}(t,u) = \frac{\partial}{\partial u} \mathbb{P}^i\{\widetilde{Q}(t) = j, Q(t) = k, V(t) \leq u\} \qquad (4.153)$$

be its transition density function, which is assumed to be differentiable, where $\mathbb{P}^i$

denotes the probability measure induced by the initial value i of the queueing process $Q(0-) = Q_0$. For the associated functionals will use the notation $\mathbb{E}^i[\cdot]$. Since we will deal with limiting distributions, we drop the superscript i in $\pi_{jk}^{(i)}(u,t)$. Let $\mathfrak{D}_{t,u}$ be the differential operator defined as

$$\mathfrak{D}_{t,u}f(t,u) = \left(\frac{\partial}{\partial t} - \frac{\partial}{\partial u}\right)f(t,u). \tag{4.154}$$

Model with Finite Waiting Room $w < \infty$. The Kolmogorov differential equations in $\pi_{jk}(t,u)$ are as follows:

$$\mathfrak{D}_{t,u}\pi_{jk}(t,u) = -g(k)\mu_j\pi_{jk}(t,u) + g(k+1)\mu_j\pi_{j,k+1}(t,u),$$
$$0 \le k \le \min(j, m+w-1),\ j \in E, \tag{4.155}$$

$$\mathfrak{D}_{t,u}\pi_{k-1,k}(t,u) = -g(k)\mu_{k-1}\pi_{k-1,k}(t,u) + g(k+1)\mu_j\pi_{.k-1}(t,0)a_{k-1}(u),$$
$$j = k-1, k = 1,\ldots,m+w, \tag{4.156}$$

$$\mathfrak{D}_{t,u}\pi_{m+w,m+w}(t,u) = -m\mu_{m+w}\pi_{m+w,m+w}(t,u) + \pi_{.m+w}(t,0)a_{m+w}(u),$$
$$j = k = m+w, \tag{4.157}$$

where $\pi_{.k}(t,u)$ is the marginal density of the process $(Q(t), V(t))$ and $g(k) = \min(k,m)$. Consider the following limiting densities and their Laplace transforms:

$$\pi_{jk}(u) = \lim_{t\to\infty}\pi_{jk}(t,u),\ \pi_{.k}(u) = \lim_{t\to\infty}\pi_{.k}(t,u), \tag{4.158}$$

$$\pi_{jk}^*(\theta) = \int_0^\infty \pi_{jk}(u)e^{-\theta u}du,\ Re(\theta) \ge 0,\ j,k \in E. \tag{4.159}$$

Now, letting $t\to\infty$ and then applying the Laplace transform to the above system (4.155-4.157) we will arrive at a system of functional equations in which by letting $\theta \downarrow 0$ we have:

$$g(k)\mu_j\pi_{jk} = g(k+1)\mu_j\pi_{j,k+1} - \pi_{jk}(0),\ 0 \le k \le \min(j,m+w-1),\ j \in E,\tag{4.160}$$

$$g(k)\mu_{k-1}\pi_{k-1,k} = \pi_{.k-1}(0) - \pi_{k-1,k}(0),\ j = k-1, k = 1,\ldots,m+w, \tag{4.161}$$

$$m\mu_{m+w}\pi_{m+w,m+w} = \pi_{.m+w}(0) - \pi_{m+w,m+w}(0),\ j = k = m+w, \tag{4.162}$$

where $\pi_{jk} = \lim_{\theta\downarrow0}\pi_{jk}^*(\theta)$ is the stationary marginal distribution of $(\widetilde{Q}(t), Q(t))$. Since

$$\sum_{j=k-1}^{m+w}\pi_{jk} = \pi_{.k}\ (\text{in notation, }\pi_k) \tag{4.163}$$

is obviously the stationary distribution of the queueing process, we have from system (4.160-4.162), by summation in j that

$$g(k)\pi_k = g(k+1)\pi_{k+1} + \frac{1}{\mu_{k-1}}(1-\delta_{k,0})\pi_{.k-1}(0) -$$
$$\sum_{j=k-1}^{m+w}\frac{1}{\mu_j}(1-\delta_{j,-1})\pi_{jk}(0),\ k = 0,\ldots,m+w, \tag{4.164}$$

$$m\pi_{m+w} = \sum_{j=m+w-1}^{m+w}\frac{1}{\mu_j}(\pi_{.j}(0) - \pi_{j,m+w}(0)), \tag{4.165}$$

where δ is Kronecker delta. Finally, summing up the above equation in $k = 0,\ldots, n-1$ for $n = 1,\ldots,m+w$ we have

$$g(n)\pi_n = \sum_{k=0}^{n-1}\sum_{j=k-1}^{m+w}\frac{1-\delta_{j,-1}}{\mu_j}\pi_{jk}(0) - (1-\delta_{n,1})\sum_{k=0}^{n-2}\frac{1}{\mu_k}\pi_{.k}(0),$$

$$n = 1,\dots,m+w. \tag{4.166}$$

Denote by $\pi_{j\cdot}(t,u) = \sum_{k=0}^{j+1} \pi_{jk}(t,u)$ the transition marginal density of the process $(\widetilde{Q}(t),V(t))$. Now, due to $\pi_{j\cdot}(0) = \pi_{\cdot j}(0)$ (thus, in notation, $\pi_j(0)$), by simple transformations it can be shown that (4.165) and (4.166) for $n = m+w$ are identical. Equations (4.166) should be augmented by $\pi e = 1$, where $\pi = (\pi_0,\dots,\pi_{m+w})$ and e is the $(m+w+1) \times 1$ matrix with all entries 1.

It remains to find the unknown probabilities $\pi_{jk}(0)$ and $\pi_k(0)$ which may also be of independent interest. Let $p = (p_0,\dots,p_{m+w})$ be the stationary distribution of the embedded Markov chain Q, $p\vec{a}$ be the scalar product of vectors p and $\vec{a}$ (the mean interarrival period in the steady state), and p_{jk} be the transition probability of Q. The following proposition is due to Dshalalow [236] considering Q as a semi-regenerative process and making use of the main convergence theorem for semi-regenerative processes.

Proposition 4.19. *It holds true that*

$$\pi_{jk}(0) = \frac{p_j p_{jk}}{p\vec{a}}, \ j,k \in E, \tag{4.167}$$

$$\pi_k(0) = \frac{p_k}{p\vec{a}}, \ k \in E. \tag{4.168}$$

Equations (4.166-4.168) yield the following main result of this section.

Theorem 4.20. *The stationary distribution* $\pi = (\pi_0,\dots,\pi_{m+w})$ *of the queueing process in the queueing system* $G_0,\dots,G_{m+w}/M_0,\dots,M_{m+w}/m/w$ *exists and can be expressed in terms of the stationary distribution of the embedded Markov chain* $p = (p_0,\dots,p_{m+w})$ *and the transition probabilities* (p_{jk}):

$$p\vec{a}g(n)\pi_n = \sum_{k=0}^{n-1} \sum_{j=k-1}^{m+w} \frac{1-\delta_{j,-1}}{\mu_j} p_j p_{jk} - (1-\delta_{n,1})\sum_{k=1}^{n-1} \frac{1}{\mu_{k-1}} p_{k-1},$$

$$n = 1,\dots,m+w, \tag{4.169}$$

$$\pi_0 = 1 - \sum_{j=1}^{m+w} \pi_j, \tag{4.170}$$

where $g(n) = \min(n,m)$.

For $\mu_j = \mu$, $j \in E$, (4.169) leads to

$$\pi_n = \frac{p_{n-1}}{p\vec{a}\mu g(n)}, \ n = 1,\dots,m+w. \tag{4.171}$$

Assuming further that $a_i = a$, $i \in E$, we get

$$\pi_n = \frac{p_{n-1}}{a\mu g(n)}, \ n = 1,\dots,m+w, \tag{4.172}$$

which coincides with the classical formula of Takács [277] but is somewhat more general, since in the last formula we restrict ourselves to the equalities of the first moments only; the PDF's $\vec{A}$ may still differ. Formula (4.171) was established in Dshalalow [224] and formula (4.169) is due to Dshalalow [236].

Remark 4.21. Theorem 4.10 can formally be applied to a system with an infinite buffer, $G_0,\dots/M_0,\dots/m/\infty$, if in formulas (4.169) and (4.170) we set $w = \infty$.

This result is due to Dshalalow [228].

4.5.2 Embedded queueing process

The embedded queueing process was studied in Dshalalow [236] under the following restrictions:

$$A_0(x) = \ldots = A_{h-1}(x),\ \mu_0 = \ldots = \mu_{h-1}, \tag{4.173}$$

$$A_h(x) = \ldots = A_{m+s-1}(x),\ \mu_h = \ldots = \mu_{m+s-1}, \tag{4.174}$$

$$A_{m+s}(x) = \ldots = A_{m+w}(x) = A(x),\ \mu_{m+s} = \ldots = \mu_{m+w} = \mu, \tag{4.175}$$

where $0 \leq h \leq m,\ 0 \leq s \leq w$. For $h = 0$, we set $A_0(x) = \ldots = A_{m+s-1}(x)$ (same with μ's). If $h = m$ and $s = 0$, we set $A_h(x) = A(x)$ and $\mu_h = \mu$.

Although the stationary distribution p exists without these restrictions and can be obtained numerically from $p = pP$, $pe = 1$, an explicit and analytically tractable solution is possible only under the above or similar assumptions whose interpretation is as follows. The system has two control levels. The first control level allows one to monitor the input and service under a given number of busy channels. The second control level is turned on when all channels are busy and a few customers are waiting. The "best" values of h and s can be subject to optimization of a relevant objective function.

Theorem 4.22. *Under restrictions (4.173-4.175) the stationary distribution $p = (p_0,\ldots,p_{m+w})$ satisfies the following formulas:*

$$\sigma^{-1}p_j = \begin{cases} \displaystyle\sum_{i=j}^{h-1}(-1)^{i-j}\binom{i}{j}E_i, & j = 0,\ldots,h-1 \\[2mm] \displaystyle\sum_{i=j}^{m-1}(-1)^{i-j}\binom{i}{j}U_i, & j = h,\ldots,m-1 \\[2mm] b_{m+s-j}, & j = m,\ldots,m+s-1 \\[2mm] c_{m+w-j}, & j = m+s,\ldots,m+w \end{cases} \tag{4.176}$$

where

$$\sigma^{-1} = E_0 + U_{h0} + \sum_{j=0}^{s}b_j + \sum_{j=0}^{w-s-1}c_j, \tag{4.177}$$

b_j *and* c_j *are the coefficients of the power series* $b(u) = \sum_{j\geq 0}b_j u^j$, *and* $c(u) = \sum_{j\geq 0}c_j u^j$, *which can be obtained from the respective Taylor series expansions of the functions*

$$B(u) = c_{w-s} + \frac{\alpha(m\mu(1-u)) - u}{\alpha_h(m\mu_h(1-u)) - u}\sum_{i\geq 1}c_{w-s+i}u^i, \tag{4.178}$$

$$C(u) = \frac{\alpha(m\mu(1-u))(1-u)}{\alpha(m\mu(1-u)) - u}, \tag{4.179}$$

both analytic at 0, with

$$\alpha_j(\theta) = \int_0^\infty e^{-\theta x} A_j(dx). \tag{4.180}$$

$$U_{hr} = \begin{cases} \sum_{i=h}^{m-1} U_i \Delta_{ih}^r, \ r = 0,\dots,h-1 \\ U_r, \ r = h,\dots,m-1 \end{cases} \tag{4.181}$$

$$\Delta_{jk}^r = \begin{cases} \sum_{i=k+1}^{j} (-1)^{j-i} \binom{i}{r}\binom{j}{i}, & k = 0,\dots,j-1 \\ 0, & k \geq j \\ \quad where\ 0 \leq r \leq i \leq j,\ j = 0,1,\dots \end{cases} \tag{4.182}$$

$$U_r = d_r^h \sum_{j=r+1}^{m} W_j/(\alpha_j^h d_j^h - 1),\ r = h,\dots,m-1, \tag{4.183}$$

$$d_r^k = \begin{cases} 1, & r = 0 \\ \prod_{i=1}^{r} \dfrac{\alpha_i^k}{1-\alpha_i^k}, & r = 1,\dots,m+w, \end{cases} \tag{4.184}$$

$$\alpha_r^k = \alpha_k(r\mu_k), \tag{4.185}$$

$$E_r = d_r^0 \sum_{j=r}^{h-1} \frac{(1-\alpha_{j+1}^h)U_{hj+1} - \alpha_{j+1}^h U_{hj} + W_{j+1}}{\alpha_{j+1}^0 d_j^0},\ r = 0,\dots,h-1, \tag{4.186}$$

$$W_r = \binom{m}{r}\left[b_{s+1}\alpha_m^h - \sum_{j=0}^{s-1} b_{s-j} S_{jr}^h - \sum_{j=s}^{w-1} c_{w-j} S_{jr} - S_{w-1,r} \right],\ r = 0,\dots,m, \tag{4.187}$$

$$S_{jr}^k = \begin{cases} \alpha_r^k(\frac{m}{m-r})^{j+1} - \sum_{i=0}^{j+1} (\frac{m}{m-r})^{j+1-i} q_i^k,\ r = 0,\dots,m-1 \\ 0,\ r = m,\dots,m+w, \end{cases} \tag{4.188}$$

$$q_r^j = \int_0^\infty e^{-m\mu_j x} \frac{(m\mu_j x)^r}{r!} A_j(dx). \tag{4.189}$$

(*Upper indices in S are dropped whenever they are in α and q.*)

Observe that while the above formulas are rather cumbersome and can be rivaled by the direct evaluation $p = pP$, $pe = 1$, similar results applied to the system $G_0,\dots/M_0,\dots/m/\infty$ (see Dshalalow [228]) have no alternative.

4.6 CLOSED SYSTEMS AND GENERAL MACHINE REPAIR PROBLEMS

A queueing system is said to be *closed* if a servicing facility serves only a group of permanent customers that from time to time require service. Another name for such a system is a model with a *finite source*. Because of the need of these systems in engineering and technology, the research in this area, as well as terminology bifurcated in two directions: reliability and queueing. In the former they use the term an (*unreliable*) *machine* instead of a *customer*, and a *repairman* instead of a *server*. A connection between closed and multi-channel queues was first observed in Takács [277] and then fully explored in Dshalalow [224,234-236].

A basic closed queue, or, equivalently, *machine repair problem* or *process of servicing machines*, consists of $m + 1$ "unreliable" machines serviced by a single repairman. Each of the continuously working machines is subject to occasional breakdowns independent of each other and of the repairman's work. If all machines are intact the repairman is idle, but as soon as any of the machines fails the repairman services it immediately. Broken down machines line up in the order of their failures. Originally, basic systems were "Markovian," i.e., with both failures and repairs exponentially distributed, in notation, $M/M/m + 1/1$. Takács [277], for the first time, systematically studied machine repair problems $M/G/m + 1/1$ with generally distributed service times and obtained elegant formulas for the queueing process. The research in this area is still of interest and more new papers with various modifications of the classical machine repair problems appear in the queueing literature: [233,237-245].

Dshalalow in [224] and [236] proposed machine repair problems with available reserve machines ready to replace any failed machines. A machine is *reserve* if it is idle at the beginning, but it immediately replaces any broken down working machine. In general, working and reserve machines are indistinguishable. That means any reserve machine can become a working machine, and once failed, will be refurbished and then take its place in the main or reserve facility. Dshalalow in [236] considered two models, with and without repairmans vacations.

Model 1 refers to a system with m working and $w + 1$ reserve machines and a single repairman. As soon as a working machine breaks down it lines up for repair and is replaced by any of the reserve machines if available. A fully repaired machine gets the status of a "working machine" and begins to work as long as the total number of intact machines is less than m. Otherwise, a machine becomes reserve regardless of whether it was previously working or reserve. When all working and reserve machines are intact the repairman rests, but he is ready to restore his service as soon as any new (working) machine breaks down. Note that in this and the following model 2, only machines that are working can break down.

Model 2 refers to a system with m working and w reserve machines and a single repairman, who leaves on vacation when all machines (the system) become intact. A vacation segment is not interrupted if in the meanwhile some machines fail. Upon his return, the repairman inspects the system and if all machines are still intact he leaves on vacation again and he does so until upon his return at least one broken machine is waiting. Under these assumptions this would be a standard exhaustive system with multiple vacations. However, once applied to a repair machine problem it is subject to the following modification. We assume that each time the repairman leaves the system he returns with a "brand new" machine, which replaces one broken machine if available or one of the reserve machines, otherwise. [An older machine is "discarded."] In the latter case, there is no change in the status of the system. In the first case, the system can become intact each time the repairman replaces only one broken machine, which would generate another vacation segment. Although this machine repair problem with vacations may appear somewhat special, it serves as an intermediate system between model 1 and multi-channel queues in terms of establishing a duality relation. This will be discussed below.

In both models we assume state dependent service, repairman vacations, and machine failures, and the models are formalized as follows.

Model 1. Let $I^1(t)$ be the number of intact and $W^1(t)$ the number of working machines at time t. Obviously, $W^1(t) = \min\{m, I^1(t)\}$. Let $\tau_1, \tau_2, \ldots$ be the successive instants of the completion of machine repairs. Let $I_n^1 = I^1(\tau_n -)$. Therefore, the process $I^1(t)$ is with the state space $\{0, \ldots, m+w+1\}$ and the embedded process I_n^1 is with the state space $\{0, \ldots, m+w\}$. Assume that the nth repair time of a machine is distributed according to the PDF $A_i(x) \in \{A_0(x), \ldots, A_{m+w}(x)\}$ if $I_n^1 = i$. Given $I_n^1 = i$, each of the $W^1(t)$ machines in the interval (τ_n, τ_{n+1}) can break down (independently of each other and of the repairman) exponentially with rate $\mu_i \in \{\mu_0, \ldots, \mu_{m+w}\}$.

Model 2. Let $I^2(t)$ be the total number of intact and $W^2(t)$ the number of working machines at time t. In this case, $I^1(t) \in \{0, \ldots, m+w\}$. Analogously, $W^2(t) = \min\{m, I^2(t)\}$. Let T_n be the completion time of the nth repair. The system is said to *regenerate* if it becomes "intact," i.e., the number of intact machines becomes $m+w$. Suppose that at time T_n the system regenerates and therefore the repairman leaves on vacation. At time T_{n+1}, the repairman returns from the first vacation segment with a new machine, which replaces one of the broken machines if extant, thereby decreasing the number of broken machines by one. Consequently, $I^1(T_{n+1}) = \min\{I^1(T_{n+1}) + 1, m+w\}$. The repairman continues his sequence of vacation segments until upon his return and after one replacement there are still one or more broken machine, lined up, and then the repairman begins processing them one by one. We denote $I_n^2 = I^2(T_n -)$. The state space of the embedded process I_n^2 is $\{0, \ldots, m+w\}$.

Note that the state $\{m+w\}$ applies to some of the instants of repairman returns from vacations during which no new machines broke, and therefore such states of I_n^2 are possible due to vacation policy only. [In contrast with model 2, in model 1, the maximum state, $\{m+w+1\}$ of the process $I^1(t)$, can never be reached by the embedded process I_n^1.] Now, given $I_n^2 = i$, the length of the interval $[T_n, T_{n+1})$ is distributed as $A_i(x) \in \{A_0(x), \ldots, A_{m+w}(x)\}$, whether it is a vacation segment or repair time. Failures of working machines occur as in model 1.

A parallel between model 2 and the $G_0, \ldots, G_{m+w}/M_0, \ldots M_{m+w}/m/w$ system.

Model 2 is identical to the multi-channel $G_0, \ldots, G_{m+w}/M_0, \ldots M_{m+w}/m/w$ system described in the previous section, whose notation will be used unchanged. Suppose that at time T_n a customer departs from a source and arrives at the system at time T_{n+1}. It either enters one of the free channels available, occupies the waiting room, or leaves the system unprocessed if there were $m+w$ customers in the system on his arrival. So, while in the $G_0, \ldots, G_{m+w}/M_0, \ldots M_{m+w}/m/w$ system at time T_n a customer departs from the source, at the same time in the finite source model 2 the repairman starts a new service or goes on vacation; both completions take place at time T_{n+1}. If in the interval $[T_n, T_{n+1})$ a working machine breaks down, at the same time a customer is processed by one of the channels in $G_0, \ldots, G_{m+w}/M_0, \ldots M_{m+w}/m/w$ and leaves the system. When the $(n+1)$th arrival occurs (at time T_{n+1}), the queue in $G_0, \ldots, G_{m+w}/M_0, \ldots, M_{m+w}/m/w$ is increased by one unless there have been $m+w$ customers. In the latter case the $(n+1)$th customer is lost. If in the closed system, T_{n+1} is the $(n+1)$th repair completion, the number of intact machines increases by one. Otherwise, if T_{n+1} is the instant of the repairman's return from a vacation, the number of intact machines also increases by one unless there have been $m + w$ intact machines just prior to his return. Consequently, the processes $I^2(t)$ and $Q(t)$ are identical.

The following will be to establish a formal relationship between model 1 and model 2, as it turns out to be more vivid than the relationship between model 1 and the queueing system $G_0,\ldots,G_{m+w}/M_0,\ldots M_{m+w}/m/w$.

A relation between models 1 and 2. First observe that the processes $I^1(t)$ and $I^2(t)$ are semi-regenerative relative to the sequences $\tau = \{\tau_0 = 0, \tau_1,\ldots\}$ and $T = \{T_0 = 0, T_1,\ldots\}$, that $I^1 = \{I^1_n; n = 0,\ldots\}$ and $I^2 = \{I^2_n; n = 0,\ldots\}$ are embedded Markov chains, and that (τ, I^1) and (T, I^2) are their respective Markov renewal processes. Since idle periods of the repairman in model 1 are distributed exponentially, the Markov chains I^1 and I^2 are stochastically equivalent.

Let $\mathfrak{B}_n$ be the nth busy period and $\mathfrak{I}_n$ be the idle period immediately following $\mathfrak{B}_n$. Let $c(t)$ denote the counting process associated with the point process $\mathfrak{B}_n$. The process $I^1(t)$ on its busy period and process $I^2(t)$ are stochastically equivalent, i.e.,

$$\mathbb{E}^i[g(I^1(t)) \mid \mathfrak{I}_{c(t)} > t\} = \mathbb{E}^i[g(I^2(t))] \tag{4.190}$$

for any Borel measurable function g, and

$$\mathbb{P}^i\{I^1(t) \in A \mid \mathfrak{I}_{c(t)} > t\} = \frac{\mathbb{P}^i\{I^1(t) \in A, I^1(t) \in \{0,\ldots,m+w\}\}}{\mathbb{P}^i\{I^1(t) \in \{0,\ldots,m+w\}\}}$$

$$= \begin{cases} 0, & A = \{m+w+1\} \\[2mm] \dfrac{\mathbb{P}^i\{I^1(t) \in A\}}{1 - \mathbb{P}^i\{I^1(t) = m+w+1\}}, & A \subseteq \{0,\ldots,m+w\}. \end{cases} \tag{4.191}$$

Consequently, (4.190) and (4.191) yield that

$$\mathbb{P}^i\{I^1(t) = k\} = [1 - \mathbb{P}^i\{I^1(t) = m+w+1\}]\mathbb{P}^i\{I^2(t) = k\}. \tag{4.192}$$

Let $\tilde{I}^1(t)$ be the minimal semi-Markov process relative to the sequence τ (of successive repair completions). [Note that $\tilde{I}^1(t)$ preserves the information about the number of intact machines just before the last repair completion prior to t, and it "continuously interpolates" the embedded Markov chain.] Then, we have

$$\mathbb{P}^i\{I^1(t) = m+w+1\} = \mathbb{P}^i\{\tilde{I}^1(t) = m+w, \mathfrak{I}_{c(t)} \leq t\}. \tag{4.193}$$

The last equation is due to the following arguments. That $\tilde{I}^1(t) = m+w$ means t belongs to a repair cycle following the last repair completion at $\mathfrak{I}_{c(t)}$ (which is also the start of an idle period following busy period $\mathfrak{B}_{c(t)}$). This cycle consists of the idle period $[\mathfrak{I}_{c(t)}, \mathfrak{B}_{c(t)+1})$ and the repair period lasting from $\mathfrak{B}_{c(t)+1}$ to the next completion at $\tau_{C(t)+1}$, where $C(t) = C([0,t])$ is the counting processes associated with τ. In other words, over the period $[\mathfrak{I}_{c(t)}, \tau_{C(t)+1})$, the value of the semi-Markov process $\tilde{I}^1(t)$ is $m+w$, while the state $\{m+w+1\}$ of $I^1(t)$ takes place only over the idle period lasting from $\mathfrak{I}_{c(t)}$ through $\mathfrak{B}_{c(t)+1}$. Note that by the definition of $c(t)$, $t \in [\mathfrak{B}_{c(t)}, \mathfrak{B}_{c(t)+1})$ and hence either $t \in [\mathfrak{B}_{c(t)}, \mathfrak{I}_{c(t)})$ or $t \in [\mathfrak{I}_{c(t)}, \mathfrak{B}_{c(t)+1})$.

Now, from the theory of alternating renewal processes (*cf.* Cox [257], section 7.2, formula (6)) [i.e., renewal processes in which inter-renewal times are convolutions of two or more independent r.v.'s alternating through upcoming inter-renewal segments] we know that the probability of time t to fall into an interval of the "first type" has the limiting distribution $p_1 = \dfrac{r_1}{r_1 + r_2}$, where r_1 and r_2 are the mean lengths of the intervals of the first and the second type, respectively. Although repair cycles do not form an alternating renewal process we can "extract" them from the entire point process of all events and "stick them together." In other words, given $\{\tilde{I}^1(t) = m + w\}$, t really runs an alternating renewal process of repair cycles $[\mathfrak{I}_{c(t)}, {}^\tau C(t) + 1)$ including idle periods and initial repair periods. Therefore, given that t does belong to such a cycle, i.e., $\tilde{I}^1(t) = m + w$, we have that

$$\lim_{t \to \infty} \mathbb{P}^i\{\mathfrak{I}_{c(t)} \le t \mid \tilde{I}^1(t) = m + w\} = \frac{\dfrac{1}{\mu_{m+w}m}}{\dfrac{1}{\mu_{m+w}m} + a_{m+w}}$$

$$= \frac{1}{a_{m+w}\mu_{m+w}m + 1}. \tag{4.194}$$

On the other hand, by the ergodic theorem for semi-Markov processes ($cf.$ Çinlar [255], p. 342),

$$\lim_{t \to \infty} \mathbb{P}^i\{\tilde{I}^1(t) = m + w\} = \frac{p_{m+w}r_{m+w}}{p\vec{r}}, \tag{4.195}$$

where

$$r_k = \mathbb{E}^k[\tau_1] = \begin{cases} a_k, \; k = 0,\dots,m+w-1 \\ a_{m+w} + \dfrac{1}{\mu_{m+w}m}, \; k = m+w \end{cases} \tag{4.196}$$

is the mean first repair cycle and $\vec{r} = (r_0,\dots,r_{m+w})^{\mathrm{T}}$, and $p = (p_0,\dots,p_{m+w})$ is the stationary distribution of the embedded Markov chain, I_n^1 which (as has been mentioned) is identical to the stationary distribution of the embedded Markov chain I_n^2.

Now, by (4.193-4.195),

$$\pi_{m+w+1}^1 = \frac{p_{m+w}}{p\vec{a}\mu_{m+w}m + p_{m+w}}, \tag{4.197}$$

where

$$\pi_k^1 = \lim_{t \to \infty} \mathbb{P}^i\{I^1(t) = k\}, \; k = 0,\dots,m+w+1. \tag{4.198}$$

Finally, (4.192) yields that

$$\pi_k^1 = (1 - \pi_{m+w+1}^1)\pi_k, \; k = 0,\dots,m+w, \tag{4.199}$$

where $\pi_k = \lim_{t \to \infty} \mathbb{P}^i\{I^1(t) = k\}$ has been found in the previous section and the probability π_{m+w+1}^1 satisfies formula (4.197), so that we have

$$\pi_k^1 = \pi_k \frac{p\vec{a}\mu_{m+w}m}{p\vec{a}\mu_{m+w}m + p_{m+w}}, \; k = 0,\dots,m+w. \tag{4.200}$$

(4.200) and (4.197) give the stationary distribution of the process of intact machines for model 1.

Example 4.23. Let $\mu_k = \mu$, $k = 0,\dots,m+w$. Then, combining (4.200) with formula (4.171) and by (4.197), we have

$$\pi_k^1 = \frac{m}{g(k)} \frac{p_{k-1}}{p\vec{a}\mu m + p_{m+w}}, \; k = 1,\dots,m+w+1, \tag{4.201}$$

and

$$\pi_0^1 = 1 - \sum_{k=1}^{m+w+1} \pi_k^1. \tag{4.202}$$

If in formula (4.201) we set $w = 0$ and replace m (machines) by $m+1$ we arrive at a state dependent version of the classical machine repair problem investigated by Takács [277]:

$$\pi_k^1 = \frac{m+1}{k} \frac{p_{k-1}}{p\vec{a}\mu(m+1) + p_m}, k = 1,\ldots,m+1, \tag{4.203}$$

$$\pi_0^1 = 1 - \sum_{k=1}^{m+1} \pi_k^1. \tag{4.204}$$

To have (4.203) reduce to Takács' formula we set $p\vec{a} = 1$ (which, in particular, is the case when $a_k = a$, $k = 0,\ldots,m$):

$$\pi_k^1 = \frac{m+1}{k} \frac{p_{k-1}}{\mu(m+1) + p_m}, k = 1,\ldots,m+1. \tag{4.205}$$

Note that the method of construction of "dual models" gives rise to investigation of other modifications of repair machines problems, such as those with "true vacations."

4.7 OPEN PROBLEMS AND FUTURE RESEARCH DIRECTIONS

1) It is worthwhile to study a system with w-policy, mentioned in the introduction as case 4 under "workload dependence." This system will rival quorum systems and can be analyzed by using first excess level techniques. Combined with D-policy, w-policy can specify an interesting hysteresis scenario for workload.
2) The ic- and sd-dependent system discussed in the beginning of Subsection 4.2.1 can be generalized for batch arrivals and batch service.
3) The system with random server capacity treated in Subsection 4.2.4 can be modified to an $M^X/G_N^{r,R}/1$ type by using similar techniques.
4) An interesting extension of single server D-policy systems would be one with bulk arrivals and batch service with first excess level analysis as an option.
5) One can consider state dependent multi-channel systems with unreliable servers by using results of Section 4.6 for machine repair problems.

ACKNOWLEDGEMENT. I would like to thank Gary Russell for his very careful reading the final draft of this chapter, and numerous editorial suggestions throughout the manuscript, especially the bibliography.

BIBLIOGRAPHY

I. Single Server Systems

A) Queue Length Dependence

Input rate (continuous dependence, upon service completions), discouraged arrivals, finite waiting room

[1]   Abolnikov, L. and Dshalalow, J.H., On a multilevel controlled bulk queue-

ing system $M^X/G^{r,R}/1$, *J. Appl. Math. Stoch. Anal.*, **5**:3 (1992), 237-260.

[2]   Abolnikov, L. and Dshalalow, J.H., Ergodicity conditions and invariant probability measure for an embedded Markov chain in a controlled bulk queueing system with a bilevel service delay discipline. Part I, *Appl. Math. Let.*, **5**:4 (1992), 25-27.

[3]   Abolnikov, L. and Dshalalow, J.H., Ergodicity conditions and invariant probability measure for an embedded Markov chain in a controlled bulk queueing system with a bilevel service delay discipline. Part II, *Appl. Math. Let.*, **5**:5 (1992), 15-18.

[4]   Abolnikov, L., Dshalalow, J.H., and Dukhovny, A.M., On some queue length controlled stochastic processes, *J. Appl. Math. Stoch. Anal.*, **3**:4 (1990), 227-244.

[5]   Abolnikov, L., Dshalalow, J.H., and Dukhovny, A.M., On stochastic processes in a multilevel bulk queueing system, *Stoch. Anal. Appl.*, **10**:2 (1992), 155-179.

[6]   Abolnikov, L., Dshalalow, J.H. and Dukhovny, A.M., First passage processes in queueing system $M^X/G^r/1$ with service delay discipline, *Intern. J. Math. and Math. Sci.*, **17**:3 (1994), 571-586.

[7]   Baba, Y., The $M^X/G/1$ queue with finite waiting room, *J. Oper. Res. Soc. Japan*, **27**:3 (1984), 260-272.

[8]   Bocharov, P.P., *Finite Capacity Single Server Queues*, People Friendship University Press, Moscow 1985.

[9]   Dshalalow, J.H., *Warteschlangensysteme mit Feedback*, Thesis of Doct. Diss. at Technische Universität Berlin, Dept. Math., Berlin, W. Germany, 1983.

[10]   Dshalalow, J.H., On a first passage problem in general queueing systems with multiple vacations, *J. Appl. Math. Stoch. Anal.*, **5**:2 (1992), 177-192.

[11]   Dshalalow, J.H., Single-server queues with controlled bulk service, random accumulation level and modulated input, *Stoch. Anal. Appl.*, **11**:1 (1993), 29-41.

[12]   Dshalalow, J.H., First excess level analysis of random processes in a class of stochastic servicing systems with global control, *Stoch. Anal. Appl.*, **12**:1 (1994), 75-101.

[13]   Dshalalow, J.H. and Russell, G., On a single-server queue with fixed accumulation level, state dependent service, and semi-Markov modulated input flow, *Intern. J. Math. and Math. Sci.*, **15**:3 (1992), 593-600.

[14]   Dshalalow, J.H. and Tadj, L., A queueing system with random server capacity and multiple control, *Queueing Sys.*, **14** (1993), 369-384.

[15]   Goel, L.R., Heterogeneous queueing with arrivals depending on queue length, *Math. Operationsforsch. Stat.*, **7**:6 (1976), 945-952.

[16]   Gupta, S.K., Queues with hyper-Poisson input and exponential service time distribution with state dependent arrival and service rates, *Oper. Res.*, **15** (1967), 847-856.

[17]   Haddidi, N., Busy period of queues with state dependent arrival and service rates, *J. Appl. Prob.*, **11** (1974), 842-848.

[18]   Hoorn, M.H. van, *Algorithms and Approximations for Queueing Systems*, CWI Tracts, Amsterdam 1984.

[19]   Knessl, C., Matkovsky, B., Schuss, Z., and Tier, C., On the performance of state-dependent single server queues, *SIAM J. Appl. Math.*, **46**:4 (1986), 657-697.

[20]   Knessl, C., Matkovsky, B., Schuss, Z., and Tier, C., System crash in a finite capacity $M/G/1$ queue, *Stoch. Models*, **2**:2 (1986), 171-201.

[21]   Martin, M. and Artalejo, J.R., Analysis of an $M/G/1$ queue with two types of impatient units, *Adv. Appl. Prob.*, **27** (1995), 840-861.

[22]   Natvig, B., On a queueing model where potential customers are discouraged by queue length, *Scand. J. Stat.*, **2** (1975), 34-42.

[23]   Niu, S.-C. and Cooper, R.B., Transform-free analysis of $M/G/1/K$ and related queues, *Math. Oper. Res.*, **18**:2 (1993), 486-510.

[24]   Shanthikumar, J.G. and Chandra, M.J., Application of level crossing analysis to discrete state processes in queueing systems, , *Nav. Res. Log. Quart.*, **29**:4 (1982), 593-608.

[25]   Shanthikumar, J.G. and Sumita, U., On the busy-period distributions of $M/G/1/K$ queues with state-dependent arrivals and FCFS/LCFS-P service disciplines, *J. Appl. Prob.*, **22** (1985), 912-919.

[26]   Sharma, O.P. and Maheswar, V.R., Transient behavior of a simple queue with discouraged arrivals, *Optimization*, **27** (1993), 283-291.

[27]   Schellhaas, H., Computation of the state probabilities in $M/G/1$ queues with state dependent input and state dependent service, *OR Spektrum*, **5** (1983), 223-228.

[28]   Schellhaas, H., Computation of the state probabilities in a class of semi-regenerative queueing models, in: *Semi-Markov Models, Theory and Applications* (ed.: J. Jansen), Plenum Press, New York 1986.

[29]   Tijms, H.C. and Duyn Schouten, F., van der, Inventory control with two switch-over levels for a class $M/G/1$ queueing systems with variable arrival and service rate, *Stoch. Proc. Appl.*, **6** (1978), 213-222.

[30]   Tijms, II.C. and van Hoorn, M.H., Algorithms for the state probabilities and waiting times in single server queueing systems with random and quasi-random input and phase-type service times, *OR Spektrum*, **2** (1981), 145-152.

[31]   Tijms, H.C. and van Hoorn, M.H., Computational methods for single-server and multi-server queues with Markovian input and general service times, 71-98, in: *Applied Probability-Computer Science: The Interface, Vol. 2* (ed.: R.L. Disney and T.J. Ott), Birkhäuser, Boston 1982.

Service rate and batch size (continuous dependence and upon service completions)

[32]   Abolnikov, L. and Dshalalow, J.H., Feedback queueing systems; duality principle and optimization, *Autom. Rem. Contr.*, **39** (1978), 11-20.

[33]   Abolnikov, L. and Dshalalow, J.H., On a multilevel controlled bulk queueing system $M^X/G^{r,R}/1$, *J. Appl. Math. Stoch. Anal.*, **5**:3 (1992), 237-260.

[34]   Abolnikov, L. and Dshalalow, J.H., Ergodicity conditions and invariant probability measure for an embedded Markov chain in a controlled bulk queueing system with a bilevel service delay discipline. Part I, *Appl. Math. Let.*, **5**:4 (1992), 25-27.

[35]   Abolnikov, L. and Dshalalow, J.H., Ergodicity conditions and invariant probability measure for an embedded Markov chain in a controlled bulk queueing system with a bilevel service delay discipline. Part II, *Appl. Math. Let.*, **5**:5 (1992), 15-18.

[36]   Abolnikov, L., Dshalalow, J.H., and Dukhovny, A.M., On some queue length controlled stochastic processes, *J. Appl. Math. Stoch. Anal.*, **3**:4

(1990), 227-244.

[37] Abolnikov, L., Dshalalow, J.H., and Dukhovny, A.M., On stochastic process-
      es in a multilevel bulk queueing system, *Stoch. Anal. Appl.*, **10**:2 (1992),
      155-179.

[38] Abolnikov, L., Dshalalow, J.H., and Dukhovny, A.M., First passage process-
      es in queueing system $M^X/G^r/1$ with service delay discipline, *Intern. J.
      Math. and Math. Sci.*, **17**:3 (1994), 571-586.

[39] Alimov, D., On a queueing model with variable service rate, *Izv. Acad. Sci.
      Turkm. SSR Ser. Fiz.-Mat.*, No. 1 (1980), 3-6.

[40] Conolly, B.W., Some applications of the theory of infinite capacity service
      systems to a single server system with linearly state dependent service, *J.
      Appl. Prob.*, **8** (1971), 202-207.

[41] Conolly, B.W., The generalized state-dependent queue: the busy period
      Erlangian, *J. Appl. Prob.*, **11** (1974), 618-623.

[42] Conway, R. and Maxwell, W., A queueing model with state dependent
      service rate, *The J. Industr. Eng.*, **12**:2 (1961), 130-136.

[43] Deb, R., Optimal control of batch service queues with switching costs, *Adv.
      Appl. Prob.*, **8** (1976), 177-194.

[44] Delbrouck, L.E.N., A feedback queueing system with batch arrivals, bulk
      service and queue-dependent service time, *J. Assoc. Comp. Mach.*, **17**:2
      (1970), 314-323.

[45] Dshalalow, J.H., *Warteschlangensysteme mit Feedback*, Thesis of Doct.
      Diss. at Technische Universität Berlin, Dept. Math., Berlin, W. Germany,
      1983.

[46] Dshalalow, J.H., A single-server queue with random accumulation level, *J.
      Appl. Math. Stoch. Anal.*, **4**:3 (1991), 203-210.

[47] Dshalalow, J.H., On a first passage problem in general queueing systems
      with multiple vacations, *J. Appl. Math. Stoch. Anal.*, **5**:2 (1992), 177-192.

[48] Dshalalow, J.H., Single-server queues with controlled bulk service, random
      accumulation level and modulated input, *Stoch. Anal. Appl.*, **11**:1 (1993),
      29-41.

[49] Dshalalow, J.H., First excess level analysis of random processes in a class of
      stochastic servicing systems with global control, *Stoch. Anal. Appl.*, **12**:1
      (1994), 75-101.

[50] Dshalalow, J.H. and Russell, G., On a single-server queue with fixed accu-
      mulation level, state dependent service, and semi-Markov modulated input
      flow, *Intern. J. Math. and Math. Sci.*, **15**:3 (1992), 593-600.

[51] Dshalalow, J.H. and Tadj, L., A queueing system with random server
      capacity and multiple control, *Queueing Sys.*, **14** (1993), 369-384.

[52] Dshalalow, J.H. and Tadj, L., On applications of first excess level random
      processes to queueing systems with random server capacity and capacity
      dependent service time, *Stoch. and Stoch. Reports*, **45** (1993), 45-60.

[53] Fakinos, D., The single server queue with service depending on queue size
      and with the preemptive resume last-come-first-served queue discipline, *J.
      Appl. Prob.*, **24** (1987), 758-767.

[54] Fakinos, D., The $G/G1$ (LCFS/P) queue with service depending on queue
      size, *Europ. J. Oper. Res.*, **59** (1992), 303-307.

[55] Fedegruen, A. and Tijms, H.C., Computation of the stationary distribution
      of the queue size in an $M/G/1$ queueing system with variable service rate,
      J. Appl. Prob., **17** (1980), 515-522.

[56]   Gupta, S.K., Queues with hyper-Poisson input and exponential service time distribution with state dependent arrival and service rates, *Oper. Res.*, **15** (1967), 847-856.

[57]   Gupta, S.K., On bulk queues with state dependent parameters, *J. Oper. Res. Soc. Jap.*, **9**:2 (1967), 69-79.

[58]   Haddidi, N., Busy period of queues with state dependent arrival and service rates, *J. Appl. Prob.*, **11** (1974), 842-848.

[59]   Haddidi, N. and Conolly, B.W., On the improvement of the operational characteristics of single server queues by means of a queue length dependent mechanism, *Appl. Stat.*, **18** (1969), 229-240.

[60]   Harris, C.M., Queues with state-dependent stochastic service rates, *Oper. Res.*, **15** (1967), 117-130.

[61]   Harris, C.M., A queueing system with multiple service time distributions, *Nav. Res. Logist. Quart.*, **14** (1967), 231-239.

[62]   Harris, C.M., On queues with state-dependent Erlang service, *Nav. Res. Logist. Quart.*, **18**:1 (1971), 103-110.

[63]   Harris, C.M. and Marchal, W.G., State dependence in $M/G/1$ server-vacation models, *Oper. Res.*, **36**:4 (1988), 560-565.

[64]   Hoorn, M.H. van, *Algorithms and Approximations for Queueing Systems*, CWI Tracts, Amsterdam 1984.

[65]   Ivnitskiy, V.A., A stationary regime of a queueing system with parameters dependent on the queue length and with nonordinary flow, *Eng. Cyb.* (1974), 85-90.

[66]   Jo, K. and Stidham, J., Jr., Optimal service rate control of $M/G/1$ queueing systems using phase methods, *Adv. Appl. Prob.*, **15** (1983), 616-637.

[67]   Knessl, C., Matkovsky, B., Schuss, Z., and Tier, C., On the performance of state-dependent single server queues, *SIAM J. Appl. Math.*, **46**:4 (1986), 657-697.

[68]   Loris-Teghem, J. and Manya, N., Analysis of a queueing system with group arrivals and state dependent service times, related to a stochastic continuous-review (s,S) inventory model, *Europ. J. Oper. Res.*, **11** (1982), 82-92.

[69]   Matias, M.B., A queueing system with a queue length dependent service, *Revista Ci. Mat. Sér. A*, **3** (1972), 19-32.

[70]   Muh, D.C.R., A bulk queueing system under N-policy with bilevel service delay discipline and start-up time, *J. Appl. Math. Stoch. Anal.*, **6**, No. 4 (1993), 359-384.

[71]   Muh, D.C.R., On a Class of N-Policy Multilevel Control Queueing Systems, Doctoral Thesis, Florida Tech, Melbourne, FL 1994.

[72]   Neuts, M.F., A general class of bulk queues with Poisson input, *Ann. Math. Stat.*, **38** (1967), 759-770.

[73]   Schäl, M., The analysis of queues with state-dependent parameters by Markov renewal processes, *Adv. Appl. Prob.*, **3** (1971), 155-175.

[74]   Shanthikumar, J.G., On a single-server queue with state-dependent service, *Nav. Res. Logist. Quart.*, **26**:2 (1979), 305-309.

[75]   Shakhbazov, A.A., A queueing system with warm-up and queue length dependent service time, *Izv. Acad. Sci. AsSSR Ser. Phys.-Tech. Math.*, No. 1 (1974), 32-35.

[76]   Schellhaas, H., Computation of the state probabilities in $M/G/1$ queues with state dependent input and state dependent service, *OR Spektrum*, **5** (1983), 223-228.

[77] Schellhaas, H., Computation of the state probabilities in a class of semi-regenerative queueing models, in: *Semi-Markov Models, Theory and Applications* (ed.: J. Jansen), Plenum Press, New York 1986.

[78] Suzuki, T. and Ebe, M., Decision rules for the queueing system $M/G/1$ with service depending on queue-length, *Memoirs of the Defense Acad. Japan*, **7**:3 (1967), 1263-1273.

[79] Tadj, L., On a bulk queueing system with random server capacity and multiple control, *Eng. Simul.*, **15**:1 (1993), 3-10.

[80] Tijms, H.C. and Duyn Schouten, F., van der, Inventory control with two switch-over levels for a class $M/G/1$ queueing systems with variable arrival and service rate, *Stoch. Proc. Appl.*, **6** (1978), 213-222.

[81] Tijms, H.C. and van Hoorn, M.H., Algorithms for the state probabilities and waiting times in single server queueing systems with random and quasi-random input and phase-type service times, *OR Spektrum*, **2** (1981), 145-152.

[82] Tijms, H.C. and van Hoorn, M.H., Computational methods for single-server and multi-server queues with Markovian input and general service times, 71-98, in: *Applied Probability-Computer Science: The Interface, Vol. 2* (ed.: R.L. Disney and T.J. Ott), Birkhäuser, Boston 1982.

[83] Welch, P.D., On a generalized $M/G/1$ queueing process in which the first customer of each busy period receives exceptional service, *Oper. Res.*, **12** (1964), 736-752.

[84] Pakes, A.G., On the busy period of the modified $GI/G/1$ queue, *J. Appl. Prob.*, **10** (1973), 192-197.

Input and service rate with multilevel hysteretic control

[85] Bahary, E. and Kolesar, P., Multilevel bulk service queues, *Oper. Res.*, **20**:2 (1972), 406-420.

[86] Dukhovny, A.M., On queueing systems with bilevel hysteretic feedback, 129-146, in *Operations Research Methods and Reliability in System Analysis*, Kiev 1976.

[87] Gebhard, R.F., A queueing process with bilevel hysteretic service rate control, *Nav. Res. Log. Quart.*, **14**:1 (1967), 55-67.

[88] Loris-Teghem, J., Hysteretic control of an $M/G/1$ queueing system with two service time distributions and removable server, in: *Point Processes and Queueing Problems*, Colloquia Mathematica Societatis János Bolyai, Hungary, **24** (1978), 291-305.

[89] Lu, F. and Serfozo, R., $M/M/1$ queueing decision processes with monotone hysteretic optimal policies, *Oper. Res.*, **32**:5 (1984), 1116-1132.

[90] Teghem, J., Jr., On uniform hysteretic policies in a queueing system with variable service rates, *Cah. Centr. d'Etud. Rech. Opér.*, **21**:2 (1979), 121-126.

[91] Yadin, M. and Naor, P., On queueing systems with variable service capacities, *Nav. Res. Log. Quart.*, **14**:1 (1967), 43-53.

Idle and vacation periods disciplines (q-, T-, and E-limited disciplines)

[92] Abolnikov, L. and Dshalalow, J.H., On a multilevel controlled bulk queueing system $M^X/G^{r,R}/1$, *J. Appl. Math. Stoch. Anal.*, **5**:3 (1992), 237-260.

[93] Abolnikov, L. and Dshalalow, J.H., Ergodicity conditions and invariant probability measure for an embedded Markov chain in a controlled bulk que-

ueing system with a bilevel service delay discipline. Part I, *Appl. Math. Let.*, **5**:4 (1992), 25-27.

[94] Abolnikov, L. and Dshalalow, J.H., Ergodicity conditions and invariant probability measure for an embedded Markov chain in a controlled bulk queueing system with a bilevel service delay discipline. Part II, *Appl. Math. Let.*, **5**:5 (1992), 15-18.

[95] Abolnikov, L., Dshalalow, J.H., and Dukhovny, A.M., On some queue length controlled stochastic processes, *J. Appl. Math. Stoch. Anal.*, **3**:4 (1990), 227-244.

[96] Abolnikov, L., Dshalalow, J.H., and Dukhovny, A.M., On stochastic processes in a multilevel bulk queueing system, *Stoch. Anal. Appl.*, **10**:2 (1992), 155-179.

[97] Abolnikov, L., Dshalalow, J.H., and Dukhovny, A.M., First passage processes in queueing system $M^X/G^r/1$ with service delay discipline, *Intern. J. Math. and Math. Sci.*, **17**:3 (1994), 571-586.

[98] Abolnikov, L., Dshalalow, J.H., and Dukhovny, A.M., Stochastic analysis of a controlled bulk queueing system with continuously operating server: continuous time parameter queueing process, *Stat. Prob. Let.*, **16** (1993), 121-128.

[99] Abolnikov, L., Dshalalow, J.H., and Dukhovny, A.M., A multilevel controlled bulk queueing system with vacationing server, *Oper. Res. Letters*, **13** (1993), 183-188.

[100] Abolnikov, L., Dshalalow, J.H., and Dukhovny, A.M., Stochastic analysis of a controlled bulk queueing system with continuously operating server: continuous time parameter queueing process, *Stat. Prob. Letters*, **16** (1993), 121-128.

[101] Babitsky, A.V., $M/G/1$ vacation model with limited service discipline and hybrid switching-on policy, *Math. Probl. Engin.* (to appear).

[102] Chaudhry, M.L. and Templeton, J.G.C., The queueing system $M/G^B/1$ and its ramifications, *Europ. J. Oper. Res.*, **6** (1981), 56-60.

[103] Chaudhry, M.L., Gupta, U.C., and Madill, B.R., Computational aspects of bulk-service queueing system with variable capacity and finite space: $M/G^Y/1/N + B$, *J. Oper. Res. Soc. Jap.*, **34**:4 (1991), 404-421.

[104] Chaudhry, M.L., Madill, B.R., and Briére, G., Computational analysis of steady-state probabilities of $M/G^{a,b}/1$ and related nonbulk queues, *Queueing Sys.*, **2** (1987), 93-114.

[105] Dshalalow, J.H., A single-server queue with random accumulation level, *J. Appl. Math. Stoch. Anal.*, **4**:3 (1991), 203-210.

[106] Dshalalow, J.H., On a first passage problem in general queueing systems with multiple vacations, *J. Appl. Math. Stoch. Anal.*, **5**:2 (1992), 177-192.

[107] Dshalalow, J.H., Single-server queues with controlled bulk service, random accumulation level and modulated input, *Stoch. Anal. Appl.*, **11**:1 (1993), 29-41.

[108] Dshalalow, J.H., First excess level analysis of random processes in a class of stochastic servicing systems with global control, *Stoch. Anal. Appl.*, **12**:1 (1994), 75-101.

[109] Dshalalow, J.H. and Russell, G., On a single-server queue with fixed accumulation level, state dependent service, and semi-Markov modulated input flow, *Intern. J. Math. and Math. Sci.*, **15**:3 (1992), 593-600.

[110] Dshalalow, J.H. and Tadj, L., A queueing system with random server

capacity and multiple control, *Queueing Sys.*, **14** (1993), 369-384.

[111] Dshalalow, J.H. and Tadj, L., On applications of first excess level random processes to queueing systems with random server capacity and capacity dependent service time, *Stoch. and Stoch. Reports*, **45** (1993), 45-60.

[112] Dshalalow, J.H. and Yellen, J., Bulk input queues with quorum and multiple vacations, *Math. Probl. Engin.*, **2**:2 (1996), 95-106.

[113] Easton, G.D. and Chaudhry, M.L., The queueing system $E_k/M^{a,b}/1$ and its numerical analysis, *Comp. Oper. Res.*, **9** (1982), 197-205.

[114] Federguen, A. and So, K.C., Optimality of threshold policy in single-server queueing systems with server vacations, *Adv. Appl. Prob.*, **23** (1991), 388-405.

[115] Gold, H. and Tran-Gia, P., Performance analysis of a batch service queue arising out of manufacturing system modeling, *Queueing Sys.*, **14** (1993), 413-426.

[116] Harris, C.M. and Marchal, W.G., State dependence in $M/G/1$ server-vacation models, *Oper. Res.*, **36**:4 (1988), 560-565.

[117] Heyman, D.P., The T-policy for the $M/G/1$ queue, *Manag. Sci.*, **23** (1977), 775-778.

[118] Jacob, M.J. and Madhusoodanan, T.P., Transient solution for a finite capacity $M/G^{a,b}/1$ queueing system with vacations to the server, *Queueing Sys.*, **2** (1987), 381-386.

[119] Jacob, M.J., Krishnamoorthy, and Madhusoodanan, T.P., Transient solution of a finite capacity $M/G^{a,b}/1$ queueing system, *Nav. Res. Log.*, **35** (1988), 437-441.

[120] Kambo, N.S. and Chaudhry, M.L., Distribution of the busy period for the bulk service queueing system $E_k/M^{a,b}/1$, *Comp. Oper. Res.*, **9** (1982), 86-1 to 86-7.

[121] Kasahara, S., Takine, T., Takahashi, Y., and Hagesawa, T., Analysis of an $SPP/G/1$ system with multiple vacations and E-limited service discipline, *Queueing Sys.*, **14** (1993), 349-367.

[122] Kasahara, S., Takagi, H., Takahashi, Y., and Hagesawa, T., $M/G/1/K$ system with push-out scheme under vacation policy, *J. Appl. Math. Stoch. Anal.*, **9**:2 (1996), 143-157.

[123] Keilson, J. and Servi, L., Oscillating random walk models for $GI/G/1$ vacation systems with Bernoulli schedules, *J. Appl. Prob.*, **23** (1986), 790-802.

[124] Keilson, J. and Servi, L., Blocking probabilities for $M/G/1$ vacation systems with occupancy level dependent schedules, *Oper. Res.*, **37**:1 (1989), 134-140.

[125] Kella, O., Optimal control of the vacation scheme in an $M/G/1$ queue, *Oper. Res.*, **38**:4 (1990), 724-728.

[126] Lee, T.T., $M/G/1/N$ queue with vacation time and limited service discipline, *Perform. Eval.*, **9** (1989), 181-190.

[127] Neuts, M.F., A general class of bulk queues with Poisson input, *Ann. Math. Stat.*, **38** (1967), 759-770.

[128] Neuts, M.F., Queues solvable without Rouche's theorem, *Oper. Res.*, **27**:4 (1979), 767-781.

[129] Neuts, M.G., The $M/G/1$ queue with a limited number of admissions or a limited admission period during each service time, *Stoch. Mod.*, **1**:3 (1985), 361-391.

[130] Neuts, M.F., Generalizations of the Pollaczek-Khinchine integral equations

in the theory of queues, *Adv. Appl. Prob.*, **18** (1986), 952-990.

[131]   Ooki, M., Fukagawa, Y., Murakami, S., and Yoshida, S., Analysis of non-preemptive priority queues with service disciplines depending on the number of waiting messages, *Asia-Pacif. J. Oper. Res.*, **10**:1 (1993), 57-78.

[132]   Tadj, L., On a bulk queueing system with random server capacity and multiple control, *Eng. Simul.*, **15**:1 (1993), 3-10.

[133]   Teghem, J., Jr., Optimal control of queues: removable servers [Tutorial paper XIX], *Belgian J. Oper. Res. Stat. Comp. Sci.*, **25**:2-3 (1985), 99-128.

[134]   Teghem J., Jr., Control of the service process in a queueing system, *Europ. J. Oper. Res.*, **23** (1986), 141-158.

[135]   Zhang, Z. and Vickson, R.G., A simple approximation for mean waiting time in $M/G/1$ queue with vacations and limited service discipline, *Oper. Res. Let.*, **13** (1993), 21-26.

Busy period disciplines (N-policy, vacations, and start-up periods)

[136]   Baker, K., A note on operating policies for the queue $M/M/1$ with exponential startups, *INFOR*, **11**:1 (1973), 71-72.

[137]   Borthakur, A., Medhi, J., and Gohain, R., Poisson input queueing system with startup time and under control-operating policy, *Compt. Oper. Res.*, **14**:1 (1987), 33-40.

[138]   Chae, K.C. and Lee, H.W., $M^X/G/1$ vacation models with N-policy: heuristic interpretation of the mean waiting time, *J. Oprnl. Res. Soc.*, **46** (1995), 258-264.

[139]   Dshalalow, J.H., Excess level processes in queueing, in: *Advances in Queueing* (ed. by Dshalalow J.D.), 243-262, CRC Press, Boca Raton, FL 1995.

[140]   Federguen, A. and So, K.C., Optimality of threshold policy in single-server queueing systems with server vacations, *Adv. Appl. Prob.*, **23** (1991), 388-405.

[141]   Heyman, D.P., Optimal operating policies for $M/G/1$ queueing systems, *Oper. Res.* **16** (1968), 362-382.

[142]   Hong, J.W. and Lie, C.H., Mean waiting time analysis of cyclic server system under N-policy, *J. Kor. Oper. Res. Soc.*, **18**:3 (1993), 51-63.

[143]   Jaiswal, N.K. and Simha, P.S., Optimal operating policies for the finite-source queueing process, *Oper. Res.*, **20**:3 (1972), 698-707.

[144]   Kramer, M., Stationary distributions in a queueing system with vacation times and limited service, *Queueing Sys.*, **4** (1989), 57-68.

[145]   Kubat, P. and Servi, L.D., Cyclic service queues with very short service times, *European J. Opnl. Res.*, **53** (1991), 172-188.

[146]   LaMaire, R.O., $M/G/1/N$ vacation model with varying E-limited service discipline, *Queueing Sys.*, **11** (1992), 357-375.

[147]   Lee, D-S., A two-queue model with exhaustive and limited service discipline, *Stoch. Mod.*, **12**:2 (1996), 285-305.

[148]   Lee, H.-S. and Srinivasan, M.M., Control policies for the $M^X/G/1$ queueing system, *Mgt. Sci.*, **35**:6 (1989), 708-721.

[149]   Lee, H.-S., Optimal control of the $M^X/G/1/K$ queue with multiple server vacations, *Comput. Oper. Res.*, **22**:5 (1995), 543-552.

[150]   Lee, H.W., Lee, S.S., and Chae, K.C., Operating characteristics of $M^X/G/1$ queue with N-policy, *Queueing Sys.*, **15** (1994), 387-399.

[151]   Lee, H.W., Lee, S.S., Park, J.O., and Chae, K.C., Analysis of $M^X/G/1$ queue with N-policy and multiple vacations, *J. Appl. Prob.*, **31** (1994), 467-

496.

[152] Lee, H.W., Lee, S.S., and Chae, K.C., A fixed-size batch service queue with vacations, *J. Appl. Math. Stoch. Anal.*, **9**:2 (1996), 205-219.

[153] Lee, S.S., Lee, H.W., Yoon, S.H., and Chae, K.C., Batch arrival queue with N-policy and single vacation, *Compt. Oper. Res.*, **22**:2 (1995), 173-189.

[154] Medhi, J. and Templeton, J.G.C., A Poisson input queue under N-policy and with general start up time, *Compt. Oper. Res.*, **19**:1 (1992), 35-41.

[155] Muh, D.C.R., A bulk queueing system under N-policy with bilevel service delay discipline and start-up time, *J. Appl. Math. Stoch. Anal.*, **6**, No. 4 (1993), 359-384.

[156] Muh, D.C.R., On a Class of N-Policy Multilevel Control Queueing Systems, Doctoral Thesis, Florida Tech, Melbourne, FL 1994.

[157] Neuts, M.F., Generalizations of the Pollaczek-Khinchine integral equations in the theory of queues, *Adv. Appl. Prob.*, **18** (1986), 952-990.

[158] Park, J.O. and Lee, H.W., Optimal strategy in N-policy system with early set-up, *J. Opnl. Res. Soc.* (to appear).

[159] Ramaswami, R. and Servi, L., The busy period of the $M/G/1$ vacation model with a Bernoulli schedule, *Stoch. Mod.*, **4**:3 (1988), 507-521.

[160] Rubin, I. and Zhang, Z., Switch-on policies for communications and queueing systems, in: *Proceedings of the Third International Conference on Data Communication*, 329-339, Elsevier, North-Holland, Amsterdam, 1988.

[161] Shanthikumar, J.G., Optimal control of an $M/G/1$ priority queue via N-control, *Amer. Journ. Math. Manag. Sci.*, **1** (1981), 191-212.

[162] Takagi, H., Time-dependent process of $M/G/1$ vacation models with exhaustive service, *J. Appl. Prob.*, **29** (1992), 418-429.

[163] Takine, T. and Hagesawa, T., A note on $M/G/1$ vacation systems with waiting time limits, *Adv. Appl. Prob.*, **22** (1990), 513-518.

[164] Talman, A.J.J., A simple proof of the optimality of the best N-policy in the $M/G/1$ queueing control problem with removable server, *Statistica Neerl.*, **32** (1979), 143-150.

[165] Teghem, J., Jr., Optimal control of queues: removable servers [Tutorial paper XIX], *Belgian J. Oper. Res. Stat. Comp. Sci.*, **25**:2-3 (1985), 99-128.

[166] Teghem J., Jr., Control of the service process in a queueing system, *Europ. J. Oper. Res.*, **23** (1986), 141-158.

[167] Welch, P.D., On a generalized $M/G/1$ queueing process in which the first customer of each busy period receives exceptional service, *Oper. Res.*, **12** (1964), 736-752.

[168] Yadin, M. and Naor, P., Queueing systems with removable service station, *Oper. Res. Quart.*, **14** (1963), 393-405.

Hysteresis ((v,N)-policy)

[169] Loris-Teghem, J., Hysteretic control of an $M/G/1$ queueing system with two service time distributions and removable server, in: *Point Processes and Queueing Problems*, Colloquia Mathematica Societatis János Bolyai, Hungary, **24** (1978), 291-305.

[170] Loris-Teghem, J., Imbedded and non-imbedded stationary distributions in a finite capacity queueing system with removable server, *Cah. Centr. d'Etud. Rech. Opér.*, **26**:1-2 (1984), 87-94.

[171] Teghem, J., Jr., Optimal control of queues: removable servers [Tutorial paper XIX], *Belgian J. Oper. Res. Stat. Comp. Sci.*, **25**:2-3 (1985), 99-128.

[172] Teghem J., Jr., Control of the service process in a queueing system, *Europ. J. Oper. Res.*, **23** (1986), 141-158.

B) Workload Dependence

Limited virtual waiting time

[173] Afanas'eva, L.G., Existence of a limit distribution in queueing systems with bounded sojourn time, *Theor. Veroyat. Primenen.*, **10**:3 (1965), 570-578.

[174] Baccelli, F., Boyer, P., and Hebuterne, G., Single-server queues with impatient customers, *Adv. Appl. Prob.*, **16** (1984), 887-905.

[175] Barrer, D.Y., Queueing with impatient customers and ordered service, *Oper. Res.*, **5** (1957), 650-656.

[176] Daley, D.J., Single-server queueing systems with uniformly limited queueing time, *J. Austr. Math. Soc.*, **4** (1964), 489-505.

[177] Daley, D.J., General customer impatience in the queue $GI/G/1$, *J. Appl. Prob.*, **2** (1965), 186-205.

[178] De Kok, A.G. and Tijms, H.C., A queueing system with impatient customers, *J. Appl. Prob.*, **22** (1985), 688-696.

[179] Dijk, N.M., van, Queueing systems with restricted workload: an explicit solution, *J. Appl. Prob.*, **27** (1990), 393-400.

[180] Gavish, B. and Schweitzer, P.J., The Markovian queue with bounded waiting time, *Manag. Sci.*, **23** (1977), 1349-1357.

[181] Hassin, R. and Haviv, M., Equilibrium strategies for queues with impatient customers, *Oper. Res. Let.*, **17** (1995), 41-45.

[182] Haugen, R.B. and Skogan, E., Queueing systems with stochastic time out, *IEEE Trans. Comm.*, **28**:12 (1980), 1984-1989.

[183] Hokstad, P., A single server queue with constant service time and restricted accessibility, *Manag. Sci.*, **25**:2 (1979), 205-208.

[184] Neuts, M.G., The $M/G/1$ queue with a limited number of admissions or a limited admission period during each service time, *Stoch. Mod.*, **1**:3 (1985), 361-391.

[185] Neuts, M.F., Generalizations of the Pollaczek-Khinchine integral equations in the theory of queues, *Adv. Appl. Prob.*, **18** (1986), 952-990.

[186] Takács, L., A single-server queue with limited virtual waiting time, *J. Appl. Prob.*, **11** (1974), 612-617.

[187] Takine, T. and Hagesawa, T., A note on $M/G/1$ vacation systems with waiting time limits, *Adv. Appl. Prob.*, **22** (1990), 513-518.

[188] Yurkevich, O.M., On multi-server queueing systems with limited waiting time, *Eng. Cybern.* (in Russian), **5** (1970), 50-58.

Input rate

[189] Knessl, C., Matkovsky, B., Schuss, Z., and Tier, C., Asymptotic analysis of a state-dependent $M/G/1$ queueing system, *SIAM J. Appl. Math.*, **46**:3 (1986), 483-505.

[190] Knessl, C., Matkovsky, B., Schuss, Z., and Tier, C., Distribution of the maximum buffer content during a busy period for state-dependent $M/G/1$ queues, *Stoch. Mod.*, **3**:2 (1987), 191-226.

[191] Knessl, C., Matkovsky, B., Schuss, Z., and Tier, C., Busy period distribution in state-dependent queues, *Queueing Sys.*, **2** (1987), 285-305.

[192] Knessl, C., Matkovsky, B., Schuss, Z., and Tier, C., A state-dependent

$GI/G/1$ queue, *Europ. J. Appl. Math.*, **5** (1994), 217-241.

Service rate
[193] Callahan, J.R., A queue with waiting time dependent service times, *Nav. Res. Log. Quart.*, **20**:2 (1973), 321-324.
[194] Cohen, J.W., On the optimal switching level for an $M/G/1$ queueing system, *Stoch. Proc. Appl.*, **4** (1976), 297-316.
[195] Knessl, C., Matkovsky, B., Schuss, Z., and Tier, C., Asymptotic analysis of a state-dependent $M/G/1$ queueing system, *SIAM J. Appl. Math.*, **46**:3 (1986), 483-505.
[196] Knessl, C., Matkovsky, B., Schuss, Z., and Tier, C., Distribution of the maximum buffer content during a busy period for state-dependent $M/G/1$ queues, *Stoch. Mod.*, **3**:2 (1987), 191-226.
[197] Knessl, C., Matkovsky, B., Schuss, Z., and Tier, C., Busy period distribution in state-dependent queues, *Queueing Sys.*, **2** (1987), 285-305.
[198] Knessl, C., Matkovsky, B., Schuss, Z., and Tier, C., A state-dependent $GI/G/1$ queue, *Europ. J. Appl. Math.*, **5** (1994), 217-241.
[199] Posner, M., Single-server queues with service time dependent on waiting time, *Oper. Res.*, **21** (1973), 610-616.
[200] Tijms, H.C., On a switch-over policy for controlling the workload in a queueing system with two constant service rates and fixed switch over costs, *Zeitschrift für Oper. Res.*, **21** (1977), 19-32.

Busy period discipline (D-policy)
[201] Balachandran, K.R., Queue length dependent priority queues, *Manag. Sci.*, **17** (1971), 463-471.
[202] Balachandran, K.R., Control policies for a single server system, *Manag. Sci.*, **19** (1973), 1013-1018.
[203] Balachandran, K.R. and Tijms, H., On the D-policy for the $M/G/1$ queue, *Manag. Sci.*, **21** (1975), 1073-1076.
[204] Boxma, O.J., Note on a control problem of Balachandran and Tijms, *Manag. Sci.*, **22** (1976), 916-917.
[205] Dshalalow, J.H., Excess level processes in queueing, in: *Advances in Queueing* (ed. by Dshalalow J.D.), 243-262, CRC Press, Boca Raton, FL 1995.
[206] Li, J. and Niu, S-C., The waiting time distribution for the $GI/G/1$ queue under D-policy, *Prob. Engineer. Inform. Sci.*, **6** (1992), 287-308.
[207] Rubin, I. and Zhang, Z., Switch-on policies for communications and queueing systems, in: *Proceedings of the Third International Conference on Data Communication*, 329-339, Elsevier North-Holland, Amsterdam, 1988.
[208] Teghem, J., Jr., Optimal control of queues: removable servers [Tutorial paper XIX], *Belgian J. Oper. Res. Stat. Comp. Sci.*, **25**:2-3 (1985), 99-128.
[209] Teghem J., Jr., Control of the service process in a queueing system, *Europ. J. Oper. Res.*, **23** (1986), 141-158.

C) Interdependence of Arrival and Service

[210] Borst, S.C., Boxma, O.J., and Combé, M.B., An $M/G/1$ queue with customer collection, *Stoch. Models*, **9**:3 (1993), 341-371.
[211] Chao, X., Monotone effect of dependency between interarrival and service times in a simple queueing system, *Oper. Res. Let.*, **17** (1995), 47-51.

[212] Conolly, B.W. and Choo, Q.H., The waiting time process for a generalized correlated queue with exponential demand and service, *SIAM J. Appl. Math.*, **37** (1979), 263-175.

[213] Hadidi, N., Queues with partial correlation, *SIAM J. Appl. Math.*, **40** (1981), 467-475.

[214] Hadidi, N., Further results on queues with partial correlation, *Oper. Res.*, **33** (1985), 203-209.

[215] Langaris, C., Busy period analysis of a correlated queue with exponential demand and service, *J. Appl. Prob.*, **24** (1987), 476-485.

[216] Mitchell, C.R. and Paulson, A.S., $M/M/1$ queues with interdependent arrival and service processes, *Nav. Res. Log. Quart.*, **26** (1979), 47-56.

[217] Niu, S.-C., On queues with dependent interarrival and service times, *Nav. Res. Log. Quart.*, **28**:3 (1981), 497-501.

II. Multiserver and Closed Systems

Queue length and number of servers state dependent (arrival and service rates)

[218] Antelava, N.I. and Arsenishvili, G.L., A multi-channel queue with a variable arrival rate, *Reports of Acad. Sci. Georg. SSR*, **90**:3 (1978), 529-531.

[219] Bhat, U.N., The queue $GI/M/3$ with service rate depending on the number of busy servers, *Ann. Math. Stat.*, **36** (1965), 1081.

[220] Bhat, U.N., The queue $GI/M/2$ with service rate depending on the number of busy servers, *Ann. Inst. Stat. Math., Tokyo*, **18** (1966), 211-221.

[221] Cheah, J.Y. and Smith, J.M., Generalized $M/G/C/C$ state dependent queueing models and pedestrian traffic flows, *Queueing Sys.*, **15** (1994), 365-386.

[222] Dshalalow, J.H., Many-server feedback queueing systems, *Eng. Cybern.*, **16** (1978), 78-88.

[223] Dshalalow, J.H., *Warteschlangensysteme mit Feedback*, Thesis of Doct. Diss. at Technische Universität Berlin, Dept. Math., Berlin, W. Germany, 1983.

[224] Dshalalow, J.H., On the multiserver queue with finite waiting room and controlled input, *Adv. Appl. Prob.*, **17** (1985), 408-423.

[225] Dshalalow, J.H., On a multi-channel transportation loss system with controlled input and controlled service, *J. Appl. Math. Simul.*, **1**:1 (1987), 41-55.

[226] Dshalalow, J.H., Infinite channel queueing system with controlled input, *Math. Oper. Res.*, **12** (1987), 665-677.

[227] Dshalalow, J.H., Multi-channel queueing systems with infinite waiting room and stochastic control, Techn. Rep. MA-0188, Florida Tech (1988), 1-32.

[228] Dshalalow, J.H., Multichannel queueing systems with infinite waiting room and stochastic control, *J. Appl. Prob.*, **26** (1989), 345-362.

[229] Dshalalow, J.H., On a multi-channel queue with state dependent input flow and interruptions, *J. Appl. Math. Simul.*, **2**:3 (1989), 199-204.

[230] Gupta, S.K., On bulk queues with state dependent parameters, *J. Oper. Res. Soc. Jap.*, **9**:2 (1967), 69-79.

[231] Liu, L., Kashyap, B.R.K., and Templeton, J.G.C., The service $M/M^R/\infty$ with impatient customers, *Queueing Sys.*, **2** (1987), 363-372.

[232] Reynolds, J.F., The stationary solution of a multiserver queueing model

with discouragement, *Oper. Res.*, **16**:1 (1968), 64-71.

Machine repair problems and finite source (closed systems)

[233] Albright, S.C., Optimal maintenance-repair policies for the machine repair problem, *Nav. Res. Log. Quart.*, **27** (1980), 17-27.

[234] Dshalalow, J.H., On a duality principle in processes of servicing machines with double control, *J. Appl. Math. Simul.*, **1**:3 (1988), 245-251.

[235] Dshalalow, J.H., Duality principle in double controlled processes of servicing machines with reserve replacement, *Stoch. Proc. Appl.*, **35** (1990), 193.

[236] Dshalalow, J.H., On single-server closed queue with priorities and state dependent parameters, *Queueing Sys.*, **8** (1991), 237-254.

[237] Goheen, L.C., On the optimal operating policy for the machine repair problem when failure and repair times have Erlang distribution, *Oper. Res.*, **25**:3 (1977), 484-492.

[238] Jaiswal, N.K. and Simha, P.S., Optimal operating policies for the finite-source queueing process, *Oper. Res.*, **20**:3 (1972), 698-707.

[239] Knessl, C., Matkovsky, B., Schuss, Z., and Tier, C., The two repairman problem: a finite source $M/G/1$ queue, *SIAM J. Appl. Math.*, **47** (1987), 367-397.

[240] Sztrik, J., On the finite-source $\vec{G}/M/r$ queue, *European J. Opnl. Res.*, **20** (1985), 261-268.

[241] Wang, K.-H., Profit analysis of the machine repair problem with a single service station subject to breakdowns, *J. Opl. Res. Soc.*, **41**:12 (1990), 1153-1160.

[242] Wang, K.-H., Profit analysis of the $M/M/R$ machine repair problem with spares and server breakdowns, *J. Opl. Res. Soc.*, **45**:5 (1994), 539-548.

[243] Wang, K.-H. and Wu, J.-D., Cost analysis of the $M/M/$R machine repair problem with spares and two modes of failure, *J. Opl. Res. Soc.*, **46** (1995), 783-790.

[244] Wilson, J.G. and Benmerzouga, A., Optimal m-failure policies with random repair time, *Oper. Res. Let.*, **9** (1990), 203-209.

[245] Winston, W., Optimal control of discrete and continuous time maintenance system with variable service rates, *Oper. Res.*, **25**:2 (1977), 259-268.

Idle and vacation periods disciplines (q-, T-, and E-limited disciplines), variable number of channels

[246] Bell, C., Optimal operation of an $M/M/2$ queue with removable servers, *Oper. Res.*, **28**:5 (1980), 1189-1204.

[247] Liu, L., Kashyap, B.R.K., and Templeton, J.G.C., The service $M/M^R/\infty$ with impatient customers, *Queueing Sys.*, **2** (1987), 363-372.

[248] Neuts, M.F. and Nadarajan, R., A multiserver queue with thresholds for the acceptance of customers into service, *Oper. Res.*, **30**:5 (1982), 948-960.

[249] Rhee, H.-K. and Sivazlian, B.D., Distribution of the busy period in a controllable $M/M/2$ queue operating under the triadic $(0,K,N,M)$ policy, *J. Appl. Prob.*, **27**:2 (1990), 425-432.

[250] Romani, Un modelo de la teoria de colas con numero variable de canales, *Trab. Est.*, **8** (1957), 175-189.

[251] Winston, W., Optimality of monotonic policies for multiple server exponential queueing systems with state-dependent arrival rates, *Oper. Res.*, **26**:6 (1978), 1089-1094.

III. General Surveys, Monographs, and Related Theory

[252]   Abolnikov, L. and Dshalalow, J.H., A first passage problem and its applications to the analysis of a class of stochastic models, *J. Appl. Math. Stoch. Analysis*, 5:1 (1992), 83-98.

[253]   Abolnikov, L. and Dukhovny, A., Markov chains with transition delta-matrix: ergodicity conditions, invariant probability measures and applications, *J. Appl. Math. Stoch. Anal.*, 4, No. 4 (1991), 335-355.

[254]   Chaudhry, M.L. and Templeton, J.G.C., *A First Course in Bulk Queues*, J. Wiley, New York, 1983.

[255]   Çinlar, E., *Introduction to Stochastic Processes*, Prentice Hall, Englewood Cliffs, NJ, 1975.

[256]   Cooper, R.B., *Introduction to Queueing Theory*, CEEPress, The George Washington University, Washington, D.C. 1990.

[257]   Cox, D.R., *Renewal Theory*, Methuen, London 1962.

[258]   Crabill, T., Gross, D., and Magazine, M., A classified bibliography of research on optimal design and control of queues, *Oper. Res.*, **25** (1977), 219-232.

[259]   Doshi, B.T., Queueing systems with vacations - a survey, *Queueing Sys.*, **1** (1986), 29-66.

[260]   Doshi, B., Single-server queues with vacations, in: *Stochastic Analysis of Computer and Communication Systems*, (ed.: Takagi, H.), Elsevier Sci. Publishers B.V. (North-Holland) 1990, 217-265.

[261]   Dshalalow, J.H., On modulated random measures, *J. Appl. Math. Stoch. Anal.*, 4:4 (1991), 305-312.

[262]   Dshalalow, J.H., On applications of Little's formula, *J. Appl. Math. Stoch. Anal.*, 6:3 (1993), 271-275.

[263]   Dshalalow, J.H., On termination time processes, in *Studies in Applied Probability*, Edited by J. Galambos and J. Gani, Essays in honor of Lajos Takács, *J. Appl. Prob.*, Special Volume **31A** (1994), 325-336.

[264]   Dshalalow, J.H., First excess levels of vector processes, *J. Appl. Math. Stoch. Anal.*, 7:3 (1994), 457-464.

[265]   Dshalalow, J.H., Excess level processes in queueing, in: *Advances in Queueing* (ed. by Dshalalow J.D.), 243-262, CRC Press, Boca Raton, FL 1995.

[266]   Dshalalow, J.H. (ed.), *Advances in Queueing*, CRC Press, Boca Raton, FL 1995.

[267]   Dshalalow, J.H., Ergodic theorems for modulated stochastic processes, in: *Proc. World Congress of Nonlinear Analysts '92* (ed. V. Lakshmikantham), 1745-1755, Walter de Greyter, Berlin 1996.

[268]   Gnedenko, B.V. and Kovalenko, I.N., *An Introduction to the Theory of Queues*, Nauka, Moscow (1966) 60-83.

[269]   Gnedenko, B.V. and Kovalenko, I.N., *Introduction to Queueing Theory*, 2nd ed., Revised and Supplemented, Birkhäuser, Boston 1989.

[270]   Gross, D. and Harris, C., *Fundamentals of Queueing Theory*, 2nd Edition, Wiley, New York 1985.

[271]   Knessl, C. and Tier, C., Applications of singular perturbation methods in queueing, in: *Advances in Queueing* (ed. by J.H. Dshalalow), 311-336, CRC Press, Boca Raton, FL 1995.

[272]   Kulkarni, V.G. and Liang, H.M., Retrial queues revisited, IN THIS BOOK.

[273] Medhi, J., *Recent Developments in Bulk Queueing Models*, Wiley Eastern Limited, New Delhi, 1984.

[274] Neuts, M.F., *Matrix-Geometric Solutions in Stochastic Models - an Algorithmic Approach*, The John Hopkins University Press, Baltimore, 1981.

[275] Neuts, M.F., *Structured Stochastic Matrices of M/G/1 Type and Their Applications*, Marcel Dekker, New York, 1989.

[276] Takács, L., A telefon-forgalom elméletének néhány valószinüségszámitási kérdéséröl (On some probability problems in teletraffic) (Hung.), *Magyar Tud. Akadémia Mat. Fiz. Oszt. Közl.*, **8**, (1958) 151-210.

[277] Takács, L., *Introduction to the Theory of Queues*, Oxford Univ. Press, New York 1962.

Part II

Telecommunications and Computer Networks

Chapter 5

Queueing analysis of polling models: progress in 1990-1994

Hideaki Takagi

ABSTRACT Polling models generally refer to systems of multiple queues served by nondedicated servers with rules that allocate the servers to the queues. The basic models consist of separate queues with independent Poisson inputs served by a single server in cyclic order. Significant progress in the analysis of these and extended systems, which began over thirty years ago, has enriched queueing theory as well as contributed to the system performance evaluation techniques in such fields as computers, communications, manufacturing, and transportation. This chapter surveys research activities during recent years.

CONTENTS

5.1 INTRODUCTION

A basic *polling model* is a system of multiple queues attended by a single server in cyclic order. The term *polling* originated with the *polling* data link control scheme in which the central computer interrogates each terminal on a multidrop communication line to find whether or not it has data to transmit. The addressed terminal transmits data, and the computer then examines the next terminal. This was an application of a polling model studied in the early 1970s. Situations represented by polling models and their variations appear not only in computers and communications but also in other fields of engineering, such as manufacturing (e.g., a patrolling machine repairman, assembly work on a carousel, and an automated guided vehicle) and transportation systems (e.g., traffic lights at an intersection, mail delivery, and elevators). The ubiquitous application is not surprising because the cyclic allocation of the server (resource) is a natural and simple way for fair arbitration to multiple queues (requesters). Therefore, polling models in various settings have been studied by many researchers since the late 1950s, focusing on the applications to technologies emerging at each period. The reader is referred to a monograph [152] and surveys [42, 55, 95, 127, 140, 151, 154, 155, 158] for the developments in the analysis (an early stage of), optimization, and applications of polling models prior to 1990. Recently, a special issue (Volume 11, No. 1-2, 1992, edited by O.J. Boxma and H. Takagi) of *Queueing Systems* was devoted to polling models, and Volume 35 for stochastic modeling of telecommunication systems (1992, edited by P. Nain and K.W. Ross) of *Annals of Operations Research* devoted a section to polling models. Conti et al. [72] surveyed some latest applications of polling models to the performance evaluation of metropolitan area networks.

The aim of this chapter is to highlight some progress in the analysis and optimization of polling models witnessed during recent years. This is done in Section 5.3; Section 5.2 provides a description of basic polling models and a summary of their analysis results available before 1990. In Section 5.4, possible research topics in the future are suggested. The references include original papers published in journals and conference proceedings as well as those in the stage of preprints and technical reports, dated during 1990-1994, that happen to have come under my notice.

5.2 BASIC MODELS AND SUMMARY OF PROGRESS BEFORE 1990

Let us describe basic polling models and summarize the progress in their analysis before 1990. To do so, we introduce some notation common to all models. The number of queues in the system is denoted by N. Queues are indexed by i, $i = 1, 2, \ldots, N$, in the order of server movement. In continuous-time systems, we assume a Poisson process for the arrival of customers at rate λ_i for queue i. The Laplace-Stieltjes transform (LST) of the probability distribution function (PDF), the mean, and the second moment of the service time of a customer at queue i are denoted by $B_i^*(s)$, b_i, and $b_i^{(2)}$, respectively. The total load offered to the system is then given by

$$\rho = \sum_{i=1}^{N} \rho_i; \quad \rho_i := \lambda_i b_i. \tag{5.1}$$

The LST of the PDF, the mean, and the variance of the time needed by the server to switch from queue i to queue $i+1$ are denoted by $R_i^*(s)$, r_i, and δ_i^2, respectively. The switchover times times are independent of the arrival and service processes. The mean and the variance of the total switchover time are then given by

$$R := \sum_{i=1}^{N} r_i; \quad \Delta^2 := \sum_{i=1}^{N} \delta_i^2. \tag{5.2}$$

The mean polling cycle time, generally independent of the queue index i, is denoted by $E[C]$. The LST of the PDF and the mean of the waiting time W_i of a customer at queue i are denoted by $W_i^*(s)$ and $E[W_i]$, respectively. A system in which the arrival and service processes in all queues and switchover times are independent of queue index i is called *symmetric*. For a symmetric system, we let $\rho_i = \rho/N = \lambda b$, and omit subscript i from other variables.

In the presentation of this section, we restrict ourselves to basic models and leave out references for the sake of conciseness. Readers who are interested in the variations of the basic models and the references are referred to Takagi [152, 154, 155].

5.2.1 Single-buffer systems

A system in which each queue can accommodate at most one customer at a time is called a *single-buffer system*. Those customers that arrive to find the buffer occupied are lost. This system can model a patrolling machine repairman and an interactive transaction processing system in a computer shared by multiple users.

For a symmetric single-buffer system in which the service time is a constant b, closed-form expressions for the mean cycle time $E[C]$ and the mean waiting time $E[W]$ are available:

$$E[C] = R + E[Q]b \tag{5.3}$$

and

$$E[W] = (N-1)b - \frac{1}{\lambda} + \frac{NR}{E[Q]} \tag{5.4}$$

where

$$E[Q] = \frac{N \sum_{n=0}^{N-1} \binom{N-1}{n} \prod_{j=0}^{n} \left[e^{\lambda(R+jb)} - 1 \right]}{1 + \sum_{n=1}^{N} \binom{N}{n} \prod_{j=0}^{n-1} \left[e^{\lambda(R+jb)} - 1 \right]} \tag{5.5}$$

is the mean number of customers served in a polling cycle.

For an asymmetric single-buffer system, we define the *station time* ω_k for the kth visited queue as a time interval consisting of the switchover time from queue $k-1$ to queue k and the possible service time at queue k, where the visit number k is incremented one by one at every visit to all queues. The LST of the PDF for the joint distribution of N successive station times $\omega_{k-N+1}, \ldots, \omega_k$ is defined by

$$\Omega_k^*(s_1, \ldots, s_N) := E\left[\exp\left(-\sum_{j=1}^{N} \omega_{k-N+j} s_j \right) \right], \tag{5.6}$$

which satisfies the equations

$$\Omega_k^*(s_1, \ldots, s_N) = R_{k-1}^*(s_N + \lambda_k)[1 - B_k^*(s_N)]\Omega_{k-1}^*(0, s_1 + \lambda_k, \ldots, s_{N-1} + \lambda_k)$$
$$+ R_{k-1}^*(s_N)B_k^*(s_N)\Omega_{k-1}^*(0, s_1, \ldots, s_{N-1}) \quad k = 1, 2, \ldots. \tag{5.7}$$

The LST of the PDF and the mean of the waiting time at the kth visited queue are then given by

$$W_k^*(s) = \frac{\lambda_k[I_k^*(s) - I_k^*(\lambda_k)]}{(\lambda_k - s)[1 - I_k^*(\lambda_k)]}; \quad E[W_k] = \frac{E[I_k]}{1 - I_k^*(\lambda_k)} - \frac{1}{\lambda_k}, \tag{5.8}$$

where $I_k^*(s)$ is the LST of the PDF of the intervisit time I_k for the kth visited queue, which is given by

$$I_k^*(s) = \Omega_{k-1}^*(0, s, \ldots, s)R_{k-1}^*(s). \tag{5.9}$$

From (5.7), we can get a set of $N(2^{N-1} - 1)$ linear equations for the same number of unknowns that are used to obtain $W_k^*(s)$. This number has been reduced to $2^N - 1$ by appropriately augmenting parameters in $\Omega_k^*(\cdot)$.

Taking the limit $N \to \infty$ with ρ and R fixed at finite values, we obtain a *continuous polling model*. In this limit, the server travels around a circle on which customers arrive uniformly. If the service time is constant, we have

$$I^*(s) = e^{-sR}\left(\frac{1-\rho}{1 - \rho e^{-sR}} \right)^{R/b}; \quad W^*(s) = \frac{1 - I^*(s)}{E[I]s}. \tag{5.10}$$

The following mean values are available for systems with generally distributed service times:

$$E[C] = E[I] = \frac{R}{1-\rho}; \quad E[W] = \frac{R}{2(1-\rho)} + \frac{\rho b^{(2)}}{2b(1-\rho)}. \tag{5.11}$$

5.2.2 Infinite-buffer systems

In a system in which any number of customers can be accommodated without loss at each queue, we usually deal with four basic disciplines with respect to the rule by which the server leaves the queue. In an *exhaustive service* system, the server continues to serve each queue until it empties. Customers that arrive at the queue being served are also served in the current service period. In a *gated service* system, the server serves only those customers that were found in a queue when it visited the queue. Those that arrive at the queue during its service period are set aside to be served in the next round of polling. In a *k-limited service* system, each queue is served until either it empties or k customers are served, whichever occurs first. (The k-limited service system is subdivided into *exhaustively limited service* system and the *gatedly limited service* system, depending on whether the served k messages include those that arrive during the service period.) In a *k-decrementing service* system, each queue is served until either it empties or the queue size decreases to k less than that found at the polling instant.

For a broad class of infinite-buffer systems, including the four basic models mentioned above, the mean cycle time is simply given by

$$E[C] = \frac{R}{1-\rho}. \tag{5.12}$$

For symmetric systems, we have the following results for the mean waiting time:

$$E[W]_{\text{exhaustive}} = \frac{\delta^2}{2r} + \frac{N\lambda b^{(2)} + r(N-\rho)}{2(1-\rho)} \tag{5.13a}$$

$$E[W]_{\text{gated}} = \frac{\delta^2}{2r} + \frac{N\lambda b^{(2)} + r(N+\rho)}{2(1-\rho)} \tag{5.13b}$$

$$E[W]_{1-\text{limited}} = \frac{\delta^2}{2r} + \frac{N\lambda b^{(2)} + r(N+\rho) + N\lambda\delta^2}{2(1-\rho-N\lambda r)} \tag{5.13c}$$

$$E[W]_{1-\text{decrementing}} = \frac{\delta^2}{2r} + \frac{N\lambda b^{(2)}(1-\lambda r) + (r+\lambda\delta^2)(N-\rho)}{2[1-\rho-\lambda r(N-\rho)]}. \tag{5.13d}$$

For asymmetric systems, the exact computation of the mean waiting time at each queue is available for exhaustive and gated service systems. For an exhaustive service system, the mean waiting time $E[W_i]$ at queue i can be expressed in terms of the moments of the intervisit time I_i as

$$E[W_i] = \frac{\lambda_i b_i^{(2)}}{2(1-\rho_i)} + \frac{E[(I_i)^2]}{2E[I_i]}. \tag{5.14}$$

We have

$$E[I_i] = \frac{R(1-\rho_i)}{1-\rho}; \quad E[(I_i)^2] = (E[I_i])^2 + \delta_{i-1}^2 + \frac{1-\rho_i}{\rho_i} \sum_{\substack{j=1 \\ (j \neq i)}}^{N} r_{ij}, \tag{5.15}$$

where r_{ij} is the covariance of the station times ω_i and ω_j for queues i and j, respectively. For the exhaustive service system, the station time ω_i is defined as the time interval between successive instants at which the server leaves queue $i-1$ and queue i. The set $\{r_{ij}; i, j = 1, 2, \ldots, N\}$ is computed by solving the following set of equations:

$$r_{ij} = \frac{\rho_i}{1-\rho_i} \left(\sum_{m=i+1}^{N} r_{jm} + \sum_{m=1}^{j-1} r_{jm} + \sum_{m=j}^{i-1} r_{mj} \right) \quad j < i \qquad (5.16a)$$

$$r_{ij} = \frac{\rho_i}{1-\rho_i} \left(\sum_{m=i+1}^{j-1} r_{jm} + \sum_{m=j}^{N} r_{mj} + \sum_{m=1}^{i-1} r_{mj} \right) \quad j > i \qquad (5.16a)$$

$$r_{ii} = \frac{\delta_{i-1}^2}{(1-\rho_i)^2} + \frac{\lambda_i b_i^{(2)} E[I_i]}{(1-\rho_i)^3} + \frac{\rho_i}{1-\rho_i} \sum_{\substack{j=1 \\ (j \neq i)}}^{N} r_{ij}. \qquad (5.16c)$$

Note that equation (5.14) has a form of *decomposition property* for an M/G/1 queue with *generalized vacations*. Although there are $O(N^2)$ equations in (5.16), an alternative set of $O(N)$ equations to yield the mean waiting times is also available. A similar set of equations holds for a gated service system. An exact analysis of a 1-limited or 1-decrementing service system with two queues is possible by reducing the determination of unknowns to a boundary value problem on a complex plane. A number of approximation techniques have been proposed for other limited and decrementing service systems. While the individual mean waiting times may not be available, their load-weighted sum, called the *pseudoconservation law*, has been derived for systems with the four basic disciplines and a mixture thereof. The extensions of the pseudoconservation law in the recent years are surveyed in Subsection 5.3.3.

Major variations of the above-mentioned models, which needed certain ingenious treatment, were: (5.1) systems with zero switchover times, (5.2) systems with non-cyclic polling order, (5.3) tandem queues served by a single server, and (5.4) discrete-time systems. In particular, non-cyclic polling order included *deterministic*, as in elevators, *probabilistic* (*random* and *Markovian routing*), and *state-dependent* (e.g., a *greedy server*). Many minor variations were also brought in by incorporating batch Poisson arrivals, modified service disciplines, priorities within each queue, and so on.

5.3 PROGRESS IN 1990–1994

In this section, we overview several topics investigated recently.

5.3.1 Single-buffer systems

In a single-buffer system, each queue can accommodate only one customer at a time. A customer that arrives to find the queue nonempty is blocked and lost. One may add a feature of *Bernoulli feedback* so that the customer whose service is completed returns to the queue with probability p_i, where $0 \leq p_i < 1$, without losing analytical tractability. If the system is asymmetric and the service times are generally distributed, an exact analysis of this system requires the solution of a set of $O(2^N)$ linear equations. Such an analysis was given by Takine et al. [167] for a system with nonzero switchover times, and by Takine et al. [165] and Takine and Hasegawa [164] for a system with zero switchover times. Bunday et al. [54] analyze a scanning polling system with constant service and switchover times with Bernoulli feedback; see also Bunday and Sztrik [53]. Chung et al. [68] analyze a single-buffer system with Markovian server routing. Takagi [162] studies discrete-time single buffer systems.

An interesting variation of the single-buffer system is a system consisting of many single-buffer queues and an infinite- or finite-buffer queue. This system was used to model a token ring network with a gateway or a bridge that connects the network to an external network [136, 168].

Another interesting variation is brought in by considering two classes of customers (the priority customer and the ordinary customer) that are treated differently according to the length of the preceding cycle time. While the priority customer is always served when the server arrives, the ordinary customer is served only if the preceding cycle time does not exceed the prescribed limit. This system was introduced as a model of a timed-token protocol that handles both synchronous and asynchronous traffic by Takagi [156], who analyzed a symmetric system with constant service times. The analysis was extended to an asymmetric system with general service times by Nakamura et al. [137]. Woodward [184 (sec. 7.2.2)] treats a symmetric system by the equilibrium point analysis (EPA) technique.

There is some controversy over the characteristics of the mean waiting times at lightly even-loaded queues located between two heavily loaded queues. While Takine et al. [166, 169] and Jung [102] claim that the mean waiting times at the lightly loaded queues increase in the direction of polling, Mukherjee et al. [135] present numerical results with opposite behavior, even though they consider the same model.

5.3.2 Continuous polling models

The service time was assumed to be a constant in the previous analysis of a continuous polling model on a circle. Bisdikian and Merakos [30] study the output process of this model. Coffman and Gilbert [69], who derived equation (5.10), pointed out a difficulty in obtaining the waiting time distribution when the service time has general distribution. Recently, however, Kroese and Schmidt [113] successfully analyzed a continuous polling model on a circle with general service times by means of point processes and regenerative processes in combination with stochastic integration theory.

Studies of other continuous polling models have also been tried, such as polling on a graph [13, 70, 114] and noncyclic polling in two and more dimensional space [19, 27].

5.3.3 Extension of pseudoconservation laws

An equation expressing a weighted sum of the mean waiting times $\{E[W_i]; 1 \leq i \leq N\}$, called the *pseudoconservation law*, was derived by Boxma and Groenendijk [43] for a continuous-time system in which each queue has a Poisson arrival process and any one of exhaustive, gated, 1-limited, and 1-decrementing service disciplines by unifying the results obtained before by others for similar systems with a single service discipline. It was later extended to systems with compound Poisson arrival processes by Boxma and Groenendijk [44] and Boxma [41]. An error in this extension for a queue with decrementing service discipline was corrected by

Chiarawongse and Srinivasan [65], who assumed that all the customers contained in each arriving batch belong to the same class.

Combining these results and correcting the previous errors, we can get the most general result as follows. We assume that the system has a Poisson arrival process at rate λ such that each arrival contains G_i customers for queue i, $1 \leq i \leq N$, simultaneously. The pseudoconservation law is then given by

$$\sum_{i \in E,G} \rho_i E[W_i] + \sum_{i \in L} \rho_i \left(1 - \frac{\lambda_i R}{1 - \rho}\right) E[W_i] + \sum_{i \in D} \rho_i \left[1 - \frac{\lambda_i(1 - \rho_i)R}{1 - \rho}\right] E[W_i]$$

$$= \frac{\lambda \sum_{i=1}^{N} \left[\rho g_i b_i^{(2)} + g_i^{(2)} b_i^2\right] + 2\lambda \sum_{i=2}^{N} b_i \sum_{j=1}^{i-1} g_{ij} b_j}{2(1 - \rho)}$$

$$+ \rho \frac{\Delta^2}{2R} + \frac{R\left(\rho - \sum_{i=1}^{N} \rho_i^2\right)}{2(1 - \rho)} + \frac{R \sum_{i \in G,L} \rho_i^2}{1 - \rho} + \frac{\lambda R \sum_{i \in L} g_i^{(2)} b_i}{2(1 - \rho)} \qquad (5.17)$$

$$+ \frac{\lambda R \sum_{i \in D}\left[(1 - 2\rho_i)g_i^{(2)} b_i - (\lambda g_i)^2 g_i b_i b_i^{(2)}\right]}{2(1 - \rho)},$$

where $g_i = E[G_i]$, $g_{ij} = E[G_i G_j]$ for $i \neq j$, $g_i^{(2)} = E[G_i(G_i - 1)]$, $\lambda_i = \lambda g_i$, $\rho_i = \lambda g_i b_i$, and E, G, L, and D stand for the index sets of queues with exhaustive, gated, limited, and decrementing service disciplines, respectively. Some special cases are as follows. If each arrival contains only customers for a single queue, we have $g_{ij} = 0$ for $i \neq j$, $1 \leq i, j \leq N$. Furthermore, if each arrival contains a single customer, we have $g_i = 1$ and $g_i^{(2)} = 0$ for $1 \leq i \leq N$. If the numbers of customers contained in each arrival are independent for different queues, we have $g_{ij} = g_i g_j$ for $i \neq j$, $1 \leq i, j \leq N$. If all the switchover times are zero, equation (5.17) reduces to the *conservation law* for multiclass $M^X/G/1$ systems given in Takagi [159 (sec. 3.5)].

For discrete-time systems, the pseudoconservation law, first derived by Boxma and Groenendijk [44], must be corrected in two ways. First, Bisdikian [29] points out an error in the terms not related to switchover times. Second, the term for queues with decrementing service discipline is erroneous, as is shown by Chiarawongse and Srinivasan [65]. Consequently, if the numbers Λ_i of customers that arrive at queue i in each slot are independent slot by slot, the corrected pseudoconservation law is given by

$$\sum_{i \in E,G} \rho_i E[W_i] + \sum_{i \in L} \rho_i \left(1 - \frac{\lambda_i R}{1 - \rho}\right) E[W_i] + \sum_{i \in D} \rho_i \left[1 - \frac{\lambda_i(1 - \rho_i)R}{1 - \rho}\right] E[W_i]$$

$$= \frac{\sum_{i=1}^{N} \left[\rho \lambda_i b_i^{(2)} + \lambda_i^{(2)} b_i^2\right] + 2 \sum_{i=2}^{N} b_i \sum_{j=1}^{i-1} \lambda_{ij} b_j - \rho^2}{2(1 - \rho)}$$

$$+ \rho \frac{\Delta^2}{2R} + \frac{R\left(\rho - \sum_{i=1}^{N} \rho_i^2\right)}{2(1 - \rho)} + \frac{R \sum_{i \in G,L} \rho_i^2}{1 - \rho} + \frac{R \sum_{i \in L} \lambda_i^{(2)} b_i}{2(1 - \rho)} \qquad (5.18)$$

$$+ \frac{R\sum_{i \in D}\left[(1 - 2\rho_i)\lambda_i^{(2)}b_i - \rho_i\lambda_i^2 b_i^{(2)}\right]}{2(1 - \rho)},$$

where $\lambda_i = E[\Lambda_i]$, $\lambda_{ij} = E[\Lambda_i\Lambda_j]$ for $i \neq j$, $\lambda_i^{(2)} = E[\Lambda_i(\Lambda_i - 1)]$, and $\rho_i = \lambda_i b_i$.

We note that Chang and Sandhu [60, 62] show a pseudoconservation law for a continuous-time system in which each queue has a Poisson arrival process and one of exhaustive-limited, gate-limited, and general-decrementing service disciplines. However, their result contains undetermined constants. A similar pseudoconservation law for a system with such service disciplines and compound Poisson arrival processes can also be derived [24, 187].

Pseudoconservation laws have also been extended to a continuous-time system, where each queue has Poisson arrival processes of multiple priority classes of customers and one of exhaustive, gated, or 1-limited service disciplines [84, 144], and to a similar discrete-time system [163].

5.3.4 Time-limited service systems

Limited service systems have gained much attention recently in view of the application to the performance modeling of timed-token passing protocols employed in the medium access control (MAC) layer of local and metropolitan area networks standards such as IEEE 802.4 token passing bus and ANSI/IEEE Fiber Distributed Data Interface (FDDI). In timed-token protocols, the time during which each station on the network can continue to transmit packets is limited according to a certain rule that depends on the congestion of the network. Polling models with limited service can be used to approximate the operation of timed-token protocols for which exact performance analysis seems hopeless. We note, however, that the exact result for the mean waiting time in limited service polling systems is only available for a symmetric system with 1-limited service as given in equation (5.13c).

Several approximate formulas for the mean waiting time in limited service polling systems were published before 1990; see Takagi [155]. Chang and Sandhu [61] propose an approximation technique for the mean waiting time in a polling system in which the number of customers served at queue i per visit of the server is exhaustively limited by K_i, $1 \leq i \leq N$. They use the known result for a single-queue system with server vacations

$$E[W_i] = \frac{\lambda_i b_i^{(2)}}{2(1 - \rho_i)} + \frac{E[V_i^2]}{2E[V_i]}, \tag{5.19}$$

which holds if the vacation times, V_i for queue i, (corresponding to the intervisit times in the polling system) satisfy certain conditions [74, 88]. While $E[V_i] = R(1 - \rho_i)/(1 - \rho)$ is known as in equation (5.15), $E[V_i^2]$ is estimated from the assumption that V_i is exponentially distributed [62] or it has gamma distribution in which the two parameters are determined by the approximate analysis of cycle times [63].

Leung [124] takes a different approach to consider a system in which the parameter K_i is an independent random variable (*probabilistically-limited service*). He uses a numerical technique based on the discrete Fourier transform to determine the queue size distribution (truncated with sufficient accuracy) at service completion times in each queue, and calculates the mean waiting time. LaMaire [116] studies a single-queue system with server vacations in which K_i is a random variable and the vacation time can be correlated with the value of K_i that was used for the preceding service period. When using this result in the framework of a polling model, he uses simulation to determine the distribution of K_i and a dependence on the vacation time.

In the time-token protocol, the maximum length of a service period is limited not by the number of served customers but by the sum of the service times elapsed since the start of the service period as well as during the preceding cycle time. Therefore, for the approximation of the timed-token protocol by a limited service system, the time limit M_i is somehow connected with parameter K_i. Chang and Sandhu [63] simply approximate

$$K_i = \frac{M_i + \widehat{B}_i(M_i)}{b_i}, \tag{5.20}$$

where $\widehat{B}_i(M_i)$ denotes the mean residual service time of a customer when M_i expires. This form assumes that the service of a customer is continued until its completion even if M_i expires in the middle of a service time; this option is called *asynchronous overrun* in FDDI. Karvelas and Leon-Garcia [104, 105] show a new approach to the delay analysis of symmetric timed-token networks by which the formula (5.13a) for the exhaustive service system is used with appropriately inflated service times. De Souze e Silva et al. [78] propose a numerical technique for an asymmetric time-limited service system with exponentially distributed service times and constant switchover times. The stability and delay performance of timed-token protocols are also studied by Altman [10, 11], Altman and Kofman [17], Altman and Liu [20], Conti et al. [71, 73], Genter and Vastola [92], Rubin and Wu [141], Tangemann [170-174], and Tangemann and Sauer [175, 176]. A survey of these studies is given by Conti et al. [72].

Although some interesting techniques have recently been proposed to study a single-queue, time-limited service system with server vacations, their extension to polling systems remains much unexplored.

5.3.5 Reservation schemes

In systems with exhaustive, gated, limited, and decrementing service disciplines, the number of messages served in a queue is determined only when and after the server visits the queue. Recently, a few service disciplines, according to which this number is determined prior to the time of visit, have been analyzed. Let us call such a discipline a *reservation scheme*.

A simple reservation scheme, described by Bertsekas and Gallager [25 (Subsec. 3.5.2)], is that when the server visits queue i it serves only those messages that were there when the server left queue $i-1$. Thus, the switchover time from queue $i-1$ to queue i is supposed to be used for scheduling the services at queue i. The analysis of a polling system with this discipline is similar to that of a polling system with gated service discipline.

Boxma et al. [49] study two reservation schemes: the globally-gated discipline and the *cyclic reservation multiple access* (CRMA). In a polling system with *globally-gated discipline*, there is a special queue that we designate as queue 1. When the server visits queue i, only those messages are served that were present in queue i when the server visited queue 1 most recently. Customers who arrive at queue i afterwards (even before the server reaches queue i) will be served during the next cycle. Explicit expressions for the mean waiting time and the pseudoconservation

law for a polling system with globally-gated discipline and Poisson arrival processes are available as follows:

$$E[W_i] = \frac{1}{\rho(1+\rho)} \left(1 + 2\sum_{j=1}^{i-1} \rho_j + \rho_i \right)$$

$$\cdot \left[\frac{\rho}{2(1-\rho)} \sum_{j=1}^{N} \lambda_j b_j^{(2)} + \rho\frac{\Delta^2}{2R} + \frac{\rho(1+\rho)}{2(1-\rho)}R \right] + \sum_{j=1}^{i-1} r_j \qquad (5.21a)$$

$$\sum_{i=1}^{N} \rho_i E[W_i] = \frac{\rho}{2(1-\rho)} \sum_{j=1}^{N} \lambda_j b_j^{(2)} + \rho\frac{\Delta^2}{2R} + \frac{\rho(1+\rho)}{2(1-\rho)}R + \sum_{i=2}^{N} \rho_i \sum_{j=1}^{i-1} r_j. \qquad (5.21b)$$

This model is extended to a system with server interruptions by Boxma et al. [50].

Altman et al. [16] obtain a peculiar result when the globally-gated discipline is
applied to a scanning polling system. In this system, the server first visits the
queues in one direction, that is, in the order $1, 2, \ldots, N$, serving only those customers that were present when the service to queue 1 was started. Then the server re-
verses its orientation and visits the queues in the opposite direction, namely in the
order $N, N-1, \ldots, 1$, serving only those customers that were present when the service to queue N was started upon the orientation reversal, and so on. The mean
waiting time for queue i in this system is given by

$$E[W_i] = \frac{\rho}{2(1-\rho)} \sum_{j=1}^{N} \lambda_j b_j^{(2)} + \frac{\Delta^2}{2R} + \frac{R}{1-\rho}, \qquad (5.22)$$

which is independent of i. Thus the mean waiting times are identical for all
queues exhibiting the fairness. Higher-order moments of the waiting times differ
generally.

The CRMA was originally proposed for controlling the access to high-speed
local and metropolitan area networks based on a slotted unidirectional folded bus
architecture. For modeling the CRMA, Boxma et al. [49] assume that a single
server controls the system by sending out "collectors" periodically for a cyclic tour
among the queues to collect their current service requests. After a completion of a
tour by a collector, the requests are served (once the services from previous tours
have been completed) in the order collected. The mean waiting time for each queue
can be calculated.

Khamisy et al. [109] generalize the globally-gated discipline by assuming that
there are n $(1 \leq n \leq N)$ special queues, each of which has a disjoint set of subordinate queues (these are logical relations independent of the queues indices). When
the server visits a subordinate queue, only those messages are served that were
present there when the server visited its master queue most recently. This scheme
is called the *synchronized gated discipline*. An extreme case in which $n = 1$ is the
globally-gated discipline, while another case in which $n = N$ is the ordinary gated
service discipline. The mean waiting times as well as the pseudoconservation law
are derived.

Lee and Sengupta [119, 120] analyze a polling system with limited service and
reservations to model the *pipeline polling protocol* for satellite communications.
Let K be a parameter, and let $k_{i,j}$ be the number of messages in queue i when the
server leaves queue i in the jth cycle. Then, the maximum number of messages
served in queue i in the $j + 1$th cycle is given by

$$\max[1, \min(K, k_{i,j})]. \tag{5.23}$$

At each cycle, at least one message and at most K messages can be served for each queue. An exact analysis for a single-queue system with server vacations and an approximate analysis for a polling system are provided. The stability of this protocol is discussed by Chang [59].

5.3.6 Extended analysis

Various new developments in the analysis of individual systems, as well as generic methodologies, have continued to appear. Let us mention a few of them. Systems of two queues, one with exhaustive service and the other with limited or decrementing service, are studied by Ibe [99], Katayama [107], Lee [118], Ozawa [138], and Weststrate and van der Mei [183]. Kubat and Servi [115] derive the optimal server scheduling for systems with two queues and zero service times. Levy and Sidi [128] analyze exhaustive and gated service systems with compound Poisson arrivals. Approximation techniques for the mean waiting times in limited and decrementing service systems, based on pseudoconservation laws, are proposed by Balsamo [25], Casares-Giner [56], and Chang and Hwang [57, 58]. Simulation is used to compare the waiting time distributions for systems with different service disciplines by Charzinski et al. [64]. Systems in which a setup time (also called a switch-in time) is needed before starting service at each queue are analyzed by Altman et al. [12], Altman and Yechiali [22], and Gupta and Srinivasan [96]. Systems with zero switchover times, in which a setup time is needed before starting service when a message arrives in an empty system, are treated by Fuhrmann and Moon [89] and Takagi [153]. Stochastic ordering of polling systems can be found in Altman et al. [18], Boxma and Kelbert [45], Levy et al. [129], and Liu et al. [131]. Resing [139] characterizes polling systems with multitype branching processes; he notes that he was unaware of an early work by Fuhrmann [86]. An idea of ancestral line is also employed for numerical computation by Konheim and Levy [111] and Konheim et al. [112]. Tsai and Rubin [180] analyze the mean waiting time in a symmetric polling system with two priority classes of customers. Zhang and Acampora [186] propose and analyze a modified polling scheme for the indoor wireless local area networks. Harrison and Coury [98] also analyze a discrete-time polling model for a wireless multi-channel cellular network, in which the server serves several queues in one slot.

5.3.7 Stability

Stability issues in polling models are not straightforward, as they should be analyzed in the framework of multidimensional Markov chains; see Szpankowski [151] for stability criteria in multidimensional stochastic processes. Recently, Georgiadis and Szpankowski [93, 94] prove the following necessary and sufficient condition for cyclic polling systems with k_i-limited service discipline at queue i:

$$\rho < 1 \text{ and } \lambda_i < \frac{k_i(1-\rho)}{R} \text{ for all } i \in \{1, 2, \ldots, N\}. \tag{5.24}$$

While this condition was "known" for years, it lacked a formal proof. The proof is based on stochastic dominance techniques and application of Loynes' stability criteria for an isolated queue. Altman et al. [18] also provide a sufficient condition under which the vector of queue sizes is ergodic. In addition, they show that the queue sizes, station times, intervisit times, and cycle times are stochastically increasing in arrival rates, in service times, and in switchover times. Furthermore, the mean cycle time, the mean intervisit time, and the mean station time are invariant under general service disciplines and general arrival and service processes. Altman and Spieksma [21] give necessary and sufficient conditions for the existence of all the moments of station times and show the geometric convergence. Sufficient conditions for the central limit theorem and the law of iterated logarithm for the moments of station times are also given. Fricker and Jaïbi [85] and Massoulié [132] also deal with stability.

5.3.8 Numerical algorithms

The exact mean waiting times in an asymmetric exhaustive service system can be computed by solving a set of $O(N^2)$ equations in equations (5.16a)-(5.16c), which is due to Ferguson and Aminetzah [83], or another set of $O(N)$ equations derived by Sarkar and Zangwill [142] (implementation of the Sarkar-Zangwill algorithm is given in Garner [91] and Tayur and Sarkar [177]). When using the standard method for solving a set of linear simultaneous equations even the latter approach requires $O(N^3)$ operations. The situation is the same for a gated service system. Later, Srinivasan [14] and Srinivasan et al. [149] propose an algorithm that requires $O(N^2)$ operations to compute the mean waiting time at a given queue. Iterative algorithms that require $O(N \log_\rho \epsilon)$ operations, where ϵ is the required accuracy, are proposed by Konheim and Levy [111], Konheim et al. [112], and Levy [125]. Federgruen and Katalan [82] show an efficient numerical method to compute the steady state queue size distributions for exhaustive and gated service systems.

An exact computation of the mean waiting time in an asymmetric limited service system has been unavailable. Thus, Blanc [31-35] and Blanc and van der Mei [36, 37] develop the *power-series algorithm*, which can be applied to systems with Coxian distributed service and switchover times. Leung [122-124] uses discrete Fourier transforms to approximate unknown functions that appear in the analysis of limited service systems.

5.3.9 Networks of queues

An open network of queues served by a single server in cyclic order is considered by Sidi and Levy [145] and Sidi et al. [146]. This is a generalization of the basic polling system for the case in which customers, after their service completion at queue i, move to queue j with probability p_{ij}, or leave the network with probability $1 - \sum_{j=1}^{N} p_{ij}$. Levy and Sidi [127] mention the modeling of selective-repeat ARQ protocol and distributed algorithm on token ring networks and robotics systems with staged jobs as potential applications of this network. Not only the queue size and the waiting time at each queue, but also the expected delay of customers who follow a specific route in the network, can be calculated for gated and exhaustive service systems. Altman and Yechaili [23] analyze a closed network of queues served by a single server in cyclic order. Katayama [106] analyzes a tandem of two queues (with Bernoulli feedback at the first queue) with customers of multiple classes served cyclically by a single server with zero switchover times. Katayama [108] also considers a tandem of two queues, with gated service and vacations for the first queue and exhaustive service for the second queue.

Walrand [182] describes dynamic optimization problems to determine, at each service completion time, which queue should be served next by a single server to minimize the given cost; for example,

$$E\left[\sum_{i=1}^{N} c_i L_i\right], \tag{5.25}$$

where L_i is the number of customers present at queue i and c_i is the cost of a customer at queue i. For networks without external arrivals, an optimization problem of this kind can be viewed as a *multi-armed bandit problem* because the system evolves only at those queues which are attended by the server.

5.3.10 Finite systems

The state space of a polling system is finite if the number of customers involved in the system is finite or if the capacity of the system is finite. Let us call the former a *finite-population system*, and call the latter a *finite-capacity system*.

A finite-population system is a closed system in which each customer alternates a period of being in the queue (including a period of being served) and a period of being in the source. From the analogy with the finite-population systems in priority queues [101 (sec. III.1), 161 (sec. 4.7)], we may consider two models of the polling system with a finite population that differ with respect to the population constraint: a multiple finite-source model and a single finite-source model. In the *multiple finite-source model*, the source of customers for each queue is associated only with that queue; a customer whose service has been completed always returns to its original source. The single-buffer system discussed in Subsections 5.2.1 and 5.3.1 is a special case of the multiple finite-source model in which the population size for each queue is one. In the *single finite-source model*, there is only a single source of customers, each of which selects one of the queues at random when it leaves the source.

Multiple and single finite-source models may be associated with the *flow control* and *congestion avoidance* mechanisms in computer communications networks. Namely, a multiple finite-source model, in which the population size is fixed for each queue, corresponds to the *window flow control* [77 (sec. 4.2)]. A single finite-source model, in which the total population size is a single constraint, corresponds to the *isarithmic congestion avoidance* scheme [77 (sec. 4.3)].

A multiple finite-source model is analyzed by Choi and Trivedi [66] and Ibe and Trivedi [100], using a technique called *stochastic Petri nets* under the assumption (which is essential in this technique) that both service and switchover times are exponentially distributed. A similar model with generally distributed service and switchover times is analyzed by Takagi [160] by using the analysis of an M/G/1//N system with server vacations.

A finite-capacity polling system with exhaustive, gated, and 1-limited service is analyzed by Takagi [157] by considering the Markov chain for the set of queue sizes embedded at the times when the server visits and leaves each queue. The server vacation time for each queue is calculated by conditioning on this Markov chain. The queue size at each queue is then obtained by utilizing the existing analysis of an M/G/1/K queue with server vacations. The independence assumption implicitly involved in Takagi [157] for the vacation time and the busy period is dropped by Kofman [110]. Jung [102] and Jung and Un [103] provide a different technique, called *virtual buffering*, to analyze a finite-capacity polling system with exhaustive service. They introduce a virtual buffer of infinite capacity for each queue when the server is on vacation for that queue. When the server comes to the queue, the messages that exceed the capacity of the real buffer are removed from the queue. Based on this model, a Markov chain for the set of queue sizes at the instant of time when the server visits each queue is studied, and the duration of a vacation is obtained. Tran-Gia [179] gives an approximate analysis of a discrete-time, 1-limited service system with a finite capacity and renewal arrival process. Lang and Bosch [117] present a similar technique for the analysis of a continuous-time, k-limited service system with a finite capacity and Poisson arrival process.

5.3.11 Multiserver systems

Polling systems with multiple servers are studied in a series of papers by Ajmone Marsan et al. [1-9] using a technique of *generalized stochastic Petri nets* (GSPNs). In this technique, once a model is described in terms of GSPN primitives, all possible marking states are mapped into a state space of a Markov chain, whose steady state distribution is computed numerically. The performance measures are then calculated. While several earlier studies of multiserver systems were all approximate, the method of Ajmone Marsan et al. can provide exact numerical values for customers' waiting times. The restrictions include exponential distributions for customers' interarrival times, service times, and the server switchover times and finite-capacity queues so as to obtain finite state spaces and a limited size of models (in terms of the numbers of queues, servers, and buffers) that can be solved with acceptable time and space complexity. A unique parameter in multiserver systems is the maximum number of servers that can attend the same queue simultaneously. Some servers may pass others at some queues (just like an empty bus on a street passes another at a crowded bus stop). These features that are difficult to handle analytically can be easily incorporated in the GSPN model.

The numerical results presented by Ajmone Marsan et al. [1-9] show that, when the total service rate is fixed, the mean waiting times can be greatly reduced by using multiple servers, particularly when the switchover times are long and the system is loaded (this situation corresponds to the large distances and high data rates in metropolitan area networks). For some values of the system parameters, the same can also happen for the customer response time (waiting time plus service time), but to a lesser degree, due to the fact that the increase in the number of servers at a fixed service rate implies an increase in the mean service time. An interesting behavior of an asymmetric system with two servers observed by Ajmone Marsan et al. [6] is that a heavily loaded queue monopolizes one server while lightly loaded queues share the other server.

Multiserver systems are also studied by Borst [38] and van der Mei and Borst [181], using a different approach.

5.3.12 Effects of switchover times

Traditionally, the analysis of systems with zero switchover times has been independent from that of systems with nonzero switchover times [74]. A reason for the technical difficulty in relating the two is as follows. In a system with zero switchover times, the server executes an infinite number of cycles in any finite period during which the system is empty. This implies that the mean cycle time is zero. On the other hand, the analysis of systems with nonzero switchover times is based on the evaluation of variables averaged over a cycle time. Recently, however, a few works have appeared to relate the analysis for the two systems.

As a special case, if all the switchover times are constant, we can derive from equations (5.16a)-(5.16c) another set of equations with respect to $\hat{r}_{ij}: = r_{ij}/R$ which can be used to calculate the ratio $E[(I_i)^2]/E[I_i]$ for $E[W_i]$ in equation (5.14). We can then obtain a set of equations for a system with zero switchover times by letting $R{\rightarrow}0$. This idea is given by Choudhury [67] and Levy and Kleinrock [126]. More recently, Cooper et al. [75] found the following results. For an exhaustive service system,

$$E[W_i] = E[W_i^0] + \frac{R(1 - \rho_i)}{2(1 - \rho)}, \tag{5.26a}$$

where W_i^0 is the waiting time in the corresponding system with zero switchover times and modified service-time variances

$$\widetilde{b}_i^{(2)} = b_i^{(2)} + \frac{\delta_{i-1}^2}{R}\frac{1 - \rho}{\lambda_i}. \tag{5.26b}$$

For a gated service system,

$$E[W_i] = E[W_i^0] + \frac{R(1 + \rho_i)}{2(1 - \rho)}, \tag{5.27a}$$

where W_i^0 is the waiting time in the corresponding system with zero switchover times and modified service-time variances

$$\widetilde{b}_i^{(2)} = b_i^{(2)} + \frac{\delta_i^2}{R}\frac{1 - \rho}{\lambda_i}. \tag{5.27b}$$

Equations (5.26a) and (5.27a) are also derived by Fuhrmann [87] in a special case in which the switchover times are constant. We note that the modification of the service-time variances as in equations (5.26b) and (5.27b) was used by Ferguson and Aminetzah [83] when they derived the pseudoconservation laws. The results of Fuhrmann [87] and Cooper et al. [75] are extended to transforms (and therefore to higher moments) of the waiting times by Srinivasan et al. [150].

5.3.13 Optimal server routing/action

When we have control over the server movement, we can optimize its routing (a sequence of choices of the queues to serve) and its action (service discipline) in each queue at given decision epochs in order to minimize a given objective function. Such *stochastic optimization problems* may be classified into static, semi-dynamic, and dynamic ones.

In the *static optimization*, decisions are made once and for all prior to the operation on the system by selecting controllable parameters. For the static optimization of service disciplines when the polling order is given (not necessarily cyclic), Levy et al. [129] prove that the exhaustive service discipline dominates all other disciplines (under the assumption that the server does not wait idling at a queue) with respect to the total amount of unfinished work in the system at any time. Boxma et al. [46-48] (see also Boxma [42]) investigate the statically optimized polling order table (ratio of occurrences of all queues in the table, size of the table, and the order within the table) that minimizes

$$\sum_{i=1}^{N} \rho_i E[W_i] \tag{5.28}$$

for an exhaustive or gated service system. Interestingly, they show that the *golden ratio policy* provides a good heuristic for determining a good visit order in the table. An optimization that specifies not only the polling order but also the starting time of each polling is considered by Borst et al. [40].

In the *semi-dynamic optimization*, the decision for the polling order is made at the beginning of each polling cycle based on the knowledge of the system process until that time. Browne and Yechiali [51] formulate a *Markov decision process* to consider the minimization of the cycle time given the set of queue sizes $\{L_1,...,L_N\}$ at the beginning of a polling cycle. According to them, the expected cycle time is minimal if the server visits queues in the order of increasing values of L_i/λ_i (this rule is independent of the service times) in an exhaustive and gated service system with zero switchover times. This result is extended by Fabian and Levy [80, 81] for the cycle time maximization. A similar optimization is considered for single-buffer systems by Browne and Yechiali [52]. Yechiali [185] summarizes the results including those for systems with binomial-gated and Bernoulli-gated disciplines.

Liu et al. [131] address *dynamic optimization* problems that determine the server's routing and action so as to stochastically minimize the unfinished work and the number of customers in the system at all times. They establish the following results: When the server is at a nonempty queue, it should neither idle nor switch until that queue is empty (exhaustive and greedy discipline). For symmetric systems, (*i*) when the system has become empty, the server should stay idling at the last visited queue (patient policy); and (*ii*) when the server has emptied a queue, it should subsequently visit the queue with the largest queue size (stochastically largest queue policy). (The claim (*ii*) is also confirmed by Miyoshi et al. [134].) The cyclic routing is optimal when the only available information is the previous decision. (Prior to this work, Liu and Nain [130] consider a particular polling system arising from the videotext system, and identify optimal scheduling policies depending on the amount of information available to the controller. Towsley et al. [178] prove that the largest queue policy minimizes the number of customers lost when the queues have finite and equal buffer capacities.) A similar dynamic optimization is studied by Ajmone Marsan et al. [9] for systems with multiple servers. One of their findings is that the cyclic polling order provides a very good performance, often superior to some idealistic schemes, in particular when systems with a large number of queues and a small number of servers are heavily loaded (high arrival rates and long switchover times). This is because the cyclic polling follows the minimum distance Hamiltonian tour of queues. For single-buffer systems, Harel and Stulman [97] find the optimal parameter d of a *horizon server* such that it acts like a greedy server within d nearest queues, and acts like a cyclic server when all the d nearest queues are empty.

An optimal buffer capacity system is studied by Birman et al. [28], Gail et al. [90], and Sasaki [143]. It relates to *deterministic optimization problems*, in which the gradual (i.e., non-instantaneous) arrival process is given and the server routing is *longest queue first*, *least time to reach bound*, or *partially gated* (only those messages with completed arrivals by polling instants are served).

5.4 OPEN PROBLEMS AND FUTURE RESEARCH DIRECTIONS

A strong thrust for the progress in mathematical theory comes from the need for application. One of the recent application fields that motivated the study of polling models was media access protocols in local and metropolitan area networks. Another incentive is the establishment of mathematical foundations, such as stability, stochastic decomposition, and boundary value problems. Also, there can be innumerable results from the "Cartesian product" of solved models (combinations of variations). Certain developments made in the recent years are overviewed in Section 5.3. It is likely that many more models will be explored along the existing lines.

Some new directions, though challenging, may be suggested. First, we can investigate real systems and build models with new features. For example, the server in switchover may slow down if there are customers to serve in the next queue, as is usual in transportation and other systems accompanying mechanical movements. If there are no customers to serve in the whole system, the server may go to a home base like elevators or stop at the last served queue (called a *homing server*, a *dormant server*, or a *patient server*, as opposed to a *roving server* in the conventional model). Such models are studied recently by Ajmone Marsan et al. [6], Blanc and van der Mei [37], Borst [39], Eisenberg [79], and Srinivasan and Gupta [148]. Note that the implementation of the homing or patient server requires the information about the whole system, which may not be easy to obtain for certain applications. Polling models with non-Poisson arrival processes should also be given attention in view of vigorous modeling efforts for broadband ISDN systems. They include modulated Markov Poisson processes, linear input models [76, 121], gradual input processes [28, 90, 143], and "Cruz-type" processes [14, 15]. Embedding polling models in a network of queues or in other total system models is also noteworthy. Optimization of the server routing/action is much unexplored because of its analytical intractability. As an alternative, Matsumoto [133] proposes a method of controlling the server by a *neural network* in an asymmetric system so that the mean waiting time is minimized.

ACKNOWLEDGEMENTS

This work is supported in part by the Telecommunications Advancement Foundation. The author wants to thank Dr. E. Altman, Professors O.J. Boxma, R.B. Cooper, and M.M. Srinivasan, and Mr. Gary Russell (an editorial assistant to Professor Jewgeni H. Dshahalow) for their valuable comments on the draft version of this text.

BIBLIOGRAPHY

[1]   Ajmone Marsan, M., de Moraes, L.F., Donatelli, S. and Neri, F., Analysis of symmetric nonexhaustive polling with multiple servers, *IEEE INFOCOM '90*, San Francisco, California, June 3-7 (1990), 284-295.

[2]   Ajmone Marsan, M., de Moraes, L.F., Donatelli, S. and Neri, F., Cycles and waiting times in symmetric exhaustive and gated multiserver multiqueue systems, *IEEE INFOCOM '92*, Florence, Italy, May 6-8 (1992), 2315-2324.

[3]   Ajmone Marsan, M., Donatelli, S. and Neri, F., GSPN models of multiserv-
        er multiqueue systems, *The Third International Workshop on Petri Nets
        and Performance Models*, Kyoto, Japan, December 11-13 (1989), 19-28.

[4]   Ajmone Marsan, M., Donatelli, S. and Neri, F., GSPN models of Markovian
        multiserver multiqueue systems, *Perf. Eval.* **11**:4 (1990), 227-240.

[5]   Ajmone Marsan, M., Donatelli, S. and Neri, F., Multiserver multiqueue sys-
        tems with limited service and zero walk time, *IEEE INFOCOM '91*, Bal
        Harbour, Florida, April 9-11 (1991), 1178-1188.

[6]   Ajmone Marsan, M., Donatelli, S., Neri, F. and Fantini, G., Some numeri-
        cal results on finite-buffer Markovian multiserver polling systems, *IEEE Int.
        Telecomm. Symp.*, Rio de Janiero, Brazil (1994).

[7]   Ajmone Marsan, M., Donatelli, S., Neri, F. and Rubino, U., GSPN models
        of random, cyclic, and optimal 1-limited multiserver multiqueue systems,
        ACM SIGCOMM' 91 Conference, Comm. Arch. & Protocols, Zürich, Swit-
        zerland, Sept. 3-6 (1991), 69-80.

[8]   Ajmone Marsan, M., Donatelli, S., Neri, F. and Rubino, U., On the con-
        struction of abstract GSPNs: an exercise in modeling, *Proc. of the Fourth
        International Workshop on Petri Nets and Perf. Models*, Melbourne,
        Australia, Dec. 2-5 (1991), 2-17.

[9]   Ajmone Marsan, M., Donatelli, S., Neri, F. and Rubino, U., Good and bad
        dynamic polling orders in symmetric single buffer Markovian multiserver
        multiqueue systems, *IEEE INFOCOM '93*, San Francisco, California,
        March 30 - April 1 (1993), 176-185.

[10]   Altman, E., Analysis of timed-token ring protocols, Report RR 1570,
        INRIA, Dec. 1991.

[11]   Altman, E., Analyzing timed-token ring protocols using the power series al-
        gorithm, In: *The Fund. Role of Teletraffic in the Evol. of Telecomm. Net-
        works - ITC* 14, (ed. by J. Labetoulle and J.W. Roberts), Elsevier Science
        B.V., Amsterdam **B** (1994), 961-971.

[12]   Altman, E., Blanc, H., Khamisy, A. and Yechiali, U., Gated-type polling
        systems with walking and switch-in times, *Stochastic Models* **10**:4 (1994),
        741-763.

[13]   Altman, E. and Foss, S., Polling on a graph with general arrival and service
        time distribution, Report No. 1992, INRIA (1993).

[14]   Altman, E., Foss, S., Riehl, E.R. and Stidham, S., Performance bounds and
        pairwise stability for generalized vacation and polling systems, Research
        Report UNC/OR TR93-8, Univ. of NC at Chapel Hill (1993).

[15]   Altman, E., Foss, S., Riehl, E.R. and Stidhman Jr., S., Sample path
        analysis of token rings, In: *The Fund. Role of Teletraffic in the Evol. of
        Telecommunication Networks - ITC* 14, (ed. by J. Labetoulle and J.W.
        Roberts), Elsevier Science B.V., Amsterdam **B** (1994), 811-820.

[16]   Altman, E., Khamisy, A. and Yechiali, U., On elevator polling with global-
        ly gated regime, *Queueing Sys.* **11**:1-2 (1992), 85-90.

[17]   Altman, E. and Kofman, D., Sample path bounds for performance measures
        of token rings, Research Report, Telecom Paris (1993).

[18]   Altman, E., Konstantopoulos, P. and Liu, Z., Stability monotonicity and in-
        variant quantities in general polling systems, *Queueing Sys.* **11**:1-2 (1992),
        35-57.

[19]   Altman, E. and Levy, H., Queueing in space, *Adv. Appl. Prob.* **26**:4 (1994),
        1095-1116.

[20] Altman, E. and Liu, Z., Improving the stability characteristics of asynchronous traffic in FDDI token rings, In: *High Speed Networks and Their Performance*, (ed. by H.G. Perros and Y. Viniotis), North-Holland (1994), 441-460.

[21] Altman, E. and Spieksma, F., Geometric ergodicity and moment stability of station times in polling systems, Tech. Report No. TW-92-09, Univ. of Leiden, The Netherlands (1992).

[22] Altman, E. and Yechiali, U., Cyclic Bernoulli polling, *Zeitschrift für Opn. Res.* **38**:1 (1993), 55-76.

[23] Altman, E. and Yechiali, U., Polling in a closed network, *Prob. in the Eng. and Inform. Sciences* **8**:3 (1994), 327-343.

[24] Baba, Y., Analysis of batch arrival cyclic service multiqueue systems with limited service discipline, *J. of the Opns. Res. Soc. of Japan* **34**:1 (1991), 93-104.

[25] Balsamo, S., Analysis of asymmetric polling systems with limited service: waiting time approximation (1990), preprint.

[26] Bertsekas, D. and Gallager, R., *Data Networks*, second edition, Prentice-Hall, New Jersey 1992.

[27] Bertsimas, D.J. and van Ryzin, G., A stochastic and dynamic vehicle routing problem in the Euclidean plane, *Opns. Res.* **39**:4 (1991), 601-615.

[28] Birman, A., Gail, H.R., Hantler, S.L., Rosberg, Z. and Sidi, M., An optimal service policy for buffer systems, *J. ACM* (1991), (to appear).

[29] Bisdikian, C., A note on the conservation law for queues with batch arrivals, *IEEE Trans. on Comm.* **41**:6 (1993), 832-835. Correction in **42**:1 (1994), 178.

[30] Bisdikian, C. and Merakos, L., Output process from a continuous token-ring local area network, *IEEE Trans. on Comm.* **40**:12 (1992), 1796-1799.

[31] Blanc, J.P.C., A numerical approach to cyclic-service queueing models, *Queueing Sys.* **6**:2 (1990), 173-188.

[32] Blanc, J.P.C., Cyclic poling systems: limited service versus Bernoulli schedules, Tech. Report FEW 442, Tilburg University, The Netherlands (1990).

[33] Blanc, J.P.C., The power-series algorithm applied to cyclic polling systems, *Comm. in Statistics - Stoch. Models* **7**:4 (1991), 527-545.

[34] Blanc, J.P.C., Performance evaluation of polling systems by means of the power-series algorithm, *Annals of Opns. Res.* **35**:1-4 (1992), 155-186.

[35] Blanc, J.P.C., An algorithmic solution of polling models with limited service discipline, *IEEE Trans. on Comm.* **40**:7 (1992), 1152-1155.

[36] Blanc, J.P.C. and van der Mei, R.D., Optimization of polling systems: computation of derivatives by means of power-series algorithm, In: *Performance '93*, (ed. by G. Iazeolla and S. Lavenberg), Rome, Italy, Sept. 29-Oct. 1 (1993), 385-398.

[37] Blanc, J.P.C. and van der Mei, R.D., The power-series algorithm applied to polling systems with a dormant server, In: *The Fund. Role of Teletraffic in the Evol. of Telecommunication Networks - ITC* 14, (ed. by J. Labetoulle and J.W. Roberts), Elsevier Science B.V., Amsterdam **B** (1994), 865-874.

[38] Borst, S.C., Polling systems with multiple coupled servers, CWI Report BS-R9408, The Netherlands (1994).

[39] Borst, S.C., A pseudo-conservation law for a polling system with a dormant server, In: *The Fund. Role of Teletraffic in the Evol. of Telecommunication Networks - ITC* 14, (ed. by J. Labetoulle and J.W. Roberts), Elsevier

Science B.V., Amsterdam **B** (1994), 729-742.

[40]   Borst, S.C., Boxma, O.J., Harink, J.H.A. and Huitema, G.B., Optimization of fixed time polling schemes, *Telecomm. Sys.* **3** (1994), 31-59.

[41]   Boxma, O.J., Workloads and waiting times in single-server systems with multiple customer classes, *Queueing Sys.* **5**:1-3 (1989), 185-214.

[42]   Boxma, O.J., Analysis and optimization of polling systems, In: *Queueing, Perf. and Control in ATM (ITC* 13), (ed. by J.W. Cohen and C.D. Pack), Elsevier Science Publishers B.V., Amsterdam (1991), 173-183.

[43]   Boxma, O.J. and Groenendijk, W.P., Pseudo-conservation laws in cyclic-service systems, *J. Appl. Prob.* **24**:4 (1987), 949-964.

[44]   Boxma, O.J. and Groenendijk, W.P., Waiting times in discrete-time cyclic-service systems, *IEEE Trans. on Comm.* **COM-36**:2 (1988), 164-170.

[45]   Boxma, O.J. and Kelbert, M., Stochastic bounds for a polling system, *Annals of Opns. Res.* **48**:1-4 (1994), 295-310.

[46]   Boxma, O.J., Levy, H. and Weststrate, J.A., Optimization of polling systems, In: *Perf. '90*, (ed. by P.J.B. King, I. Mitrani, and R.J. Pooley), Elsevier Science Publishers B.V., Amsterdam (1990), 349-361.

[47]   Boxma, O.J., Levy, H., and Weststrate, J.A., Efficient visit frequencies for polling tables: minimization of waiting cost, *Queueing Sys.* **9**:1-2 (1991), 133-162.

[48]   Boxma, O.J., Levy, H. and Weststrate, J.A., Efficient visit orders for polling systems, *Perf. Eval.* **18**:2 (1993), 103-123.

[49]   Boxma, O.J., Levy, H. and Yechiali, U. Cyclic reservation schemes for efficient operation of multiple-queue single-server systems, *Annals of Opns. Res.* **35**:1-4 (1992), 187-208.

[50]   Boxma, O.J., Weststrate, J.A. and Yechiali, U., A globally gated polling system with server interruptions, and applications to the repairman problem, *Prob. in Eng. and Inform. Sci.* **7**:2 (1993), 187-208.

[51]   Browne, S. and Yechiali, U., Dynamic priority rules for cyclic-type queues, *Adv. in Appl. Probl.* **21**:2 (1989), 432-450.

[52]   Browne, S. and Yechiali, U., Dynamic scheduling in single server multi-class service systems with unit buffers, *Naval Res. Logistics* **38**:3 (1991), 383-396.

[53]   Bunday, B.D. and Sztrik, J., The maintenance of bi-directionally patrolled machines, *I.M.A. J. of Math. Appl. in Bus. and Ind.* **3** (1992), 377-386.

[54]   Bunday, B.D., Sztrik, J. and Tapsir, R.B., A heterogeneous scan service polling model with single-message buffer, In: *Perf. of Distr. Syst. and Integ. Comm. Networks*, (ed. by T. Hasegawa, H. Takagi, and Y. Takahashi), Elsevier Science Publishers B.V., Amsterdam (1992), 99-111.

[55]   Campbell, G.M., Cyclic queueing systems, *Euro. J. of Opnl. Res.* **51**:2 (1991), 155-167.

[56]   Casares-Giner, V., A pseudoconservation law based approximate analysis for cyclic service queueing system with decrementing service discipline, In: *Teletraffic and Datatraffic in a Period of Change, ITC-13*, (ed. by A. Jensen and V.B. Iversen), Elsevier Science Publishers B.V., Amsterdam (1991), 749-754.

[57]   Chang, C.-J. and Hwang, L.-C., Analysis for a general service nonexhaustive polling system with pseudoconservation law, Technical Report, National Chiao Tung University, Taiwan, December (1991).

[58]   Chang, C.-J. and Hwang, L.-C., Analysis of a general service nonexhaustive polling system using a heuristic combination method and pseudo-conserva-

tion law, *Perf. Eval.* (1993), (to appear).

[59]   Chang, K.C., Stability conditions for a pipeline polling scheme in satellite communications, *Queueing Sys.* **14**:3-4 (1993), 339-348.

[60]   Chang, K.C. and Sandhu, D., Pseudo-conservation laws in cyclic-server, multiqueue systems with a class of limited service policies, *IEEE INFOCOM '90*, San Francisco, California, June 3-7 (1990), 260-267.

[61]   Chang, K.C. and Sandhu, D., Mean waiting time approximations in cyclic-service systems with exhaustive limited service policy, *IEEE INFOCOM '91*, Bal Harbour, Florida, April 9-11 (1991), 1168-1177. Also in *Perf. Eval.* **15**:1 (1992), 21-40.

[62]   Chang, K.C. and Sandhu, D., Pseudo-conservation laws in cyclic-service systems with a class of limited service policies, *Annals of Opns. Res.* **35**:1-4 (1992), 209-229.

[63]   Chang, K.C. and Sandhu, D., Delay analysis of token-passing protocols with limited token holding times, *IEEE INFOCOM '92*, Florence, Italy, May 6-8 (1992), 2299-2305.

[64]   Charzinski, J., Renger, T. and Tangemann, M., Simulative comparison of the waiting time distributions in cyclic service polling systems with different service strategies, In: *The Fund. Role of Teletraffic in the Evol. of Telecommunication Networks - ITC* 14, (ed. by J. Labetoulle and J.W. Roberts), Elsevier Science B.V., Amsterdam **A** (1994), 719-728.

[65]   Chiarawongse, J. and Srinivasan, M.M., On pseudo-conservation laws for the cyclic server system with compound Poisson arrivals, *Opns. Res. Letters* **10**:8 (1991), 453-459.

[66]   Choi, H. and Trivedi, K.S., Approximate performance models of polling systems using stochastic Petri nets, *IEEE INFOCOM '92*, Florence, Italy, May 6-8 (1992), 2306-2314.

[67]   Choudhury, G.L, Polling with a general service order table: Gated service, *IEEE INFOCOM '90*, San Francisco, California, June 3-7 (1990), 268-276.

[68]   Chung, H., Un, C.K. and Jung, W.Y., Performance analysis of Markovian polling systems with single buffers, *Perf. Eval.* **19**:4 (1994), 303-315.

[69]   Coffmann Jr., E.G. and Gilbert, E.N., A continuous polling system with constant service times, *IEEE Trans. on Info. Theory* **IT-33**:4 (1986), 584-591.

[70]   Coffman Jr., E.G. and Stolyar, A., Continuous polling on graphs, *Prob. in the Eng. and Inform. Sciences* **7**:2 (1993), 209-226.

[71]   Conti, M., Gregori, E. and Lenzini, L., A vacation model for interconnected FDDI networks, *ACM 1992 Computer Science Conf.*, Kansas City, Missouri, March 3-5 (1992), 9-16.

[72]   Conti, M., Gregori, E. and Lenzini, L., Metropolitan area networks (MANs): protocols, modeling and performance evaluation, In: *Perf. Eval. of Comp. and Communication Sys., Joint tutorial papers of Perf. '93 and Sigmetrics '93*, (ed. by L. Donatiello and R. Nelson), Lecture Notes in Computer Science **729**, Springer-Verlag, Berlin (1993), 81-120.

[73]   Conti, M., Gregori, E. and Lenzini, L., Design and analysis of a medical communication system based on FDDI, *Microcomputer Appl.* (Special issue on Medical Applications) **12**:3 (1993), 118-127.

[74]   Cooper, R.B., Queues served in cyclic order: Waiting times, *The Bell System Tech. J.* **49**:3 (1970), 399-413.

[75]   Cooper, R.B., Niu, S.-C. and Srinivasan, M.M., A decomposition theorem

for polling models: the switchover times are effectively additive, *Opns. Res.* (1992), (to appear).

[76] Daganzo, C.F., Some properties of polling systems, *Queueing Sys.* **6**:2 (1990), 137-154.

[77] Davies, D.W., Barber, D.L.A., Price, W.L., and Solomonides, C.M., *Computer Networks and Their Protocols*, John Wiley and Sons, New York 1979.

[78] de Souza e Silva, E., Gail, H.R. and Muntz, R.R., Polling systems with server timeout, *IEEE/ACM Trans. Networking* (1993), (to appear).

[79] Eisenberg, M., The polling system with a stopping server, *Queueing Sys.* **18**:3,4 (1994), 387-431.

[80] Fabian, O. and Levy, H., Polling system optimization through dynamic routing policies, *IEEE INFOCOM '93*, San Francisco, California, March 30-April 1 (1993), 194-200.

[81] Fabian, O. and Levy, H., Pseudo-cyclic policies for multi-queue single server systems, *Annals of Opns. Res.* **48**:1-4 (1994), 127-152.

[82] Federgruen, A. and Katalan, Z., Approximating queue size and waiting time distributions in general polling systems, *Queueing Sys.* **18**:3,4 (1994), 353-386.

[83] Ferguson, M.J. and Aminetzah, Y.J., Exact results for nonsymmetric token ring systems, *IEEE Trans. on Commun.* **COM-33**:3 (1985), 223-231.

[84] Fournier, L. and Rosberg, Z., Expected waiting times in cyclic service systems under priority disciplines, *Queueing Sys.* **9**:4 (1991), 419-439.

[85] Fricker, C. and Jaïbi, R., Monotonicity and stability of periodic polling models, *Queueing Sys.* **15**:1-4 (1994), 211-238.

[86] Fuhrmann, S.W., Performance analysis of a class of cyclic schedules, Bell Lab. Tech. Memo. 59531-810316.01TM, March 16 (1981).

[87] Fuhrmann, S.W., A decomposition result for a class of polling models, *Queueing Sys.* **11**:1-2 (1992), 109-120.

[88] Fuhrmann, S.W. and Cooper, R.B., Stochastic decompositions in the M/G/1 queue with generalized vacations, *Opns. Res.* **33**:5 (1985), 1117-1129.

[89] Fuhrmann, S.W. and Moon, A., Queues served in cyclic order with an arbitrary start-up distribution, *Naval Res. Logistics* **37**:1 (1990), 123-133.

[90] Gail, H.R., Grover, G., Guérin, R., Hantler, S.L., Rosberg, Z. and Sidi, M., Buffer size requirements under longest queue first, In: *Perf. of Dist. Systems and Integ. Commun. Networks*, (ed. by T. Hasegawa, H. Takagi and Y. Takahashi), Elsevier Science Publishers B.V., Amsterdam (1992), 413-424. Also in *Perf. Eval.* **18**:2 (1993), 133-140.

[91] Garner, G.M., Mean and variance of waiting time for nonsymmetric cyclic queueing systems, Tec. memorandum 59482-880600-00TM, AT&T Bell Lab. (1988).

[92] Genter, W.L. and Vastola, K.S., Delay analysis of the FDDI synchronous data class, *IEEE INFOCOM '90*, San Francisco, California, June 3-7 (1990), 766-773.

[93] Georgiadis, L. and Szpankowski, W., Stability of token passing rings, *Queueing Sys.* **11**:1-2 (1992), 7-33.

[94] Georgiadis, L. and Szpankowski, W., Stability criteria for yet another multidimensional distributed system, Revised version, Report No. 1996, INRIA (1993).

[95] Grillo, D., Polling mechanism models in communication systems - some ap-

plication examples, In: *Stoch. Anal. of Computer and Commun. Systems,* (ed. by H. Takagi), Elsevier Science Publishers B.V., Amsterdam (1990), 659-698.

[96]   Gupta, D. and Srinivasan, M.M., Polling systems with state-dependent set-up times, Presentation at the ORSA/TIMS Phoenix (1993) meeting.

[97]   Harel, A. and Stulman, A., Polling greedy and horizon servers on a circle, *Opns. Res.* (1990), (to appear).

[98]   Harrison, P.G. and Coury, S., Waiting time distribution in a class of wireless multichannel local area networks, (1994), preprint.

[99]   Ibe, O.C., Analysis of polling systems with mixed service disciplines, *Commun. in Statistics - Stoch. Models* **6**:4 (1990), 667-689.

[100]   Ibe, O.C. and Trivedi, K.S., Stochastic Petri net models of polling systems, *IEEE J. on Selected Areas in Commun.* **8**:9 (1990), 1649-1657.

[101]   Jaiswal, N.K., *Priority Queues*, Academic Press, New York 1968.

[102]   Jung, W.Y., Analysis of finite-capacity polling systems based on virtual buffering and lumped modeling, Ph.D. dissertation, Report CRT-T-9109, Commun. Res. Lab., Korea Adv. Inst. of Science and Tech., Seoul, Korea, June (1991).

[103]   Jung, W.Y. and Un, C.K., Analysis of a finite-buffer polling system with exhaustive service based on virtual buffering, *IEEE Trans. on Commun.* **42**:12 (1991), 3144-3149.

[104]   Karvelas, D. and Leon-Garcia, A., A general approach to the delay analysis of symmetric token passing networks, *IEEE INFOCOM '91*, Bal Harbour, Florida, April 9-11 (1991), 181-190.

[105]   Karvelas, D. and Leon-Garcia, A., Delay analysis of various service disciplines in symmetric token passing networks, *IEEE Trans. on Commun.* **41**:9 (1993), 1342-1355.

[106]   Katayama, T., Analysis of a multi-class service tandem queueing model with feedback attended by a single server, In: *Teletraffic and Datatraffic in a Period of Change, ITC*-13, (ed. by A. Jensen and V.B. Iversen), Elsevier Science Publishers B.V., Amsterdam (1991), 743-748.

[107]   Katayama, T., Performance analysis for a two-class priority queueing model with general decrementing service, *IEICE Trans. on Commun.* **E75-B**:12 (1992), 1301-1307.

[108]   Katayama, T., Analysis of a cyclic-service tandem queue with a gate and vacations, *Symposium on Performance Models for Info. Commun. Networks*, Hakone, Japan, January 19-21 (1994), 321-332.

[109]   Khamisy, A., Altman, E. and Sidi, M., Polling systems with synchronization constraints, *Annals of Opns. Res.* **35**:1-4 (1992), 231-267.

[110]   Kofman, D., Blocking probability, throughput and waiting time in finite capacity polling systems, *Queueing Sys.* **14**:3-4 (1993), 385-411.

[111]   Konheim, A.G. and Levy, H., Efficient analysis of polling systems, *IEEE INFOCOM '92*, Florence, Italy, May 6-8 (1992), 2325-2331.

[112]   Konheim, A.G., Levy, H. and Srinivasan, M.M., Descendant set: An efficient approach for the analysis of polling systems, *IEEE Trans. on Commun.* **42**:2-4 (1994), 1245-1253.

[113]   Kroese, D.P. and Schmidt, V., A continuous polling system with general service times, *The Annals of Appl. Probability* **2**:4 (1992), 906-927.

[114]   Kroese, D.P. and Schmidt, V., Single-server queues with spatially distributed arrivals, *Queueing Sys.* **17**:1,2 (1994), 317-345.

[115] Kubat, P. and Servi, L.D., Cyclic service queues with very short service times, *Euro. J. of Opnl. Res.* **53**:2 (1991), 172-188.

[116] LaMaire, R.O., An M/G/1 vacation model of an FDDI station, *IEEE J. on Selected Areas in Commun.* **9**:2 (1991), 257-264.

[117] Lang, M. and Bosch, M., Performance analysis of finite capacity polling systems with limited-M service, In: *Teletraffic and Datatraffic in a Period of Change, ITC*-13, (ed. by A. Jensen and V.B. Iversen), Elsevier Science Publishers, Amsterdam (1991), 731-735.

[118] Lee, D.-S., Analysis of a cyclic server queue with the exhaustive and limited service discipline, (1994), preprint.

[119] Lee, D.-S. and Sengupta, B., A reservation based cyclic server queue with limited service, *Perf. Eval. Rev.* (1992 ACM SIGMETRICS and Performance '92), **20**:1 (1992), 70-77.

[120] Lee, D.-S. and Sengupta, B., An approximate analysis of a cyclic server queue with limited service and reservations, *Queueing Sys.* **11**:1-2 (1992), 153-178.

[121] Lee, T.Y.S., Naive control of a queueing network, (1993), preprint.

[122] Leung, K.K., Waiting time distributions for token-passing systems with limited-one service via discrete Fourier transforms, *IEEE INFOCOM '90*, San Francisco, California, June 3-7 (1990), 1111-1118.

[123] Leung, K.K., Waiting time distribution for token passing systems with limited-k service via discrete Fourier transforms, In: *Performance '90*, (ed. by P.J.B. King, I. Mitrani and R.J. Pooley), Elsevier Science Publishers, Amsterdam (1990), 333-347.

[124] Leung, K.K., Cyclic-service systems with probabilistically-limited service, *IEEE J. on Selected Areas in Commun.* **9**:2 (1991), 185-193.

[125] Levy, H., A note on the complexity of Swartz's method for calculating the expected delay in non-symmetric cyclic polling systems, *Opns. Res. Letters* **10**:6 (1991), 363-368.

[126] Levy, H. and Kleinrock, L., Polling systems with zero switch-over periods: A general method for analyzing the expected delay, *Perf. Eval.* **13**:2 (1991), 97-107.

[127] Levy, H. and Sidi, M., Polling systems: Applications, modeling, and optimization, *IEEE Trans. on Commun.* **38**:10 (1990), 1750-1760.

[128] Levy, H. and Sidi, M., Polling systems with simultaneous arrivals, *IEEE Trans. on Commun.* **39**:6 (1991), 823-827.

[129] Levy, H., Sidi, M. and Boxma, O.J., Dominance relations in polling systems, *Queueing Sys.* **6**:2 (1990), 155-172.

[130] Liu, Z. and Nain, P., Optimal scheduling in some multi-queue single server systems, *IEEE Trans. on Autom. Cont.* **37**:2 (1992), 247-252.

[131] Liu, Z., Nain, P. and Towsley, D., On optimal polling policies, *Queueing Systems* **11**:1-2 (1992), 59-83.

[132] Massoulié, L., Stability of a non-Markovian polling systems, Laboratoire des Signaux et Systèmes, CNRS (1993), preprint.

[133] Matsumoto, Y., On optimization of polling policy represented by neural network, (1994), preprint.

[134] Miyoshi, N., Takahashi, Y. and Hasegawa, T., Discount optimal server assignment policy for symmetric polling systems, *Perf. Models for Info. Commun. Networks*, Numazu, Japan, January 21-23 (1993), 152-162.

[135] Mukherjee, B., Kwok, C.K., Lantz, A.C. and Melody Moh, W.-H.L.,

Comments on "Exact analysis of asymmetric polling systems with single buffers", *IEEE Trans. on Commun.* **38**:7 (1990), 944-946.

[136] Murata, M. and Takagi, H., Performance of token ring networks with a finite capacity bridge, *Computer Networks and ISDN Systems* **24**:1 (1992), 45-64.

[137] Nakamura, K., Takine, T., Takahashi, Y. and Hasegawa, T., Analysis of an asymmetric polling model with cycle-time constraint, In: *Arch. and Perf. Issues of High-Capacity Local and Metropolitan Area Networks*, (ed. by G. Pujolle), NATO ASI Series F., Springer-Verlag, Berlin (1990), 493-508.

[138] Ozawa, T., Alternating service queues with mixed exhaustive and K-limited services, *Perf. Eval.* **11**:3 (1990), 165-171.

[139] Resing, J.A.C., Polling systems and multitype branching processes, *Queueing Sys.* **13**:4 (1993), 409-426.

[140] Rubin, I. and Baker, J.E., Media access control for high-speed local area and metropolitan area communication networks, In: *Proc. of the IEEE*, **78**:1 (1990), 168-203. Reprinted in *Adv. in Local and Metropolitan Area Networks*, (ed. by W. Stallings), IEEE Computer Society Press (1994), 96-131.

[141] Rubin, I. and Wu, C.-H., Analysis of an M/G/1 queue with vacations and its application to FDDI asynchronous timed-token service system, *IEEE Global Telecommun. Conf. (GLOBECOM '86)*, Orlando, Florida, Dec. 6-9 (1992), 1630-1634.

[142] Sarkar, D. and Zangwill, W.I., Expected waiting time for nonsymmetric cyclic queueing systems - Exact results and applications, *Management Science* **35**:12 (1989), 1463-1474.

[143] Sasaki, G.H., Input buffer requirements for round robin polling systems, *Perf. Eval.* **18**:3 (1993), 237-261.

[144] Shimogawa, S. and Takahashi, Y., A note on the pseudo-conservation law for a multi-queue with local priority, *Queueing Sys.* **11**:1-2 (1992), 145-151.

[145] Sidi, M. and Levy, H., Customers routing in polling systems, In: *Performance '90*, (ed. by P.J.B. King, I. Mitrani and R.J. Pooley), Elsevier Science Publishers B.V., Amsterdam (1990), 319-331.

[146] Sidi, M., Levy, H. and Fuhrmann, S.W., A queueing network with a single cyclically roving server, *Queueing Sys.* **11**:1-2 (1992), 121-144. A corrigendum in "Correction to equation (5.6) in the paper: A queueing network with a single cyclically roving server", *Queueing Sys.* **16**:1-2 (1994), 193.

[147] Srinivasan, M.M., Waiting time variances in continuous-time polling systems, Technical Report No. 91-37, The Univ. of Michigan, Dept. of Ind. and Opns. Eng., Ann Arbor, Michigan (1991).

[148] Srinivasan, M.M. and Gupta, D., When should a roving server be patient?, (1993), preprint.

[149] Srinivasan, M.M., Levy, H. and Konheim, A.G., The individual station technique for the analysis of continuous-time polling systems, Tech. Report No. 91-36, The Univ. of Michigan, Dept. of Ind. and Opns. Eng., Ann Arbor, Michigan (1991).

[150] Srinivasan, M.M., Niu, S.-C. and Cooper, R.B., Relating polling models with nonzero and zero switchover times, *Queueing Sys.* (1993), (to appear).

[151] Szpankowski, W., Towards computable stability criteria for some multidimensional stochastic processes, In: *Stoch. Anal. of Computer and Commun. Systems*, (ed. by H. Takagi), Elsevier Science Publisher B.V.,

Amsterdam (1990), 131-172.

[152]   Takagi, H., *Analysis of Polling Systems*, The MIT Press, Cambridge, Massa-chusetts 1986.

[153]   Takagi, H., Queueing analysis of polling systems with zero switchover times, TRL Research Report, TR87-0034, IBM Japan, Ltd., Tokyo (1987).

[154]   Takagi, H., Queueing analysis of polling models, *ACM Comp. Surveys* **20**:1 (1988), 5-28.

[155]   Takagi, H., Queueing analysis of polling models: An update, In: *Stoch. Anal. of Computer and Commun. Systems*, (ed. by H. Takagi), Elsevier Science Publishers B.V., Amsterdam (1990), 267-318.

[156]   Takagi, H., Effects of the target token rotation time on the performance of a timed-token protocol, In: *Performance '90*, (ed. by P.J.B. King, I. Mitrani, and R.J. Pooley), Elsevier Science Publishers B.V., Amsterdam (1990), 363-370.

[157]   Takagi, H., Analysis of finite capacity polling systems, *Adv. in Appl. Probab.* **23**:2 (1991), 373-387.

[158]   Takagi, H., Application of polling models to computer networks, *Computer Networks and ISDN Systems* **22**:3 (1991), 193-211.

[159]   Takagi, H., *Queueing Analysis: A Foundation of Performance Evaluation, Vol. 1: Vacation and Priority Systems, Part 1*, Elsevier Science Publishers B.V., Amsterdam 1991.

[160]   Takagi, H., Analysis of an $M/G/1//N$ queue with multiple server vacations, and its application to a polling model, *J. of the Opns. Res. Soc. of Japan* **35**:3 (1992), 300-315.

[161]   Takagi, H., *Queueing Analysis: A Foundation of Performance Evaluation, Vol. 2: Finite Systems*, Elsevier Science Publishers B.V., Amsterdam 1993.

[162]   Takagi, H., Analysis of single-buffer polling models for time-slotted communication protocols, In: *Local and Metropolitan Communication Systems - LAN & MAN*, (ed. by T. Hasegawa, G. Pujolle, H. Takagi and Y. Takahashi), Chapman and Hall (1995).

[163]   Takahashi, Y. and Krishna Kumar, B., On the pseudo-conservation law for discrete-time multi-queue systems with priority disciplines, *Symp. on Perf. Models for Info. Commun. Networks*, Hakone, Japan, January 19-21 (1994), 178-188.

[164]   Takine, T. and Hasegawa, T., A cyclic-service finite source model with round-robin scheduling, *Queueing Sys.* **11**:1-2 (1992), 91-108.

[165]   Takine, T., Takagi, H., Takahashi, Y. and Hasegawa, T., Analysis of asymmetric single-buffer polling and priority systems without switchover times, *Perf. Eval.* **11**:4 (1990), 253-264.

[166]   Takine, T., Takahashi, Y. and Hasegawa, T., Exact analysis of asymmetric polling system with single buffers, *IEEE Trans. on Commun.* **36**:10 (1988), 1119-1127.

[167]   Takine, T., Takahashi, Y. and Hasegawa, T., Average message delay of an asymmetric single-buffer polling system with round-robin scheduling of services, In: *Modelling Tech. and Tools for Comp. Perf. Eval.*, (ed. by D. Potier and R. Puigjaner), Plenum Publishing Corp., New York (1989), 179-187.

[168]   Takine, T., Takahashi, Y. and Hasegawa, T., Modeling and analysis of a single-buffer polling system interconnected with external networks, *INFOR* **28** (1990), 160-177.

[169] Takine, T., Takahashi, Y. and Hasegawa, T., Further results on asymmetric polling systems with single buffers, (1990), preprint.

[170] Tangemann, M., A mean value analysis for throughputs and waiting times of the FDDI timed token protocol, In: *Teletraffic and Datatraffic in a Period of Change, ITC*-13, (ed. by A. Jensen and V.B. Iversen), Elsevier Science Publishers B.V., Amsterdam (1991), 173-179.

[171] Tangemann, M., Timer threshold dimensioning and overload control in FDDI networks, *IEEE INFOCOM '92*, Florence, Italy, May 6-8 (1992), 363-371.

[172] Tangemann, M., Mean waiting time approximations for symmetric and asymmetric polling systems with time-limited service, *Proc. 7, GI/ITG Fach. Messung, Modellierung und Bewertung von Rechen- und Kommunikationssystemen*, Aachen, Germany, September 21-23 (1993).

[173] Tangemann, M., Modelling and analysis of high-speed local area networks with time-limited token passing media access protocols, Ph.D. Thesis, Inst. of Commun. Switching and Data Technics, Univ. of Stuttgart, Germany (1994).

[174] Tangemann, M., Mean waiting time approximation for FDDI, In: *The Fund. Role of Teletraffic in the Evol. of Telecommunication Networks - ITC* 14, (ed. by J. Labetoulle and J.W. Roberts), Elsevier Science Publishers B.V., Amsterdam **B** (1994), 973-984.

[175] Tangemann, M. and Sauer, K., Performance analysis of the FDDI media access control protocol, *Fourth International Conf. on Data Commun. Systems and their Perf.*, Barcelona, Spain, June 20-22 (1990), 32-43.

[176] Tangemann, M. and Sauer, K., Performance analysis of the timed token protocol of FDDI and FDDI-II, *IEEE J. on Selected Areas in Commun.* **9**:2 (1991), 271-278.

[177] Tayur, S. and Sarkar, D., Implementation of an efficient algorithm to calculate waiting times in a nonsymmetric cyclic queueing system, Internal Memorandum 45313-880822-01IM, AT&T Bell Lab. (1988).

[178] Towsley, D., Fdida, S. and Santoso, H., Design and analysis of flow control protocols for metropolitan area networks, In: *High-Capacity Local and Metropolitan Area Networks*, (ed. by G. Pujolle), Springer (1991), 471-492.

[179] Tran-Gia, P., Analysis of polling systems with general input process and finite capacity, *IEEE Trans. on Commun.* **40**:2 (1992), 337-344.

[180] Tsai, Z. and Rubin, I., Mean delay analysis for a message priority-based polling scheme, *Queueing Sys.* **11**:3 (1992), 223-240.

[181] van der Mei, R.D. and Borst, S.C., Analysis of multiple-server polling systems by means of the power-series algorithm, CWI Report BS-R9410, The Netherlands (1994).

[182] Walrand, J., Queueing networks, In: *Handbooks in Opns. Res. and Mgmt. Sci., Vol. 2: Stoch. Models*, (ed. by D.P. Heyman and M.J. Sobel), Elsevier Science Publishers B.V., Amsterdam (1990), 519-603.

[183] Weststrate, J.A. and van der Mei, R.D., Waiting times in a two-queue model with exhaustive and Bernoulli service, *ZOR-Mathematical Methods of Operations Research* **40** (1994), 289-303.

[184] Woodward, M.E., *Communication and Computer Networks. Modelling with Discrete-Time Queues*, IEEE Computer Society Press, Los Alamitos, California 1994.

[185] Yechiali, U., Optimal dynamic control of polling systems, In: *Queueing,*

Perf. and Control in ATM (*ITC*-13), (ed. by J.W. Cohen and C.D. Pack), Elsevier Science Publishers B.V., Amsterdam (1991), 205-217.

[186] Zhang, Z. and Acampora, A.S., Performance of a modified polling strategy for broadband wireless LANs in a harsh fading environment, *Telecommun. Sys.* **1**;3 (1993), 279-294.

[187] Zhigang, A., Mean waiting delay for a token ring network with batch arrivals and limited service policies, (1993) preprint.

Chapter 6
Product-form loss networks

Hisashi Kobayashi and Brian L. Mark

ABSTRACT In a loss system customers arrive in attempt to seize some of the available system resources. At the time of arrival, a customer finding insufficient available resources leaves the system, hence the term *loss system*. Otherwise, the customer seizes the required amount of resources for a random holding time and then leaves the system, thereby making available the previously held resources. In this chapter we investigate a class of loss systems known as *product-form loss networks*, which are generalizations of the classical Erlang and Engset loss models. Loss networks can provide a mathematical model for various resources sharing systems including circuit-switched networks and other types of connection-oriented communication networks.

We develop basic properties of loss networks related to the product-form stationary distribution by generalizing the classical loss models. We find analogies between loss networks and the well-studied product-form queueing networks, which suggest that the parallel methods of analysis can be applied to both types of models. Blocking probabilities and other quantities of interest for loss networks are expressible in terms of a normalization constant, which is analogous to the normalization constant of queueing networks. Evaluation of the normalization constant usually poses significant computational challenges even for simple loss networks with special structure. We discuss state-of-the-art computational methods and asymptotic approximations for evaluating the normalization constant and blocking probabilities in product-form loss networks. In conclusion we discuss open problems and related areas of research.

CONTENTS

6.1 INTRODUCTION

In his 1963 seminal paper, J.R. Jackson [17] introduced a class of queueing

0-8493-8076-6/97/$0.00+$.50

networks that possess a simple product-form stationary state distribution. Jackson's model consists of a system of $\cdot/M/1$ queues together with a Markov chain governing customer routing between queues. Subsequently, Baskett, Chandy, Muntz, and Palacios [3] enlarged the class of product-form queueing networks by incorporating multiple customer classes and including the additional service disciplines of processor sharing, last come first serve with preemptive resume, and infinite-server stations. They showed not only that the stationary distribution of queue lengths has a product form, but also that it depends on the service time distributions only through their means; i.e., the product-form state distribution is *insensitive* to the form of the service time distribution.

In this chapter we investigate a class of models called *loss networks*, which share the product-form and insensitivity properties of Jackson networks. In fact, there are several analogies between loss networks and queueing networks, and often parallel methods find application to both types of networks. The Jackson network may be viewed as a generalization of the $M/M/1$ queue whereby several queues form a network with customer routing between queues. On the other hand, a loss network is a generalization of the classical loss model that was studied by Erlang in the 1930's in the context of blocking in telephone exchanges. As we shall see, the loss network arises from the Erlang model by introducing multiple customer and server classes and allowing customers to have multiple servers simultaneously.

While the topic of queueing networks has been an active research area since the work of Jackson, more recently there has been an increasing interest in loss networks since they are finding new applications to modeling and analysis of new telecommunications and computer systems. In the literature, loss networks are often called *circuit-switched networks* since they provide accurate models for such networks. In a circuit-switched network, a typical connection is set up by reserving one *circuit* on each link along a path from a source to a destination. These circuits are held throughout the duration of a call and then released simultaneously. If a new call cannot be established due to lack of sufficient available circuits, the call is rejected, hence the name *loss network*. In this chapter we will use much of the terminology for circuit-switched networks when we discuss loss networks because of a close analogy between them. By contrast, *packet-switched networks* are more appropriately modeled by queueing networks in which each queueing station represents a packet switch and the links represent the rules for routing packets among the switches.

The focus of this chapter is on loss networks that possess a product-form stationary state distribution. Recently there has also been interest of loss networks with state-dependent or alternative routing in order to optimize performance in circuit-switched networks. Such networks do not have a product-form stationary distribution and are beyond the scope of this chapter. However, the product-form loss network provides a basis for considering loss networks with various control strategies. See, for example, [20, 27] and references contained therein for work on alternative routing in loss networks.

The remainder of the chapter is organized as follows. In Section 6.2, we characterize the loss network as a generalization of the Erlang loss model and derive several of its qualitative properties, such as insensitivity and reversibility. Section 6.3 focuses on the analysis of quantitative properties associated with loss networks, in particular, blocking probabilities and generating functions. Analogous to queueing networks, many of the quantities of interest in loss networks can be expressed in terms of a normalization constant (also called the *partition function*).

In Section 6.4, we discuss recursive algorithms for computing normalization constants and blocking probabilities for loss stations. The recursive algorithms are efficient for loss stations of small-to-moderate size, but require excessive computing time and lose numerical accuracy as the number of servers becomes large. In Section 6.5, we discuss an alternative approach to computing the normalization constant: direct numerical inversion of the associated generating function. This approach, made viable by recent progress in numerical transform inversion techniques [1, 2], holds promise for efficiently solving larger and more general loss networks than the recursive methods. In the realm of large loss networks, asymptotic approximations become attractive in terms of computational work and accuracy. Section 6.6 deals with asymptotic approximation of an inversion integral for loss stations. The approximation techniques discussed here do not extend readily to more general loss networks. However, by considering the equilibrium behavior of the general loss network in the limit of large scale, asymptotic approximations for blocking probabilities can be obtained. Section 6.7 covers this class of asymptotic approximations, including the reduced load approximation and its generalizations. Finally, Section 6.8 summarizes this chapter and discusses related open problems and future research in the area of loss networks. The Appendices contain derivations of some results discussed in the main text and a glossary listing some of the more important notations used throughout the chapter.

6.2 CHARACTERIZATION AND BASIC PROPERTIES

6.2.1 Generalized loss stations

We use the general term *station* to denote an entity that provides service to arriving customers. A station consists of a number of *servers* and possibly a *waiting room*. A *loss station* is characterized by having a finite number of servers and no waiting room. An arriving customer either begins service immediately or is rejected due to the lack of a sufficient number of available servers. By contrast, a *queueing station* has infinite waiting room; no customer is rejected.

A loss station may be viewed as an infinite-server queueing station with a *truncated* state space. If a stationary Markov process satisfies the *partial balance* equations on a subset of the original state space, the truncated process defined on the smaller state space will have an equilibrium distribution of the same form, up to a normalization constant. More concretely, let $X(t)$ be a stationary Markov process defined on a countable state space $\mathcal{S}$ with equilibrium distribution $\pi(x)$, $x \in \mathcal{S}$ and transition rates $q(x, x')$, $x, x' \in \mathcal{S}$. A subset $A \subset \mathcal{S}$ is said to be *partially balanced* (cf. [41]) if the partial balance equation

$$\pi(x) \sum_{x' \in A} q(x, x') = \sum_{x' \in A} \pi(x') q(x', x), \quad \forall x \in A$$

holds on A. Partial balance implies that $\pi(x)$ would remain a stationary distribution in A (up to renormalization) if the process were "truncated" to the space A, i.e., transitions in and out of A are set to zero. Conversely, if $\pi(x)$ is the stationary distribution of the truncated (to A) version of $X(t)$, then A is partially balanced.

The classical Erlang loss model may be denoted as an $M/M/C(0)$ queue (see Figure 1); i.e., a loss station with C servers where arriving customers form a

Poisson process with rate λ and each customer occupies a server for an exponentially distributed holding time with mean $1/\mu$. Let $n(t)$ be the number of busy servers at time t. Then the stationary distribution for $n(t)$ is given by:

$$\pi_C(n) \triangleq \lim_{t\to\infty} P(n(t) = n) = \frac{a^n}{n!} \left[\sum_{i=0}^{C} \frac{a^i}{i!} \right]^{-1}, \quad n = 0, 1, \ldots, C, \qquad (6.1)$$

where $a = \lambda/\mu$ is the *offered load*. As $C\to\infty$, $\pi_C(n)$ approaches the Poisson distribution with parameter a, the stationary distribution of an infinite-server (IS) station or $M/M/\infty$ queue. Alternatively, equation (6.1) may be viewed as a truncated version of the stationary distribution for an IS station. The offered load a is given in units of *erlangs*.

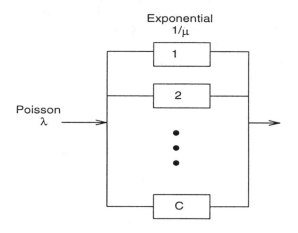

Figure 1. Erlang model.

In the classical Engset loss model, which may be denoted as an $M(N)/M/C(0)$ queue, the arrival process is generated by a finite source model with N sources (see Figure 2).

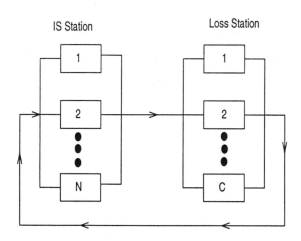

Figure 2. Engset station.

Each source consists of a customer who waits for an exponentially distributed integer-generation time at the IS station, arrives at the loss station where it either acquires a server for an exponentially distributed holding time or is blocked, and then begins a new cycle. The arrival rate to the station has the state dependent form $\lambda(n) = (N - n)\nu$, where ν is the parameter of an exponentially distributed inter-generation time. For the case $N > C$, the stationary distribution for the number of busy servers, $n(t)$, is given by

$$\pi_C(N, n) = \binom{N}{n} b^n \left[\sum_{i=0}^{C} \binom{N}{i} b^i \right]^{-1}, \quad n = 0, 1, \ldots, C, \qquad (6.2)$$

where $b = \nu/\mu$ is the offered load per customer. If $N \leq C$, then an arriving customer will always find at least one free server and the Engset station is equivalent to an $M(N)/M/\infty$ queue with stationary distribution

$$\pi_C(N, n) = \binom{N}{n} p^n (1 - p)^{N-n}, \quad n = 0, 1, \ldots, N, \qquad (6.3)$$

where $p = \nu/(\nu + \mu)$ is the stationary probability that a given customer is in service at the IS station.

It has long been known that the stationary distribution for the Erlang model given by (6.1) remains true for an arbitrary holding time distribution (cf. [36]). In other words, the stationary product-form distribution for the Erlang model is *insensitive* to the holding time distribution and depends only on the mean $1/\mu$. Thus, we define an *Erlang station* to be an $M/G/C(0)$ queue. Insensitivity with respect to holding time distribution applies also to the IS and Engset models. Furthermore, the product-form distribution of the Engset model is insensitive to the form of the inter-generation time (cf. [13]); i.e., the time that a given customer waits before arriving at the station for service need not be exponentially distributed. Thus, we define an *Engset station* as a $G(N)/G/C(0)$ queue. As we shall see, the insensitivity property also holds for the more general case of a loss network.

Let us generalize the classical models by introducing the set $\mathcal{R} = \{1, \ldots, R\}$ of customer classes with multiple server acquisition and general holding times. Customers of class $r \in \mathcal{R}$ seek to acquire A_r servers for a generally distributed holding time with mean $1/\mu_r$. The *generalized IS station*, denoted $\cdot/G_{\mathcal{R}}/\infty$, has an infinite number of servers. The *generalized loss station*, denoted $\cdot/G_{\mathcal{R}}/C(0)$, has C servers with no waiting room; an arriving customer of class r is *blocked* if there are less than A_r available servers.

Two multiclass source models are of particular interest. The *multiclass Poisson source* (or *infinite source*), denoted by $M_{\mathcal{R}}$, consists of independent Poisson processes with rates λ_r, $r \in \mathcal{R}$. The *offered load* from class r customers is given by $a_r = \lambda_r/\mu_r$, which may be interpreted as the average number of class r customers in service at an IS station. The *multiclass finite source*, denoted $G_{\mathcal{R}}(N_{\mathcal{R}})$, consists of independent finite sources with populations N_r, $r \in \mathcal{R}$, and generally distributed inter-generation times with means $1/\nu_r$. The offered load from a class r customer is given by $b_r = \nu_r/\mu_r$. For the finite source model, the stationary probability that a given class r customer is in service at an IS station is given by

$$p_r = \nu_r/(\nu_r + \mu_r) = b_r/(1 + b_r). \tag{6.4}$$

With these multiclass source models we have *generalized* Erlang and Engset loss stations which we denote by $M_{\mathcal{R}}/G_{\mathcal{R}}/C(0)$ and $G_{\mathcal{R}}(N_{\mathcal{R}})/G_{\mathcal{R}}/C(0)$, respectively. Letting the number of servers $C \to \infty$, we obtain generalized IS stations. The IS stations $M_{\mathcal{R}}/G_{\mathcal{R}}/\infty$ and $G_{\mathcal{R}}(N_{\mathcal{R}})/G_{\mathcal{R}}/\infty$ are simple examples of *open* and *closed* queueing networks, respectively. The generalized loss and IS stations may be characterized by a state process $\mathbf{n}(t) = (n_r(t) \in \mathcal{R})$, where $n_r(t)$ denotes the number of class r customers in the station at time t. For a generalized Engset station with C servers and population vector given by $\mathbf{N} = (N_r : r \in \mathcal{R})$, the set of feasible states is

$$\mathcal{S}(\mathbf{N}, C) = \left\{ \mathbf{n} \in \mathbb{Z}_+^R \sum_{r \in \mathcal{R}} : A_r n_r \leq C; \mathbf{n} \leq \mathbf{C} \right\}. \tag{6.5}$$

Denote the equilibrium state distribution by

$$\pi_C(\mathbf{N}, \mathbf{n}) = \lim_{t \to \infty} P(\mathbf{n}(t) = \mathbf{n}).$$

For a Poisson source model, the set of feasible states and the stationary distribution are denoted by $\mathcal{S}(C)$ and $\pi_C(\mathbf{n})$, respectively.

The generalized loss stations have many of the properties associated with stations found in queueing networks. A queueing station is said to be *quasireversible* if its state process $\mathbf{n}(t)$ is a stationary Markov process with the property that the state at an arbitrary time t_0 is independent of: (*i*) the arrival times of class r customers after time t_0; and (*ii*) the departure times of class r customers prior to time t_0. The property of quasireversibility was introduced by Kelly [21] to characterize a wide class of queueing stations which, together with certain rules governing customer routing, give rise to product-form queueing networks. We extend this definition to loss stations by assuming the convention that the departure process includes both customers who successfully complete service and those who are blocked and do not receive service. A closely related property is reversibility. A stochastic process $n(t)$ is *reversible* if it is statistically identical to the time-reversed process $n^R(t) = n(\tau - t)$, where τ is an arbitrary time origin. For a stationary Markov process, reversibility holds if and only if its stationary distribution satisfies the *detailed balance equations*. The following theorem gives a characterization of the generalized Erlang station in terms of these properties (see Appendix A for a simple proof).

Theorem 6.1. *The generalized Erlang station is quasireversible and its state-process $\mathbf{n}(t)$ is a reversible Markov process with stationary distribution given by*

$$\pi_C(\mathbf{n}) = \frac{1}{G(C)} \prod_{r \in \mathcal{R}} \frac{a_r^{n_r}}{n_r!}, \quad \mathbf{n} \in \mathcal{S}(C), \tag{6.6}$$

where $a_r = \lambda_r/\mu_r$ and $G(C)$ is the normalization constant defined by

$$G(C) \triangleq \sum_{\mathbf{n} \in \mathcal{S}(C)} \prod_{r \in \mathcal{R}} \frac{a_r^{n_r}}{n_r!}.$$

Note that the stationary distribution reduces to that of the Erlang station when there is a single customer class. By using the quasireversibility property of the Erlang station, we give an analogous characterization for the generalized Engset station.

Theorem 6.2. *For the generalized Engset loss system, $\mathbf{n}(t)$ is a reversible Markov process with stationary distribution:*

$$\pi_C(\mathbf{N}, \mathbf{n}) \frac{1}{G(\mathbf{N}, C)} \prod_{r \in \mathcal{R}} \binom{N_r}{n_r} b_r^{n_r}, \quad \mathbf{n} \in \mathcal{S}(\mathbf{N}, C), \tag{6.7}$$

where $b_r = \nu_r / \mu_r$, $r \in \mathcal{R}$ and $G(\mathbf{N}, C)$ is the normalization constant defined by

$$G(\mathbf{N}, C) \triangleq \sum_{\mathbf{n} \in \mathcal{S}(\mathbf{N}, C)} \prod_{r \in \mathcal{R}} \binom{N_r}{n_r} b_r^{n_r}.$$

Proof. This loss system can be viewed as a two-station closed queueing network consisting of a (generalized) IS station in tandem with a loss station (see Figure 2). With Poisson source models, both stations are quasireversible and possess the insensitivity property with respect to service time distributions. Hence, the stationary distribution of the tandem connection has the form

$$\pi(\mathbf{n}^1, \mathbf{n}^2) \propto \pi_1(\mathbf{n}^1) \pi_2(\mathbf{n}^2),$$

where $\mathbf{n}^1$ and $\mathbf{n}^2$ represent the number of customers of each class at the IS station and the loss station, respectively, and π_i, $i = 1, 2$ denote the marginal distributions of the two stations. Hence,

$$\pi(\mathbf{n}^1, \mathbf{n}^2) \propto \prod_{r \in \mathcal{R}} \frac{(\lambda_r / \mu_r)^{n_r^1}}{n_r^1!} \frac{(\lambda_r / \nu_r)^{n_r^2}}{n_r^2!}.$$

By making the identification $\mathbf{N} - \mathbf{n} = \mathbf{n}^1$ and $\mathbf{n} = \mathbf{n}^2$ and applying the state truncation property with the closed network constraint $\mathbf{n}^1 + \mathbf{n}^2 = \mathbf{N}$, we obtain that the stationary distribution of the original system has the form

$$\pi(\mathbf{N}, \mathbf{n}) \propto \prod_{r \in \mathcal{R}} \binom{N_r}{n_r} \left(\frac{\nu_r}{\mu_r} \right)^{n_r}$$

which, upon normalization, yields the result (6.7). Reversibility can be established by showing that the distribution given by (6.7) satisfies the detailed balance equations. $\square$

We observe that the Poisson source model may be considered as a limiting case of the finite source model as the population size increases towards infinity. Consider the generalized Engset loss station in a regime where for each $r \in \mathcal{R}$,

$$N_r \to \infty, \quad b_r \to 0,$$

such that $N_r b_r = a_r$ remains constant. In this regime,

$$\binom{N_r}{n_r} b_r^{n_r} \to \frac{a_r^{n_r}}{n_r!}, \quad r \in \mathcal{R}.$$

Hence,

$$\pi_C(\mathbf{N}, \mathbf{n}) \to \pi_C(\mathbf{n}),$$

i.e., the stationary distribution of the Engset station approaches that of an Erlang station.

6.2.2 Open, closed, and mixed loss networks

Let us now classify the servers of a loss station by means of the set $\mathfrak{J} = \{1,\ldots,J\}$ of server *types*. There are C_j servers of type $j \in \mathfrak{J}$ such that $\sum_{j \in \mathfrak{J}} C_j = C$. Thus, the servers of the loss station are specified by a vector $\mathbf{C} = (C_j : j \in \mathfrak{J})$. An arriving class r customer seeks to simultaneously acquire A_{jr} servers of type j (see Figure 3). This generalized loss station with multiple server types is denoted by $\cdot/G_{\mathfrak{J},\mathfrak{R}}/C_{\mathfrak{J}}(0)$. Letting $\mathbf{C} \to \infty$, we obtain a generalized IS station with multiple server types.

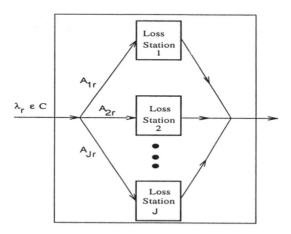

Figure 3. Generalized loss station with multiple server types.

This station operation provides a model for the operation of a *circuit-switched network* under a fixed routing with the following interpretation: each server represents a *circuit* and $\mathfrak{J}$ is a set of links, where link j consists of C_j circuits. The customer represents *calls* to the network seeking to establish connections or *routes*, which are represented by customer classes in the loss station model. A *route* consists of a path of links from a source to a destination node. The nodes or switches of a circuit-switched network are not explicitly taken into account in the loss network model, but they are implicitly specified by the set of routes. A call on a route holds a fixed number of circuits on each link on the path associated with the route. The notion of route subsumes the notion of call class and we use the symbol $\mathfrak{R}$ to denote the set of all routes in the loss network. A route r call requires A_{jr} circuits on each link $j \in \mathfrak{J}$. Thus, in our terminology, a *loss network* is a generalized loss station with multiple server types and we denote it by $\cdot/G_{\mathfrak{J},\mathfrak{R}}/C_{\mathfrak{J}}(0)$. Letting the link capacities $\mathbf{C} \to \infty$, we obtain an infinite-server network $\cdot/G_{\mathfrak{J},\mathfrak{R}}/\infty$.

As depicted in Figure 3, the loss network may be viewed as a system of generalized loss stations $\cdot/G_{\mathfrak{R}}/C(0)$. An arriving call at a particular route seeks to simultaneously acquire servers from a subset of the loss stations. Each of the loss stations corresponds to a server type. In the loss network context, a single link corresponds to a generalized loss station.

We introduce source models for a loss network as follows. We partition the routes of a loss network into the set of open routes and the set of closed routes, denoted by $\mathcal{R}_O$ and $\mathcal{R}_C$, respectively, with $\mathcal{R} = \mathcal{R}_O \cup \mathcal{R}_C$. The call arrival process to an open route p is Poisson with rate λ_p, while that for a closed route s is a finite source of population N_s with mean inter-generation time $1/\nu_s$. All call arrival processes are assumed to be independent. An *open loss network*, as depicted in Figure 4, consists entirely of open routes, while a *closed loss network* consists entirely of closed routes. A *mixed loss network*, as shown in Figure 5, has both kinds of routes and we denote it by $M_{\mathcal{R}_O}, G_{\mathcal{R}_C}(N_{\mathcal{R}_C})/G_{\mathcal{J},\mathcal{R}}/C_{\mathcal{J}}(0)$.

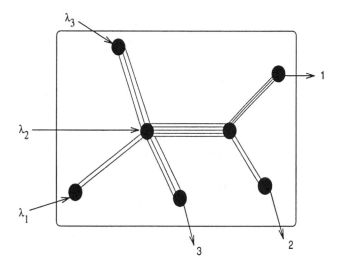

Figure 4. Open loss network.

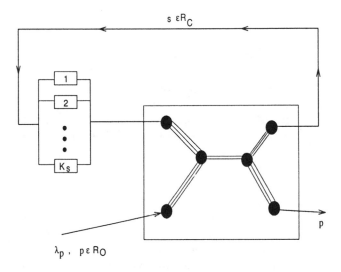

Figure 5. Mixed loss network.

A mixed loss network generalizes both the Erlang and Engset stations in the previous section. Denote its state process by $\mathbf{n}(t) = [n_r(t) : r \in \mathcal{R}]$. Analogous to

Theorem 6.1, an open loss network is quasireversible. We summarize the properties of the stationary state distribution for the mixed loss network as follows.

Theorem 6.3. *The state process of the mixed loss network is a reversible Markov process with steady state distribution given by*

$$\pi_C(\mathbf{N},\mathbf{n}) = \frac{1}{G(\mathbf{N},\mathbf{C})} \prod_{p \in \mathcal{R}_O} \frac{a_p^{n_p}}{n_p!} \prod_{s \in \mathcal{R}_C} \binom{N_s}{n_s} b_s^{n_s}, \quad \mathbf{n} \in \mathcal{N}(\mathbf{C}), \qquad (6.8)$$

where $b_s = \nu_s/\mu_s$, $s \in \mathcal{R}_C$,

$$\mathcal{I}(\mathbf{N},\mathbf{C}) = \{\mathbf{n} \in \mathbb{Z}_+^R : \mathbf{A}\mathbf{n} \le \mathbf{C}; n_s \le N_s, s \in \mathcal{R}_C\}, \qquad (6.9)$$

and

$$G(\mathbf{N},\mathbf{C}) = \sum_{\mathbf{n} \in \mathcal{I}(\mathbf{N},\mathbf{C})} \prod_{p \in \mathcal{R}_O} \frac{a_p^{n_p}}{n_p!} \prod_{s \in \mathcal{R}_C} \binom{N_s}{n_s} b_s^{n_s}. \qquad (6.10)$$

Open and closed routes of a loss network may be viewed as analogs to open and closed subchains in a queueing network with multiple customer classes [3, 32]. To illustrate this analogy, consider a queueing network consisting of the set $\mathcal{N} = \{1,\ldots,N\}$ of queueing stations and the set $\mathcal{R}$ of customer classes. Customer routing is governed by a first-order Markov chain $\mathcal{M}$, with the transition probability matrix P (often called the *switching matrix*). In general, $\mathcal{M}$ is decomposable into L irreducible subchains $\mathcal{M}_1,\ldots,\mathcal{M}_L$ (cf. [32]). The subchains are of two types: the set $\mathcal{J}_O$ of *open* subchains and the set $\mathcal{J}_C$ of *closed* subchains. Open subchains, $l \in \mathcal{J}_O$, are driven by independent Poisson arrival processes of rates λ_l., with $p_{(0,c)}^{(n,c)} = p_{(n,c)}^{(0,c)} = 0$ for all $(n,c) \in \mathcal{M}_l$. The links and calls of a loss network may be regarded as analogs, respectively, to the stations and the customers of the queueing network. The open routes $p \in \mathcal{R}_O$ are driven by independent Poisson processes of rates λ_p. In a closed route $s \in \mathcal{R}_C$, the population is held constant at K_l.. However, in a loss network, the notion of route subsumes a further classification of calls (e.g., calls of class c on route r) since a call does not change its class membership during its lifetime. This contrasts with the subchain concept in a queueing network, where a customer may change its class membership after completing service at a station within the subchain. In a subchain, a customer receives service (after possibly waiting for service) and then moves to a next station and changes its class membership according to subchain transition probabilities. On the other hand, the route concept involves simultaneous occupancy of circuits from a fixed set of links.

6.3 BLOCKING FORMULAE AND GENERATING FUNCTIONS

The main performance measure for a loss network is the blocking or loss probability. The analogous measure of interest in queueing networks is the mean sojourn time. The remainder of this chapter focuses on computing or approximating blocking probabilities in loss networks. In this section, we define and obtain expressions for blocking probabilities in loss networks. The blocking probabilities are expressed in terms of the stationary state distribution of the loss network. In particular, the blocking probabilities can be expressed in terms of *normalization constants* (also called *partition functions*). Usually (as in the previous section) the state is specified by a vector $\mathbf{n} = (n_r : r \in \mathcal{R})$, giving the number of calls carried on each route. We refer to this state description as the *route state*. An alternative state descriptor, the *link state*, is the vector $\mathbf{m} = (m_j : j \in \mathcal{J})$, which gives the number of calls carried on each link. We examine the relations between the two state descriptors and look at generating functions of the link state and the normalization constant. The generating function representation of the normalization constant is the basis for several methods of calculating the blocking probabilities.

6.3.1 Blocking probabilities

The term *blocking probability* may refer to either *time congestion* or *call congestion*. The *time congestion* for a given class of calls is defined as the proportion of time that the network state is such that no new calls of the same class can be accepted. *Call congestion* is the proportion of call arrivals in a given class that are blocked. Since the state process of a loss network is ergodic, the call congestion and time congestion are stationary probabilities. For Poisson arrivals, there is no ambiguity in the use of the term *blocking probability*, since it follows from the PASTA property (cf. [42]) that the stationary state distribution seen by an arriving call is the same as the stationary distribution at an arbitrary point in time. Hence, in this case, the call congestion and the time congestion are equal.

Consider, for example, an Erlang loss station with C servers and an offered load of a erlangs. The time congestion and call congestion are both equal to the stationary probability that all C servers are busy. The blocking probability is given by the celebrated Erlang B loss formula:

$$E(C,a) \triangleq \frac{a^C}{C!} G(C)^{-1} = \pi_C(C), \tag{6.11}$$

where $G(C) = \sum_{i=0}^{C} \frac{a^i}{i!}$ is the normalization constant. Now consider an Engset station with C servers, a population of N customers, and the offered load b. No blocking can occur if $N \leq C$. If $N > C$, the time congestion is given by the Engset loss formula

$$T(N,C,b) \triangleq \binom{N}{C} b^n G(N,C)^{-1} = \pi_C(N,C), \tag{6.12}$$

where $G(N,C) = \sum_{i=1}^{C} \binom{N}{i} b^i$ is the normalization constant. To determine the call congestion, note that the expected number of customers arriving to the station when there are n busy servers is given by

$$(N-n)\nu \pi_C(N,n).$$

The expected number of arrivals per unit time is given by

$$\sum_{n=0}^{C} (N-n)\nu \pi_C(N,n).$$

Hence, the call congestion is given by

$$\widetilde{T}(N,C,b) = \frac{(N-C)\nu \pi_C(N,C)}{\sum_{n=0}^{C}(N-n)\nu \pi_C(N,n)} = \binom{N-1}{C} b^C G(N-1,C)^{-1}$$

$$= T(N-1, C, b). \tag{6.13}$$

The last equality shows that the call congestion for an Engset station with population N is equal to the time congestion for an Engset station with population $N-1$.

For a mixed loss network, the time congestion on route $r \in \mathcal{R}$, denoted by L_r, can be expressed as:

$$L_r = 1 - \frac{G(\mathbf{N}, \mathbf{C} - \mathbf{A}_r)}{G(\mathbf{N}, \mathbf{C})}. \tag{6.14}$$

Denote the call congestion on route $r \in \mathcal{R}$ by $\tilde{L}_r$. Since the open routes are driven by Poisson arrival processes, it follows from the PASTA property that

$$L_r = \tilde{L}_r, \quad r \in \mathcal{R}_0. \tag{6.15}$$

The proportion of arriving calls over a closed route $s \in \mathcal{R}_C$ that find the system in state $\mathbf{n}$ is

$$a_s(\mathbf{N}, \mathbf{n}) = \frac{(N_s - n_s)\nu_s \pi_C(\mathbf{N}, \mathbf{n})}{\sum_{\mathbf{n}' \in \mathcal{N}(\mathbf{N}, \mathbf{C})} (N_s - n'_s)\nu_s \pi_C(\mathbf{N}, \mathbf{n}')}.$$

Using (6.8) and performing some algebraic manipulations yields the simple result

$$a_s(\mathbf{N}, \mathbf{n}) = \pi_C(\mathbf{N} - \mathbf{e}_s, \mathbf{n}),$$

where $\mathbf{e}_s$ denotes the R-dimensional unit vector with a one in the entry corresponding to route s. This is a generalization of the well-known property for the quasi-random input model (cf. [22]). The call congestion for route s calls is then given by

$$\tilde{L}_s = a_s(\mathbf{N}, \mathbf{C}) = 1 - \frac{G(\mathbf{N} - \mathbf{e}_s, \mathbf{C} - \mathbf{A}_s)}{G(\mathbf{N} - \mathbf{e}_s, \mathbf{C})}. \tag{6.16}$$

The expressions for L_r and $\tilde{L}_r$, $r \in \mathcal{R}$, given above, may be viewed as generalizations of the classical Erlang and Engset loss formulas.

The blocking probability for class r customers at a generalized Erlang station is given by the following *generalized Erlang loss formula*:

$$\varepsilon_r(C, \mathbf{a}) \triangleq 1 - \frac{G(C - A_r)}{G(C)}, \tag{6.17}$$

where $\mathbf{a} = (a_r : r \in \mathcal{R})$ is the vector of offered loads. The time congestion for class s customers in a generalized Engset station may be expressed as

$$\mathcal{T}_s(\mathbf{N}, C, \mathbf{a}) \triangleq 1 - \frac{G(\mathbf{N}, C - A_s)}{G(\mathbf{N}, C)} \tag{6.18}$$

while the call congestion is given by

$$\tilde{\mathcal{T}}_s(\mathbf{N}, C, \mathbf{a}) \triangleq 1 - \frac{G(\mathbf{N} - \mathbf{e}_s, C - A_s)}{G(\mathbf{N} - \mathbf{e}_s, C)}. \tag{6.19}$$

The blocking probabilities are expressed succinctly in terms of normalization constants. The normalization constant is a sum of product-form terms which is generally difficult to compute directly. As $C \to \infty$, $G(\mathbf{N}, C)$ increases monotonically to the normalization constant for an IS network:

$$G(\mathbf{N}, \infty) = \exp\left(\sum_{r \in \mathcal{R}_O} a_r\right) \prod_{s \in \mathcal{R}_s} (1 + b_s)^{N_s}. \tag{6.20}$$

Later in the chapter, we will discuss efficient computational algorithms and approximations for calculating the normalization constant and blocking probabilities of a loss network.

6.3.2 Link state distribution

Rather than characterizing the loss network in terms of route populations, an alternative description is provided by the link occupancies. Let $m_j(t)$ denote the number of occupied circuits on link j at time t. Then the process $\mathbf{m}(t) = (m_j(t): j \in \mathcal{J})$ is related to the state process $\mathbf{n}(t)$ as follows:

$$\mathbf{m}(t) = \mathbf{A}\mathbf{n}(t).$$

Denote the stationary distribution of $\mathbf{m}(t)$ by $p_C(\mathbf{N}, \mathbf{m}) = \lim_{t \to \infty} P(\mathbf{m}(t) = \mathbf{m})$. The link state distribution $p_C(\mathbf{N}, \mathbf{m})$ and the route state distribution $\pi_C(\mathbf{N}, \mathbf{n})$ are related as follows:

$$p_C(\mathbf{N}, \mathbf{m}) = \sum_{\mathbf{n}: \, \mathbf{An} = \mathbf{m}} \pi_C(\mathbf{N}, \mathbf{n}), \quad \mathbf{m} \in \mathcal{M}(\mathbf{N}, \mathbf{C}),$$

where

$$\mathcal{M}(\mathbf{N}, \mathbf{C}) = \{\mathbf{m} \geq 0 \colon \mathbf{An} = \mathbf{m}, \mathbf{n} \in \mathcal{G}(\mathbf{N}, \mathbf{C})\}.$$

This gives,

$$p_C(\mathbf{N}, \mathbf{m}) = \frac{1}{G(\mathbf{N}, \mathbf{C})} \sum_{\mathbf{n}: \, \mathbf{An} = \mathbf{m}} \prod_{r \in \mathcal{R}} \frac{a_r^{n_r}}{n_r!}, \quad \mathbf{m} \in \mathcal{M}(\mathbf{N}, \mathbf{C}).$$

The normalization constant $G(\mathbf{N}, \mathbf{C})$ is the same as that defined earlier in terms of the route state distribution. In terms of the link state distribution, we have

$$G(\mathbf{N}, \mathbf{C}) = \sum_{\mathbf{m} \in \mathcal{M}(\mathbf{N}, \mathbf{C})} p_C(\mathbf{N}, \mathbf{m}).$$

The loss probability for a class on route $r \in \mathcal{R}$ can be expressed in terms of the link state distribution:

$$L_r = \sum_{\mathbf{m} \in \mathcal{M}_r(\mathbf{C})} p_C(\mathbf{m}),$$

where

$$\mathcal{M}_r(\mathbf{N}, \mathbf{C}) = \{\mathbf{m} \in \mathcal{M}(\mathbf{N}, \mathbf{C}) \colon \mathbf{m} + \mathbf{A}_r \not\leq \mathbf{C}\}$$

is the set of link states such that a new call on route r cannot be accepted.

The link state probabilities may be computed directly via the following recursion:

$$m_j p_C(\mathbf{N}, \mathbf{m}) = \sum_{r \in \mathcal{R}_O} A_{jr} a_r p_C(\mathbf{N}, \mathbf{m} - \mathbf{A}_r)$$

$$+ \sum_{s \in \mathcal{R}_C} A_{js} N_s b_s \sum_{k=1}^{N_s} (-b_s)^{k-1} p_C(\mathbf{N}, \mathbf{m} - k\mathbf{A}_s), \quad j \in \mathcal{J}. \tag{6.21}$$

This recursion is a generalization of a recursion for generalized Erlang stations, first reported by Kaufman [19] and Roberts [33] (see Appendix B). An alternative form of this recursion, involving the two arguments $\mathbf{m}$ and $\mathbf{N}$ is

$$m_j p_C(\mathbf{N}, \mathbf{m}) = \sum_{r \in \mathcal{R}_O} A_{jr} a_r p_C(\mathbf{N}, \mathbf{m} - \mathbf{A}_r)$$

$$+ \sum_{s \in \mathcal{R}_C} A_{js} N_s b_s p_C(\mathbf{N} - \mathbf{e}_s, \mathbf{m} - k\mathbf{A}_s), \quad j \in \mathcal{J}. \tag{6.22}$$

In practice, the recursion (6.21) would usually be preferable to (6.22), since the recursion in two arguments involves more computation and significantly higher storage requirements. However, as we shall see later on, the practical applicability of recursion (6.21) appears to be limited to networks of special structure, in particular, loss stations.

6.3.3 Generating functions

The probability generating functions (pgf's) for the stationary route state distribution of the mixed loss network are defined by

$$\pi_C^*(\mathbf{N}, \mathbf{z}) \triangleq E[\mathbf{z}^\mathbf{n}] = \sum_{\mathbf{n} \in \mathcal{S}(\mathbf{N}, \mathbf{C})} \pi_C(\mathbf{N}, \mathbf{n}) \mathbf{z}^\mathbf{n}, \tag{6.23}$$

where $\mathbf{z}^\mathbf{n} \triangleq \prod_{r \in \mathcal{R}} z_r^{n_r}$. We may write the generating function in the form

$$\pi_C^*(\mathbf{N}, \mathbf{z}) = \frac{1}{G(\mathbf{N}, \mathbf{C})} \sum_{\mathbf{n} \in \mathcal{S}(\mathbf{N}, \mathbf{C})} \prod_{p \in \mathcal{R}_O} \frac{(a_p z_p)^{n_p}}{n_p!} \prod_{s \in R_C} \binom{N_s}{n_s} (b_s z_s)^{n_s}. \tag{6.24}$$

Letting $\mathbf{C} \rightarrow \infty$ and using (6.20), it follows that the pgf for the route-state distribution of an IS network is given by

$$\pi_\infty^*(\mathbf{N}, \mathbf{z}) = \exp\left\{ \sum_{p \in \mathcal{R}_O} a_p(z_p - 1) \right\} \prod_{s \in \mathcal{R}_C} \left(\frac{1 + b_s z_s}{1 + b_s} \right)^{N_s}. \tag{6.25}$$

Note that $\pi_\infty^*(\mathbf{N}, \mathbf{z})$ is just the product of the marginal pgf's of the Poisson random variables n_p, $p \in \mathcal{R}_O$ and the binomial random variables n_s, $s \in \mathcal{R}_C$.

Define the pgf for the link state distribution $p_C(\mathbf{m})$ by

$$p_C^*(\mathbf{N}, \mathbf{z}) = E[\mathbf{z}^\mathbf{m}],$$

where $\mathbf{z}^\mathbf{m} = \prod_{j \in \mathcal{J}} z_j^{m_j}$ (for convenience, we shall sometimes suppress the population vector $\mathbf{N}$ in the notation). The pgf for the link state distribution of an IS network has a simple form (see Appendix B):

$$p_\infty^*(\mathbf{N}, \mathbf{z}) = \sum_{\mathbf{m} \geq 0} p_\infty(\mathbf{N}, \mathbf{m}) \mathbf{z}^\mathbf{m} = G(\mathbf{N}, \infty) \prod_{r \in \mathcal{R}_O} \exp(a_r \mathbf{z}^{\mathbf{A}_r}) \cdot \prod_{s \in \mathcal{R}_C} (1 + b_s \mathbf{z}^{\mathbf{A}_s})^{N_s}, \tag{6.26}$$

where $G(\mathbf{N}, \infty)$ is the normalization constant for an IS network:

$$G(\mathbf{N}, \infty) = \exp\left(-\sum_{r \in \mathcal{R}_O} a_r \right) \cdot \prod_{s \in \mathcal{R}_C} (1 + b_s)^{N_s}. \tag{6.27}$$

The pgf $p_\infty^*(\mathbf{N}, \mathbf{z})$ provides an alternative means for deriving the recurrence (6.21). The partial derivative of $p_\infty^*(\mathbf{N}, \mathbf{z})$ with respect to $\mathbf{z}$ is given by

$$\frac{\partial p_\infty^*(\mathbf{N}, \mathbf{z})}{\partial z_j} = \sum_{\mathbf{m} \geq 0} m_j p(\mathbf{N}, \mathbf{m}) \mathbf{z}^{\mathbf{m} - \mathbf{e}_j} = \sum_{\mathbf{m} \geq 0} (m_j + 1) p(\mathbf{N}, \mathbf{m} + \mathbf{e}_j) \mathbf{z}^\mathbf{m}. \tag{6.28}$$

Using the above expression for the pgf, the above partial derivative can also be written as

$$\frac{\partial p_\infty^*(\mathbf{N}, \mathbf{z})}{\partial z_j} = K \frac{\partial}{\partial z_j} \left(\exp(\sum_{r \in \mathcal{R}_O} a_r \mathbf{z}^{\mathbf{A}_r}) + \sum_{s \in \mathcal{R}_C} N_s \ln(1 + b_s \mathbf{z}^{\mathbf{A}_s}) \right)$$

$$= K p_\infty^*(\mathbf{N}, \mathbf{z}) (\sum_{r \in \mathcal{R}_O} a_r A_{jr} \mathbf{z}^{\mathbf{A}_r - \mathbf{e}_j})$$

$$+ \sum_{s \in \mathcal{R}_C} N_s b_s A_{js} \mathbf{z}^{\mathbf{A}_s - \mathbf{e}_j} (\sum_{k \geq 1} (-b_s)^{k-1} \mathbf{z}^{(k-1)\mathbf{A}_s}) \tag{6.29}$$

Equating coefficients of $\mathbf{z}^{\mathbf{n}}$ in (6.28) and (6.29) yields

$$(m_j + 1) p(\mathbf{N}, \mathbf{m} + \mathbf{e}_j) = \sum_{r \in \mathcal{R}_O} a_r A_{jr} p(\mathbf{N}, \mathbf{m} + \mathbf{e}_j - \mathbf{A}_r)$$

$$+ \sum_{s \in \mathcal{R}_C} N_s b_s A_{js} \sum_{k \geq 1} (-b_s)^{k-1} p(\mathbf{N}, \mathbf{m} + \mathbf{e}_j - k\mathbf{A}_s).$$

Substituting $\mathbf{m}$ for $\mathbf{m} + \mathbf{e}_j$ gives the recurrence (6.21). Note that we have dropped the subscript $\mathbf{C}$ in the link state distribution $p(\mathbf{N}, \mathbf{m})$ since the recurrence applies both to the IS network and the case of finite $\mathbf{C}$.

Define the generating function of the normalization constant $G(\mathbf{N}, \mathbf{C})$ with respect to the J-dimensional vector $\mathbf{C}$ by

$$G^*(\mathbf{N}, \mathbf{z}) = \sum_{\mathbf{C} \geq 0} G(\mathbf{N}, \mathbf{C}) \mathbf{z}^{\mathbf{C}}. \tag{6.30}$$

The generating function $G^*(\mathbf{N}, \mathbf{z})$ has a simple product form (see Appendix B):

$$G^*(\mathbf{N}, \mathbf{z}) = \prod_{j \in \mathcal{J}} \left(\frac{1}{1 - \mathbf{z}_j} \right) \exp \left(\sum_{r \in \mathcal{R}_O} a_r \mathbf{z}^{\mathbf{A}_r} \right) \prod_{s \in \mathcal{R}_C} (1 + b_s \mathbf{z}^{\mathbf{A}_s})^{N_s}. \tag{6.31}$$

From (6.26) and (6.31), we see that $G^*(\mathbf{N}, \mathbf{z})$ and $p_\infty^*(\mathbf{N}, \mathbf{z})$ are related as follows:

$$G^*(\mathbf{N}, \mathbf{z}) \propto \left(\prod_{j \in \mathcal{J}} \frac{1}{1 - z_j} \right) p_\infty^*(\mathbf{N}, \mathbf{z}).$$

In the case $J = 1$ (i.e, for a loss stations), the generating function $G^*(\mathbf{N}, z)$ is a formal power series in variable z. The normalization constant $G(\mathbf{N}, C)$ can be expressed in terms of $G^*(\mathbf{N}, z)$ as follows:

$$G(\mathbf{N}, C) = [z^C] G^*(\mathbf{N}, z) = \frac{1}{C!} \frac{d^C}{dz^C} G^*(\mathbf{N}, z) \Big|_{z=0}, \tag{6.32}$$

where $[z^C]$ denotes the operator which extracts the coefficient of z^C in a formal power series. Taking z to be a complex variable, $G^*(\mathbf{N}, z)$ is an analytic function in z. By Cauchy's integral formula [37] we can write

$$G(\mathbf{N}, C) = \frac{1}{2\pi i} \int_{C_r} \frac{G^*(\mathbf{N}, z)}{z^{C+1}} dz, \tag{6.33}$$

where C_r is a circle in the complex plane with radius $r < 1$. Representations (6.32) and (6.33) suggest methods of computing the normalization constant, which will be discussed in the following sections. In the multidimensional case,

$$G(\mathbf{N}, \mathbf{C}) = [\mathbf{z}^{\mathbf{C}}] G^*(\mathbf{z}) = \frac{1}{\mathbf{C}!} \frac{d^{\mathbf{C}}}{d\mathbf{z}^{\mathbf{C}}} G^*(\mathbf{N}, \mathbf{z}) \Big|_{\mathbf{z}=0},$$

where $\mathbf{C}! = C_1! \ldots C_j!$ and

$$\frac{d^{\mathbf{C}}}{d\mathbf{z}^{\mathbf{C}}} = \frac{\partial^{C_1 + \ldots + C_J}}{\partial z_1^{C_1} \ldots \partial z_J^{C_J}}.$$

The multidimensional version of Cauchy's integral formula leads to the representation

$$G(\mathbf{N}, \mathbf{C}) = \frac{1}{(2\pi i)^J} \oint_\Gamma \frac{G^*(\mathbf{N}, \mathbf{z})}{\mathbf{z}^{\mathbf{C}+\mathbf{1}}} \, d\mathbf{z}, \tag{6.34}$$

where $\mathbf{1}$ is the vector of all ones and Γ is a simple closed contour about the origin.

6.4 RECURSIVE ALGORITHMS

6.4.1 Classical models

The blocking probability for an Erlang loss station is given by the Erlang B loss formula in equation (6.11). A brute-force evaluation of the Erlang formula involves the summation of $O(C)$ factorial terms. For large C, this can lead to numerical instability. The normalization constant may be written as follows:

$$G(C) = \sum_{i=0}^{C} \frac{a^i}{i!} = G(C-1) + \frac{a^C}{C!}.$$

Dividing both sides by $a^C/C!$ and recalling the Erlang B formula, we obtain

$$E(C,a)^{-1} = \frac{C}{a} E(C-1,a)^{-1} + 1, \quad C = 1, 2, \dots. \tag{6.35}$$

With the initial condition $E(0,a) = 1$, (6.35) is an efficient recursion for computing $E(C,a)$ which requires $O(C)$ work. Recursion (6.35) also holds for non-integer values of C when $E(C,a)$ is extended to real values of C via an integral representation (cf. [18]). An analogous recursion can be obtained for the Engset model. Write the normalization constant as

$$G(N,C) = \sum_{i=0}^{C} \binom{N}{i} b^i = G(N, C-1) + \binom{N}{C} b^C.$$

Dividing both sides by $\binom{N}{C} b^C$ we get

$$T(N,C,b)^{-1} = \frac{C}{b(N-C+1)} T(N, C-1, b)^{-1} + 1. \tag{6.36}$$

With the initial condition $T(N,0,b) = 1$, (6.36) yields an $O(C)$ recursion for computing $T(N,C,b)$ for a fixed value of N and intermediate values of C.

Alternatively, $G(N,C)$ can be written as

$$G(N,C) = \sum_{i=0}^{C} \binom{N}{i} b^i = \sum_{i=0}^{C} \left\{ \binom{N-1}{i} + \binom{N-1}{i-1} \right\} b^i,$$

which suggests the following recursion in N and C:

$$G(N,C) = G(N-1,C) + bG(N-1,C-1). \tag{6.37}$$

With the initial condition $G(N,0) = 1$ for all N, (6.37) provides a recursion for computing $G(N,C)$ for all intermediate values of N and C. The blocking probability for the Engset station is then given by

$$T(N,C,b) = 1 - \frac{G(N,C-1)}{G(N,C)}. \tag{6.38}$$

In practice, the normalization constants $G(N-1,C)$ and $G(N,C)$ may be large, which could lead to numerical instabilities in taking their ratio. It is therefore of interest to develop a recursion directly in terms of $T(N,C,b)$, analogous to (6.36). Dividing both sides of (6.37) by $G(N,C)$ gives

$$1 = \frac{G(N-1,C)}{G(N,C)} + b\frac{G(N-1,C-1)}{G(N,C)}. \tag{6.39}$$

Defining

$$B(N,C) = \frac{G(N,C-1)}{G(N,C)}, \quad F(N,C) = \frac{G(N-1,C)}{G(N,C)},$$

we can write (6.39) in two ways:

$$1 = F(N,C) + bB(N,C)F(N,C-1) \tag{6.40}$$

and

$$1 = F(N,C) + bB(N-1,C)F(N,C). \tag{6.41}$$

From (6.40) and (6.41), the following recursion in $B(N,C)$ can be obtained:

$$B(N,C) = \frac{B(N-1,C)(1 + bB(N-1,C-1))}{1 + bB(N-1,C)}. \tag{6.42}$$

The blocking probability is then given by

$$T(N,C,b) = 1 - B(N,C). \tag{6.43}$$

Recursion (6.42) was obtained by Pinsky, Conway, and Liu [31]. The computational and storage requirements for the recursion are both $O(NC)$.

6.4.2 Generalized loss stations

The recursions given by (6.35), (6.36), and (6.42) are efficient methods of computing blocking probabilities for Erlang and Engset stations if the state distributions are not desired. Unfortunately, these recursions do not allow direct extensions to the generalized loss stations. It is worthwhile to note that the recursion (6.37) can be extended to the generalized Engset station as follows:

$$G(\mathbf{N},C) = G(\mathbf{N} - \mathbf{e}_s, C) + b_s G(\mathbf{N} - \mathbf{e}_s, C - A_s), \quad s \in \mathfrak{R}, \tag{6.44}$$

with the initial condition $G(\mathbf{N},0) = 1$ for all $\mathbf{N}$. However, like recursion (6.22), (6.44) is a recursion in two arguments and may have high computation and storage requirements in practical problems.

Let us specialize recurrence (6.21) to the case of generalized loss stations. For a generalized Erlang station, (6.21) becomes

$$mp(m) = \sum_{r \in \mathfrak{R}} a_r p(m - A_r), \tag{6.45}$$

subject to the normalization condition $\sum_{m=0}^{C} p(m) = 1$. This recursion was obtained by Kaufman [19] and Roberts [33]. The computational effort is $O(RC)$ and the storage requirement is $O(C)$. The blocking probabilities are given by

$$B_r = \sum_{i=0}^{A_r} p(C - i), \quad \in \mathfrak{R}. \tag{6.46}$$

If just the blocking probabilities are desired, only the values

$$\pi(C-i), \quad i = 0, \ldots, \max_{r \in \mathcal{R}} A_r$$

need to be stored. For the generalized Engset model, recurrence (6.21) reduces to

$$mp(\mathbf{N}, m) = \sum_{r \in \mathcal{R}} A_r N_r b_r \sum_{k=1}^{N_r} (-b_r)^{k-1} p(\mathbf{N}, m - kA_r), \tag{6.47}$$

subject to the normalization requirement $\sum_{m=0}^{C} p(\mathbf{N}, m) = 1$. The computational effort is $O(RC \max_{r \in \mathcal{R}} N_r)$ and the storage requirement is $O(C)$. The blocking probabilities are given as in (6.46).

An alternative recursion can be obtained by considering the generating function for the normalization constant. Specializing (6.31) for the generalized Erlang station, we have

$$G^*(z) = \frac{1}{1-z} \cdot \prod_{r \in \mathcal{R}} \exp(a_r z^{A_r}). \tag{6.48}$$

Let $G_r^*(z)$ and $G_r(C)$, respectively, denote the generating function and normalization constant for a loss station with the set of customer classes $\{1, \ldots, r\}$. In particular, $G_R^*(z) = G^*(z)$ and $G_R(C) = G(C)$. Then we can write

$$G^*(z) = G_{R-1}^*(z) \cdot \exp(a_R z^{A_R}). \tag{6.49}$$

It follows that the normalization can be written as

$$\begin{aligned}
G_R(c) &= [z^C] G_{R-1}^*(z) \cdot \exp(a_R z^{A_R}) \\
&= \sum_{l=0}^{C} [z^{C-l}] G_{R-1}^*(z) \cdot [z^l] \exp(a_R z^{A_R}).
\end{aligned} \tag{6.50}$$

Clearly, $[z^{C-l}] G_{R-1}^*(z) = G_{R-1}(C-l)$, and

$$[z^l] \exp(a_R z^{A_R}) = \begin{cases} \dfrac{a_R^i}{i!}, & \text{if } iA_R = l \text{ for some } i \\ 0, & \text{otherwise.} \end{cases} \tag{6.51}$$

Substituting these expressions into (6.49) we deduce the following relation:

$$G_r(C) = \sum_{i=0}^{\lfloor C/A_r \rfloor} G_{r-1}(C - iA_r) \cdot \frac{a_r^i}{i!}, \quad r = 2, \ldots, R. \tag{6.52}$$

This recursion was obtained by Kaufman [19] and is similar to the Buzen recursion for computing normalization constants in closed queueing networks [8].

A similar recursion can be derived for the normalization constant of the generalized Engset station. The associated generating function is given by

$$G^*(\mathbf{N}, z) = \frac{1}{1-z} \cdot \prod_{s \in \mathcal{R}} (1 + b_s z^{A_s})^{N_s}. \tag{6.53}$$

Following an argument similar to the above, we obtain the recursion[1]

$$G_r(C) = \sum_{i=0}^{\lfloor C/A_r \rfloor} G_{r-1}(C - iA_r) \cdot \binom{N_r}{i} b_r^i, \quad r = 2, \ldots, R. \tag{6.54}$$

---

[1] For convenience, here we suppress the population vector in the notation for the normalization constant.

This recursion was obtained by Kogan [25], who also gives an efficient implementation of the algorithm using two arrays.

Kogan's method can be applied to mixed loss stations with both open and closed customer classes. Denote the two arrays by $\mathbf{a}[k]$ and $\mathbf{b}[k]$, $k = 0, 1, \ldots, C$. The basic outline of the algorithm is given below:

Initialize:

    1. $l \leftarrow 1$

    2. $\mathbf{a}[k] \leftarrow G_1(k), \quad k = 0, 1, \ldots, C$

Repeat:

    3. $l \leftarrow l + 1$

    4. $\mathbf{b}[k] \leftarrow G_l(k), \quad k = 0, \ldots, C$

    5.   if $l = R$ then stop: $G(C) = \mathbf{b}[C]$

              else $\mathbf{a}[k] \leftarrow \mathbf{b}[k], \quad k = 0, 1, \ldots, C$.

In the initialization step, the normalization constant value for a loss station with capacity k and only class 1 customers is computed and stored in $\mathbf{a}[k]$, for $k = 0, 1, \ldots, C$. If class 1 is open, the computation in step 2 is

$$\mathbf{a}[k] \leftarrow \sum_{i=0}^{\lfloor \frac{k}{A_1} \rfloor} a_1^i, \quad k = 0, 1, \ldots, C. \tag{6.55}$$

If class 1 is closed, the required computation is

$$\mathbf{a}[k] \leftarrow \sum_{i=0}^{\min(\lfloor \frac{k}{A_1} \rfloor, N_1)} \binom{N_1}{i} b_1^i, \quad k = 0, 1, \ldots, C. \tag{6.56}$$

Carrying out these computations is equivalent to finding the normalization constants for classical loss stations with capacities $[k/A_1]$. The computational effort is $O(C/A_1)$ and $O(N_1 C/A_1)$ for open and closed customer classes, respectively. The computation in step 4 is

$$\mathbf{b}[k] \leftarrow \sum_{i=0}^{\lfloor \frac{k}{A_l} \rfloor} a_l^i \mathbf{a}[k - iA_l], \quad k = 0, 1, \ldots, C, \tag{6.57}$$

if l is closed and

$$\mathbf{b}[k] \leftarrow \sum_{i=0}^{\min(\lfloor \frac{k}{A_l} \rfloor, N_l)} \binom{N_l}{i} b_l^i \mathbf{a}[k - iA_l], \quad k = 0, 1, \ldots, C, \tag{6.58}$$

if l is open. These computations can be carried out efficiently in a recursive manner. The computational effort is $O(C^2/A_l^2)$ for open l and $O(C^2 N_l/A_l^2)$ for closed l.

6.4.3 More general loss networks

The algorithms that have been discussed so far are efficient methods for computing blocking probabilities in the classical and generalized loss stations for small to moderate values of C, R, and N_r, $r \in \Re$. Unfortunately, for large values of these parameters, the recursive methods can take an excessive computation time and suffer from loss of accuracy. For loss networks of general topology, one can develop a recursive method from recurrence relation (6.21), but such an algorithm would be computationally impractical for most realistic networks. For an open loss network with all link capacities equal to C, the recursive method requires $O(RC^J)$ of work. Moreover, the storage requirement is $O(C^J)$ and the recursion is susceptible to problems of numerical accuracy. The applicability of recursive algorithms to loss networks appears to be restricted to loss stations and other networks of special topological structure. Tsang and Ross [38], for example, develop algorithms for *tree networks* (see Subsection 6.5.2).

6.5 NUMERICAL EVALUATION OF THE INVERSION INTEGRAL

Recently, there has been much interest in efficient algorithms for performing numerical inversion of generating functions. Abate and Whitt [1, 2] present an algorithm based on the Fourier-series method to numerically invert probability generating functions. Versions of this algorithm have been applied to the transient M/G/1 queue [12], computing the normalization constants of closed queueing networks [11], and computing the normalization constants of resource sharing models analogous to the loss networks discussed here. The results obtained by Choudhury, Leung, and Whitt [11] using the numerical transform inversion technique indicate that the method often can be more efficient than the recursive algorithms discussed in the previous section and is potentially applicable to a larger class of loss networks.

6.5.1 One-dimensional case

In the one-dimensional case, equation (6.33) gives the integral representation of the normalization constant[2]

$$G(C) = \frac{1}{2\pi i} \int_{C_r} \frac{G^*(z)}{z^{C+1}} dz.$$

With the change of variables $z = re^{i\theta}$, the above integral can be rewritten as

$$G(C) = \frac{1}{2\pi r^C} \int_{-\pi}^{\pi} G^*(re^{i\theta}) e^{-i\theta C} d\theta.$$

Using a step approximation to the integral with lC equally spaced samples taken over the interval $[-\pi, \pi]$, we have

$$G(C) \approx \frac{1}{2r^C lC} \sum_{k=-lC}^{lC} G^*(re^{i\pi k/lC}) e^{-\pi k/lC}.$$

In fact, an expression for the error involved in the approximation can be obtained via the Poisson summation formula (cf. [2]). Including the error term, we have

$$G(C) = \frac{1}{2r^C lC} \sum_{k=-lC}^{lC} G^*(re^{i\pi k/lC}) e^{-i\pi k/lC} - \epsilon_A, \qquad (6.59)$$

where the *aliasing error* is given by

$$\epsilon_A = \sum_{n=1}^{\infty} G(C + 2nlC) r^{2nlC}.$$

If the normalization constants $G(C)$ are bounded by a known constant K (e.g.,

---

[2]For convenience, we suppress the population vector $\mathbf{N}$ and write $G(C)$ for $G(\mathbf{N}, C)$, etc.

if $G^*(z)$ is a pgf and the $G(C)$ are probabilities), then the aliasing term can be bounded by

$$|\epsilon_A| \leq K_1 \frac{r^{2lC}}{1 - r^{2lC}}.$$

In general, however, the normalization constants can grow without bound and the aliasing error ϵ_A can be arbitrarily large. In order to control the aliasing error, the normalization constants are scaled by means of two parameters α and β as follows:

$$\widetilde{G}(C) = \alpha \beta^C G(C).$$

The corresponding *scaled generating function* is given by

$$\widetilde{G}^*(z) = \alpha G^*(\beta z).$$

With the scaled generating function, the aliasing error becomes

$$\widetilde{\epsilon}_A = \sum_{n=1}^{\infty} \widetilde{G}(C + 2nlC)r^{2nlC}.$$

Since the blocking probabilities are given as ratios of normalization constants, it is of interest to control the *relative aliasing error* defined by

$$\epsilon_R = \frac{\widetilde{\epsilon}_A}{\widetilde{G}(C)}.$$

This relative error is bounded by

$$|\epsilon_R| \leq \sum_{n=1}^{\infty} \left| \frac{\widetilde{G}(C + 2nlC)}{\widetilde{G}(C)} \right| r^{2nlC} = \alpha^{2lC} \sum_{n=1}^{\infty} \left| \frac{G(C + 2nlC)}{G(C)} \right| r^{2nlC}.$$

The parameters α and β are chosen in order to keep the relative aliasing error small and yet avoid numerical underflow or overflow. In practice, a choice of α and β must be made heuristically, since the values of $G(C)$ are not known *a priori*.

The basic inversion formula (6.59) involves a sum of $2lC$ terms. In order to accelerate the convergence of the sum, Choudhury, Leung, and Whitt [10] propose truncating some of the terms involved in the summation. The basic idea of truncation is to identify the terms which dominate the sum in the inversion formula. The inversion formula consists of a weighted sum of values of the generating function at points along the circle of radius r in the complex plane. As the capacities C grow, the generating function tends to become unevenly distributed along the circle. By identifying the relative maximum points, the computation of the sum can be reduced significantly by including only the points around the maxima that have non-negligible relative amplitudes. In general, the location of the maximum points are not known, so that heuristics based on these ideas are used to truncate the summation in the inversion formula. Similar to the scaling strategy, truncation is first applied to a tractable model such as the Erlang loss station, and the basic approach is extended heuristically to handle more general models.

6.5.2 Multidimensional case

Consider a general J-dimensional generating function $G^*(z)$. Define a sequence of partial generating functions as follows:

$$G^*_{(j)}(\mathbf{z}_j; \mathbf{C}_{j+1}) = \sum_{C_1=0}^{\infty} \cdots \sum_{C_j=0}^{\infty} G(\mathbf{C}) \prod_{i=1}^{j} z_i^{C_i}, \quad 1 \leq j \leq J,$$

where $\mathbf{z} = (z_1, \ldots, z_j)$ and $\mathbf{C}_j = (C_j, C_{j+1}, \ldots, C_J)$. By convention, take $\mathbf{z}_0$ and C_{J+1} to be null vectors. Note that $\mathbf{C} = \mathbf{C}_1$ and $\mathbf{z} = \mathbf{z}_J$. The partial generating function $G^*_{(j)}(\mathbf{z}_j, \mathbf{C}_{j+1})$ is j-dimensional in the variables $z_1, \ldots, z_j$ and in particular, $G^*_{(J)}(\mathbf{z}) = G^*(\mathbf{z})$. The jth and $(j-1)$th partial generating functions are related by

$$G^*_{(j)}(\mathbf{z}_j; \mathbf{C}_{j+1}) = \sum_{C_j = 0}^{\infty} G^*_{(j-1)}(\mathbf{z}_{j-1}; \mathbf{C}_j) z_j^{C_j},$$

i.e., $G^*_{(j)}(\mathbf{z}_j, \mathbf{C}_{j+1})$ is the generating function of $G^*_{(j-1)}(\mathbf{z}_{j-1}, \mathbf{C}_j)$ in variable z_j.

Applying the Cauchy integral formula (6.33) to $G^*_{(j)}(\mathbf{z}_j, \mathbf{C}_{j+1})$ gives

$$G^*_{(j-1)}(\mathbf{z}_{j-1}; \mathbf{C}_j) = \frac{1}{2\pi i} \int_{C_{r_j}} \frac{G^*_{(j)}(\mathbf{z}_j; \mathbf{C}_{j+1})}{z_j^{C_j+1}} dz_j, \qquad (6.60)$$

where C_{r_j} denotes a circle of radius r_j about the origin.

The numerical inversion algorithm proceeds by applying the one-dimensional inversion formula recursively. In order to compute the right-hand side of (6.60) for $j = 1$, the inversion formula must be applied recursively for $j = 2, \ldots, J$. For $j = J$, the right-hand side of (6.60) can be evaluated explicitly as $G^*_{(J)}(\mathbf{z}_J, \mathbf{C}_J) = G(\mathbf{C})$. The one-dimensional numerical inversion formula at the jth step is

$$G^*_{(j-1)}(\mathbf{z}_{j-1}, \mathbf{C}_j)$$

$$= \frac{1}{2 l_j C_j r_j^{C_j}} \sum_{k = -l_j C_j}^{l_j C_j - 1} G^*_{(j)}(\mathbf{z}_{j-1}; r_j e^{i\pi k/(l_j C_j)}; \mathbf{C}_{j+1}) e^{-i\pi k/l_j} - \epsilon_j, \quad (6.61)$$

where the aliasing error is given by

$$\epsilon_j = \sum_{n=1}^{\infty} G^*_{(j-1)}(\mathbf{z}_{j-1}; C_j + 2n l_j C_j, \mathbf{C}_{j+1}).$$

Applying (6.61) recursively gives a nested sum expression for the normalization constant:

$$G(\mathbf{C}) = \prod_{j \in \mathfrak{J}} \frac{1}{2 l_j C_j r_j^{C_j}} \sum_{k_1 = -l_1 C_1}^{l_1 C_1 - 1} \cdots$$

$$\sum_{k_J = -l_J C_J}^{l_J C_J - 1} G^*(r_1 e^{i\pi k_1/l_1 C_1}, \ldots, r_J e^{i\pi k_J/l_J C_J}) \exp\left(-i\pi \sum_{j \in \mathfrak{J}} k_j/l_j\right). \qquad (6.63)$$

Thus, the normalization constant can be computed using a nested sequence of loops. As in the one-dimensional case, practical limits of numerical precision require us to choose parameters α_j and β_j to scale the normalization constant at the jth inversion step. For large C_j, truncation can usually be applied to reduce the number of terms in the jth nested sum.

In many cases, the overall dimension of the inversion can be reduced by exploiting the structure of the generating function. Suppose, for example, that the generating function can be written as a product of M factors

$$G^*(\mathbf{z}) = \prod_{i=1}^{M} G_i(\widehat{\mathbf{z}}_i), \tag{6.64}$$

where $\widehat{\mathbf{z}}_i \subseteq \{z_1, \ldots, z_J\}$ and $\widehat{\mathbf{z}}_i \cap \widehat{\mathbf{z}}_j = \emptyset$ for $i = j$. Then, the nested sum expression (6.63) can be written as a product of M independent nested sums of smaller dimension. The dimension of the overall inversion carried out in this manner is then the maximal cardinality $D = \max_{1 \le i \le M} |\widehat{\mathbf{z}}_i|$, of the sets $\widehat{\mathbf{z}}_i$.

More generally, a dimension reduction can be accomplished by choosing a subset $\widehat{\mathbf{z}}$ containing d of the J variables z_j and treating these variables as constants. Then, the generating function in the $J - d$ remaining variables is written as a product of m factors, where no two factors have any of the variables (excluding the subset $\widehat{\mathbf{z}}$) in common. The normalization constant can then be written as a sequence of d nested sums corresponding to the inversion with respect to the variables in $\widehat{\mathbf{z}}$, followed by a product of M nested sums corresponding to each of the factors. Let D be the maximum number of variables appearing in one of the factors. Then the dimension of the overall inversion is effectively $d + D$. Ideally, one can choose a set of d variables to minimize the resulting dimension $d + D$. The problem of choosing a best set of d variables can be formulated as a graph problem which is NP-hard (see [10]). In practice, the variables can often be chosen by inspection or by enumerating subsets of increasing size d.

As an example, consider a so-called *tree network* (cf. [38]). In a tree network, links $1, \ldots, J - 1$ are called *access* links and link J is called the *common link*. There are $R = J - 1$ routes and the rth one requires A_r circuits from link r and link J. Arrivals are assumed to be Poisson with intensity a_r for route r. The generating function corresponding to the tree network is given by

$$G^*(\mathbf{z}) = \prod_{j \in \mathcal{J}} \left(\frac{1}{1 - z_j} \right) \cdot \exp \left(\sum_{r \in \mathcal{R}} a_r z_r z_J \right).$$

In this case we choose the variable z_j, regard it as a constant, and write the generating function as the following product of factors:

$$G^*(\mathbf{z}) = \frac{1}{1 - z_J} \prod_{j=1}^{J-1} \left(\frac{1}{1 - z_j} \right) \exp(a_j z_j z_J).$$

Hence, it is clear that the overall dimension of the inversion can be reduced to 2, independent of J.

6.5.3 Computational complexity

The computational effort required in the recursive inversion procedure is exponential in the dimension of the generating function J. Assume that each link in the loss network has capacity C. Then the recursive numerical inversion procedure requires $O(C^J)$ evaluations of the normalization constant. Each evaluation involves $O(R)$ work. The effort required to do scaling is insignificant in comparison. Hence, the overall computational complexity of the general numerical inversion algorithm applied to the loss network is $O(RC^J)$. For large C, the algorithm is of practical use only if J is relatively small.

For large C, a truncation can be applied at each inversion step, effectively reducing the parameter C to $\bar{C} \ll C$. By exploiting the structure in particular loss networks, a dimension reduction can be used to reduce J to $\bar{J} \ll J$. Thus, by applying both the truncation and dimension reduction, the computational complexity for particular loss network models can be reduced to $O(R\bar{C}^{\bar{J}})$. The storage requirement of the algorithm is $O(R)$. For the special case of generalized loss stations (both Erlang and Engset) we have $J = 1$. Thus, the computational effort is $O(RC)$ for moderate values of C and $O(R\bar{C})$ for large values of C for which truncation can effectively be applied. We have already seen that the computational effort required for the tree network is $O(R\bar{C}^2)$.

6.6 ASYMPTOTIC APPROXIMATION OF THE INVERSION INTEGRAL

For generalized loss stations, $G^*(z) = \sum_{C=0}^{\infty} G(C)z^C$ is an analytic function in a disk D centered at the origin. By Cauchy's integral formula, the normalization constant $G(C)$ can be expressed as a contour integral

$$G(C) = \frac{1}{2\pi i} \oint_{\Gamma} \frac{G^*(z)}{z^{C+1}} dz, \qquad (6.65)$$

where Γ is a simple closed contour in D containing the origin.

In this section we shall discuss asymptotic approximations for the normalization constants and blocking probabilities of generalized loss stations in limiting regimes where $C \to \infty$. The approximations are based on well-established techniques in the asymptotic approximation of integrals (cf. [26, 43, 16]). Our discussion is based on the results of Kogan [25] and Mitra and Morrison [28] who obtain asymptotic approximations for the generalized Engset model. Extensions to more general loss networks involves additional complications in the asymptotic approximation of multidimensional integrals and is an interesting area of future research (see Section 6.8). We mention that asymptotic approximations have also been applied to integral representations for normalization constants in certain types of queueing networks (cf. Birman and Kogan [5]).

6.6.1 Saddle point method

The inversion integrals we shall be concerned with in this section have the form

$$I(C) = \int_{\Gamma} q(z)e^{Cf(z)} ds, \qquad (6.66)$$

where $q(z)$ and $f(z)$ are analytic functions in a domain containing the contour Γ and C is a large positive parameter. The *saddle point method* is a technique involving derivation of asymptotic expansions for integrals of the form (6.66) in the regime $C \to \infty$. By Cauchy's theorem, it is possible to deform continuously the contour Γ to a new contour Γ', homotopic to Γ, such that the value of the integral is unchanged. In the saddle point method, the contour Γ' is chosen to satisfy the following conditions [16, 43]:

(i) $\max_{z \in \Gamma'} \mathrm{Re} f(z)$ is attained at a finite number of points $z_1, \ldots, z_k$, which are either *saddle points* of $f(z)$ or endpoints of Γ'.

(ii) $\mathrm{Im} f(z)$ is constant on Γ'.

These properties define Γ' as a *saddle point contour*.

A saddle point of $f(z)$ is any point z_0 satisfying

$$f'(z_0) = 0. \tag{6.67}$$

z_0 is a *saddle point of order n* $(n \geq 1)$ if

$$f'(z_0) = f''(z_0) = \ldots = f^{(n)}(z_0) = 0, \ f^{(n+1)}(z_0) \neq 0.$$

Consider the representation

$$f(z) = u(x,y) + iv(x,y),$$

where $z = x + iy$ and u, v are real functions. From the Cauchy-Riemann equations, $f'(z) = u_x - iu_y$. Thus, if $z_0 = x_0 + iy_0$ satisfies (6.67), then

$$u_x(x_0, y_0) = u_y(x_0, y_0) = 0.$$

Since u is a harmonic function, the last equations and the maximum modulus theorem imply that (x_0, y_0) must be a saddle point of $u(x,y)$. Hence, z_0 is a saddle point of $f(z)$. An equivalent form of condition (ii) in the definition of a saddle point contour, that will be useful in the sequel, is [16]:

> (ii') The points $z_1, \ldots, z_k$ at which $\max_{z \in \Gamma'} \mathrm{Re} f(z)$ is attained are either saddle points or endpoints of Γ'. Furthermore, Γ' cannot be deformed into a domain of smaller values of $\mathrm{Re} f(z)$ in all neighborhoods of these points, i.e.,
>
> $$f(z_i) = \max_{z \in \Gamma'} \mathrm{Re} f(z) = \min_{\widetilde{\Gamma}} \min_{z \in \widetilde{\Gamma}} \ \mathrm{Re} f(z), \quad i = 1, \ldots k, \tag{6.68}$$
>
> where the minimum is assumed over all contours $\widetilde{\Gamma}$ such that the integrals over Γ and $\widetilde{\Gamma}$ are equal.

Given a saddle point contour Γ', the asymptotic form of $I(C)$ can be obtained by applying the so-called *principle of localization* due to Laplace. The basic idea behind Laplace's method is that if the integrand attains a sharp maximum at a point x_0, then (assuming x_0 is unique) the integral is approximately equal to an integral over a small neighborhood of this point. More generally, let us define the *contribution* of a point z_i, at which $M = \max_{z \in \Gamma'} f(z)$ is attained, as the integral

$$V_{z_i}(C) = \int_{U(z_i) \cap \Gamma'} q(z)e^{Cf(z)}dz, \tag{6.69}$$

where $U(z_i)$ is a small neighborhood of point z_i. The asymptotic form of $I(C)$ is given by the summation in the contributions of the points $z_1, \ldots, z_k$:

$$I(C) = \sum_{i=1}^{k} V_{z_i}(C) + O(\exp(C(M-c))), \quad C \to \infty, \tag{6.70}$$

where $c > 0$. The last term in (6.70) is exponentially small in comparison with the contributions $V_{z_i}(C)$. The following formula gives the contribution of an interior simple (i.e., of order one) saddle point z_0 [16]:

$$V_{z_0}(C) = e^{Cf(z_0)} \sqrt{\frac{-2\pi}{Cf''(z_0)}} \left[q(z_0) + O(C^{-1})\right]. \tag{6.71}$$

6.6.2 Application to generalized Engset station

The generating function for the normalization constant of the generalized Engset station is given by (6.53):

$$G^*(z) = \frac{1}{1-z} \prod_{s \in \mathcal{R}} \left(1 + b_s z^{A_s}\right)^{N_s}. \tag{6.72}$$

Using (6.65), the normalization constant can be expressed as

$$G(\mathbf{N}, C) = \frac{1}{2\pi i} \oint_{\Gamma} h(z) e^{Cf(z)} dz, \tag{6.73}$$

where

$$f(z) = \sum_{s \in \mathcal{R}} \gamma_s \log(1 + b_s z^{A_s}) - \log z, \quad h(z) = \frac{1}{z(1-z)}, \tag{6.74}$$

and $\gamma_s = N_s/C$. We shall consider the limiting regime in which $C \to \infty$ and $N_s \to \infty$ such that γ_s remains constant for each $s \in \mathcal{R}$. In (6.73), we take the contour Γ as a circle about the origin oriented in the counter-clockwise direction.

We seek an asymptotic approximation to the integral representation for normalization constant given by (6.73). To apply the saddle point method, we look for saddle points of $f(z)$.

Lemma 6.4. $f(z)$ *has a unique saddle point* z_0 *on the real axis. Furthermore,* z_0 *is a simple saddle point, positive, and minimizes* $f(z)$ *on the positive real axis.*

Proof. Differentiating the expression for $f(z)$ in (6.74), we obtain

$$f'(z) = \sum_{s \in \mathcal{R}} \frac{\gamma_s b_s A_s z^{A_s - 1}}{1 + b_s z^{A_s}} - \frac{1}{z}.$$

Let $g(z) = z f'(z)$. Note that $g(0) = -1$ and that $g(z)$ increases monotonically for $z > 0$ to the limit

$$g(\infty) = \lim_{z \to \infty} g(z) = \sum_{s \in \mathcal{R}} \gamma_s A_s.$$

We may assume that

$$\sum_{s \in \mathcal{R}} N_s A_s > C,$$

since otherwise no blocking would occur. Hence, $g(\infty) > 1$. By the continuity and monotonicity of $g(z)$, there must be a unique point $z_0 > 0$ satisfying $g(z_0) = f(z_0) = 0$, or equivalently,

$$\sum_{s \in \mathcal{R}} \frac{\gamma_s b_s A_s z_0^{A_s}}{1 + b_s z_0^{A_s}} = 1. \tag{6.75}$$

To see that z_0 is a simple saddle point, note that

$$(z_0)^2 f''(z_0) = \frac{\gamma_s b_s A_s^2 z_0^{A_s}}{(1 + b_s z_0^{A_s})^2} > 0. \tag{6.76}$$

This inequality also establishes that z_0 minimizes $f(z)$ on the positive real axis. $\quad\square$

Let $\Gamma' = \{ \,|z| = z_0 \}$ be the circle of radius z_0, taken in the counter-clockwise direction. We would like to use Γ' as a saddle point contour for (6.73).

Lemma 6.5. *The contour* Γ' *satisfies properties* (i) *and* (ii') *defining a saddle point contour.*

Proof. Without loss of generality, we may assume that A_s, $s \in \mathcal{R}$ have no common factors. Writing $z = re^{i\theta}$, we have

$$\operatorname{Re} f(z) = \operatorname{Re} f(re^{i\theta}) = \sum_{s \in \mathcal{R}} \gamma_s \log(1 + b_s r^{A_s} e^{iA_s \theta}) - \log r$$

$$\leq \sum_{s \in \mathcal{R}} \gamma_s \log(1 + b_s r^{A_s}) - \log r = f(r), \tag{6.77}$$

with equality if and only if θ is a multiple of 2π. Inequality (6.77) implies that

$$\max_{|z| = r} \mathrm{Re} f(z) = f(r), \tag{6.78}$$

and in particular, when $r = z_0$, that z_0 is the unique value that maximizes $\mathrm{Re} f(z)$ on the circle Γ'. This establishes property (i).

From Lemma 6.4,

$$f(z_0) = \min_{r > 0} f(r). \tag{6.79}$$

Combining (6.78) and (6.79), we have

$$f(z_0) = \max_{z \in \Gamma'} \mathrm{Re} f(z) = \min_{r > 0} \max_{|z| = r} \mathrm{Re} f(z), \tag{6.80}$$

which establishes property (ii). $\qquad\Box$

Since $h(z)$ has a pole at $z = 1$, the location of the saddle point z_0 with respect to $z = 1$ is critical in applying the saddle point method. If $z_0 < 1$, Γ' can be used directly as a saddle point contour. If $z_0 > 1$, the pole at $z = 1$ must be taken into account. Finally, if $z_0 \approx 1$, the result (6.71) is no longer valid and a different asymptotic approximation must be used. Alternatively, Mitra and Morrison [28] develop a uniform asymptotic approximation for the integral (6.73) which will be discussed in the next subsection.

The value of z_0 is related to the traffic load at the loss station. Define the *normalized load* by[3]

$$\rho = \sum_{s \in \mathcal{R}} \gamma_s p_s A_s. \tag{6.81}$$

Note that $\rho > 1$ implies

$$\sum_{s \in \mathcal{R}} N_s p_s A_s > C. \tag{6.82}$$

The left-hand side of (6.82) represents the average number of servers requested at the loss station. Hence, if $\rho > 1$ the station is said to be *overloaded*. Similarly, if $\rho < 1$, the loss station is said to be *underloaded*. Finally, if $\rho \approx 1$, the station is *critically loaded*. A more precise characterization of the critically loaded regime is the condition [28]:

$$\rho = 1 - \frac{\delta}{\sqrt{C}}, \tag{6.83}$$

where $\delta > 0$ is a constant.

Proposition 6.6. *The generalized Engset station is*
(a)   *overloaded if and only if $z_0 < 1$,*
(b)   *underloaded if and only if $z_0 > 1$,*
(c)   *critically loaded if and only if $z_0 \approx 1$.*

Proof. As in the proof of Lemma 6.4, let $g(z) = z f'(z)$. Note that $g(0) = -1$ and $g(z)$ is monotonically increasing for $z > 0$. We have $g(z_0) = f'(z_0) = 0$ and also

$$g(1) = f'(1) = \sum_{s \in \mathcal{R}} \frac{\gamma_s b_s A_s}{1 + b_s} - 1 = \rho - 1.$$

Hence, $z_0 < 1$ if and only if $g(1) > 0$, or equivalently, $\rho > 1$. Similarly, observe that $z_0 > 1$ if and only if $\rho < 1$ and $z_0 \approx 1$ if and only if $\rho \approx 1$. $\qquad\Box$

In the overloaded regime, $z_0 < 1$ and Γ' is a saddle point contour for (6.73).

---

[3] Recall from (6.4) that p_s is the stationary probability that a class s source s customer is in service at an IS station.

Hence we can apply the result (6.71) to obtain the following asymptotic form for $G(\mathbf{N}, C)$:

$$G(\mathbf{N}, C) = \frac{e^{Cf(z_0)}}{z_0(1 - z_0)\sqrt{2\pi C f''(z_0)}}[1 + O(C^{-1})], \quad C \to \infty, \qquad (6.84)$$

From (6.76), the second derivative of $f(z)$ at $z = z_0$ is given by

$$f''(z_0) = \frac{\gamma_s b_s A_s^2 z_0^{A_s - 2}}{(1 + b_s z_0^{A_s})^2} > 0. \qquad (6.85)$$

In the underloaded regime, $z_0 > 1$ and the pole of $h(z)$ at $z = 1$ lies inside the contour Γ'. Hence, the integral must also include the contribution from the residue at $z = 1$. Using (6.71) for the contribution of the saddle point z_0, we obtain the following asymptotic form for the underloaded case:

$$G(\mathbf{N}, C) = e^{Cf(1)} + \frac{e^{Cf(z_0)}}{z_0(1 - z_0)\sqrt{2\pi C f''(z_0)}}[1 + O(C^{-1})], \quad C \to \infty. \qquad (6.86)$$

In the critically loaded regime, $z_0 \approx 1$, so that the saddle point z_0 lies close to the pole at $z = 1$. The result (6.71) is not valid in this regime. Kogan [25] obtains the following asymptotic form for the critically loaded case:

$$G(\mathbf{N}, C) = \tfrac{1}{2}e^{Cf(1)}\mathrm{erfc}\{\mathrm{sgn}\,(1 - z_0)\sqrt{C[f(1) - f(z_0)]}\}$$

$$+ \frac{e^{Cf(z_0)}}{\sqrt{1\pi d C f''(z_0)}}\left[\frac{1}{z_0} + \frac{f'''(z_0)}{f''(z_0)} + O(C^{-1})\right]. \qquad (6.87)$$

Approximations of the form (6.87) for the case where a saddle point lies near a singularity are discussed by Wong [43]. They are also used in the context of queueing networks by Birman and Kogan [5].

6.6.3 Uniform asymptotic approximations and blocking probabilities

The asymptotic approximations for $G(\mathbf{N}, C)$ discussed in the previous subsection are specialized to three regimes, depending on the location of the saddle point z_0. Accuracies of these approximations depend on the value of z_0. Hence, it is desirable to have an asymptotic approximation which is *uniform* in z_0. Mitra and Morrison [28] derive a uniform asymptotic approximation (UAA) for $G(\mathbf{N}, C)$ based on techniques developed by Bleistein [6]. Their approximation is given as follows:

$$G(\mathbf{N}, C) = V[1 + O(C^{-1})], \qquad (6.88)$$

where

$$V = \tfrac{1}{2}e^{Cf(1)}\mathrm{erfc}\{\mathrm{sgn}\,(1 - z_0)\sqrt{C[f(1) - f(z_0)]}\}$$

$$+ \frac{e^{Cf(z_0)}}{\sqrt{2\pi C}}\left\{\frac{1}{z_0(1 - z_0)\sqrt{f''(z_0)}} - \frac{\mathrm{sgn}\,(1 - z_0)}{\sqrt{1[f(1) - f(z_0)]}}\right\}. \qquad (6.89)$$

For $z_0 \approx 1$, i.e., $|1 - z_0| \ll 1$, the UAA result (6.88) reduces to the approximation in (6.87). More importantly, the UAA result is uniformly valid in all three traffic loading regimes, including the case $z_0 \to 1$ [28].

Using the asymptotic form (6.88), uniform asymptotic approximations for the blocking probabilities at a generalized Engset station can be obtained. Recall that the *time congestion* for a class s customer is given by (cf. (6.18))

$$\mathcal{T}_s = \frac{G(\mathbf{N},C) - G(\mathbf{N}, C - A_s)}{G(\mathbf{N},C)}. \tag{6.90}$$

Expressing the numerator $G(\mathbf{N},C) - G(\mathbf{N}, C - A_s)$ as a contour integral, we can write

$$\mathcal{T}_s = \frac{1}{G(\mathbf{N},C)2\pi i} \oint_{\Gamma} (1 - z^{A_s})h(z)e^{Cf(z)}dz, \tag{6.91}$$

where $h(z)$ and $f(z)$ are defined in (6.74). Note that the pole of $h(z)$ at $z = 1$ is canceled by the zero at $z = 1$ from the term $1 - z^{A_s}$. Hence we can obtain an asymptotic form for the integral in (6.91) using the saddle point contour Γ', irrespective of the location of the saddle point z_0. Applying the result (6.71) for the contribution of z_0 and substituting the asymptotic form (6.88) for $G(\mathbf{N},C)$ in (6.91) yields the following UAA for the time congestion of class s customers:

$$\mathcal{T}_s = \frac{e^{Cf(z_0)}(1 - z_0^{A_s})}{V\sqrt{2\pi C f''(z_0)}z_0(1 - z_0)}[1 + O(C^{-1})]. \tag{6.92}$$

Similarly, a UAA can be obtained for the *call congestion*, given by (cf. (6.19))

$$\widetilde{\mathcal{T}}_s - \frac{G(\mathbf{N} - \mathbf{e}_s, C) - G(\mathbf{N} - \mathbf{e}_s, C - A_s)}{G(\mathbf{N} - \mathbf{e}_s, C)}. \tag{6.93}$$

Expressing the numerator term $G(\mathbf{N} - \mathbf{e}_s, C) - G(\mathbf{N} - \mathbf{e}_s, C - A_s)$ as a contour integral, we may write the call congestion as:

$$\widetilde{\mathcal{T}}_s - \frac{1}{G(\mathbf{N} - \mathbf{e}_s, C)2\pi i} \oint_{\Gamma} \frac{(1 - z^{A_s})h(z)}{1 + b_s z^{A_s}}e^{Cf(z)}dz. \tag{6.94}$$

Again, the zero at $z = 1$ from the term $1 - z^{A_s}$ cancels the pole at $z = 1$ of $h(z)$. Hence, using the saddle point contour Γ', we can obtain an asymptotic form of the integral in (6.94) which is uniform in the saddle point z_0. To obtain a UAA for $\widetilde{\mathcal{T}}_s$, we need a UAA for $G(\mathbf{N} - \mathbf{e}_s, C)$, which is given as follows:

$$(1 + b_s)G(\mathbf{N} - \mathbf{e}_s, C) = \widetilde{V}_s[1 + O(C^{-1})], \tag{6.95}$$

where

$$\widetilde{V}_s = \frac{1}{2}e^{Cf(1)}\mathrm{erfc}\{\mathrm{sgn}\,(1 - z_0)\sqrt{C[f(1) - f(z_0)]}\}$$

$$+ \frac{e^{Cf(z_0)}}{\sqrt{2\pi C}}\left\{\frac{1 + b_s}{\sqrt{f''(z_0)}z_0(1 - z_0)(1 + b_s z_0^{A_s})} - \frac{\mathrm{sgn}(1 - z_0)}{\sqrt{2[f(1) - f(z_0)]}}\right\}. \tag{6.96}$$

Applying the saddle point method and the UAA (6.95) to (6.94) results in the following UAA for the call congestion of class s customers:

$$\widetilde{\mathcal{T}}_s = \frac{e^{Cf(z_0)}(1 - z_0^{A_s})(1 + b_s)}{\widetilde{V}_s\sqrt{2\pi C f''(z_0)}z_0(1 - z_0)(1 + b_s z_0^{A_s})}[1 + O(C^{-1})]. \tag{6.97}$$

Mitra and Morrison [28] obtain simpler approximations to the blocking probabilities by reducing the UAA results to the three regimes of traffic loading. These approximations are less accurate than the UAA, but yield rough estimates of the asymptotic behavior of the blocking probabilities.

Proposition 6.7. *The following are asymptotic approximations for time congestion and call congestion specialized to each of the three regimes of traffic loading:*

   1.   ***Overloaded case:*** $\rho > 1$

$$\mathcal{T}_s = \widetilde{\mathcal{T}}_s = [1 - z_0^{A_s}][1 + O(C^{-1})]. \tag{6.98}$$

2. **Underloaded case:** $\rho < 1$

$$\mathcal{T}_s = \frac{e^{C[f(1) - f(z_0)]}(1 - z_0^{A_s})}{z_0(1 - z_0)\sqrt{2\pi C f''(z_0}}[1 + O(C^{-1})], \tag{6.99}$$

$$\widetilde{\mathcal{T}}_s = \frac{e^{C[f(1) - f(z_0)]}(1 - z_0^{A_s})(1 + b_s)}{z_0(1 - z_0)(1 + b_s z_0^{A_s})\sqrt{2\pi C f''(z_0)}}[1 + O(C^{-1})]. \tag{6.100}$$

3. **Critically loaded case:** $\rho = 1 - \delta/\sqrt{C},\ \delta > 0$

$$\mathcal{T}_s = \widetilde{\mathcal{T}}_s = \frac{2A_s e^{-(\delta/\sqrt{2\sigma})^2}}{\sigma\sqrt{2\pi C}\,\mathrm{erfc}(\delta/\sqrt{2\sigma})}[1 + O(C^{-1/2})]. \tag{6.101}$$

where $\sigma^2 = 2\sum_{s \in \mathcal{R}} b_s \gamma_s (A_s/1 + b_s)^2$ *(see* [28]).

Observe that in the overloaded regime, both congestion measures tend to $1 - z_0^{A_s}$ with rate $O(C^{-1})$. In the underloaded regime, the blocking probabilities approach zero at the rate $O(C^{-1})$. Finally, for the critically loaded regime, we have that the blocking probabilities approach zero at an exponential rate in C. These conclusions can also be made by considering the nonuniform asymptotic approximations for $G(\mathbf{N}, C)$ discussed in Subsection 6.6.2 (see Kogan [25]).

6.6.4 Generalized Erlang station

As noted in Subsection 6.2.1, the Poisson source model for call arrivals can be considered as the limiting case of a finite source model as the number of sources for each class increases to infinity. In the limiting regime,

$$b_s \to 0,\ N_s \to \infty,\ \text{and } b_s N_s = a_s,\ s \in \mathcal{R},$$

the generating function (6.72) for the Engset loss station approaches the generating function for an Erlang station (cf. (6.48)):

$$G^*(z) \to G^*_\infty(z) = \frac{1}{1 - z}\exp\left(\sum_{r \in \mathcal{R}} a_r z^{A_r}\right). \tag{6.102}$$

Using Cauchy's integral formula (6.65), the normalization constant of an Erlang station can be expressed as

$$G(C) = \frac{1}{2\pi i}\oint_\Gamma h(z)e^{f_\infty(z)}dz, \tag{6.103}$$

where $h(z)$ is defined in (6.74) and

$$f_\infty(z) = \sum_{r \in \mathcal{R}} \alpha_r z^{A_r} - \log z. \tag{6.104}$$

Here we consider the limiting regime

$$a_r \to \infty,\ C \to \infty,\ \text{and } \alpha_r = a_r/C,\ r \in \mathcal{R}.$$

Asymptotic approximations for $G(C)$ and blocking probabilities can be obtained using procedures similar to those discussed for the Engset station. It can be shown that there exists a unique positive-valued saddle point z_0 of order one which satisfies the equation

$$\sum_{r \in \mathcal{R}} \alpha_r A_r z^{A_r} = 1. \tag{6.105}$$

For the Erlang loss station, the normalized traffic loading is defined as

$$\rho_\infty = \sum_{r \in \mathcal{R}} \alpha_r A_r. \tag{6.106}$$

As in the Engset case, the location of the saddle point with respect to the pole at $z = 1$ of $h(z)$ divides the system behavior into three regimes of traffic loading: overloaded ($z_0 < 1, \rho_\infty > 1$), underloaded ($z_0 > 1, \rho_\infty < 1$), and critically loaded ($z_0 \approx 1, \rho_\infty \approx 1$). Asymptotic approximations specialized to the three regimes and uniform asymptotic approximations can be obtained in a manner analogous to that for the Engset station.

More directly, the asymptotic results for the Engset station can be carried over to the Erlang case by considering the following limit regime (cf. [28]):

$$\gamma_s \to \infty, \; b_s \to 0 \text{ and } b_s \gamma_s = \alpha_s, \; s \in \mathcal{R},$$

where $\alpha_s = a_s/C$. From (6.88), we obtain the following UAA for $G(C)$:

$$G(C) = V_\infty[1 + O(C^{-1})], \tag{6.107}$$

where

$$V_\infty = \frac{1}{2} e^{Cf_\infty(1)} \mathrm{erfc}\{\mathrm{sgn}\,(1 - z_0)\sqrt{C[F_\infty(1) = f_\infty(z_0)]}\}$$

$$+ \frac{e^{Cf_\infty(z_0)}}{\sqrt{2\pi C}} \left\{ \frac{1}{z_0(1 - z_0)\sqrt{f_\infty''(z_0)}} = \frac{\mathrm{sgn}(1 - z_0)}{\sqrt{2[f_\infty(1) - f_\infty(z_0)]}} \right\}. \tag{6.108}$$

Recall that both the time and call congestion measures for the Erlang station are given by the generalized Erlang formula (cf. (6.17)):

$$\varepsilon_r = \frac{G(C) - G(C - A_r)}{G(C)}, \quad r \in \mathcal{R}. \tag{6.109}$$

Applying the saddle point method to an integral representation of the numerator $G(C) - G(C - A_r)$ and using the UAA result (6.107), we obtain the following UAA for the generalized Erlang formula:

$$\varepsilon_r = \frac{e^{Cf_\infty(z_0)}(1 - z_0^{A_r})}{V_\infty \sqrt{2\pi C f_\infty''(z_0)} z_p(1 - z_0)} [1 + O(C^{-1})]. \tag{6.110}$$

As for the Engset station, the result (6.110) can be specialized to obtain simpler approximations in each of the three regimes of traffic loading.

6.7 APPROXIMATIONS FROM LARGE SCALE ASYMPTOTICS

In the preceding section we looked at asymptotic approximations of the inversion integral for a loss station in a limiting regime where the number of servers and the offered loads increase in line with each other. The computational algorithms for loss stations and networks discussed earlier require prohibitively long running times and lose accuracy as the link capacities increase. By contrast, the asymptotic approximations become more accurate as the capacities increase. In this sense, the loss networks are more tractable in large scale. Another approach to approximating blocking probabilities in the loss network is to study the equilibrium behavior of the loss network in the large scale limit (cf. [20, 41]).

6.7.2 Most probable equilibrium state

Consider the problem of determining the most probable equilibrium state of an open loss network. Formally, we can write this problem as

(P_1) maximize: $p(\mathbf{n}) = \frac{1}{G(\mathbf{C})} \prod_{r \in \mathcal{R}} \frac{a_r^{n_r}}{n_r!}$

subject to: $\mathbf{An} \leq \mathbf{C}$,

$\mathbf{n} \geq \mathbf{0}$,

$n_r \in \mathbb{Z}, \quad r \in \mathcal{R}$.

We shall assume that the offered loads a_r and the link capacities C_l are of order V, where V is a large parameter:

$$a_r = O(V), \quad r \in \mathcal{R},$$

$$C_l = O(V), \quad l \in \mathcal{J}.$$

Since $G(\mathbf{C})$ is a constant and $\log(x)$ is monotonically increasing for $x > 0$, we can replace the objective function by (P_1) by

$$\sum_{r \in \mathcal{R}} [n_r \log a_r - \log(n_r!)] \approx \sum_{r \in \mathcal{R}} n_r \left[1 + \log\left(\frac{a_r}{n_r}\right)\right] + o(V),$$

where we have used Stirling's approximation,

$$\log(n!) \sim \left(n + \tfrac{1}{2}\right) \log n - n + \tfrac{1}{2}\log 2\pi.$$

If we relax the integer requirement on n_r, $r \in \mathcal{R}$, and drop $o(V)$ term, we obtain the following convex programming problem [20, 41]:

(P) maximize: $\sum_{r \in \mathcal{R}} n_r[1 + \log\left(\frac{a_r}{n_r}\right)]$

subject to: $\mathbf{An} \leq \mathbf{C}$,

$\mathbf{n} \geq \mathbf{0}$.

Problem (P), referred to as the *primal problem*, has the following properties: the objective function is strictly concave over $\mathbf{n} \geq \mathbf{0}$, the feasible region is closed and convex, and the constraints are linear. These properties imply the existence of a unique optimum solution $\hat{\mathbf{n}}$. The associated Lagrangian form is given by

$$L(\mathbf{n}, \mathbf{y}) = \sum_{r \in \mathcal{R}} n_r \left[1 + \log\left(\frac{a_r}{n_r}\right)\right] + \sum_{l \in \mathcal{J}} y_l \left(C_l - \sum_{r \in \mathcal{R}} A_{lr} n_r\right),$$

where $y_l \geq 0$ and $l \in \mathcal{J}$ are the *Lagrange multipliers*. The solution $\hat{\mathbf{n}}$ satisfies the Kuhn-Tucker conditions:

$$\nabla_{\mathbf{n}} L(\hat{\mathbf{n}}, \hat{\mathbf{y}}) = 0, \tag{6.111}$$

$$\hat{y}_j \left(C_j - \sum_{r \in \mathcal{R}} A_{jr} \hat{n}_r\right) = 0, \quad \hat{y}_j \geq 0, \quad j \in \mathcal{J}. \tag{6.112}$$

From (6.111) we obtain

$$\hat{n}_r = a_r \exp\left(-\sum_{j \in \mathcal{J}} y_j A_{jr}\right) = a_r \prod_{j \in \mathcal{J}} \hat{p}_j^{A_{jr}}, \tag{6.113}$$

where $\widehat{p}_j = e^{-\widehat{y}_j}$. Since $y_j \geq 0$, we can interpret p_j as the *passing probability* that link j has at least one free circuit. Then,

$$\widehat{B}_j = 1 - \widehat{p}_j \tag{6.114}$$

is the *blocking probability* on link j. If $\mathbf{A}$ is a $0-1$ matrix, (6.113) suggests that in the asymptotic regime $V \to \infty$, blocking events are independent from link to link along a given route.

The passing probabilities are determined from the complementary slackness conditions (6.112) as

$$\sum_{r \in \mathcal{R}} A_{jr} a_r \prod_{k \in \mathcal{J}} \widehat{p}_k^{A_{kr}} = C_j, \quad j \in \mathcal{J}_a \tag{6.115}$$

$$\widehat{p}_j = 1, \quad l \in \mathcal{J} \backslash \mathcal{J}_a, \tag{6.116}$$

where $\mathcal{J}_a$ is the set of $j \in \mathcal{J}$ such that the inequality constraint

$$\sum_{r \in \mathcal{R}} A_{jr} \widehat{n}_r \leq C_j \tag{6.117}$$

is active (i.e., held at equality). The left-hand side of (6.117),

$$\widehat{U}_j = \sum_{r \in \mathcal{R}} A_{jr} \widehat{n}_r = \sum_{r \in \mathcal{R}} A_{jr} a_r \prod_{k \in \mathcal{J}} \widehat{p}_k^{A_{kr}}, \tag{6.118}$$

can be interpreted as the *utilization* on link j, i.e., the total number of circuits held on link j in the most probable equilibrium state.

Relations (6.115) and (6.116) uniquely determine the passing probabilities and hence also the blocking probabilities. Based on the values $\widehat{B}_j$, $j \in \mathcal{J}$, the set of links, $\mathcal{J}$, may be subdivided into three subsets:

$$\mathcal{O} = \{j \in \mathcal{J} : \widehat{B}_j > 0\}, \quad \mathcal{C} = \{j \in \mathcal{J} : \widehat{B}_j = 0, \widehat{U}_j = C_j\},$$

and

$$\mathcal{U} = \{j \in \mathcal{J} : \widehat{B}_j = 0, \widehat{U}_j < C_j\},$$

which are the set of *overloaded links*, the set of *critically loaded links*, and the set of *underloaded links*, respectively. This classification is analogous to the classification for generalized loss stations discussed earlier in the context of the asymptotic approximations. Equations (6.115) and (6.116) imply that for any overloaded link j, all circuits on link j are in use. For moderate values of V, this approximation can be poor, as noted in [20, 41]. The approximation suggestion suggested by (6.115) and (6.116) ignores the non-negligible effect of the $O(V^{-1})$ number of free circuits available on the overloaded links.

6.7.3 Hard and soft duals

Fortunately, the product-form solution (6.113) is a significant simplification. If we accept (6.113), we can still hope to get an accurate approximation by modifying or replacing the determining equations (6.111) and (6.112) for $\widehat{p}_j$. The passing probabilities $\widehat{p}_j$ also arise from the dual problem to (P). The dual Lagrangian functional is given by

$$L_D(\mathbf{y}) = \max_{\mathbf{n} \geq \mathbf{0}} L(\mathbf{n}, y) = \sum_{r \in \mathcal{R}} \widehat{n}_r + \sum_{j \in \mathcal{J}} y_i C_j$$

$$= \sum_{r \in \mathcal{R}} a_r \exp \left(-\sum_{j \in \mathcal{J}} y_j A_{jr} \right) + \sum_{j \in \mathcal{J}} y_j C_j.$$

Hence, the dual problem is

(D_1) $$\min_{\mathbf{y} \geq \mathbf{0}} L_D(\mathbf{y}).$$

Equivalently, the dual problem can be written as an unconstrained problem [41]:

(D_2) $$\min_{\mathbf{y}} \sum_{r \in \mathcal{R}} a_r \exp\left(- \sum_{j \in \mathcal{J}} y_j A_{jr}\right) + \sum_{j \in \mathcal{J}} C_j (y_j)_+,$$

where $(x)_+ = \max(0, x)$. The set of solutions to (D_2) are in one-to-one correspondence with the solutions to (6.115) and (6.116). Moreover, if $\mathbf{A}$ is of full rank, the objective function of (D_2) is strictly convex, implying the existence of a unique solution $\widehat{y}$.

Whittle [40] refers to the problem (D_2) as the *hard* dual problem and proposes a *softened* version of the dual to replace the hard solution defined by (6.115) and (6.116). The soft dual arises from considering the expectation of $\mathbf{n}$, rather than its most probable value. The expectation of n_r is given by

$$E[n_r] = \sum_{n \in \mathcal{S}(\mathbf{C})} n_r p(\mathbf{n}) = \frac{1}{G(\mathbf{C})} \sum_{\mathbf{n} \in \mathcal{S}(\mathbf{C})} n_r \prod_{r \in \mathcal{R}} \frac{a_r^{n_r}}{n_r!} = \frac{H_r(\mathbf{C})}{G(\mathbf{C})}, \qquad (6.119)$$

where

$$H_r(\mathbf{C}) = \sum_{\mathbf{n} \in \mathcal{S}(\mathbf{C})} n_r \prod_{r \in \mathcal{R}} \frac{a_r^{n_r}}{n_r!}.$$

Recall that the generating function for $G(C)$ is given by

$$G^*(\mathbf{z}) = \sum_{\mathbf{C} \geq \mathbf{0}} G(\mathbf{C}) \mathbf{z}^{\mathbf{C}} = \prod_{r \in \mathcal{R}} \exp\left(a_r \mathbf{z}^{A_r}\right) \prod_{j \in \mathcal{J}} \left(\frac{1}{1 - z_j}\right).$$

Define the generating function for $H_r(\mathbf{C})$ by

$$H_r^*(\mathbf{z}) = \sum_{\mathbf{C} \geq \mathbf{0}} H_r(\mathbf{C}) \mathbf{z}^{\mathbf{C}}.$$

Similar to the derivation of $G^*(\mathbf{z})$ in (6.31), we obtain

$$H_r^*(\mathbf{z}) = \sum_{\mathbf{C} \geq \mathbf{0}} H_r(\mathbf{C}) \mathbf{z}^{\mathbf{C}} = a_r \mathbf{z}^{A_r} \prod_{r' \in \mathcal{R}} \exp(a_{r'} \mathbf{z}^{A_{r'}}) \prod_{j \in \mathcal{J}} \left(\frac{1}{1 - z_j}\right).$$

Applying the multidimensional analog of Cauchy's integral formula, we have the following integral representations for $G(\mathbf{C})$ and $H_r(\mathbf{C})$:

$$G(\mathbf{C}) = \frac{1}{(2\pi i)^J} \int_{|z_j| = \epsilon} \frac{G^*(\mathbf{z})}{\mathbf{z}^{\mathbf{C}+1}} d\mathbf{z}, \quad H_r(\mathbf{C}) = \frac{1}{(2\pi i)^J} \int_{|z_j| = \epsilon} \frac{H_r^*(\mathbf{z})}{\mathbf{z}^{\mathbf{C}+1}} d\mathbf{z},$$

where $\mathbf{1}$ denotes the vector whose all entries are equal to 1 and $0 < \epsilon < 1$. Using these expressions in (6.119) and simplifying yields the following expression for $E[n_r]$:

$$E[n_r] = \frac{a_r \int_{|z_l| = \epsilon} \mathbf{z}^{A_r} \exp(V f(\mathbf{z})) \frac{d\mathbf{z}}{\mathbf{z}}}{\int_{|z_l| = \epsilon} \exp(V f(\mathbf{z})) \frac{d\mathbf{z}}{\mathbf{z}}}, \qquad (6.120)$$

where

$$f(\mathbf{z}) = \sum_{r \in \mathcal{R}} \frac{a_r}{V} \mathbf{z}^{A_r} - \sum_{j \in \mathcal{J}} \left(\frac{C_j}{V} \log(z_j) + \frac{1}{V} \log(1 - z_j)\right). \qquad (6.121)$$

Applying Laplace's principle of localization (see Subsection 6.6.1) to both integrals we find that for large V,

$$E[n_r] \approx a_r \mathbf{z}^{A_r} \big|_{\mathbf{z} = \widehat{\mathbf{p}}}, \qquad (6.122)$$

where $\widehat{\mathbf{p}}$ is a saddle point of $f(\mathbf{z})$, i.e., the value of $\mathbf{z} > \mathbf{0}$ that maximizes $f(\mathbf{z})$. Define

$$\widehat{y}_j = -\log(\widehat{p}_j), \quad j \in \mathcal{J}.$$

Then $\widehat{\mathbf{y}} = (\widehat{y}_j : j \in \mathcal{J})$ is the solution to the constrained problem

$$(S) \quad \min_{\mathbf{y}} \sum_{r \in \mathcal{R}} a_r \exp\left(\sum_{j \in \mathcal{J}} A_{jr} y_j\right) + \sum_{j \in \mathcal{J}} \theta_j(y_j),$$

where

$$\theta_j(y_j) = \log\left(\sum_{m=0}^{C_j} e^{my_j}\right). \tag{6.123}$$

Equation (6.123) implies that

$$\theta_j(y_j) = C_j(y_j)_+ + o(V), \tag{6.124}$$

which shows that the soft dual problem (S) tends to the hard dual problem (D_2) as $V \to \infty$. However, for moderate values of V, we would expect the soft dual to provide a more accurate approximation.

Setting the derivative of the objective function of (S) equal to zero we find that the solutions of $\widehat{\mathbf{y}}$ satisfies

$$\sum_{r \in \mathcal{R}} a_r A_{jr} \exp\left(-\sum_{j \in \mathcal{J}} \widehat{y}_j A_{jr}\right) = \frac{\partial \theta_j(\widehat{y}_j)}{\partial y_j}, \quad j \in \mathcal{J}. \tag{6.125}$$

The left-hand side of (6.125) is equal to the utilization

$$\widehat{U}_j = \sum_{r \in \mathcal{R}} A_{jr} a_r \prod_{k \in \mathcal{J}} \widehat{p}_k^{A_{kr}} = \sum_{r \in \mathcal{R}} A_{jr} \widehat{n}_r. \tag{6.126}$$

The right-hand side of (6.125) can be rewritten, using (6.123), as

$$\frac{\partial \theta_j(\widehat{y}_j)}{\partial y_j} = \frac{\sum_{m=0}^{C_j} m e^{m\widehat{y}_j}}{\sum_{m=0}^{C_j} e^{m\widehat{y}_j}} = C_j - \frac{\sum_{m=0}^{C_j} m \widehat{p}_j^m}{\sum_{m=0}^{C_j} \widehat{p}_j^m}, \quad j \in \mathcal{J}. \tag{6.127}$$

Thus, equations (6.125) are equivalent to

$$\widehat{U}_j = C_j - \frac{\sum_{m=0}^{C_j} m \widehat{p}_j^m}{\sum_{m=0}^{C_j} \widehat{p}_j^m}, \quad j \in \mathcal{J}. \tag{6.128}$$

For the case $\widehat{p}_j < 1$ (i.e., for overloaded links j), (6.128) reduces to

$$\widehat{U}_j = C_j - \frac{\widehat{p}_j}{1 - \widehat{p}_j}. \tag{6.129}$$

Relations (6.126) and (6.128) uniquely determine the passing probabilities. From (6.128), one can see that the soft dual solution corresponds to the assumption that the number of free circuits on link l has a truncated geometric distribution on $[0, 1, \ldots, C_j\}$ with parameter p_j. In terms of the link state process $\mathbf{m}(t)$, this assumption gives the following product-form stationary distributions:

$$p(\mathbf{m}) = \prod_{j \in \mathcal{J}} \frac{p_j^{C_j - m_j}}{\sum_{m=0}^{C_j} p_j^m}, \quad j \in \mathcal{J}. \tag{6.130}$$

Whittle [40] refers to relations (6.128) as the *saddle point approximation* for the blocking probabilities, since the dual problem (S) is derived from a saddle point approximation for the expectation of $\mathbf{n}$. Note that in the soft dual solution, the circuits on a given *overloaded link* are not necessarily all in use. In the limit of large V, the saddle point approximation agrees with the solution from the hard dual problem (D). But for moderate values of V, we would expect the saddle point approximation to be more accurate.

6.7.4 Reduced load approximations

We now turn to a widely used approximation method called the *Erlang fixed-point approximation*. This approximation may be viewed as an alternative way of *softening* the hard dual problem (D). The link independence conclusion is retained from the primal problem (P), but the determining relations for the passing probabilities (equations (6.115) and (6.116) for the hard dual (D_2) and equations (6.128) for the soft dual (S)) are replaced by the *Erlang fixed point equations*:

$$1 - p_j = E(C_j, U_j/p_j), \quad j \in \mathfrak{J}. \tag{6.131}$$

In effect, link j is treated as an Erlang loss station of capacity C_j with offered (Poisson) load of $\widehat{U}_j/\widehat{p}_j$ erlangs.

The fixed point equations are usually written in terms of the blocking probabilities:

$$B_j = E\left(C_j, \sum_{r \in \mathfrak{R}} A_{jr} a_r \prod_{k \neq j} (1 - B_k)^{A_{kr}}\right), \quad j \in \mathfrak{J}. \tag{6.132}$$

The second argument of the Erlang function in (6.132),

$$\rho_j = \sum_{r \in \mathfrak{R}} A_{jr} a_r \prod_{k \neq j} (1 - B_k)^{A_{kr}}, \tag{6.133}$$

can be interpreted as the *reduced load* seen by link j. Let us assume, initially, that $\mathbf{A}$ is a $0-1$ matrix, so that each call requires one circuit. In the absence of capacity constraints, the offered load of route r traffic to link j would be $A_{jr} a_r$.

At each link $k \neq j$ on route r, this stream is *thinned* by a factor of $(1 - B_k)^{A_{kr}}$ under the assumption of link independence. The summation over all routes $r \in \mathfrak{R}$ yields the aggregate reduced load ρ_j given by (6.133). The blocking probability for link j is then given by (6.132). If $\mathbf{A}$ is not a $0-1$ matrix, the fixed point equations contain a further approximation yielding that the aggregation arrival stream to link j is a simple Poisson process. However, the Erlang fixed point gives blocking probabilities that are correct in the asymptotic regime $V \to \infty$ and can be readily computed via successive substitution using (6.132).

Kelly [20] relates the Erlang fixed point equations to an extremal problem analogous to the hard dual problem (D_2):

$$(E) \qquad \min_{\mathbf{y}} \sum_{r \in \mathfrak{R}} a_r \exp\left(-\sum_{j \in \mathfrak{J}} y_j A_{jr}\right) + \theta_j(y_j),$$

where

$$\theta_j(y_j) = \int_0^{y_j} W(z, C_j). \tag{6.134}$$

Here, the function $W(x, C)$ is defined implicitly by the relation [20]

$$W(-\log(1 - E(C, a)), C) = a(1 - E(a, C)). \tag{6.135}$$

$W(x, C)$ is the mean number of carried circuits when the blocking probability is $1 - e^{-x}$. The solutions to problem (E) correspond precisely to the solutions of the Erlang fixed point equations (6.132). Since the objective function of (E) is strictly convex, it follows that there exists a unique Erlang fixed point $\widehat{\mathbf{B}}$.

The Erlang fixed-point approximation is a special case of more general *reduced load approximations* [20, 39, 9]. The reduced load approximation treats each link as a generalized Erlang loss station. Calls on routes passing through link j arriving as independent Poisson processes require A_{jr} circuits. The offered load of route r traffic to link j is computed as a reduced load under the assumption that blocking events are independent from link to link on the route. Let B_{jr} denote the steady-state blocking probability of route r calls passing through link j. In terms of the generalized Erlang loss formula,

$$B_{jr} = \varepsilon_r(C_j, \widetilde{\mathbf{a}}_j), \tag{6.136}$$

where $\widetilde{\mathbf{a}}_j = (a_{jr}: r \in \mathcal{R})$ is the vector of reduced loads defined by

$$\widetilde{a}_{jr} = a_r \prod_{k \neq j} (1 - B_{kr}). \tag{6.137}$$

Taken together, equations (6.136) and (6.137) define a set of generalized Erlang fixed-point equations. These equations satisfy the conditions of the Brouwer fixed-point theorem implying the existence of a *generalized Erlang fixed-point* $\widehat{\mathbf{B}} = (\widehat{B}_{jr}: j \in \mathcal{J}, r \in \mathcal{R})$. However, for some loss networks there may be multiple generalized Erlang fixed-points. This is in contrast to the uniqueness of the simpler Erlang fixed-point. The existence of multiple generalized fixed-points may indicate potential instabilities in the networks. (See Chung and Ross [9] for an example of a simple loss network with multiple fixed points.)

In general, fixed point equations can be solved by iterative use of (6.136) and (6.137) with an initial starting point. Like the Erlang fixed-point, the generalized Erlang fixed-point converges to the true blocking probabilities in the limiting regime $V \to \infty$ (cf. [9]). In (6.136), the evaluation of the generalized Erlang formula can be carried out using any of the various computational algorithms and approximation methods discussed previously for generalized loss stations. For example, the recursive algorithm defined by relations (6.45) and (6.46) provides an efficient method for moderate values of the capacity C. The numerical inversion algorithm can be specialized to the case of a generalized Erlang station, giving an efficient method for a wide range of values of C. The asymptotic approximations for the generalized Erlang loss station are best suited to the case of a large C.

6.8 DISCUSSION AND OPEN PROBLEMS

6.8.1 Recapitulation

The focus of this chapter has been on basic properties of loss networks and methods for calculating blocking probabilities. We view loss networks as generalizations of the classical loss models studied by Erlang and Engset. We defined open, closed, and mixed loss networks in analogy to pen, closed, and mixed queueing networks. An analogy between queueing networks and loss networks allowed us to establish, in a straightforward manner, properties of insensitivity, reversibility, and quasireversibility in loss networks. Moreover, parallel methods can often be applied to the analysis of both queueing and loss networks.

The main quantities of interest in a loss network are the stationary blocking probabilities. The blocking probabilities can be expressed in terms of ratios of normalization constants. A normalization constant is a sum of product-form terms over the network state-space. A state probability is expressed as the ratio of product-form terms over the normalization constant. An analogous normalization constant appears in the study of queueing networks. In most loss networks of practical interest, a straightforward evaluation of the normalization constant is computationally infeasible.

Several approaches to evaluating blocking probabilities are based on the integral representation of the normalization constant in (6.34). We saw that recursive algorithms for the generalized Erlang and Engset loss stations could be derived from their corresponding generating functions. An alternative approach to blocking probabilities included consideration of the link state distribution. The link state distribution of a loss network satisfies a recurrence relation. For the case of generalized loss stations, the recurrence provides the basis for an efficient method of computing the blocking probabilities. However, for more general loss networks and/or large link capacities, the recursive algorithms become impractical in terms of both computational effort and storage requirement. The integral representation of the normalization constant in equation (6.34) leads to the numerical transform inversion and asymptotic approximation approaches. The main advantage of the numerical inversion approach is its generality and applicability over a wide range of parameter values. The numerical inversion method is alternative to the recursive algorithms for moderate-sized problems, but can also be applied effectively to loss networks of larger size.

When the link capacities are large, asymptotic approaches become advantageous from the standpoint of both computation and numerical accuracy. We discussed asymptotic approximation techniques applied to the inversion integrals for generating functions of Erlang and Engset loss stations. The saddle point method, together with uniform asymptotic approximations of the normalization constant, give UAA results for the blocking probabilities, which can be specialized to three regimes of traffic loading. The large scale behavior of an open loss network is a basis for an important class of approximations for blocking probabilities, in particular, the reduced load approximations. The main simplification of these approximations is that blocking events on links are assumed to be independent. Thus, from a practical standpoint, the problem of calculating blocking probabilities in a network essentially reduces to a single-link case, i.e., a loss station.

6.8.2 Direct algorithm and numerical transform inversion

We mention that for some networks of a particular structure, a direct evaluation of the normalization constant is the most efficient approach. Recall that the normalization constant of a mixed loss network can be written as a sum of product form terms:

$$G(\mathbf{N}, \mathbf{C}) = \sum_{\mathbf{n} \in \mathcal{S}(\mathbf{N}, \mathbf{C})} \prod_{r \in \mathcal{R}} q_r(n_r),$$

where

$$q_r(n_r) = \begin{cases} \dfrac{a_r^{n_r}}{n_r!} & r \in \mathcal{R}_O \\[2mm] \dbinom{N_r}{n_r} b_r^{n_r} & r \in \mathcal{R}_C. \end{cases}$$

A direct approach to evaluating the normalization constant is to carry it out as the nested sum:

$$G(\mathbf{N}, \mathbf{C}) = \sum_{n_1 = 0}^{K_1} q_1(n_1) \sum_{n_2 = 0}^{K_2} q_2(n_2) \ldots \sum_{n_R = 0}^{K_R} q_R(n_R),$$

where the upper limits are given by

$$K_r = \begin{cases} \min_{j: A_{jr} > 0} \lfloor (C_j - \sum_{k=1}^{r-1} n_k A_{jk}) / A_{jr} \rfloor, & r \in \mathcal{R}_O \\ \min \{ N_r, \min_{j: A_{jr} > 0} \lfloor (C_j - \sum_{k=1}^{r-1} n_k A_{jk}) / A_{jr} \rfloor \}, & r \in \mathcal{R}_C. \end{cases}$$

Note that the upper limit K_r depends on the values of $n_1, \ldots, n_{r-1}$. Choudhury, Leung, and Whitt [10] discuss the direct algorithm and compare it with the numerical transform inversion approach. The direct algorithm involves a nested sum of dimension R whereas the numerical inversion requires a nested sum of dimension J. Since in most applications $J \ll R$, the direct algorithm usually is not a viable approach. However, for some special model structures, many of the summations can be performed independently of each other so that the effective dimension of the summation can be reduced to $\bar{R} \ll R$.

The success of the numerical inversion technique relies on the ability to reduce the effective dimension to $\bar{J} \ll J$. An efficient algorithm for determination of the minimal possible dimension for a given generating function remains an open problem. The problem was formulated as a graph problem in [10], which is likely NP-hard. Another one is the truncation, which effectively reduces the size parameter C to $\bar{C} \ll C$. In [10], heuristic methods for loss models were developed for truncating sums that approximate the inversion integrals. The challenge is to achieve computational savings through truncation, and yet maintain control over the errors that are introduced. The full range of applicability of the numerical inversion method awaits to be tested on practical loss network models.

6.8.3 Monte Carlo technique

An application of the Monte Carlo method to calculating the normalization constant of queueing networks and loss networks has been studied by Ross and Wang [35, 34]. The normalization constant can be written as an R-dimensional summation with constant upper limits as follows:

$$G(\mathbf{N}, \mathbf{C}) = \sum_{n_1 = 0}^{M_1} \sum_{n_2 = 0}^{M_2} \cdots \sum_{n_R = 0}^{M_R} q(\mathbf{n}),$$

where

$$M_r = \max \{ n_r : \mathbf{n} \in \mathcal{G}(\mathbf{N}, \mathbf{C}) \},$$

and

$$q(\mathbf{n}) = 1_{\mathcal{G}(\mathbf{N}, \mathbf{C})}(\mathbf{n}) \prod_{r \in \mathcal{R}} q_r(n_r).$$

Monte Carlo summation is a technique for evaluating multidimensional sums of this form. The idea is to generate a sequence of i.i.d. random vectors $\mathbf{v}^i = (v_r^i : r \in \mathcal{R})$, $i = 1, 2, \ldots$, with a common *sampling distribution* $p(\mathbf{n}) = P(\mathbf{v}^i = \mathbf{n})$ over the set

$$\Lambda = \{ 0, \ldots, N_1 \} \times \cdots \times \{ 0, \ldots, N_R \}.$$

Then the sum

$$\bar{Z}_n = \frac{1}{n} \sum_{i=1}^{n} \frac{q(\mathbf{v}^i)}{p(\mathbf{v}^i)}$$

is an unbiased and consistent estimator for $G(\mathbf{N}, \mathbf{C})$. Ross and Wang [34] discuss sampling techniques to obtain significant reductions in the variance of the Monte Carlo estimate.

The idea of Monte Carlo integration is similar to Monte Carlo summation. In Monte Carlo integration, a multidimensional integral is estimated via a sequence of random sampling vectors chosen from a probability density function. Ross and Wang [35] apply Monte Carlo integration to an integral representation of the normalization constant of a queueing network. In their experiments they found that Monte Carlo integration typically gave smaller confidence intervals than Monte Carlo summation (applied to the direct form of the queueing network normalization constant). In the context of loss networks, a natural avenue of investigation is to apply the Monte Carlo integration technique to the integral representation of the normalization constant (6.34). The integral representation is a multidimensional integral of dimension J, whereas the direct summation is of dimension R. Since for typical applications $J \ll R$, we would expect Monte Carlo integration to be a superior method in most cases.

The main advantage of the Monte Carlo approach is that the computational complexity grows polynomially with the problem size, rather than exponentially. The numerical transform inversion technique, for example, is useful only if the dimension J is small or dimension reduction can be employed such that the effective dimension is $\bar{J} \ll J$. A noteworthy approach would be to combine the numerical inversion approach with Monte Carlo summation. If dimension reduction is applied, but the dimension of an inner sum is still too large for a numerical evaluation, Monte Carlo summation can be used to carry out the inner nested summations.

6.8.4 Asymptotic approximations and error bounds for general loss networks

In our discussion of asymptotic approximations of the inversion integral we considered only the one-dimensional case, i.e., generalized loss stations. The results for generalized loss stations are of interest in themselves, but may also find use in combination with reduced load approximations to handle more general networks. An open problem is how to extend the known asymptotic integration techniques for loss stations to handle more general networks directly. In the general case, the inversion integral is multidimensional and additional assumptions may be required in order to obtain asymptotic approximations.

Recursive algorithms and a direct algorithm are exact methods in that their accuracy is limited only by the finite precision of computers. The numerical transform inversion algorithm may experience numerical underflow/overflow problems, as well as aliasing errors, but these can be controlled by a judicious choice of the relevant parameters. Monte Carlo methods provide confidence intervals that indicate the accuracy of the estimates. However, assessing the accuracy of the asymptotic approximation approaches is not as straightforward. Birman and Kogan [4] obtain error bounds for the asymptotic expansions of the normalization constant. The bounds are applicable to the asymptotic expansions for generalized loss stations. Error bounds of this nature in conjunction with the multidimensional case would be of great practical use.

6.8.5 More general arrival processes and reduced load approximations

In this chapter we considered arrival processes that are either Poisson (infinite source model) or finite-source. Routes with Poisson arrivals are called *open* while routes with finite-source populations are called *closed*. More generally, arrival rates can be state dependent, i.e., call arrivals to route r form a Poisson process with intensity $\lambda_r(n_r)$ when the number of route r calls in progress is n_r. The stationary distribution is then given by [15]

$$\pi(\mathbf{n}) = G(\mathbf{C})^{-1} \prod_{r \in \mathcal{R}} \frac{\prod_{k=1}^{n_r} \lambda_r(k-1)}{n_r! \mu_r^{n_r}}, \quad \mathbf{n} \in \mathcal{S}(\mathbf{C}),$$

where the normalization constant is given by

$$G(\mathbf{C}) = \sum_{\mathbf{n} \in \mathcal{S}(\mathbf{C})} \prod_{r \in \mathcal{R}} \frac{\prod_{k=1}^{n_r} \lambda_r(k-1)}{n_r! \mu_r^{n_r}}, \quad \mathbf{n} \in \mathcal{S}(\mathbf{C}).$$

Insensitivity with respect to the service time distribution carries over to state-dependent arrivals with minor modifications to the proof of Theorem 6.1.

The so-called Bernoulli-Poisson-Pascal (BPP) process [14, 15] is a state dependent process in which the arrival rates have the linear form

$$\lambda_r(n_r) = \alpha_r + \beta_r n_r.$$

If such a process were offered to an infinite server group, the distribution of the number of busy servers would be Bernoulli if $\beta_r < 0$ and α_r/β_r is a negative integer, Poisson if $\beta_r = 0$, and Pascal if $\beta_r > 0$. In particular, route r is open if $\beta_r = 0$. A closed route with population N_r corresponds to

$$\alpha_r = N_r \lambda_r, \quad \beta_r = -\lambda_r.$$

If such a process were offered to an infinite server group, the distribution of the number of busy servers would be Bernoulli if $\beta_r < 0$ and α_r/β_r is a negative integer, Poisson if $\beta_r = 0$, and Pascal if $\beta_r > 0$. In particular, route r is open if $\beta_r = 0$. A closed route with population N_r corresponds to

$$\alpha_r = N_r \lambda_r, \quad \beta_r = -\lambda_r.$$

If we extend the notion of *closed* route to include BPP processes with $\beta_r \neq 0$, the stationary distribution of the mixed loss network retains the product form

$$\pi(\mathbf{n}) = G(\mathbf{C})^{-1} \prod_{r \in \mathcal{R}_O} \frac{a_r^{n_r}}{n_r!} \prod_{s \in \mathcal{R}_C} \left[\frac{\beta_r}{\mu_r} \binom{\alpha_r/\beta_r - 1 + n_r}{n_r} \right], \quad \mathbf{n} \in \mathcal{S}(\mathbf{C}),$$

where we assume that α_r/β_r is a negative integer. The generating function for the normalization constant in this more general setting is

$$G^*(\mathbf{z}) = \prod_{j \in \mathcal{J}} \left(\frac{1}{1-z_j} \right) \exp\left\{ \sum_{r \in \mathcal{R}_O} a_r \mathbf{z}^{\mathbf{A}_r} \right\} \cdot \prod_{s \in \mathcal{R}_C} \left(1 - \frac{\beta_s}{\mu_s} \mathbf{z}^{\mathbf{A}_s} \right)^{-\alpha_r/\beta_r}.$$

The reduced load approximation is derived in the context of an open loss network. An interesting line of investigation would be to study analogous approximations for mixed networks where some of the routes are closed. The reduced load approximation is important because it effectively reduces the analysis of a loss network to that of a loss station. However, as some researchers have noted (cf. [29]), it can lose accuracy for certain combinations of traffic. Future research can deal with approximations of this nature and their accuracy over a wide range of traffic conditions (i.e., light, moderate, and heavy traffic).

ACKNOWLEDGEMENTS

The authors would like to thank Dr. Yaakov Kogan of Bell Laboratories and Dr. Alex Birman of IBM Research for stimulating discussions on the topic of loss networks. The authors also thank Dr. Kogan [25], Dr. Debasis Mitra of Bell Laboratories [28], and Dr. Kin K. Leung and Dr. Ward Whitt of Bell Laboratories [10, 11] for making available preprints of their papers prior to publication. Special thanks are due to Dr. John Morrison of Bell Laboratories for his detailed comments on Section 6.6, which saved the authors from numerous errors in an early draft. This work has been supported, in part, by a grant from the Ogasawara Foundation for the Promotion of Science and Technology; and by the NCIPT (National Center for Integrated Photonic Technology) through its ARPA grant. The second author has been supported by an NSERC postgraduate scholarship.

APPENDICES

Appendix A Proof of Theorem 6.1

Here we provide a proof of Theorem 1 (see also [7] for an alternative proof). We shall assume that the holding time of class r customers is represented exactly or approximately by a distribution having a rational Laplace-Stieltjes transform (LST) $\Phi_r(s)$. The theorem is, in fact, true for arbitrary distributions, but the proof would require us to consider continuous state-space Markov processes.

Using Cox's method of stages (cf. [22]) any rational LST can be expressed in the form

$$\Phi(s) = b_0 + \sum_{\phi=1}^{d} a_0 \ldots a_{\phi-1} b_\phi \prod_{i=1}^{\phi} \frac{\mu_i}{s + \mu_i}, \qquad (6.138)$$

where d is the number of *stages*, $a_i + b_i = 1$, $i = 0, 1, \ldots, d-1$, and $b_d = 1$ (see Figure A1). The ith stage is represented by a "generalized" server with an exponentially distributed service time with rate μ_i, where μ_i may be complex. Define $A_\phi = a_0 a_1 \ldots a_{\phi-1}$, $1 \leq \phi \leq d$. The mean service time μ is then given by $1/\mu = \sum_{l=1}^{d} A_l \mu_l$.

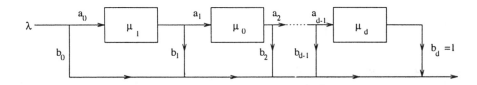

Figure A1. Cox representation of general holding time.

Now assume a d_r-state Cox representation for the LST, $\Phi_r(s)$, of class r customers. Consider the state process $\mathbf{z}(t) = (z_{r,l}(t): r \in \mathcal{R}, 1 \le l \le d_r)$, where $z_{c,l}(t)$ denotes the number of class c customers in the lth stage of service. Define the set of feasible states as $\mathcal{F}(C) = \{\mathbf{z}: \sum_{r \in \mathcal{R}} \sum_{l=1}^{d_r} A_c z_{r,l} \le C\}$ and the *blocking states*

for $c \in \mathcal{R}$ as $\mathcal{F}_r(C) = \{\mathbf{z}: \sum_{s \in \mathcal{R}} \sum_{l=1}^{d_s} A_s z_{s,l} > C - A_s\}$. Let $P_{\mathbf{z}}(\mathbf{z})$ denote the steady state distribution of $\mathbf{z}(t)$. $\mathbf{z}(t)$ is a Markov process that does not change when an arriving customer is blocked. Therefore, we introduce a *flip-flop* process $f(t) \in \{0,1\}$ that changes value each time a blocking event occurs. The joint process $\mathbf{v}(t) = (\mathbf{z}(t), f(t))$ is Markov, with equilibrium distribution $P_{\mathbf{z},f}(\mathbf{z}, f)$.

The proof proceeds by conjecturing the form of the distribution $P_{\mathbf{z},f}(\mathbf{z}, f)$ and the reverse process $\mathbf{v}_R(t) = \mathbf{v}(-t)$. In particular, we conjecture that

$$P_{\mathbf{z},f}(\mathbf{z}, f) = \tfrac{1}{2} P_{\mathbf{z}}(\mathbf{z}), \quad f \in \{0,1\}, \tag{6.139}$$

where

$$P_{\mathbf{z}}(\mathbf{z}) = P_{\mathbf{z}}(0) n! \prod_{r \in \mathcal{R}} \prod_{l=1}^{d_r} \frac{\rho_{r,l}^{z_{r,l}}}{z_{r,l}} \tag{140}$$

with $\rho_{r,l} = \lambda_r A_{r,l}/\mu_{r,l}$. The conjectured reverse process $\mathbf{v}_R(t)$ consists of independent Poisson inputs of rates λ_r, $r \in \mathcal{R}$, with the station operating as in forward time, but with the holding times realized by the Cox representation reversed in time (see Figure A2). In order to establish the validity of these assertions, it suffices to show that the following *reversed balance equation* (cf. [21]) holds:

$$P_{\mathbf{v}}(\mathbf{v}) q(\mathbf{v}, \mathbf{v}') = P_{\mathbf{v}}(\mathbf{v}') q_R(\mathbf{v}', \mathbf{v}), \tag{6.141}$$

for all $\mathbf{v}, \mathbf{v}' \in \mathcal{F}(C) \times \{0,1\}$, where $q(\cdot, \cdot)$ and $q_R(\cdot, \cdot)$ denote the transition rates of the forward and reverse processes $\mathbf{v}(t)$ and $\mathbf{v}_R(t)$, respectively.

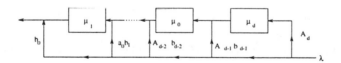

Figure A2. Cox representation in reverse time.

There are several transitions of the forward process involving a class r customer that can occur:

    (*i*)   When $\mathbf{z} \in \mathcal{F}(C)$, a class r customer in stage ϕ departs from the station entering state $\mathbf{z}'$. Then $q(\mathbf{v}, \mathbf{v}') = z_{r,\phi} \mu_{r,\phi} b_{r,\phi}$ and $q_R(\mathbf{v}', \mathbf{v}) = {}_{r,\phi} b_{r,\phi} \lambda_c$, where $\mathbf{v} = (\mathbf{z}, f)$ and $\mathbf{v}' = \mathbf{z}', f)$.

    (*ii*)   When $\mathbf{z} \in \mathcal{F}(C)$, a class r customer in stage $\phi - 1$ moves to stage ϕ entering state $\mathbf{z}'$, for $1 < \phi < d_r$. Then $q(\mathbf{v}, \mathbf{v}') = z_{r,\phi-1} \mu_{r,\phi-1} a_{r,\phi-1}$ and $q_R(\mathbf{v}', \mathbf{v}) = \mu_{r,\phi}(z_{r,\phi} + 1)$. In case $\phi = 1$, an external arrival occurs and we have $q(\mathbf{v}, \mathbf{v}') = \lambda_r a_{r,0}$ and $q_R(\mathbf{v}', \mathbf{v}) = (z_{r,1} + 1)\mu_{r,1}$.

    (*iii*) When $\mathbf{z} \in \mathcal{F}_r(C)$, a class r customer arrives and is blocked. Then $q((\mathbf{z}, f), (\mathbf{z}, 1-f)) = q_R((\mathbf{z}, 1-f), (\mathbf{z}, f)) = \lambda_r$.

It is straightforward to verify that these transition rates satisfy (6.141). This establishes that the Erlang station is quasireversible with equilibrium distribution (6.139).

By summing up $P_{\mathbf{z}}(\mathbf{z})$ over all states with $\sum_{l=1}^{d_s} z_{s,l} = n_r$, $r \in \mathcal{R}$, we obtain

$$\pi_C(\mathbf{n}) = G(C)^{-1} \prod_{r \in \mathcal{R}} \frac{a_r^{n_r}}{n_r!}, \quad \mathbf{n} \in \mathcal{G}(C), \tag{6.142}$$

where $n_r = \sum_{l=1}^{d_r} a_{r,l} = \lambda_r / \mu_r$. To establish the reversibility of $\mathbf{n}(t)$ it suffices to verify that $\pi_C(\mathbf{n})$ satisfies the detailed balance equations:

$$\lambda_r \pi_B(\mathbf{n}_r^-) = n_r \mu_r \pi_B(\mathbf{n}), \quad \mathbf{n}, \mathbf{n}_r^- \in \mathcal{G}(C), \tag{6.143}$$

where $\mathbf{n}_r^-$ denotes the state vector with one less class r customer than $\mathbf{n}$.   □

Appendix B Derivation of link state recurrence

Here we derive the link state recurrences (6.21) and (6.22) (see also [15, 30]). Consider the conditional expectation

$$E[n_r \mid \mathbf{m}] = \frac{\sum_{\mathbf{n}: \, A\mathbf{n} = \mathbf{m}} \pi_C(\mathbf{N}, \mathbf{n}) n_r}{p_C(\mathbf{N}, \mathbf{m})}. \tag{6.144}$$

Note that the stationary route-state distribution $\pi_C(\mathbf{N}, \mathbf{n})$ satisfies the detailed balance equations[4],

$$\pi(\mathbf{N}, \mathbf{n}) n_r = \pi(\mathbf{N}, \mathbf{n} - \mathbf{e}_r) a_r, \quad r \in \mathcal{R}_O, \tag{6.145}$$

$$\pi(\mathbf{N}, \mathbf{n}) n_s = \pi(\mathbf{N}, \mathbf{n} - \mathbf{e}_s)(N_s - (n_s - 1)) b_s, \quad s \in \mathcal{R}_C, \tag{6.146}$$

where $\mathbf{e}_r$ denotes the vector with unity in the rth component and zeros elsewhere. The right-hand side of (6.146) can be written as

$$\pi(\mathbf{N}, \mathbf{n})(N_s - (n_s - 1)) b_s = \pi(\mathbf{N}, \mathbf{n} - \mathbf{e}_s) N_s b_s - \pi(\mathbf{N}, \mathbf{n} - \mathbf{e}_s)(n_s - 1) b_s$$

$$= N_s b_s \sum_{k \geq 1} (-b_s)^{k-1} \pi(\mathbf{N}, \mathbf{n} - k\mathbf{e}_s), \tag{6.147}$$

where the upper limit in the summation over k is understood to be the minimum of N_s and n_s.

Summing up both sides of (6.145) over all $\mathbf{n}$ such that $A\mathbf{n} = \mathbf{m}$ yields

$$\sum_{\mathbf{n}: \, A\mathbf{n} = \mathbf{m}} \pi(\mathbf{N}, \mathbf{n}) n_r = \sum_{\mathbf{n}: \, A\mathbf{n} = \mathbf{m}} \pi(\mathbf{N}, \mathbf{n} - \mathbf{e}_r) a_r = p(\mathbf{N}, \mathbf{m} - \mathbf{A}_r) a_r, \quad r \in \mathcal{R}, \tag{6.148}$$

which gives

$$p(\mathbf{N}, \mathbf{m}) E[n_r \mid \mathbf{m}] = p(\mathbf{N}, \mathbf{m} - \mathbf{A}_r) a_r, \quad r \in \mathcal{R}. \tag{6.149}$$

Repeating this process for equation (6.147) yields:

$$\sum_{\mathbf{n}: \, A\mathbf{n} = \mathbf{m}} \pi(\mathbf{N}, \mathbf{n}) n_r = \sum_{k \geq 1} \sum_{\mathbf{n}: \, A\mathbf{n} = \mathbf{m}} \pi(\mathbf{N}, \mathbf{n} - k\mathbf{e}_s) N_s b_s (-b_s)^{k-1} \tag{6.150}$$

$$= N_s b_s \sum_{k \geq 1} (-b_s)^{k-1} p(\mathbf{N}, \mathbf{m} - k\mathbf{A}_s). \tag{6.151}$$

Multiplying both sides of (6.144) by A_{jr}, summing up over all $r \in \mathcal{R}$, and making use of (6.148) and (6.151) results in

---

[4]For convenience, we shall omit the subscript $\mathbf{C}$.

$$\sum_{r \in \mathcal{R}} A_{jr} E[n_r \mid \mathbf{m}] p(\mathbf{N}, \mathbf{m}) = \sum_{r \in \mathcal{R}_O} A_{jr} a_r p(\mathbf{N}, \mathbf{m} - \mathbf{A}_r)$$

$$+ \sum_{s \in \mathcal{R}_C} A_{js} N_s b_s \sum_{k \geq 1} (-b_s)^{k-1} p(\mathbf{N}, \mathbf{m} - k\mathbf{A}_s). \tag{6.152}$$

By noting that

$$\sum_{r \in \mathcal{R}} A_{jr} E[n_r \mid \mathbf{m}] = E[\sum_{r \in \mathcal{R}} A_{jr} n_r \mid \mathbf{m}] = E[m_j \mid \mathbf{m}] = m_j \tag{6.153}$$

we obtain the recursion

$$m_j p(\mathbf{N}, \mathbf{m}) = \sum_{r \in \mathcal{R}_O} A_{jr} a_r p(\mathbf{N}, \mathbf{m} - \mathbf{A}_r)$$

$$+ \sum_{s \in \mathcal{R}_C} A_{js} N_s b_s \sum_{k=1}^{N_s} (-b_s)^{k-1} p(\mathbf{N}, \mathbf{m} - k\mathbf{A}_s), \quad j \in \mathcal{J}. \tag{6.154}$$

We can rewrite the detailed balance equation (6.146) as

$$\pi(\mathbf{N}, \mathbf{n}) n_s = \pi(\mathbf{N}, \mathbf{n} - \mathbf{e}_s) N_s b_s, \quad s \in R_C. \tag{6.155}$$

Summing up over all $\mathbf{n}$ such that $\mathbf{A}\mathbf{n} = \mathbf{m}$ yields:

$$\sum_{\mathbf{n}:\, \mathbf{An} = \mathbf{m}} n_s \pi(\mathbf{N}, \mathbf{n}) = \sum_{\mathbf{n}:\, \mathbf{An} = \mathbf{m}} \pi(\mathbf{N} - \mathbf{e}_s, \mathbf{n} - \mathbf{e}_s) \tag{6.156}$$

$$= p(\mathbf{N} - \mathbf{e}_s, \mathbf{m} - \mathbf{A}_s) N_s b_s, \quad \mathbf{m} \geq \mathbf{A}_s. \tag{6.157}$$

Multiplying both sides of (6.144) by A_{jr}, summing up over $r \in \mathcal{R}$, and applying (6.148) and (6.157) gives

$$\sum_{r \in \mathcal{R}} A_{jr} E[n_r \mid \mathbf{m}] p(\mathbf{N}, \mathbf{m}) = \sum_{r \in \mathcal{R}_O} A_{jr} a_r p(\mathbf{N}, \mathbf{m} - \mathbf{A}_r)$$

$$+ \sum_{s \in \mathcal{R}_C} A_{js} N_s b_s p(\mathbf{N} - \mathbf{e}_s, \mathbf{m} - \mathbf{A}_s). \tag{6.158}$$

Now, applying (6.153) results in the recursion of the two arguments $\mathbf{m}$ and $\mathbf{N}$:

$$m_j p(\mathbf{N}, \mathbf{m}) = \sum_{r \in \mathcal{R}_O} A_{jr} a_r p(\mathbf{N}, \mathbf{m} - \mathbf{A}_r)$$

$$+ \sum_{s \in \mathcal{R}_C} A_{js} N_s b_s p(\mathbf{N} - \mathbf{e}_s, \mathbf{m} - k\mathbf{A}_s), \quad j \in \mathcal{J}. \tag{6.159}$$

Appendix C   Derivation of generating functions

The probability generating function for the link state distribution $p_\infty(\mathbf{N}, \mathbf{m})$ is derived as follows:

$$p_\infty^*(\mathbf{N}, \mathbf{z}) = \sum_{\mathbf{m} \geq 0} p_\infty(\mathbf{N}, \mathbf{m}) \mathbf{z}^{\mathbf{m}}$$

$$= \sum_{\mathbf{m} \geq 0} \sum_{\mathbf{An} = \mathbf{m}} \pi_\infty(\mathbf{N}, \mathbf{m}) \mathbf{z}^{\mathbf{An}}$$

$$= G \sum_{\mathbf{n} \geq 0} \prod_{r \in \mathcal{R}_O} \frac{a_r^{n_r}}{n_r!} \mathbf{z}^{\mathbf{A}_r n_r} \prod_{s \in \mathcal{R}_C} \binom{N_s}{n_s} b_s^{n_s} \mathbf{z}^{\mathbf{A}_s n_s}$$

$$= G \prod_{r \in \mathcal{R}_O} \sum_{n_r \geq 0} \frac{(a_r \mathbf{z}^{\mathbf{A}_r})^{n_r}}{n_r!} \cdot \prod_{s \in \mathcal{R}_C} \sum_{n_s \geq 0} \binom{N_s}{n_s} (b_s \mathbf{z}^{\mathbf{A}_s})^{n_s}$$

$$= G \prod_{r \in \mathcal{R}_O} \exp(a_r \mathbf{z}^{\mathbf{A}_r}) \cdot \prod_{s \in \mathcal{R}_C} (1 + b_s \mathbf{z}^{\mathbf{A}_s})^{N_s}, \tag{6.160}$$

where G is a normalization constant.

The generating function $G^*(\mathbf{N}, \mathbf{z})$ can be derived in a similar fashion as follows:

$$G^*(\mathbf{N}, \mathbf{z}) = \sum_{\mathbf{C} \geq 0} \sum_{\mathbf{An} \leq \mathbf{C}} \prod_{r \in \mathcal{R}_O} \frac{a_r^{n_r}}{n_r!} \mathbf{z}^{\mathbf{A}_r n_r} \prod_{s \in \mathcal{R}_C} \binom{N_s}{n_s} b_s^{n_s} \mathbf{z}^{\mathbf{A}_s n_s} \cdot \mathbf{z}^{\mathbf{C}}$$

$$= \sum_{\mathbf{n} \geq 0} \prod_{r \in \mathcal{R}_O} \frac{a_r^{n_r} \mathbf{z}^{\mathbf{A}_r n_r}}{n_r!} \prod_{s \in \mathcal{R}_C} \binom{N_s}{n_s} b_s^{n_s} \mathbf{z}^{\mathbf{A}_s n_s} \prod_{j \in \mathcal{J}} (1 + z_j + z_j^2 + \ldots)$$

$$= \prod_{j \in \mathcal{J}} \left(\frac{1}{1 - z_j} \right) \prod_{r \in \mathcal{R}_O} \sum_{n_r \geq 0} \frac{(a_r \mathbf{z})^{n_r \mathbf{A}_r}}{n_r!} \cdot \prod_{s \in \mathcal{R}_C} \sum_{n_s \geq 0} \binom{N_s}{n_s} (b_s \mathbf{z})^{n_s \mathbf{A}_s}$$

$$= \prod_{j \in \mathcal{J}} \left(\frac{1}{1 - z_j} \right) \exp \left(\sum_{r \in \mathcal{R}_O} a_r \mathbf{z}^{\mathbf{A}_r} \right) \prod_{s \in \mathcal{R}_C} (1 + b_s \mathbf{z}^{\mathbf{A}_s})^{N_s}. \tag{6.161}$$

Appendix D Glossary

This glossary lists some of the important notations used throughout the chapter. All vectors are defined as column vectors.

D.1 Loss stations

C	number of servers
$\mathcal{R} = \{1, \ldots, R\}$	set of all classes
$n_c, \mathbf{n} = (n_c : c \in \mathcal{R})$	number of class c calls in progress, state vector

Erlang station

M/M/C(0)	classical Erlang model
M/G/C(0)	Erlang loss station
$M_\mathcal{R} / G_\mathcal{R} / C(0)$	generalized Erlang loss station
$a_r, \mathbf{a} = (a_r : r \in \mathcal{R})$	offered load for class r calls, vector of a_r
$E(C, a)$	Erlang-B loss formula
$\varepsilon_r(C, \mathbf{a})$	generalized Erlang formula
$\pi_C(\mathbf{n})$	stationary state distribution
$G(\mathbf{C})$	normalization constant
$G_\infty^*(\mathbf{z})$	generating function for normalization constant

Engset station

M(N)/M/C(0)	classical Engset model
G(N)/G/C(0)	Engset loss station
$G_{\mathcal{R}}(N_{\mathcal{R}})/G_{\mathcal{R}}/\infty$	generalized Engset loss station
$N_s, \mathbf{N} = (N_s : s \in \mathcal{R})$	population of class s calls, vector of N_s
$b_s, \mathbf{b} = (b_s : s \in \mathcal{R})$	ratio of service time to idle time for class s calls, vector of b_s
$p_s = b_s/(1 + b_s)$	stationary probability that a class s source is in service at an IS station
$\underset{\sim}{T}(N, C, b)$	Engset loss formula for time congestion
$\tilde{G}(N, C, b)$	Engset loss formula for call congestion
$\underset{\sim}{\mathcal{T}}_s(\mathbf{N}, C, \mathbf{b})$	generalized Engset loss formula for time congestion
$\tilde{\mathcal{T}}_s(\mathbf{N}, C, \mathbf{b})$	generalized Engset loss formula for call congestion
$\pi_C(\mathbf{N}, \mathbf{n})$	stationary state distribution
$G(\mathbf{N}, C)$	normalization constant
$G^*(\mathbf{N}, \mathbf{z})$	generating function for normalization constant

D.2 Loss networks

$\mathcal{R} = \{1, \ldots, R\}$	set of routes
$\mathcal{R}_O, \mathcal{R}_C$	set of open routes, set of closed routes
$n_r, \mathbf{n} = (n_r : r \in \mathcal{R})$	number of route r calls in progress, route state vector
$N_s, \mathbf{N} = (N_s : s \in \mathcal{R}_C)$	population of closed route s, population vector for closed routes
$\mathcal{J} = \{1, \ldots, J\}$	set of links
$C_j, \mathbf{C} = (C_j : j \in \mathcal{J})$	capacity on link l (number of circuits), capacity vector
$m_j, \mathbf{m} = (m_j : j \in \mathcal{J})$	number of circuits on link l, link state vector
$a_{jr}, \mathbf{A} = [A_{jr} : j \in \mathcal{J}, r \in \mathcal{R}]$	number of circuits used by link j on route r, matrix of network requirements
$\mathbf{A}_r$	rth column of $\mathbf{A}$
$\pi_\mathbf{C}(\mathbf{N}, \mathbf{n})$	stationary route state distribution
$\pi_\mathbf{C}^*(\mathbf{N}, \mathbf{z})$	probability generating function for route state
$p_\mathbf{C}(\mathbf{N}, \mathbf{m})$	stationary link state distribution
$p_\mathbf{C}^*(\mathbf{N}, \mathbf{z})$	probability generating function for link state
$\mathcal{I}(\mathbf{N}, \mathbf{C})$	set of feasible route states
$G(\mathbf{N}, \mathbf{C})$	normalization constant
$G^*(\mathbf{N}, \mathbf{z})$	generating function for normalization constant

BIBLIOGRAPHY

[1]   Abate, J. and Whitt, W., The Fourier-series method for inverting transforms of probability distributions, *Queueing Sys.* **10** (1992), 5-88.

[2]   Abate, J. and Whitt, W., Numerical inversion of probability generating functions, *Oper. Res. Letters* **12** (1992), 245-251.

[3]   Baskett, F., Chandy, K.M., Muntz, R.R. and Palacios, F.C., Open, closed, and mixed networks of queues with different classes of customers, *Journal ACM* **22**:2 (1975), 248-260.

[4]   Birman, A. and Kogan, Y., On the accuracy of the asymptotic approxima-

tions for the partition function, *Proc. Conf. on Info. Sciences and Systems*, March 1994.

[5]   Birman, A. and Kogan, Y., Asymptotic evaluation of closed queueing networks with many stations, *Stoch. Mod.* **8** (1992), 543-563.

[6]   Bleistein, N., Uniform asymptotic expansions of integrals with stationary point near algebraic singularity, *Comm. Pure Appl. Math.* **19** (1966), 353-370.

[7]   Burman, D.Y., Lehoczky, J.P. and Lim, Y., Computational algorithms for closed queueing networks with exponential servers, *J. Appl. Prob.* **21** (1984), 850-859.

[8]   Buzen, J.P., Computational algorithms for closed queueing networks with exponential servers, *Comm. ACM* **16** (1973), 527-531.

[9]   Chung, S.-P. and Ross, K.W., Reduced load approximations for multirate loss networks, *IEEE Trans. on Comm.* **41**:8 (1993), 1222-1231.

[10]   Choudhury, G.L., Leung, K.K. and Whitt, W., An algorithm to compute blocking probabilities in multi-rate multi-class multi-resource loss models, *Adv. Appl. Prob.* **27** (1995).

[11]   Choudhury, G.L., Leung, K.K. and Whitt, W., Calculating normalization constants of closed queueing networks by numerically inverting their generating functions, *J. ACM* (1995).

[12]   Choudhury, G.L., Lucantoni, D.M. and Whitt, W., Multidimensional transform inversion with applications to the transient M/G/1 queue, *Ann. Appl. Prob.* **4** (1994), 719-740.

[13]   Cohen, J.W., The generalized Engset formulae, *Philips Telecomm. Review* **18**:4 (1957), 158-170.

[14]   Delbrouck, L.E.N., On the steady state distribution in a service facility with different peakedness factors and capacity requirements, *IEEE Trans. Commun.* **COM**-**31**:11 (1983), 1209-1211.

[15]   Dziong, Z. and Roberts, J.W., Congestion probabilities in a circuit-switched integrated services network, *Perf. Eval.* **7** (1987), 267-284.

[16]   Fedoryuk, M.V., Asymptotic methods in analysis, In: *Analysis I* (ed. by R.V. Gamkredlidze), Springer-Verlag, Berlin 1989.

[17]   Jackson, J.R., Jobshop-like queueing systems, *Mgt. Sci.* **10**:1 (1963), 131-142.

[18]   Jagerman, D.L., Some properties of the Erlang loss function, *B.S.T.J.* **53**:3 (1974), 525-551.

[19]   Kaufman, J.S., Blocking in a shared resource environment, *IEEE Trans. Comm.* **COM**-**29** (1981), 1474-1481.

[20]   Kelly, F.P., Loss networks, *Ann. Appl. Prob.* **1**:3 (1991), 319-378.

[21]   Kelly, F.P., *Reversibility and Stochastic Networks*, Wiley, Chichester 1979.

[22]   Kobayashi, H., *Modeling and Analysis: An Introduction to System Performance Methodology*, Addison-Wesley 1978.

[23]   Kobayashi, H., Birman, A. and Kogan, Y., Algorithmic analysis and congestion control of connection-oriented services in large scale communication networks, A joint research document (November, 1993).

[24]   Kobayashi, H. and Mark, B.L., On queueing networks and loss networks, *Proc. of the 28th Annual Conf. on Info. Sci. and Sys.* (1994), 794-799.

[25]   Kogan, Y., Asymptotic solution of generalized multiclass Engset model, In: *Proc. ITC 14* (ed. by J. Labetoulle and J. Roberts), Elsevier, Amsterdam **1b** (1994), 1239-1250.

[26]   Olver, F.W.J., *Introduction to Asymptotics and Special Functions*, Academic Press, New York 1974.

[27]   Mitra, D., Gibbens, R.J. and Huang, B.D., State-dependent routing on symmetric loss networks with trunk reservations-I, *IEEE Trans. on Comm.* **41**:2 (1993), 400-411.

[28]   Mitra, D. and Morrison, J.A., Erlang capacity and uniform approximations for shared unbuffered resources, *IEEE/ACM Trans. on Networking* **2**:6 (1994), 558-570.

[29]   Morrison, J.A., Loss probabilities in a simple circuit-switched network, *Adv. Appl. Prob.* **26** (1993), 456-473.

[30]   Pinsky, E. and Conway, A., Exact computation of blocking probabilities in multi-facility blocking models, In: *Proc. IFIP WG 7.3 Int. Conf. on the Perf. of Dist. Sys. and Integrated Commun. Networks* (ed. by T. Hasegawa, H. Takagi and Y. Takahashi), Kyoto, Japan (1991).

[31]   Pinsky, E., Conway, A. and Liu, W., Blocking formulae for the Engset model, *IEEE Trans. on Commun.* **42**:6 (1994), 2213-2214.

[32]   Reiser, M. and Kobayashi, H., Queueing networks with multiple closed chains: theory and computational algorithms, *IBM J. Res. Develop.* **19** (1975), 283-294.

[33]   Roberts, J.W., A service system with heterogeneous user requirements, In: *Perf. of Data Comm. Systems and their Applications* (ed. by G. Pujolle), North-Holland Publishing (1981), 423-431.

[34]   Ross, K.W. and Wang, J., Monte Carlo summation applied to product-form loss networks, *Prob. Eng. Info. Sci.* **6** (1992), 323-348.

[35]   Ross, K.W. and Wang, J., Asymptotically optimal importance sampling for product-form queueing networks, *ACM Trans. Modeling Comp. Sim.* **3**:3 (1993), 244-268.

[36]   Syski, R., *Introduction to Congestion Theory in Telephone Systems*, 2nd ed., Elsevier Science Publishers, Amsterdam 1986.

[37]   Titchmarsh, E.C., *Theory of Functions*, Oxford University Press, 2nd ed., London 1939.

[38]   Tsang, D.H.K. and Ross, K.W., Algorithms to determine exact blocking probabilities for multirate tree networks, *IEEE Trans. Comm.* **COM-38** (1990), 1266-1271.

[39]   Whitt, W., Blocking when service is required from several facilities simultaneously, *AT&T Tech. J.* **64** (1985), 1807-1856.

[40]   Whittle, P., Approximation in large-scale circuit-switched networks, *Prob. Eng. Info. Sci.* **2** (1988), 279-291.

[41]   Whittle, P., *Systems in Stochastic Equilibrium*, J. Wiley, New York 1986.

[42]   Wolff, R., *Stochastic Modeling and the Theory of Queues*, Prentice-Hall, Englewood Cliffs, New Jersey 1989.

[43]   Wong, R., *Asymptotic Approximation of Integrals*, Academic Press, Boston 1986.

Chapter 7

Sojourn time distributions in non-product-form queueing networks

Hans Daduna

ABSTRACT From a customer's point of view, the most important performance measure of a queueing network is his sojourn time. In this chapter, we unify and extend the results presently found in the literature, which provide explicitly the joint distribution of a customer's successive sojourn times over a prescribed path. The central feature for obtaining these results is the quasi overtake-free property of the path traversed by the customer. We concentrate on closed networks and sketch the cases of open and mixed networks. Our setting generalizes both the product-form networks of Kelly with general servers and the networks with overall-congestion dependent transition rates of Serfozo. We obtain equilibrium probabilities, prove an "arrival theorem," and compute joint sojourn time distributions for quasi overtake-free paths. The main conclusion is that general distributional results are rare.

CONTENTS

7.1 INTRODUCTION

Queueing networks provide a versatile set of models for complex systems in different fields of applications, such as computer networks, telecommunication networks, traffic systems, production and repair systems, migration processes, neural networks, maintenance, and logistic networks. For a short overview, see the introduction of Serfozo [35]; for special models in the computer and communication areas, see Parts III and IV of Takagi [38].

From the pioneering works of Jackson [19] and Gordon and Newell [17] on modeling and performance, analysis of such systems was mainly done by using so-called *product-form networks*, where in steady-state, the nodes behave in open networks as if they were independent or, in the closed network case, as an open system conditioned on a fixed population.

These product-form networks have been developed and refined step by step

0-8493-8076-6/97/$0.00+$.50

over the last forty years, including even more specialized modeling features; see e.g., Walrand [40]. Excluding some mini-networks and highly complicated transform results, these product-form networks can be said to be the only general class where the steady-state distributions of suitably defined Markovian state processes, which describe the networks' time development, can explicitly be written down. One main exception is given in Serfozo [34] and [35], where no product-form can be found for the equilibrium, which nevertheless is amenable to explicit computation. These models are the starting point for our derivations; we take up Serfozo's approach and generalize it in a way that allows us to deal with more details on the nodes' and the customers' behavior and to make more general assumptions on the driving processes which underlie the systems' stochastic specifications.

In Section 7.2 we define these queueing networks and compute the steady-state distribution. We concentrate here on the case of closed networks, but it will become obvious that the analysis carries over the case of open networks (see Wangler [41]) and mixed open-closed networks.

Having at hand this explicit description of the equilibrium behavior of the networks, we can solve two major problems of standard performance analysis:

(a) System-oriented performance measures, such as throughput, idling probabilities, and utilization;

(b) Mean values of customer-oriented performance measures, such as passage times, sojourn times, and travel times via Little's formula; see e.g., Reiser [31] and more advanced techniques described in Kook/Serfozo [24].

We shall be concerned with the sojourn times of customers in a network or in parts of a network, and with passage times for a customer traversing a specified path. Mean values of these quantities according to (b) can be computed under steady-state assumptions for general routes of customers.

Our interest focuses on the distribution of passage times and on the joint distribution of a customer's successive sojourn times at the nodes of a path. As it will turn out, for general paths almost nothing is known, which makes it easy to highlight the most important research problems related to the topic.

We have to look for new techniques to overcome the severe restrictions that we must put on the path's structure and the nodes' operating characteristic for successfully computing a customer's joint sojourn time distribution.

Putting it the other way round, in Section 7.5 we unify and generalize for closed networks explicitly the existing results on sojourn time distributions. Describing in this way the state of the art, our main Theorem 7.16, together with the introductory Definition 7.15, exhibit in detail the restrictions which we want to overcome. The technical term - with an intuitive interpretation - is that of a "quasi overtake-free path". A short history of the problem is included in Section 7.5, a detailed description of the problems and (partial) solutions up to 1990 is given by Boxma and Daduna [4], and a more recent survey is part of Daduna and Schassberger [12].

In Section 7.3 we prove an "arrival theorem," which provides us with a network distribution at arrival instants when a customer enters a path, as a prerequisite for our main results in Sections 7.4 and 7.5.

In Section 7.4 we prove a "splitting formula" for the joint sojourn time distribution of a customer's successive visits to the nodes of a general path. No assumptions on the topology of the path are required. We only require the first node to be semi-simple, as defined by Hemker [18] for product-form networks. This splitting formula is a result of independent interest because it clarifies some of the structure of the joint sojourn time distribution for a customer traversing a general path. In addition, it is the first step for proving the main theorem followed by computing the joint distribution for customer's sojourn times during a "*two-stations-walk*".

In Section 7.6 we comment on the difficulties that arise when sojourn time distributions on paths with overtaking are considered. In Section 7.7 we discuss open problems in the area and possible future research directions.

Two areas of research, where the discussed topic emerges in recent applications, should be mentioned without going into details:

In product and inventory systems the problem of lead time control comprises the passage time problem for items or goods forwarded through a production-inventory line. The results presented here are of interest if the production-inventory system is modelled as a queueing network. For a general problem description, see Karmakar [20]; special problems are described by Enns [15].

The introduction of ATM protocols (Asynchronous Transfer Mode) as the multiplexing technique for B-ISDN (Broadband Integrated Services Digital Networks) focused interest again on discrete time queueing networks because there exists a natural discrete time scale. An overview on the topic is given by Bae and Suda [1]. Seemingly, the first result on the End-to-End-Delay and transmission time distribution in closed form in such systems is proved in Daduna [10]. The cycle time in closed discrete time tandem systems of Bernoulli-servers is of product-form.

7.2 A CLOSED NETWORK WITH CONGESTION DEPENDENT NODES

We describe in this section the nodes of the network and how to build up from these basic entities a network of strongly coupled nodes. Roughly speaking, our nodes are of the internal structure described in detail by Kelly [21], p. 58, and the coupling is via congestion dependent service rates, which are allowed to depend on the population sizes of all nodes in the network, as described in detail by Serfozo [35], p. 159.

A node consists of an unlimited string of positions numbered $1, 2, \ldots$; whenever exactly n, $n \geq 1$, customers are present at a node, they occupy the positions 1 through n. The service mechanism is governed by three node-specific functions δ, φ, γ in the following way. If n customers are present, then an arriving customer moves independently of anything else into position k, $1 \leq k \leq n+1$, with probability $\delta(k; n+1)$, while customers from their previous positions k, $k+1, \ldots, n$ are shifted to positions $k+1$, $k+2$, $\ldots, n+1$, respectively ("*shift-protocol*"). Feedback customers are treated as new arrivals from outside the nodes.

If n customers are present, a total service effort is supplied to these customers with rate φ (this capacity function φ will be specified below), and a proportion $\gamma(k; n)$ of this effort is directed to the customer in position k, $1 \leq k \leq n$. When a customer in position k, $1 \leq k \leq n$, completes service, he leaves the node and customers previously in positions $k+1$, $k+2, \ldots, n$ move one position down ("*shift-protocol*"). Notice that from the definition we have $\sum_{k=1}^{n+1} \delta(k; n+1) = 1$, $n \geq 0$, and $\sum_{k=1}^{n} \gamma(k; n) = 1$, $n \geq 1$.

Henceforth, a node with such service mechanism is said *to be governed by the* "*general service discipline* (δ, γ)". If additionally,

$$\delta(m; n) = \gamma(m; n) \text{ for all } 1 \leq m \leq n, \ n \geq 0, \tag{7.1}$$

the service discipline (δ, γ) is said to be "*symmetric*" (Kelly [21], p. 72). We shall call such nodes "*general nodes*" and "*symmetric nodes*", respectively.

The network consists of J nodes, $J \geq 2$, numbered $1, 2, \ldots, J$. There are M distinguished customers, $M \geq 1$, traveling in the network, numbered $1, 2, \ldots, M$. At any time customer m, $1 \leq m \leq M$, is of some *type* t, $t \in T(m)$, where $\{T(1), \ldots, T(M)\}$ is a collection of pairwise disjoint countable type-sets. Each type $t \in \bigcup_{m=1}^{M} T(m)$ is associated with a fixed sequence of nodes $r(t) = [r(t, 1), r(t, 2), \ldots, r(t, S(t))]$, $r(t, s) \in \{1, \ldots, J\}$, $1 \leq s \leq S(t)$, where $2 \leq S(t) < \infty$. Customer m, of type t, travels route $r(t)$, thereafter changes to type t', and travels route $r(t')$, $t, t' \in T(m)$, and so on. Multiple visits of nodes as part of a route are not excluded.

Selection of new types is done according to a Markovian rule and independent of anything else. Customer m, $1 \leq m \leq M$, being of type t, upon completing route $r(t)$ selects his type t' with probability $p_m(t, t')$, $t, t' \in T(m)$. We assume that the stochastic matrix $p_m = (p_m(t, t'): t, t' \in T(m))$ determines an ergodic Markov chain with unique equilibrium probability $\nu_m = (\nu_m(t): t \in T(m))$, which satisfies the condition $\sum_{t \in T(m)} S(t)\nu_m(t) < \infty$.

Node j, $1 \leq j \leq J$, of the network is of the type as described above, with service discipline (δ_j, γ_j) and capacity function φ_j. Customer m, $1 \leq m \leq M$, being of type t, $t \in T(m)$, entering "*stage*" s, $1 \leq s \leq S(t)$, of his route (i.e., upon arrival at node $r(t, s) = j$) requests an amount of service which is distributed according to

    (*i*)   a node-specific $\exp(\mu_j)$-distribution, if j is not a symmetric node,

    (*ii*)   a distribution function $B_{t,s}(x)$, $x \geq 0$, depending on type and stage number.

These requests are drawn independently from one another and from anything else. To avoid notational complications and to simplify the computations in case (*ii*) we shall assume that $B_{t,s}(\cdot)$ is a finite mixture of Erlang distributions, i.e.,

$$B_{t,s}(x) = \sum_{h=1}^{H(t,s)} q_{t,s}(h) E_{\beta(t,s), h}(x), \quad x \geq 0, \tag{7.2}$$

where $q_{t,s}(h) \geq 0$, $1 \leq H(t, s) < \infty$, $q_{t,s}(H(t, s)) > 0$, $\sum_{h=1}^{H(t,s)} q_{t,s}(h) = 1$, and $E_{\beta(t,s), h}(x)$, $x \geq 1$, is the Erlang distribution function with h phases, each with mean $\beta(t, s)^{-1}$. The mean of $B_{t,s}(\cdot)$ is denoted by $\mu_{t,s}^{-1}$.

This assumption is no severe restriction because the class of mixtures of Erlang distributions is dense in the class of all distributions with support included in $[0, \infty)$. (See Schassberger [32], p. 32.) For our purposes, this assumption will allow us to find a discrete state space for a Markovian description of the network's development.

Clearly, the states of the network process must include at any time as a minimum of information the number of customers present at the different nodes. Having this in mind, we are able to specify the detailed capacity functions φ_j, $j = 1, \ldots, J$. We present the necessary information as a key lemma for determining the equilibrium of the network.

Lemma 7.1. (Serfozo [35], p. 149, 155). *Let $S(M) = \{(n_1,\ldots,n_J) \in \mathbb{N}^J : n_1 + \ldots + n_J = M\}$, and define for j, $k \in \{1,\ldots,J\}$ partial operators $T_{jk}: S(M) \to S(M)$ by $n = (n_1,\ldots,n_J) \to T_{jk}(n) = (n_1,\ldots,n_{j-1}, n_j - 1, n_{j+1},\ldots, n_{k-1}, n_k + 1, n_{k+1}, \ldots, n_J)$ if $n_j > 0$, and being undefined otherwise.*

For $f \in \{1,\ldots,J\}$ and $m \in \{0,1,\ldots,M\}$, let $S_f(M,m) = \{(n_1,\ldots,n_J) \in S(M) : n_f = M - m\}$. (In particular, we have $n \in S_f(M,0)$, if and only if $n = M \cdot e_f$, e_f being the fth unit vector.) Given any functions

$$\varphi_j : S(M) \to [0,\infty), \quad j = 1,\ldots,J,$$

the following statements are equivalent:

(a)   *There exists some $f \in \{1,\ldots,J\}$ such that*

$$\varphi_j(n) \cdot \varphi_k(T_{jf}n) \cdot \varphi_f(T_{kf}n) = \varphi_k(n) \cdot \varphi_j(T_{kf}n) \cdot \varphi_f(T_{jf}n), \tag{7.3}$$

for all $n \in S(M)$, $j,k \in \{1,\ldots,J\}$, holds.

(b)   *(7.3) holds for any $f \in \{1,\ldots,J\}$.*

(c)   *There exists an $f \in \{1,\ldots,J\}$ such that for all $n \in S(M)$ the following holds:*

If $n \in S_f(M,m)$, and $n(0), n(1),\ldots, n(m) \in S(M)$ is a sequence such that $n(h) \in S_f(M,h)$, $h = 0,1,\ldots,m$, $n = n(m)$, and for $h = 1,\ldots,m$ $n(h) = T_{fj(h)}(n(h-1))$ for some $j(h) \in \{1,\ldots,J\}$, then the product

$$\Phi(n) = \prod_{h=1}^{m} \varphi_f(n(h-1)) \cdot \varphi_{j(h)}(n(h))^{-1} \tag{7.4}$$

is independent of the specific sequence $n(1),\ldots,n(m)$. (We define an empty product to be 1.)

(d)   *The statement of (c) holds for all $f \in \{1,\ldots,J\}$.*

(e)   *The function $\Phi: S(M) \to [0,\infty)$ defined recursively by $\Phi(n) = 1$, $n \in S_f(M,0)$,*

$$\Phi(n) = \Phi(T_{jf}(n)) \cdot \varphi_f(T_{jf}(n)) \cdot \varphi_j(n)^{-1}, n \in S(M) - S_f(M,0), \tag{7.5}$$

is well defined, i.e., independent of f and the coordinate j chosen in (7.5).

The restrictions imposed on the capacity functions by (7.3) at a first glance seem to be very strong. But Serfozo [34, 35] and Kook and Serfozo [24] gave examples of practical relevance which meet (7.3) and which do not fit into the well-known families of network processes that lead to "*product-form equilibrium*". Clearly, Kelly's [21] networks, where for all $j = 1,\ldots,J$,

$$\varphi_j(n) = \varphi_j(n') \text{ if } n_j = n_j', \; n, n' \in S(M) \tag{7.6}$$

holds, meet the requirements of (7.3).

Because the celebrated BCMP-networks (see Baskett, et al. [2]) are essentially special cases of Kelly's networks, (7.6) holds for their capacity functions as well.

From the distributional assumptions and the (conditional) independence assumptions imposed on the network structure, it follows by standard arguments that the state space below obeys Markovian description of the network evolution in time. Local states x_j of node j are $x_j = e_j$, which indicates an empty node j, or $x_j = [t_{j1}, s_{j1} k_{j1}; \ldots; t_{jn_j}, s_{jn_j}, k_{jn_j}]$, which indicates that there are n_j customers present at node j, which occupy positions $1,\ldots,n_j$, and there is a customer of type t_{jh} in stage s_{jh} in position $h \in *\{1,\ldots,n_j\}$ of his route $r(t_{jh})$, requesting k_{jh}

further phases of service which are $\exp(\beta_{t_{jh}, s_{jh}})$-distributed, if j is a symmetric node; if j is not symmetric then the customer requests a residual service which is $\exp(\mu_j)$-distributed and $k_{jh} \equiv 1$; this will be omitted.

Global states of the network are $x = [x_1,\ldots,x_J]$, where x_j is a local state of node j, $j = 1,\ldots,J$, such that $\sum_{j=1}^{J} \sum_{h=1}^{n_j} 1\{t_{jh} \in T(m)\} = 1$ for all $m = 1,\ldots,M$. ($1\{\cdot\}$ denotes the indicator function.) We denote the set of all possible global states by $E(M,J)$. (The effective construction of $E(M,J)$ needs some care; it is not a product space, and in general depends on the initial configuration of the distinguished customers. For our purposes we need no further detail.)

The equilibrium behavior of the network can now be described as follows.

Theorem 7.2. *For the closed network as described above, let*

$$X(t) \in E(M,J), \quad t \geq 0,$$

denote the state of the network at time t. Then the supplemented joint queue length process $X = (X(t): t \geq 0)$ is a strong Markov process. Assuming that X is irreducible on $E(M,J)$, X is ergodic and its unique limiting and stationary distribution $\Pi = \Pi(M,J)$ on $E(M,J)$ is as follows.

Let $x = [x_1,\ldots,x_J] \in E(M,J)$, with $x_j = [t_{j1}, s_{j1}, k_{j1}; \ldots; t_{jn_j}, s_{jn_j}, k_{jn_j}]$ or $x_j = e_j$, if j is a symmetric node, or with $x_j = [t_{j1}, s_{j1}; \ldots; t_{jn_j}, s_{jn_j}]$ or $x_j = e_j$, if j is not a symmetric node, $j = 1,2,\ldots,J$. Let $n = n(x) = (n_1,\ldots,n_J)$ be the joint queue length vector of x. Then we have

$$\Pi(x) = \Phi(n) \cdot \prod_{j=1}^{J} g_j(x_j) \cdot G(M,J)^{-1}. \tag{7.7}$$

Here, $G(M,J)$ is the normalizing constant, and the functions $g_j(\cdot)$, $j = 1,\ldots,J$, are defined as

$$g_j(e_j) = 1,$$

$$g_j([t_{j1}, s_{j1}; \ldots; t_{jn_j}, s_{jn_j}]) = \prod_{h=1}^{n_j} \frac{\eta_j(t_{jh}, s_{jh})}{\mu_j}, \tag{7.8}$$

if node j is not a symmetric node, or

$$g_j([t_{j1}, s_{j1}, k_{j1}; \ldots; t_{jn_j}, s_{jn_j}, k_{jn_j}])$$

$$= \prod_{h=1}^{n_j} \left(\frac{\eta_j(t_{jh}, s_{jh})}{\eta_{t_{jh}, s_{jh}}} \cdot \sum_{i=k_{jh}}^{H(t_{jh}, s_{jh})} q_{t_{jh}, s_{jh}}(i) \right), \tag{7.9}$$

if node j is a symmetric node, where $\eta_j(t,s) = \nu_m(t) \cdot 1\{r(t,s) = j\}$, $j = 1,\ldots,J$, $s \in \{1,\ldots,S(t)\}$, $t \in T(m)$, $m \in \{1,\ldots,M\}$.

Theorem 7.2 is proved by comparing the stationary continuation of X on $\mathbb{R}$, which we again denote by $X = (X(t): t \in \mathbb{R})$ with its time reversal $\bar{X} = (\bar{X}(t): = X(-t): t \in \mathbb{R})$.

This time-reversed process describes the (equilibrium) behavior in time of a closed network with nodes $1,\ldots,J$ and distinguished customers $1,\ldots,M$ with type sets $T(m)$, $m = 1,\ldots,M$. Customer m being of type $t \in T(m)$, now travels route $\bar{r}(t) = [r(t, S(t)), r(t, S(t)-1),\ldots,r(t,1)]$, i.e., route $r(t)$ of the original network in the reversed direction. Upon completing route $\bar{r}(t)$, customer m becomes a customer of type $t' \in T(m)$ with probability $\bar{p}_m(t,t') = p_m(t',t) \cdot \nu_m(t') \nu_m(t)^{-1}$, i.e., the type selection is done according to the Markov rules determined by the transition probabilities of the Markov chain, which is the time reversal of the stationary Markov chain governed by p_m.

The nodes of the new network are of the same type as those of the original one, i.e., their behavior is determined by functions $\bar{\delta}_j$, $\bar{\gamma}_j$, $\bar{\varphi}_j$, with

$$\bar{\delta}_j = \gamma_j, \ \bar{\gamma}_j = \delta_j, \ \bar{\varphi}_j = \varphi_j, \ j = 1,\ldots,J.$$

It follows that a node is symmetric if and only if it has this property under time reversal. The distributions of the requested amount of service remain the same as in the original network. For symmetric node $\bar{r}(t,s)$ this means that customer t on stage s of route $\bar{r}(t)$ requests an amount of service which is distributed according to $\bar{B}_{t,s}(\cdot) = B_{t,S(t)-s+1}(\cdot)$.

From the definition of the time reversed process, the state spaces of X and $\bar{X}$ are the same, but we must be cautious. While for a generic triple (t,s,k) in the state description of some position, t and s have the same meaning (type of customer and stage of his actual route), the meaning of k is different for the two processes. As described above, for X, k denotes the residual number of service phases the customer requests; but for $\bar{X}$, k denotes the number of exponential service phases the customer has completed or started but not yet completed.

To distinguish X, $\bar{X}$, and the associated networks we shall use the abbreviations *"direct (time) process"*, *"direct (time) network"*, *"reversed (time) process"*, and so on.

From the description of the direct and the reversed networks, the associated transition rate matrices $Q = (q(x,y) : x,y \in E(M,J))$ and $\bar{Q} = (\bar{q}(x,y) : x,y \in E(M,J))$ are easily specified. To prove Theorem 7.2 we now apply Theorem 1.13 from Kelly [21], p. 30, which as a by-product yields that the reversed process $\bar{X}$ describes the proposed reversed network.

Proof of Theorem 7.2. We have to show that

$$\Pi(x) \cdot q(x,y) = \Pi(y) \cdot \bar{q}(y,x), x,y \in E(M,J), \ x \neq y,$$

and

$$\sum_{y \neq x} q(x,y) = \sum_{y \neq x} \bar{q}(x,y), \ x \in E(M,J).$$

The second equation is easily verified by inspection. For the first set of equations there are three different cases of transitions of X (and $\bar{X}$ in reversed direction) which we have to consider:

(i) $x = [x_1,\ldots,x_J] \to y = [y_1,\ldots,y_J]$ with $x_i = y_i$ for $i \in \{1\ldots,J\} - \{j\}$ and for a symmetric node j,

$$x_j = [t_{j1}, s_{j1}, k_{j1}; \ldots; t_{ja}, s_{ja}, k_{ja}; \ldots; t_{jn_j}, s_{jn_j}, k_{jn_j}], \text{ with } 1 \leq a \leq n_j, k_{ja} > 1,$$

$$y_j = [t_{j1}, s_{j1}, k_{j1}; \ldots; t_{ja}, s_{ja}, k_{ja} - 1; \ldots; t_{jn_j}, s_{jn_j}, k_{jn_j}].$$

(At node j, a service phase expires for the customer at position a.)
Then $\Pi(x) \cdot q(x,y) = \Pi(y) \cdot \bar{q}(y,x)$ reduces to

$$g_j(x_j) \cdot \varphi_j(n_1,\ldots,n_J) \cdot \gamma_j(a,n_j) \cdot \beta_{t_{ja}, s_{ja}}$$

$$= g_j(y_j) \cdot \varphi_j(n_1,\ldots,n_J) \cdot \gamma_j(a,n_j) \cdot \beta_{t_{ja}, s_{ja}} \cdot \left(\sum_{h=k_{ja}}^{H(t_{ja}, s_{ja})} q_{t_{ja}, s_{ja}}(h) \right)$$

$$\cdot \left(\sum_{h=k_{ja}-1}^{H(t_{ja},s_{ja})} q_{t_{ja},s_{ja}}(h) \right)^{-1}.$$

But this an obvious consequence of (7.9).

(ii) $x = [x_1,\ldots,x_J] \rightarrow y = [y_1,\ldots,y_J]$, where for a symmetric node j,

$$x_j = [t_{j1},s_{j1},k_{j1};\ldots;t_{ja},s_{ja},1;\ldots;t_{jn_j},s_{jn_j},k_{jn_j}],$$

with $1 \le a \le n_j$ and $s_{ja} < S(t_{ja})$,

$$y_j = [t_{j1},s_{j1},k_{j1};\ldots;t_{ja-1},s_{ja-1},k_{ja-1};$$
$$t_{ja+1},s_{ja+1},k_{ja+1};\ldots;t_{jn_j},s_{jn_j},k_{jn_j}],$$

and for a symmetric node $f = r(t_{ja},s_{ja}+1)$ either

$$x_f = [t_{f1},s_{f1},k_{f1};\ldots;t_{fn_f},s_{fn_f},k_{fn_f}]$$

with $n_f \ge 1$, and

$$y_f = [t_{f1},s_{f1};k_{f1};\ldots;t_{fb-1},s_{fb-1},k_{fb-1};$$
$$t_{ja},s_{ja}+1,k;t_{fb},s_{fb},k_{fb};\ldots;t_{fn_f},s_{fn_f},k_{fn_f}]$$

for some b, $1 \le b \le n_f + 1$, or $x_f = e_f$ and $y_f = [t_{ja},s_{ja}+1,k]$, and for $i \in \{1,\ldots,J\} - \{j,f\}$, $x_i = y_i$. (At node j, the service of customer t_{ja} in position a is finished. This customer proceeds to node f, which is stage $s_{ja}+1$ of his actual route.)

Then $\Pi(x) \cdot q(x,y) = \Pi(y) \cdot \bar{q}\,(y,x)$ reduces to

$$g_j(x_j)g_f(x_f)\Phi(n_1,\ldots,n_J)\varphi_j(n_1,\ldots,n_J) \cdot \gamma_j(a,n_j) \cdot \beta_{t_{ja},s_{ja}} \cdot \delta_f(b,n_f+1) \cdot q_{t_{ja},s_{ja}+1}(k)$$

$$= g_j(y_j)g_f(y_f) \cdot \Phi(n_1,\ldots,n_j-1,\ldots,n_f+1,\ldots n_J)$$

$$\cdot \varphi_f(n_1,\ldots,n_j-1,\ldots,n_f+1,\ldots,n_J) \cdot \beta_{t_{ja},s_{ja}+1} \cdot \bar{\delta}_f(b,n_f+1)$$

$$\cdot q_{t_{ja},s_{ja}+1}(k) \cdot \left(\sum_{h=k}^{H(t_{ja},s_{ja}+1)} q_{t_{ja},s_{ja}+1}(h) \right)^{-1} \cdot \bar{\delta}_j(a,n_j).$$

Applying (7.5) for the pair j,f, $\Phi(\cdot)$, $\varphi_j(\cdot)$, and $\varphi_f(\cdot)$ cancel. Inserting (7.9) and noting that $\gamma_j = \bar{\delta}_j$ and $\delta_f = \bar{\gamma}_f$ shows that the latter reduces to

$$\left(\frac{\eta_j(t_{ja},s_{ja})}{\beta_{t_{ja},s_{ja}}} \right) \cdot \beta_{t_{ja},s_{ja}} \cdot q_{t_{ja},s_{ja}}(k)$$

$$= \frac{\eta_1(t_{ja},s_{ja}+1)}{\beta_{t_{ja},s_{ja}+1}} \cdot \left(\sum_{h=k}^{H(t_{ja},s_{ja}+1)} q_{t_{ja},s_{ja}+1}(h) \right)$$

$$\cdot \beta_{t_{ja},s_{ja}+1} \cdot \left(\sum_{h=k}^{H(t_{ja},s_{ja}+1)} q_{t_{ja},s_{ja}+1}(h) \right)^{-1},$$

which holds because of $\eta_j(t_{ja},s_{ja}) = \nu_m(t_{ja}) = \eta_f(t_{ja},s_{ja}+1)$ for $t_a \in T(m)$.

(iii) $x = [x_1, \ldots, x_J] \rightarrow y = [y_1, \ldots, y_J]$, where for a symmetric node j,

$$x_j = [t_{j1}, s_{j1}, k_{j1}; \ldots; t_{ja}, S(t_{ja}), 1; \ldots; t_{jn_j}, s_{jn_j}, k_{jn_j}],$$

$$y_j = [t_{j1}, s_{j1}, k_{j1}; \ldots; t_{ja-1}, s_{ja-1}, k_{ja-1}; t_{ja+1}, s_{ja+1}, k_{ja+1}; \ldots; t_{jn_j}, s_{jn_j}, k_{jn_j}],$$

and for t_{ja}, $t' \in T(m)$ and a symmetric node $f = r(t', 1)$ either

$$x_f = [t_{f1}, s_{f1}, k_{f1}; \ldots; t_{fn_f}, s_{fn_f}, k_{fn_f}]$$

with $n_f \geq 1$, and

$$y_f = [t_{f1}, s_{f1}, k_{f1}; \ldots; t_{fb-1}, s_{fb-1}, k_{fb-1};$$
$$t', 1, k; t_{fb}, s_{fb}, k_{fb}; \ldots; t_{fn_f}, s_{fn_f}, k_{fn_f}]$$

for some b, $1 \leq b \leq n_f + 1$, or $x_f = e_r$ and $y_f = [t', 1, k]$, and for $i \in \{1, \ldots, J\} - \{f, j\}$, $x_i = y_i$. (At node j, the service of internal customer t_{ja} at position a is finished. He departs from route $r(t_{ja})$ and changes his type, now being of type t', entering node $f = r(t', 1)$.)

Then, $\Pi(x) \cdot q(x, y) = \Pi(y) \cdot \overline{q}(y, x)$ reduces to

$$g_j(x_j) g_f(x_f) \cdot \Phi(n_1, \ldots, n_J) \cdot \varphi_j(n_1, \ldots, n_J) \cdot \gamma_j(a, n_j) \cdot \beta_{t_{ja}, S(t_{ja})}$$
$$\cdot p_m(t_{ja}, t') \cdot \delta_f(b, n_f + 1) \cdot q_{t', 1}(k)$$
$$= g_j(y_j) g_f(y_f) \cdot \Phi(n_1, \ldots, n_j - 1, \ldots, n_f + 1, \ldots, n_J)$$
$$\cdot \varphi_f(n_1, \ldots, n_j - 1, \ldots, n_f + 1, \ldots, n_J)$$
$$\cdot \beta_{t', 1} \cdot \overline{\gamma}_f(b, n_f + 1) \cdot q_{t', 1}(k) \cdot \left(\sum_{h=k}^{H(t', 1)} q_{t', 1}(h) \right)^{-1} \cdot \overline{\delta}_j(a, n_j) \cdot \overline{p}_m(t', t_{ja}).$$

Applying (7.5) for the pair j, f, the terms $\Phi(\cdot)$, $\varphi_j(\cdot)$, and $\varphi_f(\cdot)$ cancel. Noting $\overline{\delta}_j = \gamma_j$ and $\overline{\gamma}_f = \delta_f$, and inserting (7.9) we obtain

$$\eta_j(t_{ja}, S(t_{ja})) \cdot p_j(t_{ja}, t') \cdot q_{t', 1}(k) = \eta_1(t', 1) \cdot \overline{p}_m(t', t_{ja}) \cdot q_{t', 1}(k).$$

But this is $\dfrac{\nu_m(t_{ja})}{\nu_m(t')} \cdot p_m(t_{ja}, t') = \overline{p}_m(t', t_{ja})$, which holds by definition of $\overline{p}_m(\cdot)$.

These three cases are exhaustive if we realize that in (ii) and (iii) the possible case $f = j$ is easier to handle, and in all cases the assumption of exponential (possibly non-symmetric) nodes f and/or j simplifies the derivations.   □

Corollary 7.3. *Let us assume that the queueing disciplines* (δ_j, γ_j), $j = 1, \ldots, J$, *are generalized in the following way:*

(i)   *Every time a new customer is inserted into the queue at, say, node j in position a, $1 \leq a \leq n_j + 1$, then the old customers at node j are redistributed over positions $1, 2, \ldots, a-1, a+1, \ldots, n_j + 1$ according to some permutation σ of $\{1, \ldots, n_j\}$. Selection of position a and permutation σ will be made with probability $\delta_j(a, \sigma; n_j)$, independently of anything else.*

(ii)   *Every time a service is completed at, say, node j in position b, $1 \leq b \leq n_j$, and the gap is closed according to the shift-protocol, then the remaining customers are redistributed over positions $1, \ldots, n_j - 1$ according to some permutation τ of $\{1, \ldots, n_j - 1\}$. Selection of τ will be made with probability $\zeta_j(\tau; n_j, b)$, independent of anything else.*

(iii)   *Whenever a service phase expires at, say, node j in position b, $1 \leq b \leq n_j$, then all customers at node j are redistributed according to some per-*

mutation τ of $\{1, \ldots, n_j\}$. Selection of τ will be made with probability $\vartheta_j(\tau; n_j, b)$, independently of anything else.
Node j is now symmetric if and only if

$$\sum_\sigma \delta_j(a, \sigma; n_j) = \gamma_j(a, n_j), \ 1 \le a \le n_j + 1, \ n_j \ge 0.$$

Let $X = (X(t): t \ge 0)$ be defined as in Theorem 7.2. Then X is ergodic with equilibrium Π given by (7.7).

Proof. The proof is rendered analogously to that of Theorem 7.2. The construction of the reversed queueing discipline becomes more involved but has no principal difficulties. □

Single server queues with permutation disciplines of Corollary 7.3 are introduced by Yashkov [42] and incorporated in Kelly's [21] product-form networks by Daduna [10]. The permutations in (*i*) and (*ii*) yield the possibility of controlling the network at departure and arrival epochs, while the permutations in (*iii*) yield a (nearly) continuous control of the system. The latter holds especially if the number of phases of the mixed Erlangian distributions is large, with each phase having a small mean, as occurs when general distributions are approximated by mixtures of Erlangs.

7.3 THE ARRIVAL THEOREM

We shall consider the network described in Section 7.2 in equilibrium, i.e., for $X = (X_t: t \in \mathbb{R})$ we have $X_t \sim \Pi$, $t \in \mathbb{R}$. Given this situation, we determine the state distribution at arrival instants for a specified customer m having a prescribed type $t \in T(M)$ at some node $j = r(t, s)$ of his route $r(t)$. Clearly, every arrival instant is a departure instant as well, with respect to the node, say i, which m visited just before. Recalling the complexity of the queueing disciplines which are represented by $\delta(\cdot), \gamma(\cdot)$, and the shift-protocol, we must be cautious about where to observe customer m at his jump instant, say t_0. If we chose a version of X, having left-continuous paths with right-hand limits, then m is at t_0 still present at node i; if we chose a version with right-continuous paths with left-hand limits, then m at t_0 is present at node j, occupying a position which is selected according to $\gamma_j(\cdot)$, and the gap at node i is closed. For reasons of symmetry it is often convenient to look at the network just "between t_0^- and t_0^+," i.e., when m has left node i, the gap is closed there, but did not yet arrive at node j.

What m then is seeing during his jump usually is called the "*disposition of the other customers.*" We shall assume henceforth that X has right-continuous paths with left-hand limits.

There are various methods to obtain the distribution of the process X at arrival instants. See, e.g., Kook and Serfozo [24], p. 235, ff., and, especially for closed networks, Lavenberg and Reiser [25], and Sevcik and Mitrani [36]. The simplest way to formulate the result is to consider limiting distributions at jump instants.

Theorem 7.4. *Consider $X = (X(t): t \ge 0)$ and $S = (S_n: n \in \mathbb{N})$, where, for fixed $t \in T(m)$, $S_n = S_n(t, s)$, $1 \le s \le S(t)$, is the nth instant when customer m being of type t arrives at node $r(t, s) =: j$.*

For $x \in E(M, J)$ such that customer m of type t on stage s of route $r(t)$ is present at node j, occupying position $a_0 \in \{1, \ldots, n_j\}$, $n_j \ge 1$, and $x = (x_1, \ldots, x_J)$, $n(x) = n = (n_1, \ldots, n_J)$, we have (for j being a symmetric server)

$$\Pi_j(t,s)(x): = \lim_{n\to\infty} Pr(X(S_n(t,s)) = x)$$

$$= \Phi(n) \cdot \varphi_j(n) \cdot \prod_{\substack{k=1 \\ k \neq j}}^{J} g_k(x_k) \cdot \prod_{h=1}^{n_j} \left(\frac{\eta_j(t_{jh}, s_{jh})}{\beta_{t_{jh}, s_{jh}}} \cdot \sum_{b=1}^{H(t_{jh}, s_{jh})} q_{t_h, s_{jh}}(b) \right)^{1\{t_{jh} \neq t\}}$$

$$\cdot \delta_j(a_0, n_j) \cdot q_{t,s}(k_{ja_0}) \cdot G_j(M,J)^{-1}, \tag{7.10}$$

where $G_j(M,J)$ is the normalizing constant, which does not depend on a_0.

The set of all such arrival states x as described above will be denoted by $E(M, J; t, s)$. *(If j is not a symmetric server, (7.10) has to be modified in an obvious way.)*

Proof. Let $(T_n : n \in \mathbb{N})$ be the sequence of jumps of $(X(t) : t \geq 0)$ and $Y = (Y(t): \; t \geq 0)$, given by $Y(t) = (X(T_{N_t - 1}), X(T_{N_t}))$, $t \geq 0$, where $(N_t : t \geq 0)$ counts the number of jumps of X in $[0, t]$. Let $(N_t^* : t \geq 0)$ denote the counting process recording the number of type t arrivals on stage s of their route at node j during $[0, t]$ and $(N_t^*(x) : t \geq 0)$ denote the counting process recording the number of type t arrivals at node j such that m, after his arrival, sees the network in state $x \in E(M, J; t, s)$. Note that m himself appears as type t in the state vector x.

By standard arguments, it follows that by using $Y(t) : t \geq 0$, which is stationary, we can express $\lim_{n\to\infty} P(X(S_n) = x)$ as

$$\frac{\displaystyle\sum_{y \in E(M,J)} \Pi(y) q(y, x)}{\displaystyle\sum_{x' \in E(M,J;t,s)} \; \sum_{y' \in E(M,J)} \Pi(y') q(y', x')}.$$

Inserting (7.7), (7.8), and (7.9), resp., and using (7.4) or (7.5) (use $f: = i$ there) yields (7.10). $\square$

The "arrival theorem" is a statement (slightly modified) similar to PASTA (Poisson-Arrivals-See-Time-Averages) in open networks, or MUSTA (Moving-Units-See-Time-Averages), as discussed by Serfozo [34], pp. 27-30. As Serfozo points out, this limiting property is the Palm property of the stationary process X with respect to (t, s)-arrivals at node j. It follows that if X is started at time 0 according to $\Pi_j(t, s)$ on $E(M, J; t, s)$ (i.e., $S_0(t, s) \equiv 0$), then the sequence $(X(S_n(t, s)) : n \in \mathbb{N})$ is stationary.

Corollary 7.5. *For customer m being of type t and for some $s \in \{1, \ldots, S(t)\}$ let $x \in E(M, J; t, s)$ denote the state after m's arrival at $r(t, s) = : j$. Let $y = ad_{t,s}(x)$ be obtained from x by deleting (t, s, k) (the information concerning m) from x and closing the gap. Then y is said to be an arrival disposition of the other customers with respect to (t, s). Let $E_{ad}(M, J; t, s)$ denote the set of all such arrival dispositions which m may see during his jump with destination $r(t, s)$. For $S_n = S_n(t, s)$, $n \in \mathbb{N}$, let*

$$X_{ad}(S_n) = ad_{t,s}(X(S_n)), \; n \in \mathbb{N},$$

denote the sequence of arrival dispositions which m observes on his jumps to $r(t, s)$. Then, for $y = (y_1, \ldots, y_J) \in E_{ad}(M, J; t, s)$ with

$$n(y) = (n_1, \ldots, n_J), \; n_1 + \ldots + n_J = M - 1$$

$$\Pi_{ad,j}(t,s)(y): = \lim_{n\to\infty} Pr(X_{ad}(S_n) = y)$$

$$= \Phi(n_1,\ldots,n_j+1,\ldots n_J)\cdot\varphi_j(n_1,\ldots,n_j+1,\ldots,n_J)$$

$$\cdot\prod_{k=1}^{J} g_k(y_k)\cdot G_j(M,J)^{-1}. \tag{7.11}$$

It should be noted that in general, there remains a trace of customer m in $\Pi_j(t,s)$ and $\Pi_{ad,j}(t,s)$, due to the general capacity function. So $\Pi_{ad,j}(t,s)$ is not the network's equilibrium if customer m is deleted. Indicating where m at j is present in $E(M,J;t,s)$ avoids further complications with state space notion which may appear if m is simply deleted. Some states which are then possible network configurations may not be allowed as dispositions. (See Walrand [40], pp. 144-145.)

Analogously, for the jumping customer m there exist departure dispositions. Here, implicitly, the node is recorded where m's jump started. For $2 \leq s \leq S(t) - 1$, this implicitly given information is equivalent to that provided by the arrival dispositions as described above. However, there are differences, e.g., if m has just selected type t and arrives at $r(t,1)$. For our purposes there will be no problems, which in general may arise if the time is reversed for the process.

To overcome possible difficulties one may define $(t,S(t);t',1)$-dispositions seen by m on leaving route $r(t)$ and starting route $r(t')$, etc.

7.4 THE SPLITTING FORMULA FOR SEMI-SIMPLE NODES

Customer routes in our networks are deterministic, given the actual type of the customer. This is a versatile routing mechanism, which by redefinition of types allows us to specify and investigate subroutes, or successive joint routes. Having this in mind, we now fix a customer and his type, say type $t \in T(M)$, of customer M, and assume that for route $r(t)$ we have $r(t,1) = 1$ and $r(t,2) = 2$.

The central object we are interested in is the joint distribution of the sojourn times of M as type t at nodes $r(t,1),\ldots,r(t,S(t))$ "in the steady state."

Let $\tau^{(n)} = (\tau_1^{(n)},\ldots,\tau_{S(t)}^{(n)})$ be the vector of successive sojourn times of customer M at nodes $r(t,1),\ldots,r(t,S(t))$, at the nth visit, where $\tau^{(0)}$ is the sojourn time of M in $r(t)$, such that the departure instant from $r(t)$, i.e., from $r(t,S(t))$, is the first such instant at 0 or thereafter.

Unfortunately, not much is known about the distribution of the $\tau^{(n)}$'s. Inspection of the existing results in the literature shows that almost all computations of joint distributions for $(\tau_1^{(0)},\ldots,\tau_{S(t)}^{(0)})$ involve trying to split the route $r(t)$ into subroutes of reduced length and then applying an induction argument.

This section is devoted to establishing a "*splitting formula*," which will be the first step to the later inductive procedure. The splitting instant will be the departure time of customer M as type t from node $1 = r(t,1)$, the moment when M's sojourn in 1 expires, and his residual route $(r(t,2) = 2, r(t,3),\ldots,r(t,S(t)))$ commences.

For technical reasons we have to shift this splitting instant to the time origin. Therefore, we consider $(X(t): t \geq 0)$ on an underlying probability space $(\Omega,\mathfrak{F},P_{ad,2})$, which guarantees that $S_0(t,2) \equiv 0$ and $ad_{t,2}(X(0)) \sim \Pi_{ad,2}(t,2)$ (according to Corollary 7.5) holds. We write expected values with respect to $P_{ad,2}$ as $E_{ad,2}[\,\cdot\,]$.

For the Laplace-Stieltjes transform (LST) of $\tau^{(0)}$, the Markov property of X yields

$$E_{ad,2}\left[\exp(-\sum_{s=1}^{S(t)}\theta_s\tau_s^{(0)})\right]$$

$$=\sum_{y\in E_{ad}(M,J;t,2)}\Pi_{ad,2}(t,2)(y)E_{ad,2}[\exp(-\theta_1\tau_1^{(0)})\mid ad_{t,2}(X(0))=y,S_0(t,2)=0]$$

$$\cdot E_{ad,2}\left[\exp(-\sum_{s=2}^{S(t)}\theta_s\tau_s^{(0)})\mid ad_{t,2}(X(0))=y,S_0(t,2)=0\right],$$

$$\theta_s\geq 0,\ s=1,\ldots,S(t). \tag{7.12}$$

In general, (7.12) can not be simplified. Our aim is to compute in the setting of a network with inter-node dependent capacity functions, explicit expressions for the joint probability of M's past sojourn time in 1 and the arrival distribution he observes during his jump from node 1 to node 2.

Under additional assumptions, (7.12) splits into a known term concerning $\tau_1^{(0)}$ and a second term concerned with $(\tau_2^{(0)},\ldots,\tau_{S(t)}^{(0)})$, a shorter path.

Although the results which will follow are more general than those hitherto found in the literature, we already have to reduce the complexity of the problem.

Definition and Assumption 7.6 (Hemker [18]). We assume node 1 to be semi-simple and independent of the other nodes. That is, there exists some number d_1, $1\leq d_1<\infty$, and real nonnegative numbers μ_{1a}, $1\leq a\leq d_1$, such that for population vector $n=(n_1,\ldots,n_J)$ we have

$$\varphi_1(n)=\varphi_1(n_1)=\sum_{a=1}^{min(d_1,n_1)}\mu_{1a}, \tag{7.13a}$$

$$\delta_1(a,n_1+1)=\begin{cases}1\ \text{for}\ a=n_1+1\\0\ \text{for}\ a\neq n_1+1,\end{cases} \tag{7.13b}$$

$$\gamma_1(a,n_1)=\begin{cases}\mu_{1a}\cdot\varphi_1(n_1)^{-1}&\text{for}\ 1\leq a\leq min(d_1,n_1)\\0&\text{for}\ a>min(d_1,n_1).\end{cases} \tag{7.13c}$$

Requested service times for all customers are exp(1)-distributed.

Semi-simple servers are generalized exponential multiservers: set $\mu_{1a}=\mu_1$, $1\leq a\leq d_1$; then there are d_1 service channels with a common waiting room which is organized according to the FCFS (First-Come-First-Served) discipline.

Due to the prescribed shift-protocol, customers in service may be served at different service channels, but as the service time distribution is memoryless, this changes neither the node's main performance measures such as throughput nor the customers' individual sojourn times, compared with a standard $\cdot/M/d_1/\infty$-system.

Corollary 7.7. *Under the assumptions of Definition 7.6, the following hold:*
(a)   *Formula (7.3) of Lemma 7.1 is equivalent to*

$$\varphi_j(n)\cdot\varphi_k(T_{j1}h)=\varphi_k(h)\cdot\varphi_j(T_{k1}n)$$

for all $n\in S(M)$ and all $j,k\in\{1,\ldots,J\}$. $\tag{7.14}$

(b)   *If $n\in S_1(M,m)$, $n_1=M-m$, then $\Phi(n)$ in (7.7), (7.10), and (7.11) may be replaced by*

$$\tilde{\Phi}(n) = \left(\prod_{h=1}^{m} \varphi_{j(h)}(n(h))^{-1} \right) \cdot \left(\prod_{a=1}^{n_1} \varphi_1(a) \right)^{-1}, \qquad (7.15)$$

where $n(0), n(1), \ldots, n(m)$ is a sequence of states as prescribed in Lemma 1(c). (If this replacement is done it has to be performed in the normalizing constants as well.)

(c)   *For $n \in S(M)$ and $j \in \{2, \ldots, J\}$,*

$$\tilde{\Phi}(n) \cdot \varphi_1(n_1) = \tilde{\Phi}(n_1 - 1, n_2, \ldots, n_{j-1}, n_j + 1, n_{j+1}, \ldots, n_J)$$

$$\cdot \varphi_j(n_1 - 1, n_2, \ldots, n_{j-1}, n_j + 1, n_{j+1}, \ldots, n_J). \qquad (7.16)$$

Proof of (b). Write (7.7) as

$$\Pi(x) = \left(\Phi(n) \cdot \left(\prod_{a=1}^{M} \varphi_1(a)^{-1} \right) \prod_{j=1}^{J} g_j(x_j) \right) \cdot \left(G(M,J) \left(\prod_{a=1}^{M} \varphi_1(a)^{-1} \right) \right)^{-1}$$

and cancel the terms in the nominator and denominator separately.   □

For node 1 being semi-simple we now return to (7.12) and show that the methods used by Hemker [18] to compute $E_{ad,2}[\exp(-\theta_1 \tau_1^{(0)}) \mid ad_{t,2}(X(0)) = y, S_0(t,2) = 0]$, $\theta_1 \geq 0$, for classical product-form networks, apply in the present setting as well.

Lemma 7.8. *For $y \in E_{ad}(M, J; t, 2)$ with $n_1 = n_1(y)$ we have*

$$E_{ad,2}[exp(-\theta_1 \tau_1^{(0)}) \mid ad_{t,2}(X(0)) = y, S_0(t,2) = 0]$$

$$= \sum_{a=1}^{min(d_1, n_1 + 1)} \frac{\mu_1 a}{\varphi_1(n_1 + 1)} \prod_{k=a}^{n_1 + 1} \frac{\varphi_1(k)}{\varphi_1(k) + \theta}, \quad \theta \geq 0. \qquad (7.17)$$

Proof. We consider the network in reversed time with transition matrix $\bar{Q} = (\bar{q}(x, x') : x, x' \in E(M, J))$ and adapt Hemker's proof. As we have shown in Section 7.2 we see a network of similar structure. From the definition of $\bar{X}$, it follows that pathwise the past sojourn time of M in 1, given his arrival disposition when jumping from 1 to 2, is in the reversed network the future sojourn time of M in 1, given his arrival distribution when jumping from 2 to 1.

Recall that $E(M, J; t, 1)$ is the set of all network states such that customer M of type t is present at node 1, and that under time reversal, node 1 becomes a (non-symmetric) node with $\bar{\varphi}_1(n) = \varphi_1(n) = \bar{\varphi}_1(n_1), n = (n_1, \ldots, n_J)$,

$$\delta_1(a, n_1 + 1) = \gamma_1(a, n_1 + 1) = \begin{cases} \mu_{1a} \cdot \varphi_1(n_1 + 1)^{-1} & 1 \leq a \leq min(d_1, n_1 + 1) \\ 0 & a > min(d_1, n_1 + 1), \end{cases}$$

$$\bar{\gamma}_1(a, n_1) = \delta_1(a, n_1) = \begin{cases} 1 & a = n_1, \\ 0 & a \neq n_1, \end{cases}$$

and with exp(1)-distributed requests for service times.

For $x \in E(M, J; t, 1)$ and $\theta \geq 0$, denote by $\bar{f}(x; \theta)$ the LST of the residual sojourn time of M as type t in the reversed node 1 given the network's state is x.

For $(\bar{f}(x; \theta) : x \in E(M, J; t, 1))$, there exists a system of linear first entrance equations having a uniquely determined solution.

The complexity of this system can be reduced considerably if we assume that

$$\bar{f}(x; \theta) = h(a, n_1; \theta) \quad 1 \leq a \leq n_1, \quad \theta \geq 0,$$

holds, where n_1 is the number of customers present at node 1, and a denotes the position of M. Clearly, this assumption has to be justified: this will be done by the explicitly given $\overline{f}(\,\cdot\,;\theta)$. Now, abbreviating $\overline{q}(x) := \overline{q}(x,x)$ we obtain for $x \in E(M,J;\,t,1)$ with $n_1(x_1) = n_1$ and M on position a,

$$h(a,n_1;\theta) = \frac{\overline{q}(x)}{\overline{q}(x)+\theta} \cdot \left[\sum_{\substack{z \in E(M,J;t,1) \\ x_1 = z_1}} \frac{\overline{q}(x,z)}{\overline{q}(x)} \cdot h(a,n_1;\theta) \right.$$

$$+ \sum_{\substack{z \in E(M,J;t,1) \\ n_1+1 = n_1(z)}} \frac{\overline{q}(x,z)}{\overline{q}(x)} \cdot \left\{ \frac{\varphi_1(a)}{\varphi_1(n_1+1)} \cdot h(a+1,n_1+1;\theta) \right.$$

$$\left. + \frac{1-\varphi_1(a)}{\varphi_1(n_1+1)} \cdot h(a,n_1+1;\theta) \right\}$$

$$+ (1-\delta_{a,n}) \cdot \frac{\varphi_1(n_1)}{\overline{q}(x)} \cdot h(a,n_1-1;\theta) + \delta_{an} \cdot \frac{\varphi_1(n_1)}{\overline{q}(x)} \cdot 1 \right], \quad \theta \geq 0. \qquad (7.18)$$

Setting

$$h(a,n_1;\theta) = \prod_{k=a}^{n_1} \frac{\varphi_1(k)}{\varphi_1(k)+\theta}, \quad \theta \geq 0,\ 1 \leq a \leq n_1, \qquad (7.19)$$

and observing that this function satisfies

$$h(a,n_1;\theta) = \frac{\varphi_1(a)}{\varphi_1(n_1+1)} \cdot h(a+1,n_1+1;\theta)$$

$$+ \left(1 - \frac{\varphi_1(a)}{\varphi_1(n_1+1)} \right) \cdot h(a,n_1+1;\theta), \quad 1 \leq a \leq n_1,$$

(7.18) reduces to

$$h(a,n_1;\theta) = \frac{1}{\overline{q}(x)+\theta} \cdot \left[\sum_{\substack{z \in E(M,J;t,1) \\ x_1 = z_1\ \text{or} \\ n_1+1 = n_1(z)}} \overline{q}(x,z) \cdot h(a,n_1;\theta) \right.$$

$$\left. + (1-\delta_{an_1}) \cdot \varphi_1(n_1) \cdot h(a,n_1-1;\theta) + \delta_{an_1}\varphi_1(n_1) \right].$$

From $\overline{q}(x) = \left(\displaystyle\sum_{\substack{z \in E(M,J;t,1) \\ x_1 = z_1\text{or }n_1+1 = n_1(z)}} \overline{q}(z) \right) + \varphi_1(n_1)$, we finally conclude that

$$h(a,n_1;\theta)[\varphi_1(n_1)+\theta]$$

$$= (1-\delta_{an_1}) \cdot \varphi_1(n_1) \cdot h(a,n_1-1;\theta) + \delta_{an_1}\varphi_1(n_1), \quad 1 \leq a \leq n_1,$$

would then be equivalent to (7.18). This system is obviously solved by (7.19). Assuming now that M on his jump from 2 to 1 observes the other customers in disposition $y \in E_{ad}(M,J;t,2)$ with $n_j(y) = n_j$, $j = 1,\ldots,J$, his sojourn time in the reversed node has a conditional distribution given y obtained from its LST:

$$\sum_{a=1}^{\min(d_1,n_1+1)} \overline{\delta}(a,n_1+1) \cdot h(a,n_1+1,\theta)$$

$$= \sum_{a=1}^{\min(d_1,n_1+1)} \frac{\mu_{1a}}{\varphi_1(n_1+1)} \prod_{k=a}^{n_1+1} \frac{\varphi_1(k)}{\varphi_1(k)+\theta}. \qquad \square$$

As Hemker [18] pointed out (for product-form networks), we can prove along the same lines as Lemma 8 a similar result in direct time.

Lemma 7.9. *For $y \in E(M, J; t, 1)$ with $n_1 = n_1(y)$, we have*

$$E_{ad,1}\left[exp(-\theta_1 \tau_1^{(0)}) \mid ad_{t,1}(X(0)) = y, \ S_0(t,1) = 0\right]$$

$$= \sum_{a=1}^{min(d_1, n_1 + 1)} \frac{\mu_{1a}}{\varphi_1(n_1 + 1)} \prod_{k=a}^{n_1 + 1} \frac{\varphi_1(k)}{\varphi_1(k) + \theta}, \quad \theta \geq 0. \tag{7.20}$$

For the most important special case, the $\circ/M/d_1/\infty$-node 1, (7.17) and (7.20) simplify considerably:

Corollary 7.10. *Let node 1 be a multiserver node, i.e., $\mu_{1a} = \mu_1$, $a = 1, \ldots, d_1$. Then for $y \in E(M, J; t, 1)$ and $y' \in E(M, J; t, 2)$ with $n_1 = n_1(y') = n_1(y)$, we have*

$$E_{ad,1}\left[exp(-\theta_1 \tau_1^{(0)}) \mid ad_{t,1}(X(0)) = y, \ S_0(t,1) = 0\right]$$

$$= E_{ad,2}\left[exp(-\theta_1 \tau_1^{(0)}) \mid ad_{t,2}(X(0)) = y', S_0(t,2) = 0\right]$$

$$= \frac{\mu_1}{\mu_1 + \theta} \cdot \left(\frac{\mu_1 d_1}{\mu_1 d_1 + \theta}\right)^{(n_1 - d_1 + 1)_+}, \quad \theta \geq 0, \tag{7.21}$$

where $(k)_+ = max(k, 0)$.

The main result of this section follows.

Theorem 7.11. (Splitting formula). *For customer M as type t on his route $r(t)$ we have under $P_{ad,2}$ for his sojourn times $\tau_s^{(0)}$, $s = 1, \ldots, S(t)$:*

$$E_{ad,2}\left[exp\left(-\sum_{s=1}^{S(t)} \theta_s \tau_s^{(0)}\right)\right]$$

$$= \sum_{y \in E_{ad,}(M,J;t,2)} \Pi_{ad,2}(t,2)(y) \left\{ \sum_{a=1}^{min(d_1, n_1 + 1)} \frac{\mu_{1a}}{\varphi_1(n_1 + 1)} \cdot \prod_{k=a}^{n_1 + 1} \frac{\varphi_1(k)}{\varphi_1(k) + \theta_1} \right\}$$

$$\cdot E_{ad,2}\left[exp\left(-\sum_{s=2}^{S(t)} \theta_s \tau_s^{(0)}\right) \Big| ad_{t,d}(X(0)) = y, S_0(t,2) = 0\right]$$

$$\theta_s \geq 0, \ s = 1, 2, \ldots, S(t). \tag{7.22}$$

The joint distribution of $\tau^{(0)} = (\tau_1^{(0)}, \ldots, \tau_{S(t)}^{(0)})$, which is of interest for practical purposes, is that under $P_{ad,1}$. That is, M as type t at his arrival at node $1 = r(t,1)$ observes the other customers distributed according to the limiting and stationary arrival distribution. The "splitting formula" (7.22) yields an expression for the joint distribution of $\tau^{(0)}$ under $P_{ad,2}$.

The junction with the sojourn time problem under $P_{ad,1}$ is given by the observation that the sojourn time distributions are the same under any $P_{ad,j}$, $j = (t,s)$, $s = 1, \ldots, S(t)$. This holds in the general setting of Section 7.2 without assuming node 1 to be semi-simple.

Proposition 7.12. *Let $r(t,s) = j \in \{1, \ldots, J\}$ for $s \in \{2, \ldots, S(t)\}$. Then,*

$$E_{ad,1}\left[exp\left(-\sum_{s=1}^{S(t)}\theta_s\tau_s^{(0)}\right)\right] = E_{ad,j}\left[exp\left(-\sum_{s=1}^{S(t)}\theta_s\tau_s^{(0)}\right)\right].$$

Proof.

$$E_{ad,j}\left[exp\left(-\sum_{s=1}^{S(t)}\theta_s\tau_s^{(0)}\right)\right]$$

$$\lim_{n\to\infty}\sum_{y\in E_{ad}(M,J;t,1)}E_{ad,j}\left[exp\left(-\sum_{s=1}^{S(t)}\theta_s\tau_s^{(n)}\right)\Bigg|X_{ad}(S_n(t,1))=y\right]$$

$$\cdot P_{ad,j}(X_{ad}(S_n(t,1))=y)$$

$$=\sum_{y\in E_{ad}(M,J;t,1)}E_{ad,j}\left[exp\left(-\sum_{s=1}^{S(t)}\theta_s\tau_s^{(0)}\right)\Bigg|X_{ad}(S_0(t,1))=y,S_0(t,1)=0\right]$$

$$\cdot\lim_{n\to\infty}P_{ad,j}(X_{ad}(S_n(t,1))=y),$$

where we used the strong Markov property of X. Note that for $n>1$ with probability 1, $S_n(t,1)>0$, because there is only one type t customer. The result follows from Corollary 7.5 with $j=1$, by using (7.5) with $f=1$.

Thus (7.22) is a result involving the joint distribution of sojourn times of M at nodes of $r(t)$ in equilibrium. Via (7.20) and (7.21) we therefore have obtained explicitly the equilibrium sojourn time distribution of any customer in a semi-simple node which is located in the general network of Section 7.2, and we have proved that it is the same under $P_{ad,1}$ and $P_{ad,2}=P_{dd,1}$. The latter denotes the distribution of the "*departure disposition*" for M when leaving node 1. A somewhat surprising observation is that this result is **not** obvious from direct unconditioning the conditional sojourn time LST's, which are identical in both time directions (see (7.17), (7.20), and (7.21)).

The reason is that, in general, $E_{ad}(M,J;t,1)\neq E_{ad}(M,J;t,s)$, $s\neq 1$, which can be seen from Walrand's [40] example (pp. 144-145). Performing the proof of equality by unconditioning requires the establishment of a suitable bijection between the disposition spaces which indeed can be found.

If $E_{ad}(M,J;t,1)=E_{ad}(M,J;t,2)$, then from Corollary 7.5 it follows, that

$$E_{ad,1}\left[exp\left(-\sum_{s=1}^{S(t)}-\theta_s\tau_s^{(0)}\right)\right]$$

$$=\sum_{y\in E_{ad}(M,J;t,1)}\Pi_{ad,1}(t,1)(y)\left\{\sum_{a=1}^{min(d_1,n_1+1)}\frac{\mu_{1a}}{\varphi_1(n_1+1)}\prod_{k=a}^{n_1+1}\frac{\varphi_1(k)}{\varphi_1(k)+\theta_1}\right\}$$

$$\cdot E_{ad,1}\left[exp\left(-\sum_{s=2}^{S(t)}\theta_s\tau_s^{(0)}\right)\Bigg|ad_{t,2}(X(S_0(t,2)))\stackrel{'}{=}y\right]. \tag{7.23}$$

This formula is the generalization of Theorem 2 in Schassberger and Daduna [33], where general (mixed) networks were considered, and node 1 was assumed to be a multiserver node (see Corollary 7.10), proved there by completely different methods.

It should be pointed out that the "key lemma" 2.1 of Schassberger and Daduna [33] yields a formula of a structure different from our results (7.17), (7.20), (7.21); in the notation used here it was proved that for a multiserver node in product-form networks

$$E_{ad,1}[\exp(-\theta_1\tau_1^{(0)}) \mid X_{ad}(S_0(t,2)) = y]$$

$$= \frac{\mu_1}{\mu_1+\theta_1} \cdot \left(\frac{\mu_1 d_1}{\mu_1 d_1+\theta_1}\right)^{(n_1(y)+1-d_1)+}$$

holds. This is **not** an obvious corollary to formula (7.21) in Corollary 7.10.

Our first result on joint sojourn time distributions is a direct consequence of the splitting formula and Proposition 7.12 and generalizes the results on "*Two-Station-Walks*" of Schassberger and Daduna [33] (Corollary 2.1(*b*), (*c*)) and Daduna [9] (Theorem 2).

Theorem 7.13. *Let $S(t) = 2$, such that M, being of type t, performs a "two-stations-walk". Then in equilibrium the distribution of $(\tau_1^{(0)}, \tau_2^{(0)})$ is as follows.*

(a)   *If $r(t,1) \neq r(t,2)$ and both nodes are semi-simple servers, then*

$$E_{ad,1}\left[exp\left(-\theta_1\tau_1^{(0)} - \theta_2\tau_2^{(0)}\right)\right]$$

$$= \sum_{y \in E_{ad}(M,J;t,2)} \Pi_{ad,2}(t,2)(y) \cdot \prod_{j=1}^{2} \left\{ \sum_{a=1}^{min(d_j,n_j+1)} \frac{\mu_a}{\varphi_j(n_j+1)} \prod_{k=a}^{n_j+1} \frac{\varphi_j(k)}{\varphi_j(k)+\theta_j} \right\},$$

$$\theta_j \geq 0, \quad j = 1,2.$$

where $n_j = n_j(y)$, $j = 1,2$.

(b)   *If $r(t,1) = r(t,2)$ and the node is a semi-simple server, then*

$$E_{ad,1}\left[exp\left(-\theta_1\tau_1^{(0)} - \theta_2\tau_2^{(0)}\right)\right]$$

$$= \sum_{y \in E_{ad}(M,J;t,1)} \Pi_{ad,1}(t,1)(y) \prod_{j=1}^{2} \left\{ \sum_{a=1}^{min(d_1,n_1+1)} \frac{\mu_{1ja}}{\varphi_1(n_1+1)} \prod_{k=a}^{n_1+1} \frac{\varphi_1(k)}{\varphi_1(k)+\theta_j} \right\},$$

$$\theta_j \geq 0, \quad j = 1,2.$$

In another direction, our Theorem 7.13 generalizes for a "*simple route*" of two successive nodes Theorem 5.2 of Kook and Serfozo [24] (p. 233): There, both nodes are single servers and the network nodes are exponential servers, such that the joint queue length process is already Markovian. It should be noted that we do not need any assumption on the inter-node dependencies, as is required by Kook and Serfozo [24]. They assume that the path under consideration is overtake-free (see their Definition 5.1 and our Definition 7.15), which puts independence assumptions on the capacity functions that separate parts of the network with respect to the inter-node dependencies. This yields a factorization of the steady-state expressions, which we do not need for Theorem 7.13.

A further immediate consequence of Proposition 7.12, the splitting formula, and Corollary 7.10 for a specific sojourn time problem, which does not need any overtake-free properties, is the following theorem.

Theorem 7.14. *Assume customer M, being of type t, on his route $r(t)$ visits infinite server nodes only, i.e., for $j = r(t,s)$, $1 \leq s \leq S(t)$, we have $\varphi_j(n_1,\ldots,n_j, \ldots,n_J) = n_j$, $(n_1,\ldots,n_J) \in S(M)$, and the service time at node j is $exp(\mu_j)$-distributed. Then,*

$$E_{ad,1}\left[exp\left(-\sum_{s=1}^{S(t)}\theta_s\tau_s^{(0)}\right)\right] = \prod_{s=1}^{S(t)}\frac{\mu_{r(t,s)}}{\mu_{r(t,s)}+\theta_s}, \quad \theta_s \geq 0, \ s = 1,\ldots,S(t).$$

Proof: By induction.

7.5 SOJOURN TIME DISTRIBUTIONS ON OVERTAKE-FREE PATHS

How to compute joint distributions for customer's successive sojourn times at different nodes of a path in a network or how to compute the distribution of the passage time (i.e., the sum of sojourn times) is almost an unsolved problem. The early results of Reich [29, 30] and Burke [5, 6] can still be said to be the prototype for all paths in networks, where explicit distributional characterization can be obtained.

They considered a tandem of exponential single-server queues (Reich [29, 30]) under the FCFS discipline, where the first and the last nodes may be multiserver queues under FCFS (Burke [5, 6]).

The research starting from these results yielded ever more complex networks with such tandems embedded in specific ways. The central feature that guaranteed that the old results hold in a more general setting of product-form networks was the "overtake-free property," introduced by Walrand and Varaiya [39] and Melamed [27]. The story of the development until 1990 is told in Boxma and Daduna [4], where results on non-product form networks and approximation techniques are described as well.

Recently, Kook and Serfozo [24] embedded the single-server tandem system of Reich [29, 30] into the non-product-form network with Markovian joint queue length process as described by Serfozo [34, 35]. They suitably adjusted (and generalized) the notion of an overtake-free route in open or closed non-product-form networks. Their result jointly generalizes those of Walrand and Varaiya [39] and Melamed [27] for open networks (*"The sojourn times are independent exponential random variables"*), and of Kelly and Pollett [22] for closed networks. (*"The joint sojourn time distribution is a mixture of convolutions of Erlangian distributions and is of product-form."*)

Cyclic exponential queueing systems are the analogues of the open tandem systems considered by Burke and Reich. Although their equilibrium state characteristics were already known, when Reich and Burke proved the independence of a customer's successive sojourn times in a tandem, cyclic queues resisted with respect to sojourn time distribution until the beginning of the eighties. The reason is that independence of sojourn times could no longer be expected, but the solutions showed a pleasingly elegant structure, as we shall see below. The joint sojourn time distribution in an exponential cycle of FCFS-queues is, loosely speaking, of product-form, and again the notion of overtake-free paths becomes the central feature.

We now take the theorem of Kook and Serfozo [24] as the starting point for this section. For the closed networks defined in Section 7.2 we shall jointly generalize their results and those of Schassberger and Daduna [33].

Definition 7.15. Consider a network as described in Section 7.2 with the notation introduced there.

(i)   For nodes $i, j \in \{1, \ldots, J\}$ write $i\{m\}j$, if customer $m \in \{1, \ldots, M\}$ may consecutively visit node i and j in this direction without any intermediate sojourn at other nodes.

(ii)   A path $[j_1, j_2, \ldots, j_r]$, $r \geq 2$, is a sequence of nodes such that for all $s = 1, 2, \ldots, r-1$ there exists at least one customer $m(s) \in \{1, \ldots, M\}$ such that $j_s\{m(s)\}j_{s+1}$.

(iii)   For customer $m \in \{1, \ldots, M\}$, a path $[j_1, j_2, \ldots, j_r]$ is said to be of potential influence on m if for all $s = 1, \ldots, r-1$ there exists at least one customer $m(s) \in \{1, \ldots, M\} - \{m\}$ such that $j_s\{m(s)\}j_{s+1}$.

(iv)   Let $[r(t,u), r(t,u+1), \ldots, r(t,v)]$ for $t \in T(m)$ be a sequence of different nodes which constitute a connected subsection of route $r(t)$.

For each $s = u, u+1, \ldots, v-1$ denote by $B(s)$ the set of all nodes that lie on m-relevant paths from $r(t,s)$ to $r(t,s+1)$ with $r(t,s+1)$ excluded. ($B(s)$ are the nodes that can be used if influences of customers are forwarded from node $r(t,s)$ to node $r(t,s+1)$.) Further denote

$$B(v) := \{1, \ldots, J\} - \bigcup_{s=u}^{v-1} B(s).$$

(v)   A connected section $[r(t,u), \ldots, r(t,v)]$, $1 \leq u < v \leq S(t)$, of route $r(t)$ is overtake-free for type $t \in T(m)$ if the following holds:

(α)   all nodes $r(t,s)$, $u \leq s \leq v$, are distinct;

(β)   for $p, q, u \leq p < q \leq v$, every path $[j_1, \ldots, j_r]$ of potential influence on m that leads from $r(t,p) = j_1$ to $r(t,q) = j_r$ includes node $r(t,p+1)$;

(γ)   for $s \in \{u, \ldots, v\}$ the capacity functions φ_j, $j \in B(s)$, are independent of n_k, $k \notin B(s)$, i.e., for $n = (n_1, \ldots, n_J)$, $n' = (n'_1, \ldots, n'_J)$ with $n'_j = n_j$, $j \in B(s)$, it follows that $\varphi_j(n) = \varphi_j(n')$.

(vi)   For customer m of type t, the route $r(t)$ is quasi overtake-free if the following holds:

- The capacity function of node $r(t,s)$ depends only on this node's congestion, i.e, $\varphi_{r(t,s)}(n_1, \ldots n_{r(t,s)}, \ldots, n_J) = \varphi_{r(t,s)}(n_{r(t,s)})$, $1 \leq s \leq S(t)$.

There exists $u, v, 1 \leq u \leq v \leq S(t)$ such that

- $r(t,1), \ldots, r(t,u-1)$ and $r(t,v+1), \ldots, r(t,S(t))$ are exponential infinite server nodes, i.e., $\varphi_j(n_1, \ldots, n_j, \ldots, n_J) = n_j$.

- $r(t,u+1), \ldots, r(t,v-1)$ are single server exponential nodes under FCFS, i.e., $\varphi_j(n_1, \ldots, n_j, \ldots, n_J) = 1$.

- $r(t,u), r(t,v)$ are semi-simple exponential nodes (see Definition 7.6).

Note that property (v) in Definition 7.15 is of a purely topological nature for the underlying graph of the network. The dependencies over the graph's nodes originate in the routing structure and the congestion dependencies of the capacity function. This topological overtake-free property must interplay with the internal node structures to obtain the final definition. This is specified in 7.15 (vi). Shortly stated, we can express the eventual definition of a quasi overtake-free path $r(t)$ for m as: $r(t)$ is split into three subsections $[r(t,1), \ldots, r(t,u-1)]$, $[r(t,u), \ldots, r(t,v)]$, $[r(t,v+1), \ldots, r(t,S(t))]$, such that the first and the last subsections of infinite server nodes pose no problems; this is similar to the result of

Theorem 7.14 on sojourn times on a path of infinite server nodes. The subsection $[r(t, u), \ldots, r(t, S(t))]$ is the critical part of the path, which may begin and end with semi-simple nodes that are neither infinite nor single server nodes.

Although we insisted on the (quasi) overtake-free property of the path for computing the joint sojourn time distribution for m in $r(t)$, it is obvious that m can be overtaken during his sojourn on $r(t)$. Property 7.15(v) guarantees only that on the itinerary $r(t, u+1), \ldots, r(t, v-1)$, m cannot be overtaken.

It should be noted that one or two of the subsections may be empty. Our main theorem can now be stated. For simplicity, we consider customer M as being of type t in $r(t)$.

Theorem 7.16. *Consider a network as described in Section 7.2. Let customer M be of type $t \in T(M)$ and assume his route $r(t)$ to be quasi overtake-free. Denote by $\tau^{(n)} = (\tau_1^{(n)}, \ldots, \tau_{S(t)}^{(n)})$ the vector of successive sojourn times of M at nodes $r(t, 1), \ldots, r(t, S(t))$, when he visits $r(t)$ the nth time. Here, $\tau^{(0)}$ is that sequence of sojourns of M in $r(t)$ that is the first to expire at time 0 or thereafter. Then,*

$$E_{ad,1}\left[exp\left(-\sum_{s=1}^{S(t)} \theta_s \tau_s^{(0)} \right) \right] = \sum_{y \in E_{ad}(M, J; t, 1)} \Pi_{ad,1}(t, 1)(y) \cdot \left(\prod_{s=1}^{u-1} \frac{\mu(s)}{\mu(s) + \theta_s} \right)$$

$$\cdot \left(\prod_{s=u+1}^{v-1} \left(\frac{\mu(s)}{\mu(s) + \theta_s} \right)^{n_s+1} \right) \left(\prod_{s=v+1}^{S(t)} \frac{\mu(s)}{\mu(s) + \theta_s} \right)$$

$$\cdot \left(\sum_{a=1}^{min(d_u, n_u+1)} \frac{\mu_{ua}}{\varphi_u(n_u+1)} \cdot \prod_{k=a}^{n_u+1} \frac{\varphi_u(k)}{\varphi_u(k) + \theta_u} \right)$$

$$\cdot \left(\sum_{a=1}^{min(d_v, n_v+1)} \frac{\mu_{va}}{\varphi_v(n_v+1)} \cdot \prod_{k=a}^{n_v+1} \frac{\varphi_v(k)}{\varphi_v(k) + \theta_v} \right), \quad \theta_s \geq 0, \ s = 1, \ldots, S(t). \quad (7.24)$$

(*Here we used the abbreviations* $\mu(s) := \mu_{r(t,s)}, \ s = 1, \ldots, u-1, u+1, \ldots, v-1,$ $v+1, \ldots, S(t); \ d_w := d_{r(t,w)}, \mu_{wa} := \mu_{r(t,w)a}, \ \varphi_w = \varphi_{r(t,w)}, \ w = u, v; \ and \ n_s = n_{r(t,s)}(y_{r(t,s)}) = n_{r(t,s)}(y).$)

Proof. The first observation is that we have, analogous to Theorem 7.14,

$$E_{ad,1}\left[\exp\left(-\sum_{s=1}^{S(t)} \theta_s \tau_s^{(0)} \right) \right]$$

$$= \left(\prod_{s=1}^{u-1} \frac{\mu(s)}{\mu(s) + \theta_s} \right) \cdot E_{ad,u}\left[\exp\left(-\sum_{s=u}^{v} \theta_s \tau_s^{(0)} \right) \right] \cdot \left(\prod_{s=v+1}^{S(t)} \frac{\mu(s)}{\mu(s) + \theta_s} \right).$$

Redefining routes for customer M, we therefore can and will assume that $u = 1$ and $v = S(t)$ holds.

Despite the fact that nodes 1 and $S(t)$ are not necessarily single server nodes we now are in the situation of Theorem 5.2 of Kook and Serfozo [24]. But the work done in Section 7.4 on proving Lemma 7.8, the splitting formula, and Lemma 7.9 and Corollary 7.10 just provides the information necessary for applying the induction procedure of Kook and Serfozo: Setting $\theta_s = 0$, $s = 2, \ldots, S(t)$ in

(7.22) and applying Proposition 7.12 yields (7.24) for the case $S(t) = 1$. (The assumption $2 \leq S(t)$ imposed in Section 7.2 on our systems was for notational purposes only and can be dropped with the help of some additional notation.)

Now assume for ease of notation that we have proved (7.24) for all paths of length z, $1 \leq z \leq S(t) - 1$. We use Proposition 7.12 and then the splitting formula to reduce the problem of proving (7.24) to computing the sojourn time distribution in node $r(t, 1)$ and in the nodes of the path $[r(t, 2), \ldots, r(t, S(t))]$. Observing that due to 7.15(v) (β) the sets $B(1), \ldots, B(S(t))$ are disjoint, the probability $\Pi_{ad, 2}(t, 2)$, appearing in the splitting formula, can be factorized in a term concerning $B(1)$ and a term concerning $B(2) \cup \ldots \cup B(S(t)) =: \widetilde{K}$, denoted by $\Pi(B(1))$, $\Pi(\widetilde{K})$, respectively.

Renormalization of these measures to probabilities, which shall be granted for now, converts the LST of (7.22) (which we have to consider after application of Proposition 7.12) into

$$
E_{ad, 1}\left[\exp\left(-\sum_{s=1}^{S(t)} \theta_s \tau_s^{(0)} \right) \right]
$$

$$
= \widetilde{C} \cdot \left[\sum_{\widetilde{y}} \Pi(B(1))(\widetilde{y}) \cdot \left(\sum_{a=1}^{\min(d_1, n_1 + 1)} \frac{\mu_{1a}}{\varphi_1(n_1 + 1)} \cdot \prod_{k=a}^{n_1 + 1} \frac{\varphi_1(k)}{\varphi_1(k) + \theta_1} \right) \right]
$$

$$
\cdot \left[\sum_{\widetilde{x}} \Pi(\widetilde{K})(\widetilde{x}) E_{ad, 2}\left[\exp\left(-\sum_{s=2}^{S(t)} \theta_s \tau_s^{(0)} \right) \Big| ad_{t, 2}(X(0)) = (\widetilde{y}, \widetilde{x}), S_0(t, 2) = 0 \right] \right].
$$

$$
(7.25)
$$

(Here, $\widetilde{C}$ stands for the renormalization and the summations over $\widetilde{y}$, $\widetilde{x}$ are such that $(\widetilde{y}, \widetilde{x}) \in E_{ad}(M, J; t, 2)$. That means the probabilities of the arrival dispositions $(\widetilde{y}, \widetilde{x})$ seen by M at his jump instant from node 1 to node 2 can be factorized in just the right way to perform the induction.) Due to the strong Markov property of X and to 7.15(v), the overtake-free property of $r(t)$, we conclude that

$$
E_{ad, 2}\left[\exp\left(-\sum_{s=2}^{S(t)} \theta_s \tau_s^{(0)} \right) \Big| ad_{t, 2}(X(0)) = (\widetilde{y}, \widetilde{x}), S_0(t, 2) = 0 \right]
$$

is independent of $\widetilde{y}$. This enables us to interpret the second squared bracket in (7.25) as the LST of the joint sojourn time distribution for M being of type t on a path $[r(t, 2), \ldots, r(S(t))]$ in a suitably redefined network. And there the induction hypothesis applies.   □

Our introduction of deterministic routes for the types enabled us in the closed network case to escape from the necessity of marking the customer under consideration. This was performed by Kook and Serfozo [24] for the case of undistinguishable customers with probabilistic routing. Their closed model is included in our presentation by allowing $S(t) = 1$ and setting $|T(m)| = 1$, $m = 1, \ldots, M$, and all $p_m(\cdot, \cdot)$ independent of m. A more detailed derivation of a similar induction procedure may be found in Daduna [8], where passage time distributions in closed networks with random routing are computed. What we have derived here can be performed for open networks and for mixed networks as well; for product-form networks this was proved in Schassberger and Daduna [33]. For the case of open loss networks with bounded population size (Kook and Serfozo [24]), the results hold as well.

For paths of interest with single server nodes as in Kook and Serfozo [24], open networks with deterministic, type-dependent routing and with unbounded population size are considered by Wangler [41], who derived the independence of sojourn times on a path, generalizing the results of Reich [29, 30], Walrand and Varaiya [39], and Kook and Serfozo [24]. The result of Wangler [46] can be obtained from Theorem 16 as well as by a limiting procedure, letting diverge $M \to \infty$. In this way, his results can be generalized to include multiserver and semi-simple nodes.

On an overtake-free path (similarly defined as in Definition 7.15), the successive sojourn times of a customer in equilibrium are independent.

7.6 MISCELLANEOUS

In this section, we discuss examples which may shed some light onto the problem of overtaking. When dealing with open networks, and referring to the results of the previous sections, we agree upon referring to the analog of the result for open networks; see Boxma and Daduna [4] for detailed citations.

The most prominent example concerning overtaking due to the topological structure of the network is the Simon/Foley (S/F) network (Simon and Foley [37]): A Poisson-λ process feeds a three station network (Fig. 1) of exponential single server FCFS nodes.

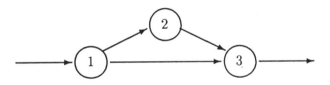

Figure 1

In equilibrium, the queue lengths behave as if they were independent. A customer traversing the network via $[1, 2, 3]$ can be overtaken when staying at node 2, so Theorem 7.16 does not apply for computing the joint sojourn time distribution of $(\tau_1^{(0)}, \tau_2^{(0)}, \tau_3^{(0)})$. From Theorem 7.13, however, it follows that on his path $[1, 2, 3]$, a customer experiences $(\tau_1^{(0)}, \tau_2^{(0)})$ and $(\tau_2^{(0)}, \tau_3^{(0)})$ as vectors with independent components. But Simon and Foley [36] proved that $(\tau_1^{(0)}, \tau_3^{(0)})$ has dependent coordinates. Moreover, Melamed [27] proved by brute-force numerical methods with careful error bounds that the customer's passage time $\tau_1^{(0)} + \tau_2^{(0)} + \tau_3^{(0)}$ is **not** distributed like the convolution of $\tau_j^{(0)}$, $j = 1, 2, 3$. Indeed, Foley and Kiessler [15] proved by using coupling arguments, that $\tau_1^{(0)}$ and $\tau_3^{(0)}$ are positively correlated.

Although Foley and Kiessler [15] explicitly point out that their methods cannot be applied to more general networks, a little bit more can be proved, using their ideas in connection with what we have computed in Sections 7.4 and 7.5.

Let us consider a general Jackson network of single server FCFS nodes (for simplification of the arguments) (Jackson [18]). Assume we can identify a subnet of the structure of Figure 1 in this network. The problem is then that the arrival stream at node 1 is not necessarily Poissonian because of possible feedback flows in the large network. Examination of the proof by Foley and Kiessler [15] shows that the starting point for their coupling argument is the explicit knowledge of the joint distribution of a customer's sojourn time $\tau_1^{(0)}$ and the disposition he observes when leaving node 1. This is explicitly given by Lemma 7.8.

Having these rather marginal results in mind, and recalling that reported simulation results indicate in open networks on feed-forward paths only positive correlations of sojourn times (Disney and Kiessler [13]), there is an obvious suggestion of what should be tried to be proved. A first attempt in this direction can be found in Daduna and Szekli [12], where more general correlation structures in Markovian networks are derived.

Another simple example, where overtaking naturally occurs, is a closed cycle of exponential single server FCFS nodes, if sojourn times are investigated which do not belong to just **one** cycle. The discussions in Kelly [22] and Boxma [3] are the best introductions to the problems connected with these systems.

Due to the assumption of single server FCFS nodes, the discussion of these examples reduced the overtaking problem to the topological structure of the customer's path in the network. Weakening this assumption poses additional difficulties as Burke [7] already stated:

In a 3-stage open tandem of

$$\text{single server - multiserver - single server}$$

all under FCFS, a customer's sojourn times $\tau_1^{(0)}$ and $\tau_3^{(0)}$ are independent.

The interpretation is: Although the topology of the network is pretty nice, the internal structure of node 2 introduces overtaking. Clearly, similar problems arise if the second node serves under another - general or symmetric - regime, which allows overtaking, e.g., processor sharing, Last-Come-First-Served, infinite server, etc.

In the light of this discussion, the following result seems to be surprising.

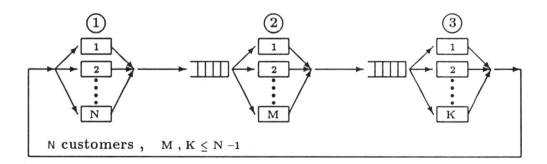

Figure 2

We have a closed tandem of exponential FCFS nodes (Figure 2). Node 1 acts as an infinite server, whereas nodes 2, 3 are multiservers where customers possibly must wait. The path [1, 2, 3] is quasi overtake-free according to Definition 7.15.

But a customer traversing the path may be overtaken by other customers repeatedly. Nevertheless, Theorem 7.16 applies, and we can explicitly compute the joint sojourn time distribution for the customer's successive visits at nodes 1, 2, 3.

Now let us consider the path [2, 3, 1]: The same holds, because via Theorem 7.16 we can compute the joint sojourn time distribution.

Finally, consider the path [3, 1, 2]: This path is **not** quasi overtake-free! It fits into the class of examples described by Burke [7], so considerable difficulties arise in computing joint sojourn time distributions. Although the set of first-entrance-time equations for the cycle time is still finite the structure of the problem is not well understood.

7.7 OPEN PROBLEMS AND FUTURE RESEARCH DIRECTIONS

As it should become clear from the main Theorem 7.16, only very little is known about the distribution of the joint sojourn time vector of a customer's sojourn times at the nodes of a prescribed path in a closed, open, or mixed network. The central feature, which yields explicit distributional results, is the overtake-freeness of the path. An intuitive description of what is done by formulating and proving this theorem can be given as follows.

Consider a cyclic sequence of exponential single server nodes under FCFS. The distribution of the cycle time in equilibrium for a customer has the LST of product-form. Then the following facts are shown:
  • the joint distribution of the customer's sojourn times when passing the cycle has a product-form LST of similar structure;
      This structure is not destroyed by the following changes and additional features:
    • the first and the last node may be substituted by multiserver exponential nodes,
    • before the first and behind the last node, infinite server nodes may be inserted into the path,
    • feed backward paths for and (very limited) feed forward paths for other customers may be allowed,
    • the network developed so far may be included into a general product-form network, or non-product form network as described in Section 7.2.

This list is a summary of sojourn time research performed in the last fifteen years - which for closed networks clarifies the gaps in our present knowledge. As pointed out, the situation for open networks is similar.

The most striking question in the open network case is that for a **simple** expression for a customer's passage time distribution in the Simon/Foley network (Figure 1) and its closed network analogue. If this can be found we would expect to find an explicit expression for the joint sojourn time distribution as well. Then it may give a hint on how to incorporate such feed forward (overtake) paths for other customers into the linear structure described above.

Another point of interest is whether multiple visits of a customer at a node may be allowed. Theorem 7.13(*b*) is a very simple case where the answer can be given, but further results beyond the classical isolated feedback node seem not to be known. A related but even more complicated problem is that of successive cycles of a customer; here, for a closed two-station network the case of two successive cycles was solved by Boxma [3] and Kelly [23], who presented some structural observations.

A set of problems that seemingly need completely different methods than those presented in Section 7.5 emerges when nodes with queueing and service regimes other than FCFS lie on the path of interest. Then, overtaking due to the internal node structure usually prevents us from proving product-form results. (See Boxma and Daduna [4], pp. 420-421, for a short review.) If these nodes are symmetric in the sense of Kelly [21] then the service times at this node need not be exponential. However, they still yield an explicit product-form (or non-product-form in the sense of Section 7.2) expression for the steady-state and arrival distributions. In general, nothing is known about passage time distributions for paths including such nodes.

Up to now, our discussion dealt with networks where the equilibrium distribution and Palm distribution, conditioned on arrivals or departures are effectively amenable. This yielded explicit sojourn time computations via conditioning, and results under Palm distributions. Deviating from the underlying Palm distribution will clearly pose interesting problems, but from our present knowledge we should not expect explicit results for the near future. Recalling the complicated structure of the transient queue length process of an isolated $M/M/1/\infty$-FCFS shows that for transient networks we cannot expect to obtain explicit queue length probabilities. It should be noted that, e.g., for passage time distributions, the conditional LST makes up a set of first-entrance-equations that do not depend on equilibrium assumptions. On the other hand, an explicit, smooth solution of this set of linear equation seems to be impossible as well.

While for the class of non-product-form networks described in Section 7.2, the steady-state distribution is (in principle) known, there are other classes where this is not the case:

- networks with blocking,
- networks with asymmetric non-exponential nodes (e.g., FCFS).

For these networks, approximation procedures have been developed - for a review, see Boxma and Daduna [4], pp. 426-442. The difficulties emerging here can be seen from the proofs in the recent paper of LeGall [26], where *the overall sojourn time in tandem queues with identical successive service times and renewal input* is computed in terms of cumbersome transform expressions. Possibly this and related results point out that research on quick and easy approximations should be a further important topic of research in this field.

ACKNOWLEDGEMENT

The author would like to thank Gary Russell for his careful reading the first draft of this paper and his valuable suggestions.

BIBLIOGRAPHY

[1]   Bae, J.J. and Suda, T., Survey of traffic control schemes and protocols in ATM networks, *Proc. IEEE* **79** (1991), 170-189.

[2]   Baskett, F., Chandy, K.M., Muntz, R.R. and Palacios, R.G., Open, closed, and mixed networks of queues with different classes of customers, *J. ACM* **22** (1975), 248-260.

[3]   Boxma, O.J., Analysis of successive cycles in a cyclic queue, In: *Performance of Computer-Communication Systems* (ed. by H. Rudin and W. Bux), North Holland, Amsterdam (1984), 293-306.

[4]   Boxma, O.J. and Daduna, H., Sojourn times in queueing networks, In: *Stoch. Anal. of Comp. and Commun. Systems* (ed. by H. Takagi), North Holland, Amsterdam (1990), 401-450.

[5]   Burke, P.J., The output of a queueing system, *Oper. Res.* **4** (1956), 699-704.

[6]   Burke, P.J., The output process of a stationary M/M/s queueing system, *Ann. Math. Stat.* **39** (1968), 1144-1152.

[7]   Burke, P.J., The dependence of sojourn times in tandem M/M/s queues, *Oper. Res.* **17** (1969), 754-755.

[8]   Daduna, H., Passage times for overtake-free paths in Gordon-Newell networks, *Adv. Appl. Prob.* **14** (1982), 672-686.

[9]   Daduna, H., On passage times in Jackson networks: Two stations walk and overtake-free paths, *Zeitschrift für Oper. Res.* **27** (1983), 239-256.

[10]   Daduna, H., *Spezielle stochastische Prozesse*, Lecture Notes, Inst. f. Math. Stochastik, Univ. Hamburg 1984.

[11]   Daduna, H., The cycle time distribution in a cycle of Bernoulli servers in discrete time, Inst. f. Math. Stochastik, Univ. Hamburg (1994), preprint No. 94-6.

[12]   Daduna, H. and Schassberger, R., Delay time distributions and adjusted transfer rates for Jackson networks, *AEÜ* **47** (Special issue on Teletraffic Theory and Engineering in Memory of Felix Pollaczek) (1993), 342-348.

[13]   Daduna, H. and Szekli, R., Positive and negative dependence in Markovian networks, *Adv. Appl. Prob.* **27** (1995), 226-254.

[14]   Disney, R. and Kiessler, P.C., *Traffic Processes in Queueing Networks, A Markov Renewal Approach*, The Johns Hopkins University Press, Baltimore 1987.

[15]   Enns, S.T., Job shop prediction and tardiness control using queueing analysis, *Int. J. Prod. Res.* **31** (1993), 2045-2057.

[16]   Foley, R.D. and Kiessler, P.C., Positive correlations in a three-node Jackson queueing network, *Adv. Appl. Prob.* **21** (1989), 241-242.

[17]   Gordon, W.J. and Newell, G.F., Closed queueing systems with exponential servers, *Oper. Res.* **15** (1967), 254-265.

[18]   Hemker, J., A note on sojourn times in queueing networks with multiserver nodes, *J. Appl. Prob.* **27** (1990), 469-474.

[19]   Jackson, J.R., Networks of waiting lines, *Oper. Res.* **5** (1957), 518-521.

[20]   Karmakar, U.S., Manufacturing lead times, order release, and capacity loading, In: *Logistics of Production and Inventory* (ed. by S.C. Graves, A.H.G. Rinnooy Kan and P.H. Zipkin) North-Holland, Amsterdam (1993), 287-329

[21]   Kelly, F.P., *Reversibility and Stochastic Networks*, Wiley, New York 1979.

[22]   Kelly, F.P. and Pollett, P.K., Sojourn times in closed queueing networks, *Adv. Appl. Prob.* **15** (1983), 638-656.

[23]   Kelly, F.P., The dependence of sojourn times in closed queueing networks, In: *Math. Comput. Perform. and Reliability* (Ed. by G. Iazeolla, P.J. Courtois, and A. Hordijk), North-Holland, Amsterdam (1984), 111-121.

[24]   Kook, K.H. and Serfozo, R.F., Travel and sojourn times in stochastic networks, *Ann. Appl. Prob.* **3** (1993), 228-252.

[25]   Lavenberg, S.S. and Reiser, M., Stationary state probabilities at arrival instants for closed queueing networks with multiple types of customers, *J. Appl. Prob.* **17** (1980), 1048-1061.

[26] LeGall, P., The overall sojourn time in tandem queues with identical succes-
      sive service times and renewal input, *Stoch. Proc. Their Appl.* **52** (1994),
      165-178.

[27] Melamed, B., Sojourn times in queueing networks, *Math. Oper. Res.* **7**
      (1982), 223-244.

[28] Melamed, B., Randomization procedures for numerical computation of de-
      lay time distributions in queueing systems, *Bull. Int. Stat. Inst.*, ISI Session
      44, Vol. II (1983), 755-775.

[29] Reich, E., Waiting times when queues are in tandem, *Ann. Math. Stat.* **28**
      (1957), 768-773.

[30] Reich, E., Note on queues in tandem, *Ann. Math. Stat.* **34** (1963), 338-341.

[31] Reiser, M., Performance evaluation of data communication systems, *Proc.
      IEEE* **70** (1982), 171-196.

[32] Schassberger, R., *Warteschlangen*, Springer-Verlag, Wien 1973.

[33] Schassberger, R. and Daduna, H., Sojourn times in queueing networks with
      multiserver nodes, *J. Appl. Prob.* **24** (1987), 511-521.

[34] Serfozo, R.F., Markovian network processes: Congestion-dependent routing
      and processing, *Queueing Sys.* **5** (1989), 5-36.

[35] Serfozo, R.F., Queueing networks with dependent nodes and concurrent
      movements, *Queueing Sys.* **13** (1993), 143-182.

[36] Sevcik, K.C. and Mitrani, I., The distribution of queueing network states at
      input and output instants, *J. ACM* **28** (1981), 358-371.

[37] Simon, B. and Foley, R.D., Some results on sojourn times in acyclic
      Jackson networks, *Man. Sci.* **25** (1979), 1027-1034.

[38] Takagi, H., *Stochastic Analysis of Computer and Communication Systems*,
      North-Holland, Amsterdam 1990.

[39] Walrand, J. and Varaiya, P., Sojourn times and the overtaking condition in
      Jacksonian networks, *Adv. Appl. Prob.* **12** (1980), 1000-1018.

[40] Walrand, J., *An Introduction to Queueing Networks*, Prentice-Hall, Engle-
      wood Cliffs 1988.

[41] Wangler, M., *Verweilzeiten und Durchlaufzeiten in offenen stochastischen
      Netzwerken mit systemzustandsabhängigen Bedienraten*, M.S. Thesis, Dep.
      of Mathematics, Univ. of Hamburg 1994.

[42] Yashkov, S.F., Properties of invariance of probabilistic models of adaptive
      scheduling in shared-use systems, *Autom. Contr. Comp. Sci.* **14** (1980), 46-
      51.

Part III

Traffic Processes

Chapter 8

Stochastic geometry models of mobile communication networks[1]

François Baccelli and Serguei Zuyev

ABSTRACT In this chapter, the authors propose a new approach for the modeling of mobile communication networks. The main tools are point processes and stochastic geometry. They show how several performance evaluation problems within this framework can actually be posed and solved by computing the mathematical expectation of certain functionals of point processes. The authors analyze models based on Poisson point processes for which analytical formulas can often be obtained. In particular, they obtain distributions of the number of mobiles with calls in progress in a base station cell and the number of such mobiles passing the boundary of the cell.

CONTENTS

8.1 INTRODUCTION

There is a rapid development of new telecommunication systems based on wireless or cellular communication networks. In contrast with stationary networks, the emitters of calls are mobile and not connected by a physical link (cable) to the corresponding station. This fact brings new problems for the performance evaluation of such systems. The main peculiarities of cellular communications are

- radio wave connections between call emitters and the ground stationary stations (base antennas);
- the problem of localization of destination (the position of a mobile at a given instant is *a priori* unknown);
- the problem of antenna association and the problem of *hand-over*, or the change of a call routing scheme when the emitter of the call in progress changes its position with respect to the ground stations.

The consideration of the first phenomenon is closely related to wave propagation studies. The state-of-the-art in this domain with relation to wireless communications is presented, e.g., in [13] with the pertinent literature quoted there. We will not be concerned with this subject here. The two last mentioned peculiarities

---

[1]Work supported by CNET through research grants 93 5 B and CTI 1B 104.

---

are due to the fact that mobile communication systems possess an additional "degree of freedom" which stems from *spatial* traffic variations in addition to the *temporal* variations present in telecommunication systems with fixed positions of emitters. We pay particular attention to this new feature making use of stochastic geometry tools to capture the space-time dynamics of mobile communication systems.

The modeling approach presented in this chapter seems relevant to strategic planning and economical studies of telecommunication networks. Generally speaking, it consists of representing the structural objects of a network (mobiles, roads, stations, etc.) as realizations of stochastic processes. All other characteristics, such as zones served by the stations, teletraffic in the network, etc., are expressed as functionals of these processes and thus depend only on the parameters of their distributions. In comparison with the current models used in strategic planning and economics of telecommunication networks, stochastic models allow the user to avoid the detailed geographic description of the network and significantly reduce the number of structural parameters of the model. This question is considered in detail in [2] and [3]. Stochastic models take into account the spatial variability of the system structure that lacks, for example, regular honeycomb lattice models of the station zones often used in mobile communication studies (cf. [13, Ch. 6]). At the same time, analytical expressions for many important characteristics are possible, which makes the analysis of the system behavior possible.

8.2 DESCRIPTION OF THE MODEL

A stochastic model of a mobile communication network should include
- a model of the stationary network;
- a model of the road system;
- a model of vehicle traffic on the roads;
- a model for the radio communication handling (allocation, capturing, release, hand-over).

There are many ways to construct the above models. In this section we propose simple parametric models which can be integrated into one global network model. The result is general enough to reflect many features of a real system and it is analytically tractable.

8.2.1 Basic cell model

The simplest probabilistic model of a telephone network is one with a single hierarchical level.

The base stations are represented through their coordinates in $\mathbb{R}^2$ plane. Each base station serves mobile telephone users located in a certain area - *cells*. Equivalently, the plane is sub-divided into cells, each of which is served by a unique station.

In this chapter, we make the assumption that subscribers are served by the closest station. This assumption is a natural first-order approximation in the case where all stations have the same power, since the closest base station antenna has usually a better radio signal reception than other stations. Note that more detailed models taking into account stations with different power levels or propagation randomness (such as physical hindrances or weather) can also be considered along these lines (see [19]).

In this case, a cell served by a station x_i is a convex polygon $\mathcal{C}_i$ known as the *Voronoi cell* with *nucleus* x_i, constructed with respect to the set $\{x_j\}$ of stations. The Voronoi cell $\mathcal{C}_i$ is the intersection of half planes H_{ij} bounded by the bisectors of the segments $[x_i, x_j]$ and containing x_i. The system of all cells forms a tessellation of the plane called the *Voronoi tessellation* (cf. [16]).

The main idea of our stochastic modeling consists of considering the configuration of the stations as realizations of a stochastic point process. The reason for this comes from the observation of real networks, where the constraints imposed on the location of base stations (tops of hills or cost constraints in towns) lead to an effective randomness for this point process. In many applications, this process can be taken as Poisson. The main advantage of a Poisson process is its simplicity. The distribution of a *Poisson process* Π is completely determined by its *intensity measure* Λ representing the mean density of points. More precisely, the number of the process points in a Borel set B obeys the Poisson distribution with parameter $\Lambda(B)$ and the number of points in disjoint Borel subsets are independent. A natural technical assumption is that the measure Λ is locally finite, e.g., $\Lambda(B) < \infty$ for any bounded Borel B. This condition guarantees that the number of points in any such set is almost surely finite. The properties of Poisson processes are presented in detail in [12].

The model, where the base stations are represented by the points of a homogeneous point process with intensity λ_s (the intensity measure here is λ_s times the Lebesgue measures in $\mathbb{R}^2$), will be referred to as *the basic Poisson cell model*.

A typical configuration of ground stations in this model is shown in Figure 1.

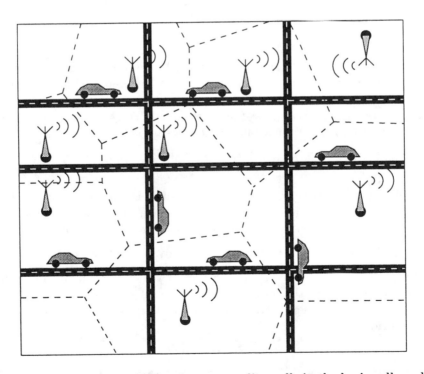

Figure 1. Ground antennas and corresponding cells in the basic cell model with Manhattan type road system.

8.2.2 Road system model

As a model for the road system, we propose the so-called *Poisson line process*, where each line is characterized by the following parameters:

- p is the distance from the line to the origin, taken positive if the line is above 0 and negative otherwise (for vertical lines, p is the abscissa of the intersection point of the line with the abscissa axis);
- α $(0 \leq \alpha < \pi)$ is the angle between the line and the abscissa axis (*inclination*);
- σ is the *type* of the road represented by the line (highway, one-way road, etc.);
- τ is the parameter defining the *traffic characteristics* of the road.

The road system is modeled by a realization of a Poisson process in the phase space $\mathbb{R} \times [0, \pi) \times \mathcal{G} \times \mathcal{T}$, where the set $E = \mathbb{R} \times [0, \pi) \times \mathcal{G}$ represents the phase space of the positions and the types of the roads and $\mathcal{T}$ is the space of traffic parameters on a road, i.e., the parameters specifying traffic characteristics on a particular road.

In the *homogeneous* case (i.e., when the distribution of the process is left invariant by shifts in the plane), the intensity measure takes the form:

$$\lambda_{rd} \, dp \, \mathcal{O}(d\alpha) Q(\alpha, d\sigma) T(\alpha, \sigma, d\tau),$$

where

- λ_{rd} is the *density of roads*,
- the probability measure $\mathcal{O}(d\alpha)$ on $[0, \pi)$ gives the *orientation distribution* of a typical road,
- $Q(\alpha, d\sigma)$ is the *distribution of the type of road* having inclination α,
- $T(\alpha, \sigma, d\tau)$ is the *distribution of the traffic parameters* on a road of type σ and inclination α.

If the orientation distribution $\mathcal{O}$ is uniform, we obtain an *isotropic* road model. If, in contrast, $\mathcal{O}$ has two atoms $\{0\}$ and $\{\pi/2\}$, we obtain the *Manhattan type* model, where the masses of the atoms represent the frequencies of the roads in each direction.

Using measure Q, one can take into account different characteristics on roads of different directions (in Manhattan the *avenues* are larger than the *streets*).

8.2.3 Traffic models

The simplest traffic model on a road is a marked Poisson point process. Let $x^{(D)}$ denote the local coordinates on a road $D = (p, \alpha, \sigma)$. Consider a Poisson process $\Theta_D(0)$ on D defined on the phase space $\mathbb{R} \times \mathbb{R}$ and driven by the intensity measure $\lambda_{tr}^{(D)} dx^{(D)} \mathcal{V}^{(D)}(dv)$. Here the parameter $\lambda_{tr}^{(D)}$ gives the *density of the traffic* and $\mathcal{V}^{(D)}(dv)$ is the *velocity distribution* of a typical vehicle (which can be positive or negative with respect to the local coordinates). All these parameters depend on the inclination and the type of the road D. A realization of this process has the following form:

$$\Theta_D(0) = \sum_j \delta_{(x_j^{(D)}, v_j)},$$

where the first term gives the position of the jth vehicle at a fixed time, say 0, and the second one gives its velocity. After a time s, the process becomes

$$\Theta_D(s) = \sum_j \delta_{(x_j^{(D)} + v_j s, v_j)}.$$

An important property of a homogeneous Poisson process is that the distribution of the process $\Theta_D(s)$ coincides with that of $\Theta_D(0)$ (this result is known as Bartlett's Theorem (cf. [12], pp. 59-60 and 49]).

To obtain a space homogeneous traffic model, one can take for space $\mathcal{T}$, the space of counting measures on $\mathbb{R}^2$ and let $T(\alpha, \sigma, \cdot)$ be the distribution of the Poisson process with intensity measure $\lambda_{tr}(\alpha,\sigma)dx \mathcal{V}(\alpha,\sigma)(dv)$ (in the previous notations $\lambda_{tr}(\alpha,\sigma) = \lambda_{tr}^{(D)}$ and $\mathcal{V}(\alpha,\sigma)(dv) = \mathcal{V}^{(D)}(dv)$). The described model is called the *isoveloxic traffic model*. The affix *iso* in this context means that the vehicles preserve their speed and do not assume *the same* speed for all vehicles.

This model is applicable when one is interested in characteristics depending on configuration of the vehicles in a fixed time instant, as for example, the number of vehicles in a fixed cell. However, the time evolution description of the system misses the phenomenon of initiation and termination of calls. This is not important in the previous situation when one may count only vehicles with a call in progress. For studying temporal characteristics like time variation of the number of communicating vehicles in a cell, the isoveloxic model needs a refinement.

Namely, we suppose that each vehicle can be in two alternative states: in a *call mode* and in a *think mode* according to whether its cellular phone is in use or not (we count only the vehicles supplied with cellular phones). We suppose then that the modes of a vehicle moving along road (p,α,σ) with speed v are driven by a Markov chain with two states. Consequently, the periods of time spent in think and call modes have exponential distributions with parameters $\kappa_0 = \kappa_0(\alpha,\sigma,v)$ and $\kappa_1 = \kappa_1(\alpha,\sigma,v)$, respectively, and the modes of different vehicles are independent. To obtain a time stationary model we choose the mode of a vehicle at time 0 to be "call" or "think" with probabilities $\theta(\alpha,\sigma,v) \overset{\text{def}}{=} \kappa_0/(\kappa_0 + \kappa_1)$ and $1 - \theta(\alpha,\sigma,v)$, respectively. The parameter θ has the meaning of an average proportion of time spent by a vehicle moving with the speed v along the road (p,α,σ) in a call mode. Usually, θ is assumed to be a constant.

8.3 ANALYTICAL RESULTS

We are interested in characterizing the distribution of variables represented as an *additive functional* of some quantities associated with roads. In the following subsection we give the definitions and the main formulas for the characteristic function of such functionals. In the upcoming subsections we show specific applications of this result.

8.3.1 Additive functionals

Let τ_D be a family of real-valued random variables with distributions T_D indexed by $D \in E = \mathbb{R} \times [0,\pi) \times \mathcal{S}$ - the phase space of the road Poisson process. Denote by $\phi_D(\eta)$ the characteristic function of the distribution T_D. Given a realization $\Delta = \Sigma_i \delta_{D_i}$ of the road process, consider a system of independent random variables $\tau_i \overset{\text{def}}{=} \tau_{D_i}$ associated with the road system. This is just another representation of the Poisson process introduced in Subsection 8.2.2.

The main quantity of interest is the additive functional $\Sigma \stackrel{\text{def}}{=} \Sigma_i \tau_i$. The independence of the system $\{\tau_i\}$ leads to the following expression for the characteristic function of the variable Σ:

$$\Phi(\eta) = \mathbf{E} \prod_{D_i \in \text{supp}\Delta} \phi_{D_i}(\eta).$$

The last formula is just an instance of the point process generating functional applied to the function $\phi_\bullet(\eta)$. Now, using the explicit expression for Poisson process generating functionals (cf. [5, p. 225]), we establish the following theorem:

Theorem 8.1.

$$\Phi(\eta) = \exp\left\{ \int_E (\phi_D(\eta) - 1)M(dD) \right\}, \tag{8.1}$$

where $M(dD) = \lambda_{rd} dp \mathcal{O}(d\alpha)Q(\alpha, d\sigma)$ is the intensity measure of the road process Δ. In particular, if the integrals below exist, then

$$\mathbf{E}\Sigma = \int_E \mathbf{E}^T \tau_D M(dD), \tag{8.2}$$

$$\mathbf{var}\,\Sigma = \int_E \mathbf{E}^T \tau_D^2 M(dD), \tag{8.3}$$

where $\mathbf{E}^T$ denotes the expectation with respect to distribution T_D.

8.3.2 The number of communicating vehicles in a cell

Let Z be a domain of $\mathbb{R}^2$ representing a base station cell and denote

$$hit(Z) \stackrel{\text{def}}{=} \{(p, \alpha) \in \mathbb{R} \times [0, p): \quad line\ (p, \alpha)\ intersects\ Z\}.$$

For $D = (p, \alpha, \sigma)$ with $(p, a) \in hit(Z)$, let τ_D be the number $\mathcal{N}(Z)$ of vehicles in communication on road (p, α) and within Z. We take $\tau_D = 0$ if $(p, \alpha) \notin hit(Z)$. As a direct application of the last theorem, we obtain the distribution of the number $\mathcal{N}(Z)$ of the vehicles in communication inside Z.

As an example, consider the refined isoveloxic traffic model introduced in Subsection 8.2.3. Fix a road (p, α, σ) and consider the process of positions of the vehicles on the road at time 0 marked with their speed v and the indicator χ of a call mode, so that $\chi = 0$ if the vehicle is in the think mode and $\chi = 1$ otherwise. Since the marks χ are chosen independently for each vehicle, the process of the communicating vehicles (being an independent thinning of a Poisson process) is again Poisson (cf. [5, p. 31]) with intensity measure $\lambda_{tr}(\alpha, \sigma)dx\theta(\alpha, \sigma, v)\mathcal{V}(\alpha, \sigma)$ (dv), where $\theta(\alpha, \sigma, v)$ is the calling rate introduced in Subsection 8.2.3.

Let $l(p, \alpha)$ be the length of the intersection of the line (p, α) with Z. The number of vehicles communicating in this part of the road has Poisson distribution with parameter $\lambda_{tr}(\alpha, \sigma)l(p, \alpha) \int \theta(\alpha, \sigma, v)\mathcal{V}(\alpha, \sigma)(dv)$. Therefore, taking

$$\phi_D(\eta) = \exp\left\{ (e^\eta - 1)\lambda_{tr}(\alpha, \sigma)l(p, \alpha) \int \theta(\alpha, \sigma, v)\mathcal{V}(\alpha, \sigma)(dv) \right\}$$

in (8.1), we obtain the characteristic function of $\mathcal{N}(Z)$.

Corollary 8.2. *In the case of an isotropic road system and constant calling probability* $\theta(\alpha,\sigma,v)\stackrel{\text{def}}{=}\theta$, *the characteristic function of the number* $\mathcal{N}(Z)$ *of communicating vehicles in zone* Z *is given by*

$$\Phi(\eta)=\exp\left\{\frac{\lambda_{rd}}{\pi}\int\limits_{\mathcal{S}}Q(d\sigma)\int\limits_{hit(Z)}(\exp\{\theta\lambda_{tr}(\sigma)l(p,\alpha)(e^{\eta}-1)\}-1)\,d\alpha dp\right\},\quad (8.4)$$

with the moments

$$\mathbf{E}\mathcal{N}(Z)=\theta\lambda_{rd}\mid Z\mid\int\limits_{\mathcal{S}}\lambda_{tr}(\sigma)Q(d\sigma),\quad (8.5)$$

and

$$\mathbf{var}\mathcal{N}(Z)=\mathbf{E}\mathcal{N}(Z)+\frac{\lambda_{rd}\theta^2}{\pi}\int\limits_{\mathcal{S}}\lambda_{tr}^2(\sigma)Q(d\sigma)\int\limits_{hit(Z)}l^2(p,\alpha)\,d\alpha dp.\quad (8.6)$$

In formulas (8.5) and (8.6) above, we have used the following elementary fact: $\int_{hit(Z)}l(p,\alpha)\,d\alpha dp$ equals π times the area $\mid Z\mid$ of domain Z.

This result allows the following interpretation. A space-stationary Poisson line process is ergodic, so the above variable $\mathcal{N}(Z)$ can be thought as of the number of communicating vehicles in a figure Z "randomly placed" on the plane with a fixed configuration of the road system. Thus, the distribution of $\mathcal{N}(Z)$ describes the spatial variations of the number of vehicles in a set of the form Z.

For dimensioning the network, it is important to know not only the moments, but also the probability of large deviations of the number of mobiles in a cell. Suppose in the sequel that the zone Z is convex.

Theorem 8.3. *Let* Z *be a convex bounded domain,* $\mid\partial Z\mid$ *and* $diam(Z)$ *being its perimeter and its diameter, respectively. Assume that* $\lambda_{tr}^*\stackrel{\text{def}}{=}\sup_{\mathcal{S}}\lambda_{tr}(\sigma)<\infty$. *Then, in the isotropic road model with a constant calling probability* θ *the following asymptotic estimate holds true:*

$$P\{\mathcal{N}(Z)\geq n\}<(\theta\lambda_{tr}^*diam(Z))^nI_n(\lambda_{rd}\mid\partial Z\mid\pi^{-1}),\quad (8.7)$$

where the function $I_n(a)$ *is given by (8.28) in Appendix.*

Proof. For the analysis of the tail behavior of the distribution of $\mathcal{N}(Z)$, rewrite its logarithmic moment generating function (see (8.4)) as

$$\log\Phi(\eta)=$$

$$\frac{\lambda_{rd}\mid\partial Z\mid}{\pi}\left[\int\limits_{hit(Z)\times\mathcal{S}}\exp\{\theta\lambda_{tr}(\sigma)l(p,\alpha)(e^{\eta}-1)\}\mid\partial Z\mid^{-1}d\alpha dp\,Q(d\sigma)-1\right].$$

$$(8.8)$$

Here we have used the fact that $\int_{hit(Z)}d\alpha dp$ equals the perimeter $\mid\partial Z\mid$ of the convex domain Z. So $\mid\partial Z\mid^{-1}d\alpha dp\,Q(d\sigma)$ is a probability measure on $hit(Z)\times\mathcal{S}$ and formula (8.8) can be interpreted as follows. Generate a number π_a having a Poisson distribution with parameter $a\stackrel{\text{def}}{=}\lambda_{rd}\mid\partial Z\mid\pi^{-1}$, then pick π_a independent roads with distribution $\mid\partial Z\mid^{-1}d\alpha dpQ(d\sigma)$, and finally, take a sum of π_a independent Poisson variables with parameter $\theta\lambda_{tr}(\sigma)l(p,\alpha)$ for a chosen road (p,α,σ). Then the distribution of this sum coincides with that of variable $\mathcal{N}(Z)$. Therefore, $\mathcal{N}(Z)$ is stochastically dominated by the sum of π_a independent Poisson distributed random variables with parameter $b\stackrel{\text{def}}{=}\theta\lambda_{tr}^*diam(Z)$.

The end of the proof is just an application of Lemma A.1.   □

Note that Lemma A.1 provides a hyper-exponential decay rate of the distribution tail of the variable $N(Z)$.

8.3.3 The number of cell border crossings

In this subsection, we study the number $\mathcal{F}_{\Delta t}(Z)$ of border crossings of a zone Z during a fixed period of time Δt. The situation here is different, because the number of vehicles entering the zone and the number of the vehicles exiting it are dependent. Indeed, the same vehicle can enter and then exit the zone if the observation time length Δt is large, the speed of a vehicle is high, or simply, the entrance and the exit points of the road are close. In practice, if a communicating vehicle is close to the cell boundary, the decision to transfer it to another antenna is taken based on many criteria, of which the geographical distance is just one factor. Hand-over cost factors imply that it is not reasonable to transfer communication to an antenna if its zone just slightly "touches" the road. Therefore, for small Δt, the crossings in distinct points of the boundary become independent (also keeping in mind the maximal speed limitations on the roads) and we can use the same machinery for finding the distribution of $\mathcal{F}_{\Delta t}(Z)$.

Theorem 8.4. *Consider the time stationary isoveloxic traffic model on the roads with a calling rate θ. Then the moment generating function of the number $\mathcal{F}_{\Delta t}(Z)$ of the border crossings of a convex zone Z during a period of time of length Δt is given by (8.1) with*

$$\phi_D(\eta) = \exp\{2(\theta c(D) + c_{01}(D))(e^\eta - 1) + c_{11}(D)(e^{2\eta} - 1)\}, \qquad (8.9)$$

and the moments

$$\mathbf{E}\mathcal{F}_{\Delta t}(Z) = \int_{\mathcal{G} \times hit(Z)} C(D)M(dD), \qquad (8.10)$$

and

$$\mathbf{var}\mathcal{F}_{\Delta t}(Z) = \int_{\mathcal{G} \times hit(Z)} [C(D) + C^2(D) + 2c_{11}(D)]M(dD), \qquad (8.11)$$

where

$$C(D) = 2[\theta c(D) + c_{01}(D) + c_{11}(D)],$$

$$M(dD) = \lambda_{rd}dp\mathcal{O}(d\alpha)Q(\alpha, d\sigma),$$

and the functions $c(D)$, $c_{01}(D)$, $c_{11}(D)$ are given by (8.12-8.16) with $D = (p, \alpha, \sigma)$.

Proof. Consider a road D that crosses a convex zone Z in two points $x_1 < x_2$ in the local coordinate system on D. Recall that the positions and velocities of vehicles at the beginning of an observation are represented as a realization of a Poisson process Θ_D in the state space $\mathbb{R}^2$ with intensity measure $\lambda_{tr}(D)dx\mathcal{V}^{(D)}(dv)$. Figure 2 shows position x and speed v of cars that cross at least one of the points x_1, x_2 during the period of time Δt. The set A_1 (respectively, A_2) there corresponds to the positions and speeds of the cars crossing point x_1 (x_2) and do not cross point $x_2(x_1)$ during the observation time Δt. The set A_0 corresponds to the positions and speeds of the cars that cross both points.

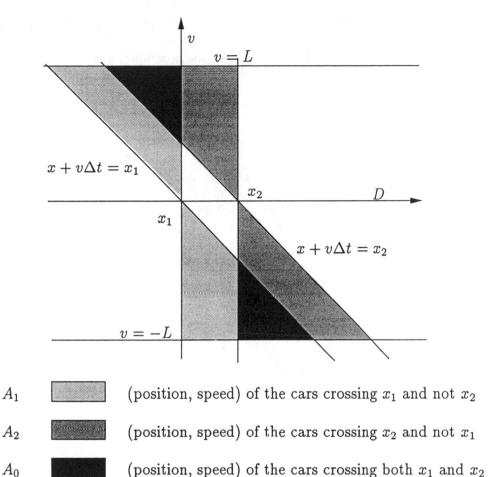

A_1 (position, speed) of the cars crossing x_1 and not x_2

A_2 (position, speed) of the cars crossing x_2 and not x_1

A_0 (position, speed) of the cars crossing both x_1 and x_2

Figure 2. Space-velocity diagram of the zone border crossing vehicles.

It is easy to see that the intensity measure of each set A_i for $i = 1, 2$ equals

$$c(D) = \lambda_{tr}(D)l(D)\mathscr{V}^{(D)}\{v\colon |v|\,\Delta t \geq l(D)\}$$
$$+ \lambda_{tr}(D)\Delta t \mathbf{E}[\,|V|\,\mathbf{1}\{|V|\,\Delta t \leq l(D)\}], \tag{8.12}$$

where $l(D) = |x_1 - x_2|$ and V is a random variable with the speed distribution $\mathscr{V}^{(D)}$. At the moment of a zone border crossing, each vehicle is in a call mode with probability θ independently of the others. Therefore, the number of vehicles in a call mode passing exactly one of the points x_1 or x_2 has a Poisson distribution with parameter $2\theta c(D)$.

Consider now a vehicle passing both points x_1 and x_2 at a speed v. The probability that this car enters the zone in a call mode and then exits it also in a call mode (after time $l(D)/v$) is given by the solution of Kolmogorov-Chapman equation for the Markov chain governing the switching mode:

$$p_{11}(D, v) = \theta^2 + \theta(1 - \theta)\exp\{-l(D)\,|v|^{-1}(\kappa_0 + \kappa_1)\}. \tag{8.13}$$

Therefore, the total number of such cars has a Poisson distribution with parameter

$$c_{11}(D) = \lambda_{tr}(D) \int p_{11}(D,v)[\,|\,v\,|\,\Delta t - l(D)]^+ \mathcal{V}^{(D)}(dv). \qquad (8.14)$$

Similarly, the number of cars entering the zone in a think mode and exiting in a call mode has a Poisson distribution with parameter

$$c_{01}(D) = \lambda_{tr}(D) \int p_{01}(D,v)[\,|\,v\,|\,\Delta t - l(D)]^+ \mathcal{V}^{(D)}(dv), \qquad (8.15)$$

where

$$p_{01}(D,v) = \theta(1-\theta)(1-\exp\{-l(D)\,|\,v\,|^{-1}(\kappa_0 + \kappa_1)\}). \qquad (8.16)$$

Of course, one can derive similar expressions for the numbers of cars entering the zone in a call mode and exiting it in a think mode.

Combining all of the above results, we see that the number of border crossings in a call mode along road D during time Δt has the distribution of the following sum of independent Poisson random variables:

$$\tau(D) = \pi_{2\theta c(D)} + \pi_{2c_{01}(D)} + 2\pi_{c_{11}}(D),$$

with moment generating function (8.9). Now the theorem follows immediately by applying Theorem 8.1.   □

Now we consider an extreme case when the observation time Δt becomes small.

Assume that the velocity distributions $\mathcal{V}^{(D)}$ have a common compact support so that the speed on any road never exceeds $L > 0$. Assume also that the traffic intensities are uniformly bounded by a constant λ_{tr}^*. Consider a road D such that the length $l(D)$ of its intersection with zone Z is at least $L\Delta t$. In this case,

$$c(D) = \lambda_{tr}(D)\Delta t \mathbf{E}\,|\,V\,| \overset{\text{def}}{=} c^I(D), \ c_{01}(D) = c_{11}(D) = 0$$

and the moment generating function of the number of border crossings in a call mode becomes

$$\phi_D^I(\eta) = \exp\{2\Delta t \theta \lambda_{tr}(D)\mathbf{E}\,|\,V\,|\,(e^\eta - 1)\}. \qquad (8.17)$$

Here the distance between the intersection points is larger than the distance any car can drive during the time Δt, so the number of active call passages of these points are independent (this explains the usage of index I in the previous formula). Introducing the notations

$$H_\Delta = \{(p,\alpha): l(p,\alpha) < \Delta t L\}$$

and $\Phi^I(\eta)$, the same as $\Phi(\eta)$ in (8.1) with $\phi_D(\eta)$ replaced by $\phi_D^I(\eta)$, we can write

$$\log\Phi(\eta) = \log\Phi^I(\eta) + \int_{\mathcal{G}\times H_\Delta} (\phi_D(\eta) - \phi_D^I(\eta))M(dD). \qquad (8.18)$$

$\Phi_{\Delta t}^I$ is the moment generating function of some random variable which we denote by $\mathcal{F}_{\Delta t}^I(Z)$.

It is easy to see that on the set $\mathcal{G}\times H_\Delta$,

$$c_{01}(D) \le \lambda_{tr}^*(\kappa_0 + \kappa_1)L(\Delta t)^2, \qquad (8.19)$$

$$c_{11}(D) \le \lambda_{tr}^* L\Delta t, \qquad (8.20)$$

$$\frac{c(D) - c^I(D)}{\lambda_{tr}(D)\Delta t} = -\mathbf{E}\,[\,|\,V\,|;\,|\,V\,|\,\Delta t \ge l(D)] + \frac{l(D)}{\Delta t}\mathcal{V}^{(D)}\{v:\,|\,v\,|\,\Delta t \ge l(D)\}, \qquad (8.21)$$

$$-\mathbf{E}\,|\,V\,| \le -\mathbf{E}[\,|\,V\,|\,;\,|\,V\,|\,\Delta t > l(D)] + \frac{l(D)}{\Delta t}\mathscr{V}^{(D)}\{v\colon |\,v\,|\,\Delta t > l(D)\} \le L.$$

$$(8.22)$$

Therefore, the second summand in the RHS of (8.18) is

$$\int_{\mathscr{I}\times H_\Delta} (\phi_D(\eta) - \phi_D^I(\eta))M(dD) = O(\Delta t)M(\mathscr{I}\times H_\Delta),$$

where $O(1)$ is uniformly bounded in η if the range of η is compact.

To estimate the last term in the product above, we use Lemma A.2 of the Appendix to obtain the following:

Corollary 7.5. *As the observation time Δt vanishes, the number of the zone border crossings in the call mode can be represented as the sum of two independent random variables*

$$\mathscr{F}_{\Delta t}(Z) = \mathscr{F}_{\Delta t}^I(Z) + \zeta_{\Delta t},$$

where $\mathscr{F}_{\Delta t}^I(Z)$ has logarithmic moment generating function

$$\log \Phi^I(\eta) = \int_{\mathscr{I}\times hit(Z)} [exp\{2\Delta t\theta\lambda_{tr}(D)\mathbf{E}^{(D)}|\,V\,|(e^\eta - 1)\} - 1]M(dD)$$

and $\zeta_{\Delta t} = O((\Delta t)^2)$ in the sense that there exists a proper limit $(\Delta t)^{-2}\zeta_{\Delta t}\overset{w}{\Rightarrow}\zeta$. In particular,

$$\mathbf{E}\mathscr{F}_{\Delta t}(Z) = 2\Delta t \int_{\mathscr{I}\times hit(Z)} \theta\lambda_{tr}(D)\mathbf{E}^{(D)}|\,V\,|\,M(dD) + O((\Delta t)^2) = \mathbf{var}\,\mathscr{F}_{\Delta t}(Z).$$

$$(8.23)$$

8.3.4 Combining the road and the basic cell models

The distributions obtained in Subsections 8.3.3 and 8.3.2 should be understood in the ergodic sense: these are distributions of characteristics of a domain Z "randomly placed" onto a road system. Returning to the basic cell model, one can look for the distribution associated with a "randomly picked" zone represented by the Voronoi tessellation.

The notion of a "random picked" point of a point process has a rigorous definition: it is the point in the origin under the *Palm distribution* of the process (cf. [5, Ch. 12] or [18, Ch. 4.4]). A Palm version of a Poisson process is obtained simply by placing an additional process point into the origin (cf. [18, p. 114]). So, denote by $\mathbf{C}_0$ the Voronoi cell with its nucleus at the origin. We study the cell $\mathbf{C}_0$'s characteristics under the product of the Palm distribution of the process Π_s of ground stations, with intensity λ_s, and the road-traffic process distribution. For simplicity, we consider in this subsection the case of an isotropic road system with a constant calling rate θ and we concentrate on the first two moments of distributions. Other cases could be treated in a similar way.

Theorem 7.6. *In an isotropic road system, the first two moments of the distribution of the number $\mathcal{N}(\mathbf{C}_0)$ of vehicles in a cell $\mathbf{C}_0$ and the number $\mathscr{F}_{\Delta t}(\mathbf{C}_0)$ of the cell $\mathbf{C}_0$ border crossings in a call mode are given by*

$$\mathbf{E}\mathcal{N}(\mathbf{C}_0) = \frac{\theta\lambda_{rd}\overline{\lambda_{tr}}}{\lambda_s},$$

$$(8.24)$$

$$\mathbf{E}\mathscr{F}_{\Delta t}(\mathbf{C}_0) = \Delta t\frac{8\theta\lambda_{rd}\overline{\lambda_{tr}\mathbf{E}|\,V\,|}}{\pi\sqrt{\lambda_s}} + O((\Delta t)^2),$$

$$(8.25)$$

$$\text{var}\mathcal{N}(\mathcal{C}_0) = \mathbf{E}\,\mathcal{N}(\mathcal{C}_0) + 0.672\frac{\theta^2\lambda_{rd}\overline{\lambda_{tr}^2}}{\lambda_s^{3/2}} + 0.280\frac{\theta\lambda_{rd}\overline{\lambda_{tr}}}{\lambda_s^2}, \qquad (8.26)$$

$$\text{var}\mathcal{F}_{\Delta t}(\mathcal{C}_0) = \mathbf{E}\mathcal{F}_{\Delta t}(\mathcal{C}_0) + 1.890\Delta t\frac{\theta\lambda_{rd}\overline{\lambda_{tr}\mathbf{E}\,|\,V\,|}}{\pi\lambda_s} + O((\Delta t)^2), \qquad (8.27)$$

where

$$\overline{\lambda_{tr}} = \int_{\mathcal{S}} \lambda_{tr}(\sigma)Q(d\sigma),$$

$$\overline{\lambda_{tr}^2} = \int_{\mathcal{S}} \lambda_{tr}^2(\sigma)Q(d\sigma),$$

$$\overline{\lambda_{tr}\mathbf{E}\,|\,V\,|} = \int_{\mathcal{S}} \lambda_{tr}(\sigma)\mathbf{E}^{(\sigma)}\,|\,V\,|\,Q(d\sigma).$$

Proof. Let $\sigma(\Pi_s)$ be the σ-algebra generated by the process Π_s. Then by (8.5),

$$\mathbf{E}\mathcal{N}(\mathcal{C}_0) = \mathbf{E}\,\mathbf{E}[\mathcal{N}(\mathcal{C}_0)\,|\,\sigma(\Pi_s)] = \theta\lambda_{rd}\mathbf{E}\,|\,\mathcal{C}_0\,|\int_{\mathcal{S}} \lambda_{tr}(\sigma)Q(d\sigma)$$

and (8.24) follows from $\mathbf{E}\,|\,\mathcal{C}_0\,| = \lambda_s^{-1}$ [15].

Similarly, (8.25) follows from (8.23), that $\int_{hit(\mathcal{C}_0)} dadp$ equals the perimeter $|\,\partial\mathcal{C}_0\,|$, and that $\mathbf{E}\,|\,\partial\mathcal{C}_0\,| = 4\lambda_s^{-1/2}$ [9].

To obtain the second moment, the following identity is useful:

$$\text{var}\xi = \text{var}\,\mathbf{E}[\xi\,|\,\sigma] + \mathbf{E}\,\text{var}[\xi\,|\,\sigma],$$

for any random variable ξ and σ-algebra σ. Taking into account that $\text{var}\,|\,\mathcal{C}_0\,| = 0.280/\lambda_s^2$ [9], (8.5) and (8.6) give

$$\text{var}\,\mathcal{N}(\mathcal{C}_0) = \mathbf{E}\mathcal{N}(\mathcal{C}_0) + 0.280\frac{\theta\lambda_{rd}\overline{\lambda_{tr}}}{\lambda_s^2} + \frac{\lambda_{rd}\overline{\theta^2\lambda_{tr}^2}}{\pi}\,\mathbf{E}\int_{hit(\mathcal{C}_0)} l^2(p,\alpha)dadp,$$

where $l(p,\alpha)$ is the length of the intersection of line (p,α) with $\mathcal{C}_0$. The last expectation can be calculated numerically (see [2, Ch. 6.6]), finally giving (8.27).

Similarly, expression (8.27) is due to (8.23) and $\text{var}\,|\,\partial\mathcal{C}_0\,| = 0.945\lambda_s^{-1}$ (cf. [16, p. 287]).   □

8.4 DISCUSSIONS AND OPEN PROBLEMS

In the end, we discuss some traffic and communication system models, which were considered previously in the literature.

The modeling of traffic flows has long history. Roughly, there are two approaches - macroscopic, where the vehicle flow is treated as a continuous fluid, and microscopic, where the behavior of individual vehicles is taken into account. Keeping in mind such peculiarities of mobile communications as switching between calling and thinking modes and the relatively low (nowadays) proportion of cars possessing radio telephones, it seems that the microscopic approach is more relevant to the nature of such communications.

In traffic science, the flow of individual vehicles on a one-lane road, along with the counting distribution, is usually characterized by its counterpart - the *headway* distribution, i.e., the distribution of time a fixed observer measures between these

two consecutive vehicles. Surely there is close relation between these two characterizations. In the case when the headways are independent, the number of vehicles passing during the observation period is just a renewal process governed by the
headway distribution (cf. [7, Ch. XI]).

In the isoveloxic model we considered, the headway distribution is exponential.
It seems the first work where a Poisson distribution for the counting process was
proposed was the paper of W.F. Adams [1]. This model shows a good fit with real
data in light traffic and in many-lanes roads (cf. a discussion in [8, p. 22-23]). Giv-
en the current density of communicating mobiles, this assumption seems justified.
A nonstationary traffic model based on the Poisson process was considered in [14].

Many other types of distributions were proposed as alternatives to the exponential headway. There are two distinguishing phenomena in a vehicle's flow that
are completely ignored by the isoveloxic model: finite length of structuring elements (vehicles) and the tendency to bunching. That two vehicles cannot be closer
than a certain distance and natural speed limitations imply the boundedness of the
flow as a function of the concentration of vehicles. This dependency is known as
the *Fundamental Diagram of Road Traffic*. F.A. Haight in [11, Ch. 3] proposes a
variety of probabilistic models that capture these phenomena. For example, an
Independent (D, M) *Traffic Queueing Model* is constructed as follows. Perform a
Bernoulli trial with the probability of success p. If the Bernoulli experiment is a
success, then choose the headway from an exponential distribution (or from the
exponential distribution shifted by the length Δ of a car; if the experiment is a
failure, then the headway is a deterministic constant exceeding Δ. Among other
possibilities, we can mention log-normal, gamma, Erlang, and Pearson distributions for headway treated in [4], and the *discrete Markov model* introduced in [6],
assuming dependency between successive headways (see also [8, pp. 24-29]).

Some remarks ought to be made on the speed distribution. One can distinguish the following cases: when a mobile follows another mobile or when it is free
from interaction with other cars. In the first case, it manifests the *free speed distribution*. In the latter, the following distributions were treated: (truncated) normal, gamma, beta, generalized Pearson type III distribution, and others (cf. [8, pp.
31-41]). As we saw in the isoveloxic model, only the mean free speed is significant
for the studied variables.

In view of the above, further developments of the presented models depend, of
course, on the nature of the studied phenomena and could be proposed in the following directions.

1. Incorporation of more elaborated traffic models. This, in particular, demands a thinner control over dependency of counting variables in the entry
   and the exit points of a road. Mixing properties of the traffic process
   should be involved here.

2. The distributions considered here described *spatial* variations of the studied
   quantities. It is interesting to characterize also their *temporal* variations.
   For example, Corollary 7.5 can be used as a basis for diffusion
   approximation for the number of communicating vehicles in a cell. Note
   that the number of cars entering and the number of cars exiting the zone
   become asymptotically independent.

3. As we already mentioned in the introduction, the association of a ground
   station is based merely on a signal power rather than on a geographical distance. In addition to the distance to the receiver, the reception power is
   strongly affected by the environment, which leads to non-homogeneous prop-

agation properties. Taking into account this random character could lead to cellular system models different from the one based on Voronoi tessellations. Some interesting signal models can be found in [13] and [10].

4. After a call is received by a ground station, it is then routed via a stationary communications network. A variety of interesting problems could be posed concerning user identification, destination allocation, routing, and hand-over scheme. A hierarchical network model introduced in [2] could serve as a starting point for modeling a multi-level stationary network.

5. A line process that we used for representation of roads possesses long range dependency. Being acceptable for local characteristics related to a particular cell, it might become non-realistic for characterization of global variables. Very often the road system of big cities resembles a web with a mixture of radial-circular and rectangular structures. In contrast, inter-urban connections are closer to random tessellations. Therefore, it seems reasonable to model urban and inter-urban road systems separately. The basic station cells are also different, being about 2 miles in size in cities and about 10-12 miles in the country.

It is impossible to list all varieties of problems arising in a such wide area as wireless communications, the area which demonstrates nowadays a real boom in its development. In this chapter, we showed how many problems could be posed and partially solved using point processes and the tools of stochastic geometry. The approach we propose gives rise to very rich and fruitful models, which we hope will interest both engineers and applied mathematicians.

APPENDIX

Lemma A.1. *Let $\{\pi_b(i)\}$ be a sequence of independent Poisson distributed random variables with parameter $b > 0$; suppose π_a is a random variable independent of $\{\pi_b(i)\}$ and Poisson distributed with parameter $a > 0$. Then,*

$$\exp\{a(e^{-b}-1)\}b^n I_n(ae^{-b}) < \mathbf{P}\left\{\sum_{i=1}^{\pi_a}\pi_b(i) \geq n\right\} < b^n I_n(a),$$

where

$$I_n(a) = \frac{1}{n!}\,\mathbf{E}\pi_a^n \sim \frac{e^{-a+n/z_0(na^{-1})}}{z_0(na^{-1})^n\sqrt{2\pi n(1+z_0(na^{-1}))}} \tag{8.28}$$

as n grows and $z_0(x)$, $x > 0$, is the unique solution of the equation $ze^z = x$.

Proof. A Poisson variable π_λ distribution tail can be written as

$$\mathbf{P}\{\pi_\lambda \geq n\} = \sum_{k=n}^{\infty}\frac{\lambda^k}{k!}e^{-\lambda} = \int_0^\lambda \frac{x^{n-1}}{(n-1)!}e^{-x}dx = \frac{\lambda^n}{n}\tilde{\gamma}(n,\lambda),$$

where

$$e^{-\lambda} < \tilde{\gamma}(n,\lambda)\overset{\text{def}}{=}n\int_0^1 y^{n-1}e^{-\lambda y}dy = \sum_{k=0}^{\infty}\frac{(-\lambda)^k n}{k!(n+k)} < 1.$$

Therefore,

$$\mathbf{P}\left\{\sum_{i=1}^{\pi_a}\pi_b(i) \geq n\right\} = \sum_{k=1}^{\infty}\mathbf{P}\{\pi_{kb} \geq n\}\frac{a^k}{k!}e^{-a} < \frac{b^n}{n!}\mathbf{E}\pi_a^n.$$

Similarly,

$$\mathbf{P}\left\{\sum_{i=1}^{\pi_a}\pi_b(i)\ge n\right\} > \exp\{a(e^{-b}-1)\}\frac{b^n}{n!}\mathbf{E}\pi^n_{ae}-b.$$

By Cauchy's formula, we can write

$$\frac{1}{n!}\mathbf{E}\pi^n_a = \frac{1}{n!}\frac{d^n}{dz^n}\mathbf{E}e^{z\pi}a = \frac{e^{-a}}{2\pi i}\oint\frac{\exp\{ae^z\}}{z^{n+1}}dz = \frac{e^{-a}}{2\pi i}\oint e^{n\psi(z)}\frac{dz}{z},$$

where $\psi(z)=an^{-1}e^z-\ln z$ and the integration is over any contour that surrounds the origin. Let z_0 be a saddle point, i.e., a solution of the equations $\psi'(z)=0$. Then the local transformation $w^2=\psi(z)-\psi(z_0)$ gives the integral

$$I_n(a)=\frac{e^{-a+n\psi(z_0)}}{2\pi i}\oint e^{nw^2}f(w)dw,$$

where

$$f(w)=\frac{1}{2}\frac{dz}{dw} \tag{8.29}$$

$$=\frac{2w}{zan^{-1}e^z-1}=\sum_{k=0}^{\infty}c_kw^k$$

and the new contour can be chosen to run as long as we wish along the imaginary axis. Therefore,

$$I_n(a)\sim\frac{e^{0a+n\psi(z_0)}}{2\pi}\sum_{k=0}^{\infty}(-1)^kc_{2k}\int_{-\infty}^{+\infty}e^{-nt^2}t^{2k}dt$$

$$=\frac{e^{-a+n\psi(z_0)}}{2\pi}\sum_{k=0}^{\infty}(-1)^kc_{2k}\frac{\Gamma(k+1/2)}{n^{k+1/2}}.$$

The first approximation reads

$$I_n(a)\sim\frac{e^{-a+n\psi(z_0)}}{2\pi}\sqrt{\frac{\pi}{n}}f(0).$$

By l'Hôpital's rule,

$$f(0)=\lim_{w\to0}\frac{2w}{zan^{-1}e^z-1}=\frac{2}{an^{-1}e^{z_0}(z_0+1)}\frac{dw}{dz}\bigg|_{z_0},$$

since $z_0an^{-1}e^{z_0}=1$. Comparing the previous expression with (8.29) we obtain

$$f(0)=\sqrt{\frac{2}{z_0+1}},$$

and finally,

$$I_n(a)=\frac{e^{-a+n\psi(z_0)}}{\sqrt{2\pi n(z_0+1)}}=\frac{e^{-a+n/z_0}}{z_0^n\sqrt{2\pi n(z_0+1)}},$$

which ends the proof. $\qquad\square$

Lemma A.2. *Let Z be a convex domain in a plane and its boundary be given by radius vectors $\mathbf{r}(\omega)$, $\omega\in[0,2\pi)$. Let γ_i be the external angle of its boundary, i.e., the angle formed by velocity vectors $\mathbf{r}'(\omega_i-0)$ and $\mathbf{r}'(\omega_i+0)$ at the points $\mathbf{r}(\omega_i)$, where these vectors are different (see Figure 3). Let $l(p,\alpha)$ be the length of the intersection of the line (p,α) with Z. Then, as Δ vanishes,*

$$\int \mathbf{1}\{(p,\alpha):0<l(p,a)<\Delta\}dp\mathbb{O}(d\alpha)$$

$$=\Delta\sum_i\sin^{-1}\gamma_i\int_0^{\gamma_i}\sin\alpha\sin(\alpha-\gamma_i)\mathbb{O}(d\alpha)+O(\Delta^2).$$

Proof. The proof is based on a simple observation that the integral over p has order Δ only when the line (p, α) cuts a singular (angular) point and is proportional to the curvature times Δ^2 in the smooth boundary points. The statement follows, since the number of singular boundary points of a convex body is at most countable (cf. [17, p. 73]). $\square$

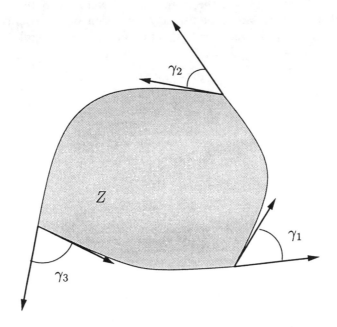

Figure 3. External angles of a planar convex domain.

ACKNOWLEDGEMENTS

The authors are grateful to Nico Temme (CWI, Amsterdam) for his help in obtaining the asymptotic expansion in the proof of Lemma A.1.

BIBLIOGRAPHY

[1]   Adams, W.F., Road traffic considered as a random series, *J. Inst. Civil Engrs.* **4** (1936), 121-130.

[2]   Baccelli, F., Klein, M., Lebourges, M. and Zuyev, S., Géomètrie aléatoire et architecture de réseaux, *Annales des Télécommunications* (1996), to appear.

[3]   Baccelli, F., Klein, M., Lebourges, M. and Zuyev, S., Stochastic geometry and architecture of communication networks, *J. Telecom. Systems* (1996), to appear. Selected Proceedings of the Third INFOCOM Telecommunications Conference.

[4]   Buckley, D.J., *Inter-Vehicle Spacing and Counting Distributions*, Ph.D. thesis, Univ. of South Wales, Australia 1965.

[5]   Daley, D.J. and Vere-Jones, D., *An Introduction to the Theory of Point Processes*, Springer-Verlag, New York 1988.

[6]   Dunne, M.C., Rothery, R.W. and Potts, R.B., A discrete Markov model of vehicular traffic, *Transport. Sci.* **2** (1968), 233-251.

[7]   Feller, W., *An Introduction to Probability Theory and its Applications*, Vol. **II**, 2nd edition, J. Wiley, New York 1971.

[8]   Gazis, D.C., Editor, *Traffic Science*, J. Wiley, New York 1974.

[9]   Gilbert, E.N., Random subdivisions of space into crystals, *Ann. Math. Statist.* **33** (1962), 958-972.

[10]   Günther, C.G., Editor, *Mobile Communications: Advanced Systems and Components*, Lecture Notes in Comp. Sci. **783**, Springer-Verlag, New York 1994.

[11]   Haight, F.A., *Mathematical Theories of Traffic Flow*, Mathematics in Science and Engineering **7**, Academic Press 1963.

[12]   Kingman, J.F.C., *Poisson Processes*, Oxford Studies in Probability, Oxford Univ. Press, Oxford, UK 1993.

[13]   Lee, W.C.Y., *Mobile Communications Design Fundamentals*, 2nd edition, Wiley Series in Telecommunications, J. Wiley, New York 1993.

[14]   Leung, K.K., Massey, W.A. and Whitt, W., Traffic models for wireless communication networks, *IEEE J. on Selected Areas in Commun.* **12**:8 (1994), 1353-1364.

[15]   Meijering, J.L., Interface area, edge length and number of vertices in crystal aggregates with random nucleation, *Philips Res. Rep.* **8** (1953), 270-290.

[16]   Okabe, A., Boots, B. and Sugihara, K., *Spatial Tesselations: Concepts and Applications of Voronoi Diagrams*, Wiley Series in Probability and Mathematical Statistics, J. Wiley, New York 1992.

[17]   Schneider, R., *Convex Bodies: The Brunn-Minkowski Theory*, Encyclopedia of Mathematics and its Applications **44**, Cambridge University Press, Cambridge 1993.

[18]   Stoyan, D., Kendall, W.S. and Mecke, J., *Stochastic Geometry and its Applications*, Wiley Series in Probability and Mathematical Statistics, J. Wiley, Chichester 1987.

[19]   Zuyev, S., Poisson power tesselations (in preparation).

Chapter 9
Fractal queueing models

Ashok Erramilli, Onuttom Narayan, and Walter Willinger

ABSTRACT *Fractal queueing* refers to contention models in which traffic processes (e.g., arrivals, service times, buffer occupancy) exhibit fluctuations or variations over a wide range of time scales. In contrast, conventional queueing theory is based on implicit assumptions (e.g., Markov property, exponential distributions) that restrict the fluctuations of the underlying traffic processes to a limited range of time scales. We motivate the need for fractal queueing models with traffic measurement studies from a variety of modern communications networks. These studies strongly suggest that actual traffic processes are consistent with "fractal" features (i.e., involving many time scales), such as long-range dependence and the infinite variance phenomenon. We then consider the problem of performing queueing analysis on the basis of fractal traffic descriptions, and illustrate how models that parsimoniously capture the empirically observed fractal traffic characteristics give rise to a much richer set of queueing responses than is known for conventional queueing theory. There is considerable scope for future research in the description, analysis, and control of fractal queueing models, and we conclude with a review of open problems.

CONTENTS

9.1 INTRODUCTION

Queueing theory has its origins in engineering problems that arise in telecommunications. In developing the famous formulas that bear his name, A.K. Erlang not only solved several problems of practical importance in telephony, but also laid the foundations of queueing theory in terms of the nature of the assumptions and analysis techniques that are routinely used to this day, even in the performance analysis of modern communications and computer systems. A case in point is the use of exponential distributions and variations thereof to describe traffic processes.

The motivation for retaining exponential distributions in the modeling of modern communications and computer systems has been to exploit their memory-less property when analyzing queueing performance; validating the exponential assumptions against actual traffic measurements has been a lesser concern. Nevertheless, queueing models have proved to be surprisingly accurate in the performance analysis of modern communications and computer systems. One reason for this success is that some of the commonly made assumptions are indeed reasonable in many practical situations; for example, measurement studies over the years have validated the use of (possibly non-homogeneous) Poisson processes to describe (first offered) call arrivals in circuit-switched voice networks. A second reason is that queueing models often predict performance accurately even when the underlying assumptions are not valid. This is the well-known robustness property of many queueing results; for example, the classical Erlang blocking formulas are known to be insensitive to the service time distribution beyond its mean. Arguably, the application of queueing models in circuit switched voice networks is one of the most successful applications of mathematical modeling in industry.

In the past twenty years, packet switched networks such as X.25, Frame Relay and ATM (Asynchronous Transfer Mode) have grown in importance. Whereas circuit switching was developed to support telephony, packet switching provides the flexibility to support communications services with a wide range of bit rate requirements, and allows the efficient support of bursty, intermittent communications (characteristic of computer networks). The advent of packet switching has motivated the development of appropriate queueing models that can be used in the analysis, design, planning, engineering, and congestion management of these networks. It has been widely recognized that traffic in packet networks is much burstier than Poisson, and this has motivated the study of more general arrival processes. As with the case of classical queueing models, the choice of these arrival processes has been driven by considerations of model tractability, which is implicitly based on the assumption that the traffic has variations or fluctuations over a (pre)specified time scale (as, for example, with the Poisson process) or over a very limited range of time scales (e.g., low order phase-type or Markovian processes). This results in several model characteristics, including (i) distributions used to describe holding times, burst lengths, sojourn times in ON/OFF ("active" and "inactive") states, etc., decay at an exponential rate asymptotically, and (ii) correlations in traffic processes (e.g., time series of arrival counts) decay at an exponential rate. The relative paucity of measurements from working packet networks (e.g., see [45] whose survey of the teletraffic literature between 1960 and 1980 revealed on the order of 5000 queueing related scientific articles and only about 50 measurements-related publications) has meant that for the most part the validity of these commonly made assumptions on traffic processes could not be verified against real data.

The situation has changed dramatically over the last five years, with the availability of high-resolution, high-quality, and high-volume traffic data from a wide range of working packet networks and services. As a result, two fundamental questions have been raised: (i) Do commonly made assumptions on traffic processes hold for real traffic? and (ii) If the assumptions are not valid, could the queueing predictions made on the basis of these assumptions nevertheless be accurate? In this paper, we will address these basic questions and illustrate how they motivate the development of a new area of research in queueing theory, *fractal queueing*.

In answer to the first question, statistical analyses of high-resolution traffic measurements have shown that in contrast to commonly made assumptions, (i) distributions of actual traffic processes typically decay more slowly ("heavy-tailed") than exponential ("light-tailed"), and (ii) correlations are commonly found to exhibit a hyperbolic ("long-range dependence") rather than an exponential ("short-range dependence") decay. Qualitatively, actual traffic processes can be said to exhibit variations on many time scales, in contrast to the limited time scale variations assumed in traditional queueing analysis and imposed by conventional traffic models. This empirical finding is in full agreement with observations made in many other branches of science and engineering (e.g., hydrology, economics, biophysics), namely that bursty phenomena typically exhibit variations over many length or time scales.

From experiences in other disciplines, bursty phenomena that involve a wide range of length or time scales pose significant challenges to modelers. A popular example is the problem of estimating the length of a coastline (see [38]); because coastlines exhibit features over a very wide range of length scales, any length measurement depends sensitively on the length of the yardstick. To this extent, the length of a coastline is arbitrary, and the coastline is better represented by the notion of a *fractal dimension*, which is a parametric representation of this dependence. A traffic analog of this example is the problem of estimating the "peakedness" (strictly speaking, the asymptotic value of the *Index of Dispersion of Counts*) of actual packet traffic; because measured traffic exhibits variability over many time scales of engineering interest, the peakedness of a given trace does not rapidly converge to a constant value (as is the case with practically all traditional traffic models), but keeps increasing with the length of the observation interval. To this extent, traditional teletraffic notions such as peakedness are arbitrary, and inapplicable to describe packet traffic. The challenge for queueing theory is then to develop models that can be used to accurately describe, analyze, and control traffic flows in emerging high-speed networks, and that exhibit certain robustness properties under constantly changing networking and services conditions.

The object of this paper is to motivate the application of fractal or self-similar models to address this challenge, review the current state-of-the-art, and point out directions for further research. Note that this is a very new and rapidly evolving area of research, and many of the papers are still in the preprint stage. As such, we have only been able to cite papers that have been forwarded to us by the authors. We regret any omissions. In Section 2, we review measurement results from a range of networks and discuss the fractal nature of measured network traffic. We then empirically show in Section 3 that the fractal traffic characteristics can have significant impacts on queueing performance and motivate the development of models that can parsimoniously describe these features. In Section 4 we review several modeling approaches that have currently been attempted, and summarize the current understanding concerning the queueing analysis of these models, along with the most relevant and significant results. The nature of the queueing behavior predicted reflects the nature of the underlying assumptions. For example, the asymptotical queue length distribution decays exponentially under very general but traditional assumptions of limited time scales of behavior for traffic processes. Even the limited analysis results on fractal queueing models have shown that a much richer set of possibilities is observed with actual traffic processes. We conclude with suggestions for future research.

9.2 THE FRACTAL NATURE OF NETWORK TRAFFIC

In the recent past, the area of traffic modeling for packet-switched networks, such as Frame Relay or ATM, has undergone significant changes, mainly due to the availability of an abundance of high-resolution, high-quality and high-volume traffic measurements from a wide range of working packet networks that carry real traffic generated by real applications. In particular, extensive statistical studies of these actual measurements have provided ample evidence that - in sharp contrast to commonly made traffic modeling assumptions - real network traffic is *fractal* in nature, i.e., exhibits fluctuations and variability over a wide range of time scales. In this section, we summarize the main findings from these recent traffic measurements studies, provide a concise mathematical framework for dealing with fractal-like traffic characteristics, and illustrate the importance of fractal traffic processes for parsimoniously modeling the complex dynamics of actual network traffic and for gaining new insights into the nature of measured packet streams.

9.2.1 Traffic modeling and traffic measurements

It is commonly accepted that traffic flows in modern broadband networks are far more complex than in traditional telephone networks, mainly due to a combination of the wide range of applications and services envisaged for these networks and of the inherent burstiness of the traffic generated by many of these applications. Moreover, in sharp contrast to the voice world, data networks are highly dynamic entities and undergo constant changes (e.g., network topology, user population, services and applications, network technologies, protocols). For example, a given WAN or Ethernet LAN, monitored just a few months apart, can experience a significant change in the traffic due to the emergence and popularity of new "killer-applications" (e.g., WWW, Mbone). The traffic volume generated by these killer-applications is a perfect example that shows exponential or even faster-than-exponential growth rates. These observations make life for the traffic modeler difficult and challenging: How can we describe such a dynamically changing world and provide adequate traffic descriptors (either at the source, application, or aggregate level)? What characteristics can be expected to be *robust* under the dynamics of "live" and evolving networks that continuously experience new services and applications? Which traffic descriptors are *simple, accurate, and useful in practice* to support the design, engineering, and operation of these networks? Which features of the traffic can be explained on *physical* grounds, hence contributing to an improved understanding of the nature of traffic in modern packet networks, and can have a significant or dominant impact on network performance?

To tackle these challenges, traffic researchers can increasingly rely on the availability of traffic measurements from working packet networks that carry real traffic generated by real applications. In the recent past, increasing volumes of empirical data have been collected and made available to researchers. This renewed interest in traffic measurement studies has been spurred by the rapid and large-scale deployment of ISDN and high-speed data networks (such as Switched Multimegabit Data Service (SMDS) and Frame Relay) and the emergence of broadband networks. Due to technological advances in the areas of data recording and data storage, traffic monitoring devices capable of recording every cell/packet/frame at OC3-speeds (i.e., 150 Mbps) for multiple hours already exist and are capable of generating data sets in the Tera-byte range. These traffic data sets, in turn, pose considerable challenges and problems for data analysts and traffic modelers in making effective and efficient use of this wealth of information for characterizing, describing, and modeling measured network traffic.

Examples of traffic measurements studies and modeling work that are based on recently collected large sets of measured network traffic include (*i*) an investigation of ISDN traffic (16 Kbps) reported in [15, 40], (*ii*) a series of articles reporting on detailed statistical analyses of high-resolution Ethernet LAN (10 Mbps) traces (see [31, 32, 61, 63]) and WAN traffic (see [11, 46, 47, 49]), (*iii*) an examination of traffic collected from working CCS subnetworks (56 Kbps) detailed in [14], and (*iv*) different attempts for dealing with variable-bit-rate (VBR) video traffic (e.g., see [5, 20, 22, 23]). Additional examples are [1, 9] and [27], which report on traffic collection and analysis efforts involving NSFNET wide-area networks, a DQDB MAN, and FASTPAC, an Australian high-speed data network, respectively. Furthermore, practically all deployed or planned ATM test beds in the United States and Europe contain significant traffic measurements components that will result in extremely high-volume data from "live" ATM networks within the next few years.

The results of these traffic measurements studies have been striking for two reasons: (1) The studies demonstrate that it is generally possible to clearly distinguish between measured packet network traffic and traffic generated by commonly assumed theoretical models, and (2) In sharp contrast to traditional traffic modeling assumptions, aggregate packet streams are statistically *self-similar* or *fractal* in nature; that is, actual network traffic looks the same when measured over time scales ranging from milliseconds to seconds and minutes and beyond. Next we specify in mathematical terms what we mean by "the fractal nature of network traffic".

9.2.2 Fractal traffic characteristics

The empirical findings observed in the above-mentioned traffic studies that measured traffic processes consistently show that variations and fluctuations over a wide range of time scales are in full agreement with observations made in many other branches of science and engineering, namely that underlying bursty phenomena typically exhibit variations over many length or time scales (e.g., see [38]). Variability or burstiness over many time scales can occur in several contexts in real traffic. We consider in the following two such concepts, namely, the *infinite variance syndrome* and *long-range dependence*, or the corresponding notions coined by Mandelbrot, i.e., the *Noah Effect* and the *Joseph Effect*, respectively. For statistical inference problems concerning the infinite variance syndrome and long-range dependence (and related concepts), see [4, 5, 28, 54, 61, 63].

9.2.2.1 The infinite variance syndrome or high variability

Many random variables of interest in traffic modeling are "heavy-tailed", which is to say that the tails of their corresponding distribution often exhibit a *hyperbolic* or *power-law* decay. More precisely, a random variable X is said to be *heavy-tailed* if

$$P[X > x] \sim L_1(x)x^{-\alpha}, \quad \text{as } x \to \infty, \tag{9.1}$$

where $\alpha > 0$, $L_1 > 0$ is a slowly varying function at infinity, i.e., $\lim_{x \to \infty} L_1(tx)/L_1(x) = 1$, for any $t > 0$, and where $a(x) \sim b(x)$ means $a(x)/b(x) \to 1$, as $x \to \infty$. Of particular interest here is the case $1 < \alpha < 2$, that is, X has finite mean but infinite variance. Mandelbrot refers to this property as the *in-*

finite variance syndrome or the *Noah Effect*. Intuitively, the infinite variance syndrome allows for compact descriptions of highly variable phenomena, i.e., of random variables that can take extreme values with non-negligible probabilities. Empirical evidence for this phenomenon in actual traffic measurements abounds and has been observed in traffic processes, such as burst lengths, sojourn times in *ON/OFF* states, packet interarrival times, resource holding times, etc. (e.g., see [14, 40, 49, 62]). This is in stark contrast to the commonly made assumption in traditional traffic modeling that random variables, such as burst lengths, sojourn times, holding times, etc. are best described in terms of *light-tailed* distributions, i.e., by distributions satisfying

$$P[X > x] \sim L_2(x)\exp(-x), \quad \text{as } x \to \infty, \qquad (9.2)$$

where L_2 is slowly varying at infinity. Light-tailed distributions are typically clustered tightly around their means, allow for only very limited fluctuations around their means, and (9.2) guarantees that all their moments are finite. This discrepancy between traditional traffic modeling theory (emphasizing light-tailed distributions) and measured traffic traces (providing strong empirical evidence for the presence of the infinite variance syndrome) can have many obvious, as well as subtle, consequences for traffic modeling and queueing performance (see Section 9.5).

9.2.2.2 Long-range dependence or persistence

As far as their time-dynamics is concerned, many empirically observed traffic processes have autocorrelation functions that span many time scales, i.e., they exhibit "long-range dependence". Formally, a covariance-stationary process $X = (X_k: k \geq 0)$ with autocorrelation function $r = (r(k), k \geq 0)$ is said to exhibit *long-range dependence* if

$$r(k) \sim L_3(k)k^{2H-2}, \quad \text{as } k \to \infty, \, 1/2 < H < 1, \qquad (9.3)$$

where L_3 is slowly varying at infinity. H is called the *Hurst parameter* and is commonly used as a measure of the degree of long-range dependence in a given time series. Mandelbrot refers to (9.3) as the *persistence phenomenon* or the *Joseph Effect*. In practically all of the recent traffic studies mentioned in Subsection 9.2.1, long-range dependence has been found to be consistent with measured traffic traces. In fact, the presence of correlations that exhibit a hyperbolic rather than an exponential decay is the major reason for the ease with which it is possible to distinguish between measured network traffic and traditional model generated traffic (see [32]). It is primarily for mathematical convenience that traditional traffic modeling considers processes that have exponentially decaying autocorrelations, i.e., are *short-range dependent:*

$$r(k) \sim L_4(k)\rho^k, \quad \text{as } k \to \infty, \, 0 < \rho < 1, \qquad (9.4)$$

where L_4 is slowly varying at infinity. Short-range dependent processes allow for only limited fluctuations or burstiness in their time-dynamics, and increasing the range over more time scales generally comes at the expense of increased model complexity. On the other hand, the empirically observed variations of traffic rates over a wide range of time scales can be captured in a compact manner via long-range dependence using the Hurst parameter.

   In the frequency domain, long-range dependence manifests itself in a spectral density that obeys a power-law near the origin; in fact, under weak regularity conditions on the slowly varying function L_3 in (9.3), X is long-range dependent iff

$$s(\omega) \sim L_5(\omega)\omega^{1-2H}, \quad \text{as } \omega\to 0, \ 1/2 < H < 1, \tag{9.5}$$

where $s(\omega) = \sum_k r(k)e^{ik\omega}$ is the spectral density of X and L_5 is slowly varying at 0 (e.g., see [4]). Property (9.5) of the spectral density is also called $1/f$-*noise phenomenon*. In contrast, short-range dependent processes are characterized by spectral densities that remain bounded for low frequencies. Note that for short-range dependent processes, $H = 1/2$. In terms of the *aggregated* processes $X^{(m)}$, which are obtained by averaging the original process X over non-overlapping blocks of size $m \geq 1$, long-range dependence can be characterized by

$$\text{var}(X^{(m)}) \sim L_6(m)m^{2H-2}, \quad \text{as } m\to\infty, \ 1/2 < H < 1, \tag{9.6}$$

where L_6 is slowly varying at infinity. Similarly, X is short-range dependent if (9.6) holds with $H = 1/2$.

Mathematically, autocorrelations that decay hyperbolically (i.e., satisfy relation (9.3)), variances of the aggregated processes that decrease more slowly than the reciprocal of the sample size (see relation (9.6)), and spectral densities that exhibit the $1/f$-noise phenomenon (i.e., see (9.5)) are different manifestations of the property that the underlying traffic process is statistically "self-similar". Formally, a covariance-stationarity packet traffic process X satisfying (9.3) is called *asymptotically second-order self-similar* with self-similarity parameter H if for all sufficiently large m, $r^{(m)}(k) \sim r(k)$, as $k\to\infty$, where $r^{(m)}$ denotes the autocorrelation function of the aggregated process $X^{(m)}$. *Exactly second-order self-similar processes* have the asymptotic proportionality strengthened to an exact equality for all k and all m. Thus, self-similarity is a well-defined mathematical concept that concerns statistical characteristics of a process X on all time scales and connects them through invariance properties using power-laws. Clearly, self-similarity seems well-suited for capturing and describing in a simple manner the empirically established feature of measured network traffic, namely that it exhibits variability and burstiness over a wide range of time scales. In the following, we will refer collectively to the infinite variance syndrome and to long-range dependence (or equivalently, $1/f$-phenomena, slowly decaying variances, or self-similarity) as the *fractal properties* of measured network traffic, using the popular notion of fractals to describe phenomena that span many time or length scales. For other approaches to characterizing the fractal nature of measured packet traffic that make explicit use of certain *fractal dimension* descriptors, we refer to [16, 19, 50].

For any empirically observed statistical characteristics, or for any assumptions made at the modeling stage, an important question, which is discussed in the following, is the physical basis or justification for these observation/assumptions. Providing such a physical explanation not only increases confidence in the empirical work and modeling assumptions, but also provides insights into the robustness of these empirical observations as network conditions and applications change.

9.2.3 On the physical nature of fractal traffic dynamics

Recent results reported in [62, 63] provide a plausible and simple explanation for the observed self-similarity of measured aggregate packet traffic in terms of the nature of the traffic generated by the individual sources/source-destination pairs that contribute to the aggregate packet stream. Developing an approach originally suggested by Mandelbrot [37] (see also [19, 57]), Willinger et al. show that the superposition of many *ON/OFF* sources, each of which exhibits the infinite variance syndrome, results in self-similar aggregate traffic. Being able to phrase the results in the well-known framework of *ON/OFF* source models (also known as "packet train models", introduced in [26]), the infinite variance property of the distribution of the sojourn time in the *ON* and/or *OFF* state is identified as the essential point of departure from traditional to self-similar traffic modeling. In this context, the infinite variance assumption on an individual *ON/OFF* source model results in highly variable *ON*- or *OFF*-periods, i.e., "train lengths" and "intertrain distances" that can be very large with non-negligible probability. In other words, the infinite variance property guarantees that each *ON/OFF* source individually exhibits fluctuations and variations that cover a wide range of time scales, and the parameter α describing the "heaviness" of the tail of the corresponding distribution in (9.1) gives a measure of the intensity of the Noah Effect. There also exists a simple relation between α and the Hurst parameter H, where the latter has been suggested in [32] as a measure of the degree of self-similarity or burstiness (or equivalently, of the Joseph Effect) of the aggregate traffic stream.

Even though independence governs the above description of the individual *ON/OFF* source model (i.e., the *ON*-periods are i.i.d., the *OFF*-periods are i.i.d., and the *ON*- and *OFF*-periods are independent from one another), an infinite variance distribution for the sojourn time in the *ON* and/or *OFF* state can be readily shown to lead to long-range dependence in the stationary "reward" process corresponding to an individual *ON/OFF* source (e.g., a reward of 1 while in the *ON* state, zero reward in the *OFF*-state). By aggregating many such sources, the dependence structure remains unchanged, but the marginal distribution becomes more Gaussian; finally, additional aggregation in time results in fractional Gaussian noise, the only exactly self-similar Gaussian process (for a more formal treatment of this derivation, see [18, 63]).

In sharp contrast to these findings, traditional traffic modeling, when cast in the framework of *ON/OFF* source models, without exception assumes finite variance distributions for the *ON*- and *OFF*-periods (e.g., exponential distribution, geometric distribution). These assumptions drastically limit the *ON/OFF* activities of an individual source, and as a result, the superposition of many such sources behaves like white noise in the sense that the aggregate traffic stream is void of any significant correlations, except possibly some in the short range. This behavior is in clear contrast with the measured network traffic described in [31].

Using actual traffic traces, Willinger et al. [62] (for earlier work, see [49]) have demonstrated that infinite variance phenomena are in full agreement with measured network traffic. They revisit the Bellcore Ethernet LAN traffic traces (see [31]) and extract from the aggregate traffic the traces generated by individual source-destination pairs. Subsequent statistical analyses of these traces (for a given hour, there are typically between 100-1000 active source-destination pairs, i.e., traffic traces) reveal that (*i*) the traffic generated by individual source-destination pairs is consistent with an *ON/OFF* model, and (*ii*) the distributions of the sojourn times in the *ON/OFF* states can be accurately described using Pareto-type distributions, which exhibit infinite variance. Thus, the Ethernet LAN data are not only consistent with self-similarity at the level of aggregate packet traffic, but they are also in full agreement with the physical explanation for self-similarity given above in terms of the fundamental characteristics of the individual source-destination pair traffics that make up the self-similar aggregate trace. Obviously, this finding helps explain some of the observed robustness features of fractal traffic. It also points out a number of potential approaches to self-similar traffic modeling, generation, and analysis.

Even though a proposed model for network traffic can be consistent with actual traffic measurements and can be fully justified on physical grounds, the main concern from a practical traffic engineering viewpoint is whether or not the statistical characteristics that distinguish the proposed model from other models for traffic are relevant and matter from a queueing performance perspective. With regard to the fractal traffic characteristics described in Subsection 9.2.2, this question is considered in the next section.

9.3 SIGNIFICANCE OF FRACTAL TRAFFIC CHARACTERISTICS

While the studies mentioned above convincingly establish the presence of variations over a wide range of time scales in actual traffic processes, the significance of these fractal characteristics to queueing performance and traffic engineering may not be clear at this point. For example, it has been argued by some that since queueing performance is determined by features in arrival processes that occur on time scales of a queueing system's busy period, long-range dependence has no practical impact and need not be incorporated into performance models. In this section, we summarize a series of simulation experiments reported in [18] that demonstrate — contrary to such arguments — the practical significance of long-range dependence for queueing performance. We also illustrate with some examples why infinite variance phenomena in traffic processes can be expected to play an increasingly important role in queueing performance of modern communications networks.

9.3.1 Importance of long-range dependence

We consider a queueing system with the following characteristics: infinite waiting room, deterministic service times, a single server, and arrivals taken from *actual* Ethernet traffic traces. The input traces consist of the measured interarrival times of actual Ethernet traffic. A 30 minute interval is representative of the time scales over which traffic in packet networks is currently measured (15-60 minutes); for example, traffic levels are typically reported at 30-minute intervals, and rate and utilization measurements over these intervals are used as baselines in "load-service" curves. In our experiments, we perform various transformations on these Ethernet traffic traces. For this purpose, we choose to work with interarrival time traces, primarily to preserve the marginal interarrival time distribution throughout the different queueing experiments. By keeping the packet interarrival time distributions of the traces used in our experiments identical, we are able to isolate the possible effects of the underlying dependence structure on queueing and separate them from those due to the distributional aspects of the interarrival times. To achieve different utilizations of the queue for a given input trace, we adjust the service time of the deterministic server, and we focus here on the average waiting time, as well as the asymptotic form of the queue length distribution as our performance measures.

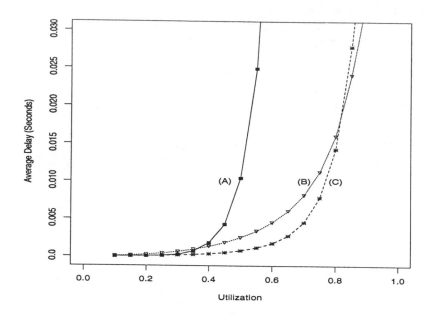

Figure 1. Average delay (in seconds) versus utilization plot,
for time series of interarrival times.

Curve (A) in Figure 1 is the average delay vs. utilization plot obtained with the original trace. As can be seen, there is a sharp rise in the average delay around 50% utilization. By way of a comparison, Figure 1 also shows (curve (B)) the delay curve predicted by the Queueing Network Analyzer (QNA). QNA uses a set of GI/G/1 approximations (based on two moment characterizations of the input traffic) that are widely applied in practice. In contrast to the delay curve obtained with the actual trace, curve (B) predicts useful capacities in excess of 80% utilization. Curve (C) in Figure 1 is the delay curve obtained by completely randomizing the sequence of interarrival times. We see that curve (C) differs substantially from curve (A), suggesting that the *best* renewal process model will be grossly in error in predicting actual performance.

Clearly, the striking discrepancy between curves (A) and (C) arises because of the dependence inherent in bursty traffic. To further investigate which aspect of the dependence structure is significant, we repeat the experiment with the input traces consisting of time series of counts, which are more in the spirit of the empirically observed LRD property observed in the traffic studies mentioned in Subsection 9.2. Figure 2 shows results obtained when the data is represented by a time series of counts over 30 millisecond intervals. The counts are randomly distributed over the 30 ms interval for the queueing simulations. Curve (A) is the delay vs. utilization curve obtained with the original trace, and while this trace is expected to be locally smoother than the original interarrival trace, there is nevertheless a sharp rise in the queueing delays at relatively low utilizations. Curve (C) is obtained with a complete shuffle of the time series of counts (i.e., by choosing an arrangement of the number of counts per 30 milliseconds at random), which results in an empirical record that is void of any correlations; once again, the discrepancy is considerable. The next experiment consists of dividing the time series of counts into blocks of size 10, and shuffling the order of the blocks. This has the effect of eliminating the dependence structure beyond 300 ms, while preserving correlations on smaller time scales. Nevertheless, the curve obtained with this trace (Curve (E)) is a poor approximation to the original. We repeat this experiment with a trace obtained by shuffling the order of the time series *within* the block of size 10.

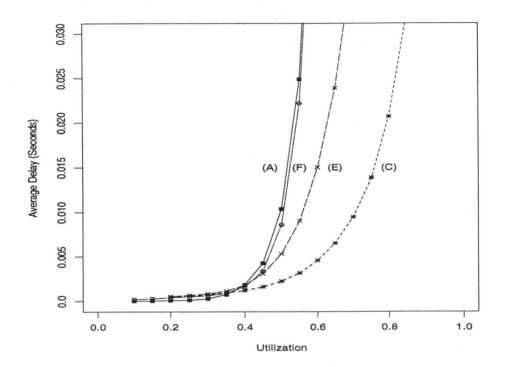

Figure 2. Average delay (in seconds) vs. utilization plot,
for time series of counts (as opposed to interarrival times).

This has the effect of changing the short-range correlations over time scales below 300 ms, while preserving correlations over longer time scales. Strikingly, the resulting plot, Curve (F), is virtually identical to the original curve. This demonstrates that several aspects of queueing performance, such as the "knee of the curve", are largely determined by long-range correlations.

Similar results are obtained with: changing block sizes; time series of interarrival counts; with other performance measures, such as queue-length distributions; other Ethernet traces; and traces from other types of packet networks. A more complete description of these empirical studies illustrating the impact of long-range dependence is given in [18]. These studies show that correlations over large time scales (from seconds to minutes and beyond) can have *measurable and practical* consequences. In Subsection 9.5, we explain these empirical results with the help of fractal queueing models, i.e., using traffic processes that parsimoniously model long-range dependence.

9.3.2 Impact of infinite variance distributions

Similarly, the presence of infinite variance distributions (or more generally, heavy-tailed phenomena) in traffic processes, such as burst lengths and resource holding times, can have significant impacts on queueing performance. For example, recall (see Subsection 9.2.3) that the empirically observed long-range dependence property of aggregate network traffic is a direct consequence of the infinite variance property displayed by the *ON/OFF* behavior of the individual sources that make up the aggregate traffic stream. Analytical results which demonstrate a queueing behavior (for individual *ON/OFF* sources as well as for the aggregate traffic) that is dramatically different depending on whether or not the *ON/OFF* behavior displays the infinite variance syndrome are described on Subsection 9.5.

For connection times (e.g., call holding times in conventional networks; see [14]) which are best described using heavy-tailed distributions, the impact on blocking performance can be significant. The Erlang B loss function and Engset finite source model are known to be insensitive to the service time distribution, beyond its mean value. Nevertheless, there are a number of reasons why these classical results will not correctly predict blocking performance: convergence to the steady state results requires on the order of 5-10 average holding times, which means that the system will not reach steady state in a time interval over which the arrival process can be considered stationary [17]. Secondly, reattempt rates can be high on account of "auto-dialing" features, and the blocking on reattempts can be very high because the state of the system evolves slowly in comparison to the reattempt rates. Over the span of an engineering period (i.e., an observation interval over which performance is assessed, and which may span several hours) arrival rates may vary, so that non-stationary effects may have to be accounted for. Thus, even though classical results are known to be robust, heavy-tailed distributions in the form of long holding times do impact blocking performance observed in networks.

A key performance measure in ATM networks is the cell loss rate caused by finite waiting rooms, or buffer overflows. However, cell losses are so bursty that the long-term average rate is believed to be an insufficient indicator of degradations in application performance. Alternately, traffic loss processes may be better described by heavy-tailed renewal processes (e.g., see [17, 36, 58]).

9.3.3 Identifying relevant time scales

We can now address the fallacy in the argument that "queueing performance should not be influenced by long-range dependence, but only by variations over a typical busy period of the queue". Empirically, in simulations with actual traffic traces, it has been observed that busy periods, like burst lengths, can span many time scales, and that there is no characteristic or "typical" length of a busy period. Intuitively, the maximum busy period places an upper "cut-off" on the time scales/lags in the autocorrelation function that need to be modeled; as the server occupancy increases, the busy periods get longer and longer, and the "relevant" time scales increase correspondingly. In particular, as the occupancy approaches 1, *all* time scales are theoretically of interest. While for a given queueing system at a particular utilization level, there exists in principle always an upper cut-off to the time scales relevant for predicting performance, these time scales cannot be estimated *a priori* as model inputs; typically, their estimation would involve solving the queue. Equally important, placing ad hoc *a priori* cut-offs on the traffic processes can lead to ill-posed problems, in the sense that predicted performance may sensitively depend on the (arbitrarily chosen) cut-offs.

Finally, there are overriding practical considerations that ultimately define the time scales of relevance. In communication networks, the mean rates, which are used as the baseline for the evaluation of performance, are estimated on the basis of arrival rate measurements on the order of 15-60 minutes. Thus, all variations in the traffic processes below these time scales are of interest. In particular, guidelines on setting occupancy thresholds, below which performance can be deemed to be acceptable, will be impacted by variations on all time scales within the occupancy measurement interval.

9.4 FRACTAL TRAFFIC DESCRIPTIONS

Traditional approaches to modeling heavy-tailed phenomena consist of representing them via mixtures of exponentials or using appropriately chosen phase-type distributions. In principle, any distribution can be approximated to an arbitrary degree by phase-type distributions (e.g., [2]). Likewise, the dependence structure in any finite time series of arrival counts can be modeled by Markov-Modulated Poisson Processes (MMPPs), or other models that are based on indexing source activity to a finite-state Markov chain. The motivation for this approach appears to be that any marked point process can be approximated by an MMPP (e.g., see [4]). While such convergence theorems are of considerable theoretical importance, they are often less useful in practice. This is because fitting heavy-tailed distributions or hyperbolically decaying autocorrelations in this manner over time scales of engineering interest typically results in highly parameterized models, which cannot be supported (in terms of inferring model parameters) in practice. For example, estimating the parameters of a mixture of a large number of exponentials to real data is known to be an ill-conditioned problem (e.g., see [36]). While there has been considerable amount of progress in computational algorithms for the analysis of large state-space Markov-based models (e.g., see [33, 34]), inferring model parameters remains largely an open problem.

However, it is possible to represent traffic processes that display fluctuations and variations over a wide range of time scales without resorting to highly parameterized models. For example, heavy-tailed phenomena can be modeled in a compact manner using distributions whose tails exhibit a power-law decay; in many cases, the performance phenomena of interest.are dominated by the properties of the tail of distribution. The *Pareto family of distributions*, for example, can parsimoniously match the power-law decay observed in many measured traffic processes (e.g., sojourn times in *ON/OFF* states, connection times):

$$P(T > t) \sim t^{-\alpha}, \quad \text{as } t \to \infty. \tag{9.7}$$

An equivalent description of *ON/OFF* sources in terms of chaotic maps is also feasible. In the chaotic map formulation, the source state is represented by a continuous variable whose evolution in discrete time is described by a low order, nonlinear dynamical system. The packet generation process is now modeled by stipulating that a source generates a batch of packets at the peak rate when the state variable is above a threshold, and is idle otherwise. Realistic *ON/OFF* behavior can now be described in terms of the parameters (small in number) of a suitably chosen map. Either of these methods enables us to model directly what appears to be a physical basis of the self-similarity observed in actual network traffic, namely the aggregation of heavy-tailed *ON/OFF* sources.

Exactly self-similar models can, by definition, parsimoniously capture autocorrelations over many time scales. The Fractional Brownian Motion (FBM) traffic arrival model, introduced in [41] can capture the second-order properties of bursty traffic processes over many time scales with three parameters:

$$A(t) = mt + \sqrt{am}Z(t), \qquad (9.8)$$

where $A(t)$ is the cumulative work up to time t, and $Z(t)$ is Fractional Brownian Motion. FBM is a Gaussian process that extends the familiar notion of Brownian Motion, characterized by independent Gaussian increments, to dependent increments with a power-law autocorrelation. Each of the three parameters in this model has a distinct physical interpretation - m is the arrival rate, the Hurst parameter H characterizes the decay of the autocorrelation function (or equivalently, the self-similar scaling factor), and a is a "peakedness" term describing the magnitude of fluctuations.

The FBM model has several obvious drawbacks in common with standard diffusion models, e.g., the "increments" can actually be negative. In heavy traffic conditions, this does not significantly impact the accuracy of model predictions. The conditions under which the FBM model can be expected to be valid in practice are described in [18] and require that (i) the range of time scales of interest for the problem coincides with the scaling region - note that the FBM model is self-similar on all time scales, whereas actual traffic processes tend to exhibit this property over a wide, but finite range of time scales, bounded by upper and lower cut-offs, (ii) the traffic is aggregated from a large number of independent users so that a purely second-order description is adequate, and (iii) the effect of flow controls on any one user is negligible.

9.5 FRACTAL QUEUEING

The general scope of engineering analysis can be divided into three distinct phases: description, analysis, and control. Most current applications of fractals are limited to the description phase, including studies on the physical bases or origins of the phenomena. Fractal models have been successfully applied to parsimoniously describe complex, bursty phenomena in many branches of science and engineering, but the resulting models are commonly known to be very difficult to analyze. As such, only few methods are available to analyze models that use fractal characterizations as inputs, and work on the control of fractal phenomena is even scarcer. Clearly, the application of fractal models to teletraffic theory and practice requires all three aspects of engineering analysis. Currently, there are few techniques that are available to routinely analyze fractal queueing models, with all the myriad possible variations that are feasible in queueing systems. This is often cited as a drawback of fractal models, though such difficulties are really representative of the problems inherent in analyzing phenomena that span many time scales. There exists, nevertheless, a limited set of analytical results that has led to significant insights into the nature of fractal queueing and has illustrated that fractal models give rise to queueing behavior that is much more versatile than what we have seen in the past based on traditional queueing models. This section summarizes some of the recent significant results in the area of fractal queueing, and readers interested in details are referred to the original papers.

9.5.1 Fractional Brownian Motion Storage Models

As indicated in the previous section, FBM models allow for a parsimonious description of long-range dependence. The analysis of queues driven by FBM arrival processes has largely been developed by Norros and his co-workers in a series of papers [7, 41, 42, 43, 56]. The starting point of these analyses is Reich's formula [6] for the backlog of work or virtual waiting time $V(t)$ in a queue with infinite waiting room, with service rate equal to C and cumulative work arrival process $A(t)$, namely,

$$V(t) = \sup_{s \leq t} (A(t) - A(s) - C(t - s)). \qquad (9.9)$$

$V(t)$ is non-negative and stationary, which follows from the stationarity of the *fractional Gaussian noise* FGN (i.e., the increment process of FBM); moreover, for $m < C$, $V(t)$ is almost surely finite [41]. Further analyses of the properties of $V(t)$ can now proceed in one of two ways: applying large deviation theory to study the complementary distribution of $V(t)$, or exploiting the scaling properties of FGN. The following derivation for the FBM storage model is due to Norros [41], though similar arguments have been used by a number of authors in other contexts. A lower bound on the tail of the complementary distribution can be obtained using the following inequality:

$$P(V(t) > x) = P(V(0) > x) \geq \max_{t \geq 0} P(A(t) > Ct + x) = \max_{t \geq 0} \bar{\Phi}\left(\frac{(C - m)t + x}{\sqrt{am}t^H}\right),$$
$$(9.10)$$

where $\bar{\Phi}(\cdot)$ is the complementary normal distribution. On the RHS, the maximum can be readily shown to be achieved at

$$t = \frac{Hx}{(1 - H)(C - m)}. \qquad (9.11)$$

From this result, it follows that the lower bound on the complementary distribution function is given by

$$P(V > x) \geq \bar{\Phi}\left(\frac{(C - m)^H x^{1 - H}}{\sqrt{am} H^H (1 - H)^{1 - H}}\right). \qquad (9.12)$$

Using well-known approximations for the tail of the normal distribution results in the following useful form for the lower bound:

$$P(V > x) \sim \frac{1}{\sqrt{2\pi}(1 + \sqrt{cx})} \exp\left(-cx^{2 - 2H}\right) \qquad (9.13)$$

with

$$c = \frac{(C - m)^{2H}}{2am\left[\left(\frac{1 - H}{H}\right)^H + \left(\frac{H}{1 - H}\right)^{1 - H}\right]^2}. \qquad (9.14)$$

For $H = 0.5$, approximating the lower bound solely by the exponential term, the resulting expression is identical to the asymptotic queue length distribution of the M/D/1 queue. For $H > 0.5$, the asymptotic queue length distribution decays as a

Weibullian or "stretched exponential", and for medium to large H-values, the discrepancies with an exponential decay predicted by traditional, short-range dependent traffic models can be substantial.

At first glance, this might appear to be a very loose lower bound. However, as Duffield and O'Connell [13] show, the bound obtained by Norros is, in fact, asymptotically tight. An intuitive explanation is provided by the "strong law of rare events", which states that rare events occur in the most likely way (see for example, [12, 59]). More formally, Duffield and O'Connell extend the results of Glynn and Whitt [21] to a very broad class of arrival processes, including those with Gaussian increments and arbitrary covariance structure. In fact, by specializing their analysis to FBM, they demonstrate that the lower bound is asymptotically tight, differing from the true asymptotic form of the distribution in at most a power-law prefactor. In principle, the results in [12] can be applied to derive bounds on the virtual waiting time for a broader class of fractal arrival processes (e.g., asymptotically self-similar, non-self similar), but we are not aware of published results on bounds that are computable.

It is also possible to derive useful performance relations for the FBM storage model without recourse to large deviations, namely by directly exploiting self-similarity or scale invariance of the increment process. Specifically, this analysis is based on the following two properties: (*i*) the distribution of the virtual waiting time is invariant as the process is observed over different time scales, i.e., $V(\alpha t)$ is identical in distribution to $V(t)$ (this follows from stationarity), and (*ii*) from the definition of the increment process, $Z(\alpha t) = \alpha^H Z(t)$. For examples using scaling analysis, see [29, 41], who derive a new class of performance results for the FBM model. These include interesting cross-over relations whereby a process with a higher H-value can, under certain conditions, have better queueing performance than a process with a lower H-value. Cross-overs are related to the fact that as the range of time scales of interest in a queueing problem decreases (e.g., increasing capacity while fixing buffer sizes), a process with higher H-values will in fact appear smoother (or less peaked) than a process with lower H-value, on sufficiently small time scales. Conversely, as the range of time scales of interest increases (as utilization levels increase), the process with higher H-values can appear burstier than a process with a lower H-value, and the performance can degrade rapidly. This variation of the "peakedness" with time scale is a key characteristic of fractal traffic processes. Self-similar models, such as the FBM model, incorporate this variation implicitly and without additional effort, in contrast to models that explicitly (and typically at the expense of added model complexity) seek to model a limited range of time scales.

The Weibullian form of the asymptotic queue length distribution, as well as the scaling properties represent the most useful results for the FBM model to date. Norros et al. [43] also analyze the exact distribution of the virtual work process. They derive the so-called Beněs formula for the virtual work; however, the results are stated in terms of an unknown function. Using a constant in place of this function in numerical studies appears to yield an upper bound very close to the Weibullian tail. Lastly, Norros examines the feasibility of short-term predictability of the arrival process [42].

Just how good are the predictions of the FBM model? Figure 3 shows a plot of $\log P(V > x)$ versus x, using trace-driven simulations (Curve (A)), at a utilization of 0.5, corresponding to the knee of Curve (A) in Figure 1. The dashed Curve (I) is the asymptotic form predicted by the FBM model.

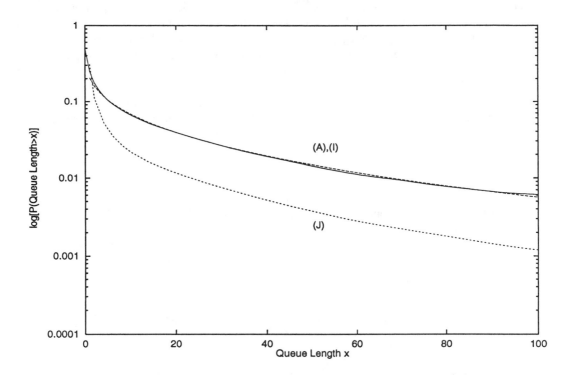

Figure 3. Complementary queue length distribution.

The values of H and a used to obtain the fit are within the confidence intervals of the parameters estimated from the trace. As can be seen, the asymptotic forms are in good agreement. For small x we expect, in general, corrections to the asymptotic form, but the asymptotic form of the curve seems to work surprisingly well even in this regime. At very large x, beyond about $x = 100$, the measured queue length distribution falls off *faster* than the dashed curve, due to limitations of the length of the simulation. In any finite set of data it is inevitable that the queue length distribution will eventually falloff faster than the predicted form for large x. Figure 3 also shows the results of a finite buffer simulation (Curve (J)), and the functional form of the queue length distribution appears to be the same as in the infinite buffer case.

The FBM model also provides insights into the sharp rise in average waiting times observed with actual traffic traces (Curve (A) in Figure 1). Using relation (9.13) to approximate the average queue length $\bar{x}$, it is readily shown that $\bar{x}$ diverges as $(1-\rho)^{-H/(1-H)}$, as $\rho \to 1$. Note that when $H = 0.5$, $\bar{x}$ behaves as $(1-\rho)^{-1}$, but for $H = 0.8$, $\bar{x}$ increases much more rapidly (as $(1-\rho)^{-4}$), which explains the sharp rise in delays observed in the simulations. Once again, this behavior is unprecedented in traditional, short-range dependent queueing models with constant or exponential service times, and shows that the use of standard queueing approximations in modern high-speed network analysis may lead to gross performance problems.

More generally, for the FBM model to be valid, the traffic should be aggregated from a large number of independent users, and the effect of network flow controls on any one user is negligible. In such a setting, the FBM model may differ from actual traffic in the nature of the short-range correlations, but as demonstrated in [18] this has a negligible effect on performance. In some environments, the number of sources that are aggregated does not appear to be large enough for a purely second-order description to be sufficient. In those cases, the reported utilizations are typically quite low; as utilizations increase, presumably as a result of increased aggregation, agreement with the FBM model can be expected to improve. Note however, that under conditions of limited aggregation, *no* purely second-order description, whether based on fractal or conventional Markov models, is expected to be valid.

9.2.5 *ON/OFF* sources

Much more drastic queueing behavior than for FBM models is obtained with single *ON/OFF* sources with heavy-tailed sojourn times in the *ON* state. This has been demonstrated in a number of ways, including analyzing the trajectories of dynamical systems that model queues driven by sources with generalized *ON/OFF* behavior. As established in [19], the queue length distributions in this setting are given by solutions of equations of the type

$$\rho(x, l) = \sum_{m=0}^{\infty} \left(\int_0^1 dy \delta((x, l) - \Phi(y, m)) \rho(y, m) \right), \qquad (9.15)$$

where $\rho(x, l)$ corresponds to the density of iterates over the source state (x) and the queue length (l). The queue length distribution $P(L > l)$ is then obtained by integrating $\rho(x, l)$ over x. For queueing to occur, the traffic generation rate in the *ON* state should exceed the service rate. Under these conditions, numerical solutions of these equations as well as heuristic arguments indicate that the queue length distribution is given by

$$P(L > l) \sim l^{-\alpha + 1}, \text{ as } l \rightarrow \infty, \qquad (9.16)$$

where α is the parameter of the Pareto distribution that governs the tail behavior of the sojourn time in the *ON* state. Thus, the queue length distribution decays as a power-law, and the rate of the decay is so slow that even the mean queue length does not exist. Using stochastic fluid models, Choudhury and Whitt [8] recently provided a complete proof of this result.

A second class of results with *ON/OFF* sources pertains to aggregates of *ON/OFF* sources. For recent results concerning the queueing behavior of the superposition of a finite number of heavy-tailed *ON/OFF* sources, see [8]. In the case of infinitely many sources, the resulting queueing behavior depends on the normalization procedure. When the service rate is scaled as the number of sources, the traffic converges to FBM. Under these conditions, an alternate derivation of relation (13) is provided by Brichet et al. [7], who directly analyze superpositions of *ON/OFF* sources. This analysis yields upper and lower bounds to the tail of the virtual work distribution; the lower bound is identical to (13), and the upper bound differs from the Weibullian in a constant prefactor.

This result might appear to contradict a recent paper by Likhanov et al. [35] (see also [3]), who analyze aggregates of *ON/OFF* sources with heavy-tailed sojourn time behavior and report a power-law decay of the queue length distributions. However, the queueing system analyzed in this case is quite different from that in [7,41]. In [35], a source in an *ON* state is assumed to instantaneously submit a burst of work at the queueing system. Given that the burst lengths are taken to be distributed as power-laws, and assuming an FCFS discipline at the burst level, an equivalence is established between this queue and an M/G/1 queue with infinite variance service times. A number of well-known results, including Pakes [44], show that in such a case, the queues decay as power-laws with unbounded average queue lengths. Note "queue length" in this case is taken to represent the number of sources that are in service. In the storage model proposed in [7,41], the queue lengths represent actual backlogs or work. The bursts from different sources can interleave, as is the case in a practical ATM multiplexer, and mapping such a queue into the M/G/1 framework is more difficult. Clearly, systems that are very different physically are bound to give rise to very different queueing asymptotics.

An alternate normalization procedure would consist of decreasing the average rate of each source under aggregation so that the resulting system is stable. In the special case when the peak rate of a single source exceeds the service rate, the queue length distributions once again decay as power-laws [56]. The analogue of a G/M/1 queue where the input is altered so as to exhibit long-range dependence in the interarrival times is considered by Resnick and Samorodnitsky [53]. By suitably choosing the arrival parameters, they show how queue size and waiting time can be made to possess infinite variance and even infinite mean. Such phenomena are non-existent in traditional queueing theory and illustrate the extent to which fractal queueing gives rise to yet unexplored queueing behavior.

9.5.3 Simulation methods

Given the relative scarcity of analytical results for self-similar traffic models, simulation studies assume a special importance. This is still an active area for research, and generation of self-similar traffic typically has to trade off simulation speed with accuracy. In particular, generating long sequences of a genuine FGN (or equivalently, FBM) process or of a truly *fractional autoregressive integrated moving average* (FARIMA) model could be prohibitive computationally (e.g., by generating from the joint distribution of the increment process). Note that FARIMA processes allow for more general autocorrelation functions than FGN models, and have been suggested in [5] as models for variable-bit-rate video traffic.

Recent investigations have focused on approximate yet fast techniques for generating self-similar traffic. One such technique is the Random Midpoint Displacement (RMD) method, which only utilizes the scaling behavior of the variance of the marginals to generate approximations to FBM (see [30]). RMD is fast, allows the rapid generation of long traces, and is known to be exact for $H = 1/2$. Accuracy can be improved with some speed trade-offs by aggregating the generated process in time. Another technique recently studied (see [48]) is based on inverting the (known) power spectrum of FBM using the Fast Fourier Transform (FFT). Fast (i.e., linear) generation methods for FBM traces are also possible using "Mandelbrot's Construction", i.e., aggregating (either stochastically generated, see [62], or using chaotic maps, see [51]) many *ON/OFF* sources; this method is tailor-made for massively parallel computers and allows the generation of realistic traffic that can easily saturate communication links with link speeds of 10-150 Mbps. As before, speed vs. accuracy trade-offs are possible by increasing the aggregation levels in space (i.e., the number of sources) or time. Pruthi [50] describes experiments which indicate that reasonable approximations to FBM are obtained by aggregating as few as 100 *ON/OFF* sources and then aggregating in time over blocks of size

100. FARIMA traces can be generated by passing FBM traces through an appropriate ARMA filter.

Another aspect of simulation deals with the difficulties inherent in estimating very low probabilities (such as cell loss probabilities) from simulations. Huang et al. [24] have extended Importance Sampling techniques to the case of FARIMA processes.

9.6 FUTURE WORK

There is considerable scope for further work in the area of fractal queueing models. Given the applicability of these models, and the many open issues, research is needed on all aspects of traffic modeling - data analysis, inference, mathematical modeling, performance analysis, and engineering methods. The following are examples of some of these open issues.

Toward routine queueing analyses of fractal models

As demonstrated in Section 9.3, leaving the Markovian framework results in model parsimony, but comes at the high price of losing the memoryless property of the exponential distribution, which forms the basis of routine analysis in traditional queueing. Fractal traffic processes also have a unique property, namely scaling; it remains to be seen if methods can be devised to exploit this feature further.

On explicit solutions for the FBM storage model

There are many aspects of the FBM storage model that are as yet unknown. It is possible that the results reviewed in this chapter exhaust all aspects of the FBM storage model that are tractable. However, there are numerous variations of the model (different performance metrics, variable service times) that are of interest, and it remains to be seen if additional results (e.g., first hitting time distributions of a fixed level, excursions above a certain level, etc.) from this model are feasible.

Beyond the FBM traffic model

While FBM appears to be a reasonable model of data traffic, other applications require more general arrival assumptions. For example, Variable-Bit-Rate (VBR) video traffic displays strong short-range correlations in addition to significant long-range dependence. Modeling the full correlation structure of VBR video traffic will require the use of asymptotically self-similar traffic models, such as FARIMA models. In addition to fitting the entire autocorrelation function, matching the marginal distributions of traffic processes can also be of practical importance (e.g., see [23]), especially if they display heavy-tailed phenomena. How to define something like an autocorrelation function for infinite variance processes is currently an active area of research (e.g., see [52, 55]).

Limited traffic aggregates

Currently, one can parsimoniously model realistic traffic under two extreme conditions: at the single source level, and in the limiting case of a large number of aggregate sources. The interim regime may not be readily susceptible to a parsimonious description, though descriptions in terms of aggregate *ON/OFF* sources may be more promising than others that would explicitly model higher-order statistics.

Further advances in the chaotic map approach to traffic modeling

The chaotic map framework has shown some promise for the rapid generation of self-similar traffic, queueing analysis, and performance of controls. However, the applicability of this approach depends heavily on methods to calculate or approximate the invariant distributions that correspond to queue length distributions. There is scope for further advances within this framework [50].

Fast generation of fractal traffic for simulations

In addition to the approaches mentioned in Subsection 9.5.3, we are aware of at least a dozen other promising techniques to generate self-similar traffic. The relative merits of these generation techniques (computational complexity and speed vs. accuracy) is a worthwhile area for future research.

From single queues to networks of queues

Most "network analysis" to date is limited to the analysis of single server queues. As demonstrated in [18] (see also [25]), long-range dependence is a very robust characteristic that is preserved as the traffic flows through queues or is subjected to short-time scale control. This insight suggests that network performance evaluation (at least to an approximate degree) with long-range dependent traffic may be feasible.

Control of fractal traffic processes

The interactions between traffic (and its fractal nature) and network controls (at the user as well as the network level) can have considerable impacts on queueing performance. This has largely not been addressed to this point. In this context, robustness issues of traffic characteristics (fractal or not) under different network protocols, under changes in network topologies and technologies, and under changes in user population and applications are of particular interest and have not yet attracted the attention they deserve.

Transient blocking analysis

The impact of "long holding times" spanning many time scales has numerous practical impacts, showing the need for queueing methods that permit the analysis of transient blocking performance in a nonstationary environment in which multiple resources can be concurrently held, and in which reattempt rates can be high.

ACKNOWLEDGEMENT

We thank Ilkka Norros of VTT Finland, Nick Duffield of AT&T Bell Labs, and Darryl Veitch and Parag Pruthi of KTH Stockholm for useful discussions, and for making available to us preprints of their work.

BIBLIOGRAPHY

[1]   Addie, R., Zuckerman, M. and Neame, T., Fractal traffic: Measurements modeling and performance, In: *Proceedings of the IEEE Infocom '95*, Boston, MA (1995), 977-984.

[2]   Altiok, T., On the phase-type approximations of general distributions, *IIE Trans.* **17** (1985), 110-116.

[3]   Anantharam, V., On the sojourn time of sessions at an ATM buffer with long-range dependent input traffic, (1995), preprint.

[4]   Asmussen, S. and Koole, G., Marked point processes as limits of Markovian arrival streams, *J. Appl. Prob.* **30** (1993), 365-372.

[5]   Beran, J., *Statistics for Long-Memory Processes*, Chapman & Hall, New York 1994.

[6]   Beran, J., Sherman, R., Taqqu, M.S. and Willinger, W., Long-range dependence in Variable-Bit-Rate video traffic, *IEEE Trans. on Communications* **43** (1995), 1566-1579.

[7]   Beneš, V.E., *General Stochastic Processes in the Theory of Queues*, Addison Wesley, New York 1963.

[8]   Brichet, F., Roberts, J.W., Simonian, A. and Veitch, D., Heavy traffic analysis of a fluid queue fed by *ON/OFF* with long-range dependence, Technical Report TD(95)03v1, COST 242 (1995).

[9]   Choudhury, G.L. and Whitt, W., Long-tail buffer-content distributions in broadband networks, (1995), preprint.

[10]   Cinotti, M., Dalle Mese, E., Giordano, S. and Russo, F., Long-range dependence in Ethernet traffic offered to interconnect DQDB MANs, University of Pisa (1994), preprint.

[11]   Cox, D.R., Long-range dependence: A review, In: *Statistics: An Appraisal* (ed. by H.A. David and H.T. David), The Iowa State University Press, Ames, Iowa (1984), 55-74.

[12]   Danzig, P., Jamin, S., Caceres, R., Mitzel, D. and Estrin, D., An empirical workload model for driving wide-area tcp/ip networks simulations, *Internetworking: Research and Experience* **3** (1992), 1-26.

[13]   Duffield, N.G., Economics of scale in queues with sources having power-law large deviation scalings, Dublin Institute of Advanced Studies, Dublin, Ireland, DIAS-APG-94-27 (1994), preprint.

[14]   Duffield, N.G. and O'Connell, N., Large deviation and overflow probabilities for the general single-serve queue with applications, In: *Proceedings of the Cambridge Philosophical Society* (1995), to appear.

[15]   Duffy, D.E., McIntosh, A.A., Rosenstein, M. and Willinger, W., Statistical analysis of CCSN/SS7 traffic data from working subnetworks, *IEEE J. Select. Areas in Communications* **12** (1994), 544-551.

[16]   Erramilli, A., Gosby, D.D. and Willinger, W., Engineering for realistic traffic: A fractal analysis of burstiness, In: *Proc. of the Bangalore Regional ITC Seminar*, Bangalore, India (1993), 155-162.

[17]   Erramilli, A. and Willinger, W., Fractal properties in packet traffic measurements, In: *Proc. of the St. Petersburg Regional ITC Seminar*, St. Petersburg, Russia (1993), 144-158.

[18]   Erramilli, A., Gordon, J. and Willinger, W., Applications of fractals in engineering for realistic traffic processes, In: *The Fundamental Role of Teletraffic in the Evolution of Telecommunications Networks* (*Proc. of ITC-14, Antibes Juan-les-Pins, France, June 1994*) (ed. by J. Labetoulle and J.W. Roberts), Elsevier, Amsterdam (1994), 35-44.

[19]   Erramilli, A., Narayan, O. and Willinger, W., Experimental queueing analysis with long-range dependent packet traffic, *IEEE/ACM Trans. on Networking* (1996), to appear.

[20]   Erramilli, A., Singh, R.P. and Pruthi, P., An application of deterministic chaotic maps to model packet traffic, *Queueing Systems* **20** (1995), 171-206.

[21]   Garrett, M.W. and Willinger, W., Analysis, modeling and generation of self-similar VBR video traffic, In: *Proc. of the ACM Sigcomm '94*, London, UK (1994), 269-280.

[22]   Glynn, P.W. and Whitt, W., Logarithmic asymptotics for steady-state tail probabilities in a single-server queue, *J. Appl. Prob.* **31** (1994), 131-156.

[23]   Heyman, D.P., Tabatabai, A. and Lakshman, T.V., Statistical analysis and simulation study of video teleconference traffic in ATM networks, *IEEE Trans. Circuits and Systems for Video Tech.* **2** (1992), 49-59.

[24]   Huang, C., Devetsikiotis, M., Lambadaris, I. and Kaye, A.R., Self-similar modeling of Variable-Bit-Rate compressed video: A unified approach, In: *Proc. of the ACM Sigcomm '95*, Cambridge, MA (1995), 114-125.

[25]   Huang, C., Devetsikiotis, M., Lambadaris, I. and Kaye, A.R., Fast simulation of self-similar traffic in ATM networks, In: *Proc. of the ICC '95* Seattle, WA (1995), 438-444.

[26]   Hwang, C.L. and Li, S.Q., On input state space reduction and buffer non-effective region, In: *Proc. of IEEE Infocom '94* (1994), 1018-1028.

[27]   Jain, R. and Routhier, S.A., Packet trains: Measurements and a new model for computer network traffic, *IEEE J. on Selected Areas in Communications* **4** (1986), 986-995.

[28]   Klivansky, S., Mukherjee, A. and Song, C., Factors contributing to self-similarity over NSFNet, Georgia Institute of Technology (1994), preprint.

[29]   Kratz, M. and Resnick, S.I., The qq-estimator and heavy tails, School of ORIE, Cornell University, Ithaca, NY (1995), preprint.

[30]   Krishnan, K.R., A new class of performance results for fractional Brownian motion traffic models, (1995), preprint.

[31]   Lau, W.-C., Erramilli, A., Wang, J.L. and Willinger, W., Self-similar traffic generation: The random midpoint displacement algorithm and its properties, In: *Proc. of the ICC '95*, Seattle, WA (1995), 466-472.

[32]   Leland, W.E., Taqqu, M.S., Willinger, W. and Wilson, D.V., On the self-similar nature of Ethernet traffic, In: *Proc. of the ACM Sigcomm '93*, San Francisco, CA (1993), 183-193.

[33]   Leland, W.E., Taqqu, M.S., Willinger, W. and Wilson, D.V., On the self-similar nature of Ethernet traffic (extended version), *IEEE/ACM Trans. on Networking* **2**:1 (1994), 1-15.

[34]   Li, S.Q. and Hwang, C.L., Queue response to input correlation functions: Discrete spectral analysis, *IEEE/ACM Trans. on Networking* **1** (1993), 522-533.

[35]   Li, S.Q. and Hwang, C.L., Queue response to input correlation functions: Continuous spectral analysis, *IEEE/ACM Trans. on Networking* **1** (1993), 678-692.

[36] Liebovitch, L.S., Testing fractal and Markov models of ion channel kinetics, *Biophysics Journal* **55** (1989), 373-377.

[37] Likhanov, N., Tsybakov, B. and Georganas, N.D., Analysis of an ATM buffer with self-similar ("fractal") input traffic, In: *Proc. of the IEEE Infocom '95*, Boston, MA (1995), 985-992.

[38] Mandelbrot, B.B., Self-similar error clusters in communication systems and the concept of conditional stationarity, *IEEE Trans. Communications Technology* **COM**-13 (1965), 71-90.

[39] Mandelbrot, B.B., Long-run linearity, locally Gaussian processes, H-spectra and infinite variances, *Intern. Econom. Review* **10** (1969), 82-113.

[40] Mandelbrot, B.B., *The Fractal Geometry of Nature*, W.H. Freeman and Co., San Francisco 1982.

[41] Mandelbrot, B.B. and Van Ness, J.W., Fractional Brownian motions, fractional noises and applications, *SIAM Review* **10** (1968), 422-437.

[42] Meier-Hellstern, K., Wirth, P.E., Yan, Y.-L. and Hoeflin, D.A., Traffic models for ISDN data users: Office automation application, In: *Teletraffic and Data Traffic in a Period of Change (Proc. of the 13th ITC, Copenhagen,* 1991) (ed. by A. Jensen and V.B. Iversen), North Holland (1991), 167-172.

[43] Norros, I., A storage model with self-similar input, *Queueing Systems* **16** (1994), 387-396.

[44] Norros, I., On the use of fractional Brownian motion in the theory of connectionless networks, *IEEE J. on Selected Areas in Communications* **13** (1995), 953-962.

[45] Norros, I., Simonian, A., Veitch, D. and Virtamo, J., A Beneš formula for the fractional Brownian storage, Technical Report TD(95)004v2, COST 242 (1995).

[46] Pakes, A.G., On the tails of waiting-time distributions, *J. of Applied Probability* **12** (1975), 555-564.

[47] Pawlita, P.F., Two decades of data traffic measurements: A survey of published results, experiences and applicability, In: *Proc. of the 12th ITC*, Torino, Italy (1988).

[48] Paxson, V., Empirically derived analytic models of wide-area TCP connections, *IEEE/ACM Transactions on Networking* **2** (1994), 316-336.

[49] Paxson, V. and Floyd, S., Wide-area traffic: The failure of Poisson modeling, In: *Proc. of the ACM Sigcomm '94*, London, UK (1994), 257-268.

[50] Paxson, V., Fast approximation of self-similar network traffic, (1995), preprint.

[51] Paxson, V. and Floyd, S., Wide area traffic: The failure of Poisson modeling, *IEEE/ACM Trans. on Networking* **3** (1995), 226-244.

[52] Pruthi, P., *An Application of Chaotic Maps to Packet Traffic Modeling*, Ph.D. Thesis, Royal Institute of Technology, Stockholm, Sweden 1995.

[53] Pruthi, P. and Erramilli, A., Heavy-tailed *ON/OFF* source behavior and self-similar traffic, In: *Proc. of the ICC '95*, Seattle, WA (1995), 445-450.

[54] Resnick, S.I., Heavy tail modeling and teletraffic data, (1995), preprint.

[55] Resnick, S.I. and Samorodnitsky, G., Performance decay in a single server exponential queueing model with long range dependence, *Operations Research*, to appear.

[56] Resnick, S. and Starica, C., Consistency of Hill's estimator for dependent data, *J. of Applied Probability* **32** (1995), 139-167.

[57] Samorodnitsky, G. and Taqqu, M.S., *Stable Non-Guassian Processes: Stochastic Models with Infinite Variance*, Chapman & Hall, New York, London 1994.

[58] Simonian, A. and Veitch, D., A storage model with long-range dependent *ON/OFF* sources with large peak rates, (1995), preprint.

[59] Taqqu, M.S. and Levy, J.B., Using renewal processes to generate long-range dependence and high variability, In: *Dependence in Probability and Statistics* (ed. by E. Eberlein and M.S. Taqqu), Progress in Prob. and Stat., Birkhauser, Boston **11** (1986), 73-89.

[60] Veitch, D., Novel methods of broadband traffic, *Proc. of Globecom* '93 (1993), Houston, TX, 1057-1061.

[61] Weiss, A., An introduction to large deviations for communication networks, *IEEE J. on Selected Areas in Communications* **13** (1995), 938-952.

[62] Willinger, W., Traffic modelling for high-speed networks: Theory versus practice, In: *Stochastic Networks* (ed. by F.P. Kelly and R.J. Williams), The IMA Volumes in Mathematics and its Applications, Springer-Verlag, New York **71** (1995), 395-409.

[63] Willinger, W., Taqqu, M.S., Leland, W.E. and Wilson, D.V., Self-similarity in high-speed packet traffic: Analysis and modeling of Ethernet traffic measurements, *Statistical Science* **10** (1995), 67-85.

[64] Willinger, W., Taqqu, M.S., Sherman, R. and Wilson, D.V., Self-similarity through high-variability: Statistical analysis of Ethernet LAN traffic at the source level, In: *Proc. of the ACM Sigcomm '95*, Cambridge, MA (1995), 100-113.

[65] Willinger, W., Taqqu, M.S., Sherman, R. and Wilson, D.V., Self-similarity through high-variability: Statistical analysis of Ethernet LAN traffic at the source level (extended version), (1995), preprint.

Chapter 10
Stochastic modeling of traffic processes

David L. Jagerman, Benjamin Melamed, and Walter Willinger

ABSTRACT Modern telecommunications networks are being designed to accommodate a heterogeneous mix of traffic classes ranging from traditional telephone calls to video and data service. Thus, traffic models are of crucial importance to the engineering and performance analysis of telecommunications systems, notably congestion and overload controls and capacity estimation.

This chapter surveys teletraffic models, addressing both theoretical and computational aspects. It first surveys the main classes of teletraffic models commonly used, and then proceeds to survey teletraffic methods for computing statistics relevant to the engineering of a teletraffic network.

CONTENTS

10.1 INTRODUCTION

Traffic is the driving force of telecommunications systems, representing customers making phone calls, transferring data files and other electronic information, or more recently, transmitting compressed video frames to a display device. The most common modeling context is queueing; traffic is offered to a queue or a network of queues, and various performance measures are calculated or estimated. These include queue length, server unitization, waiting times and traffic loss. Performance studies utilize traffic models in two basic ways: either as part of an analytical model, or to drive a discrete-event Monte Carlo simulation. Either way, traffic constitutes the common grist to the transport mechanism of the telecommunications system under study.

The material in this chapter is organized in two broad groupings. Section 10.2 surveys the main model classes of traffic streams commonly used in telecommunications. The emphasis here is on the theoretical aspects of traffic models, whether used in analytical context or Monte Carlo simulation context. Section 10.3 surveys the main teletraffic methods, designed originally to dimension and provision primarily telephone networks, subject to a prescribed grade of service. The emphasis here is on the computational aspects of the methods surveyed.

The material in each grouping is organized by and large in ascending general-

---

ity and complexity. The order of presentation also roughly tracks the corresponding chronological order of emergence of the models and methods surveyed.

10.2 MODELS OF TRAFFIC STREAMS

Simple traffic consists of single arrivals of discrete entities (packets, cells, etc.). It can be mathematically described as a *point process* [13, 10, 87], consisting of a sequence of arrival instants $T_1, T_2, \ldots, T_n, \ldots$ measured from the origin 0; by convention, $T_0 = 0$. There are two additional equivalent descriptions of point processes: *counting processes* and *interarrival time processes*. A counting process $\{N(t)\}_{t=0}^{\infty}$ is a continuous-time, non-negative integer-valued stochastic process, where $N(t) = \max\{n : T_n \leq t\}$ is the number of (traffic) arrivals in the interval $(0, t]$. An interarrival time process is a real-valued random sequence $\{A_n\}_{n=1}^{\infty}$, where $A_n = T_n - T_{n-1}$ is the length of the time interval separating the nth arrival from the previous one. The equivalence of these descriptions follows from the fact that $T_n = \sum_{k=1}^{n} A_k$, and from the equality of events

$$\{N(t) = n\} = \{T_n \leq t < T_{n+1}\} = \{\sum_{k=1}^{n} A_k \leq t < \sum_{k=1}^{n+1} A_k\}. \qquad (10.1)$$

The interarrival times, $\{A_n\}$, are assumed to form a stationary sequence, unless otherwise stated. An alternative characterization of point processes, called *stochastic intensity theory* [20, 46], is briefly discussed in Subsection 10.2.7.

Compound traffic consists of batch arrivals; that is, arrivals may consist of more than one unit at an arrival instant T_n. In order to fully describe compound traffic, one also needs to specify a real-valued random sequence $\{B_n\}_{n=1}^{\infty}$, where B_n is the (random) number of units in the batch. At a higher level of abstraction, B_n may represent some general attributes of the nth arrival, e.g., the amount of "work" associated with the nth arrival or its itinerary in a network. Such compound traffic processes, called *marked point processes* [31], are outside the scope of this paper.

Discrete-time traffic processes correspond to the case when time is slotted. Mathematically, this means that the random variables A_n can assume only integer values, or equivalently, that the random variables $N(t)$ are allowed to increase only at integer arrival instants T_n.

In addition to arrival times and batch sizes, it is often useful (and sometimes essential) to incorporate the notion of *workload* into the traffic description. The workload is a general concept describing the amount of work W_n brought to a system by the nth arriving unit; it is usually assumed independent of interarrival times and batch sizes. A typical example is the sequence of service time requirements of arrivals at a queueing system, though in queueing one usually refers to the arrival process alone as traffic. On the other hand, traffic reduces to workload description, when interarrival times are deterministic. A case in point is compressed video, also known as coded video. Video information is rarely transmitted over a network in its raw form. Rather, engineers take advantage of the considerable visual redundancy inherent in digitized pictures to compress each frame into a fraction of its original size. The compressed frames have random sizes (bit rates) which are then transported over the network and decoded at their destination. The term VBR (variable bit rate) video is used to refer to this kind of video traffic. Coded video frames (arrivals) must be delivered deterministically, every 1/30 of a second or so, for high-quality video. The workload consists of coded frame sizes (say, in bits), since a frame size is roughly proportional to its transmission time (service requirement).

The following notation will be used. The common distribution function of the A_n is denoted by $F_A(x)$. Similarly, $\lambda_A = 1/E[A_n]$ denotes the traffic rate, $\sigma_A^2 = Var[A_n]$, and $c_A = \lambda_A \sigma_A$. Unless otherwise stated, it is assumed that $0 < \sigma_A < \infty$ and that $\{A_n\}$ is simple; namely, $P\{A_n = 0\} = 0$. A traffic stream is denoted by X when a particular traffic description (via A, N, or T) is immaterial. In that case, traffic parameters may also be subscripted by X; e.g., λ_X is equivalent notation for λ_A, and similarly for other traffic parameters.

10.2.1 Renewal traffic models

This section briefly touches on renewal traffic processes and the important special cases of Poisson processes and Bernoulli processes.

Renewal models have a long history, due to their relative mathematical simplicity. In a renewal traffic process, the A_n are iid (independent, identically distributed), but their distribution is allowed to be general. Unfortunately, with few exceptions, the superposition of independent renewal processes does not yield a renewal process. Those that do, occupy a special position in traffic theory and practice. Historically, many queueing models routinely assumed a renewal offered traffic.

Renewal processes, while simple analytically, have a severe modeling drawback — the autocorrelation function of $\{A_n\}$ vanishes identically for all non-zero lags. The importance of capturing autocorrelations stems from the role of the autocorrelation function as a statistical proxy for temporal dependence in time series. Moreover, positive autocorrelations in $\{A_n\}$ can explain, to a large extent, the phenomenon of traffic burstiness. Bursty traffic is expected to dominate broadband networks, and when offered to a queueing system, it gives rise to much worse performance measures (such as mean waiting times) as compared to renewal traffic (which lacks temporal dependence); see [66] for a detailed discussion. Consequently, models that capture the autocorrelated nature of traffic are essential for predicting the performance of emerging broadband networks.

10.2.1.1 Poisson processes

Poisson models are the oldest traffic models, dating back to the advent of telephony and the renowned pioneering telephone engineer A.K. Erlang. A Poisson process [13] can be characterized as a renewal process whose interarrival times $\{A_n\}$ are exponentially distributed with rate parameter λ, that is, $P\{A_n \leq t\} = 1 - \exp(-\lambda t)$. Equivalently, it is a counting process, satisfying $P\{N(t) = n\} = \exp(-\lambda t)(\lambda t)^n/n!$, and the number of arrivals in disjoint intervals is statistically independent (a property known as *independent increments*).

Poisson processes enjoy some elegant analytical properties. First, the superposition of independent Poisson processes results in a new Poisson process whose rate is the sum of the component rates. Second, the independent increment property renders Poisson a memoryless process. This, in turn, greatly simplifies queueing problems involving Poisson arrivals. And third, Poisson processes are fairly common in traffic applications that physically comprise a large number of independent traffic streams, each of which may be quite general. The theoretical basis for this phenomenon is known as Palm's theorem [68, Vol. II, p. 582]. It roughly states that under suitable but mild regularity conditions, such multiplexed streams approach a

Poisson process as the number of streams grows, but the individual rates decrease so as to keep the aggregate rate constant. Thus, traffic on main communications arteries are commonly believed to follow a Poisson process, as opposed to traffic on upstream tributaries, which are less likely to be Poisson. However, traffic aggregation (multiplexing) need not always result in a Poisson stream; see the discussion in Subsection 10.2.6 for a counterexample.

10.2.1.2 Bernoulli processes

Bernoulli processes are the discrete-time analog of Poisson processes (time dependent and compound Bernoulli processes are defined in the natural way). Here the probability of an arrival in any time slot is p, independent of any other one. It follows that for slot k, the corresponding number of arrivals is binomial; i.e., $P\{N_k = n\} = \binom{k}{n}p^n(1-p)^{k-n}$. The time between arrivals is geometric with parameter p; i.e., $P\{A_n = j\} = p(1-p)^j$.

10.2.1.3 Phase-type renewal processes

An important special case of renewal models occurs when the interarrival times are of the so-called *phase type*. Phase-type interarrival times can be modeled as the time to absorption in a continuous-time Markov process $C = \{C(t)\}_{t=0}^{\infty}$ with state space $\{0, 1, \ldots, m\}$; here, state 0 is absorbing, all other states are transient, and absorption is guaranteed in a finite time. To determine A_n, start the process C with some initial distribution π. When absorption occurs (i.e., the process enters state 0), stop the process. The elapsed time is A_n, implying that it is a probabilistic mixture of sums of exponentials. Then, restart C with the same initial distribution π, and repeat the procedure independently to get A_{n+1}.

Phase-type renewal processes give rise to relatively tractable traffic models. They also enjoy the property that they are dense in the space of all distributions of non-negative random variables, that is, any interarrival distribution can be approximated arbitrarily close by phase-type ones; see, e.g., [3, 83].

10.2.2 Markov-based traffic models

Unlike renewal traffic models, Markov and Markov-renewal traffic models [13] introduce dependence into the random sequence $\{A_n\}$. Consequently, they can potentially capture traffic burstiness, due to nonzero autocorrelations in $\{A_n\}$.

Consider a Markov process $M = \{M(t)\}_{t=0}^{\infty}$ with a discrete state space. In this case, M behaves as follows: it stays in a state i for an exponentially distributed holding time with parameter λ_i which depends on i alone [13]; it then jumps to state j with probability p_{ij}, such that the matrix $P = [p_{ij}]$ is a probability matrix. In a simple Markov traffic model, each jump of the Markov process is interpreted as signaling an arrival, so interarrival times are exponential, their rate parameter depending on the state from which the jump occurred.

Markov models in slotted time can be defined for the process $\{A_n\}$ in terms of a Markov transition matrix $P = [p_{ij}]$ [13]. Here, state i corresponds to i idle slots separating successive arrivals, and p_{ij} is the probability of a j-slot separation, given that the previous one was an i-slot separation. Arrivals may be single, a batch of units or a continuous quantity. Batches may themselves be described by a Markov chain, whereas continuous-state, discrete-time Markov processes can model the (random) workload arriving synchronously at the system. In all cases, the Markovian structure introduces dependence into interarrival separation, batch sizes, and successive workloads, respectively.

Markov-renewal models are more general than discrete-state Markov processes, yet retain a measure of simplicity and analytical tractability. A Markov renewal process $R = \{(M_n, \tau_n)\}_{n=0}^{\infty}$ is defined by a Markov chain $\{M_n\}$ and its associated inter-jump times $\{\tau_n\}$, subject to the following constraint: the distribution of the pair (M_{n+1}, τ_{n+1}), of next state and inter-jump time, depends only on the current state M_n, but not on previous states nor on previous inter-jump times. Again, if we interpret jumps (transitions) of $\{M_n\}$ as signaling arrivals, we would have dependence in the arrival process. Also, unlike the Markov process case, the interarrival times can be arbitrarily distributed, and these distributions depend on both states straddling each interarrival interval [13].

MAP (Markovian arrival process) is a broad and versatile subclass of Markov renewal traffic processes, enjoying analytical tractability [67]. Here, the interarrival times are phase-type (see above) but with a wrinkle: traffic arrivals still occur at absorption instants of the auxiliary Markov process C, but the latter is not restarted with the same initial distribution; rather, the restart state depends on the previous transient state from which absorption had just occurred. While MAP is analytically simple, it enjoys considerable versatility; its formulation includes Poisson processes, phase-type renewal processes, and others as special cases. It also has the appealing properties that it is dense in the set of all point processes [4], and the superposition of independent MAP traffic streams results in a MAP traffic stream governed by a Markov process whose state space is the cross product of the component state spaces [67].

10.2.2.1 Markov-modulated processes

Markov-modulated models constitute an extremely important class of traffic models. The idea is to introduce an explicit notion of state into the description of a traffic stream — an auxiliary Markov process is evolving in time and its current state controls (modulates) the probability law of the traffic mechanism.

Let $M = \{M(t)\}_{t=0}^{\infty}$ be a continuous-time Markov process, with state space of $1, 2, \ldots m$ (more complicated state spaces are possible). Now assume that while M is in state k, the probability law of traffic arrivals is completely determined by k, and this holds for every $1 \leq k \leq m$. Note that when M undergoes a transition to, say, state j, then a new probability law for arrivals takes effect for the duration of state j, and so on. Thus, the probability law for arrivals is modulated by the state of M (such systems are also called *doubly stochastic*, but the term "Markov modulation" makes it clearer that the traffic is stochastically subordinated to M).

Certainly, the modulating process can be more complicated than a Markov process (so the holding times need not be restricted to exponential random variables), but such models are far less analytically tractable. For example, Markov renewal processes constitute a natural generalization of Markov-modulated processes with generally-distributed interarrival times, but those will not be reviewed here.

10.2.2.2 Markov-modulated Poisson processes

The most commonly used Markov-modulated model is the MMPP (Markov-modulated Poisson process) model, which combines the simplicity of the modulating

(Markov) process with that of the modulated (Poisson) process. In this case, the modulation mechanism simply stipulates that in state k of M, arrivals occur according to a Poisson process at rate λ_k. As the state changes, so does the rate.

MMPP models can be used in a number of ways. Consider first a single traffic source with variable rate. A simple traffic model would quantize the rate into a finite number of rates and each rate would give rise to a state in some Markov modulating process. Certainly, it remains to verify that exponential holding times of rates are an appropriate description, but the Markov transition matrix $Q = [Q_{kj}]$ of the putative M can be easily estimated from empirical data: Simply quantize the empirical data, and then estimate Q_{kj} by calculating the fraction of times that M switched from state k to state j.

As a simple example, consider a two-state MMPP model, where one state is an "on" state with an associated positive Poisson rate, and the other is an "off" state with associated rate zero (such models are also known as *interrupted Poisson* for obvious reasons). These models have been widely used to model voice traffic sources [42]; the "on" state corresponds to a talk spurt (when the speaker emits sound), and the "off" state corresponds to a silence (when the speaker pauses for a break). This basic MMPP model can be extended to aggregations of independent traffic sources, each of which is an MMPP, modulated by an individual Markov process M_i, as described above. Let $J(t) = (J_1(t), J_2(t), ..., J_r(t))$, where $J_i(t)$ is the number of active sources of traffic type i, and let $M(t) = (M_1(t), M_2(t), ..., M_r(t))$ be the corresponding vector-valued Markov process taking values on all r-dimensional vectors with nonnegative integer components. The arrival rate of class i traffic in state $(j_1, j_2, ..., j_r)$ of $M(t)$ is $j_i \lambda_i$.

10.2.2.3 Transition-modulated processes

Transition-modulated processes are a variation on the state modulation idea. Essentially, the modulating agent is a state transition rather than a state per se. However, note that a state transition can be described simply by a pair of states whose components are the one before transition and the one after it.

A transition-modulated traffic model in discrete time is described in [91]; its generalization to continuous time is straightforward. Let $M = \{M_n\}_{n=1}^{\infty}$ be a discrete-time Markov process on the positive integers. State transitions occur on slot boundaries, and are governed by an $m \times m$ Markov transition matrix $P = [P_{ij}]$. Let B_n denote the number of arrivals in slot n, and assume that the probabilities $P\{B_n = k \mid M_n = i, M_{n+1} = j\} = t_{ij}(k)$ are independent of any past state information (the parameters $t_{ij}(k)$ are assumed given). Notice that these probabilities are conditioned on transitions (M_n, M_{n+1}) of M from state M_n to state M_{n+1} during slot n. Furthermore, the number of traffic arrivals during slot n is completely determined by the transition of the modulating chain (through the parameters $t_{ij}(k)$).

Markov-modulated traffic models are a special case of Markovian transition-modulated ones: simply take the special case where the conditioning event is $\{M_n = i\}$. That is, $t_{ij}(k) = t_i(k)$ depends only on the state i of the modulating chain in slot n, but is independent of its state j in the next slot, $n+1$. Conversely, Markovian transition-modulated processes can be thought of as Markov-modulated ones, but on a larger state space. Indeed, if $\{M_n\}$ is Markov, so is the process $\{(M_n, M_{n+1})\}$ of its transitions.

As before, multiple transition-modulated traffic models can be defined, one for each traffic class of interest. The complete traffic model is obtained as the super-position of the individual traffic models. For queueing studies in discrete time, another wrinkle is the assignment of priorities to different classes, so as to order their arrivals in a buffer [91].

10.2.3 Fluid traffic models

The fluid traffic paradigm dispenses with individual traffic units. Instead, it views traffic as a stream of fluid, characterized by a flow rate (e.g., bits per second), so that a traffic count is replaced by a traffic volume.

Fluid models are appropriate to cases where individual units are numerous relative to a chosen time scale. Put differently, an individual unit is by itself of vanishingly little significance, just as one molecule more or less in a water pipeline has but an infinitesimal effect on the flow. In the B-ISDN (Broadband Integrated Services Digital Networks) context of ATM (Asynchronous Transfer Mode), all packets are fixed size *cells* of relatively short length (53 bytes); in addition, the high transmission speeds (say, on the order of gigabit/second) render the transmission impact of individual cells negligible. The analogy of a cell to a fluid molecule is a plausible one. To further highlight this analogy, contrast an ATM cell with a much bigger transmission unit, say, a compressed high-quality video frame, which consists on the order of a thousand cells. A traffic arrival stream of coded frames should be modeled as a discrete stream of arrivals, since such frames are typically transmitted at the rate of 30 frames per second. However, a fluid model is appropriate for the constituent cells.

An important advantage of fluid models is their conceptual simplicity. But important benefits will also accrue to a simulation model of fluid traffic. To see that, consider again a broadband ATM scenario. If one is to distinguish among cells, then each of them would have to count as an event. The time granularity of event processing would be quite fine, and consequently, processing cell arrivals would consume vast CPU and possibly memory resources, even on simulated time scales of minutes. A statistically meaningful simulation may often be infeasible. A fluid simulation would assume that the incoming fluid flow remains (roughly) constant over much longer time periods. Traffic fluctuations are modeled by events signaling a change of flow rate. As these changes can be assumed to happen far less frequently than individual cell arrivals, one can realize enormous savings in computing. In fact, infeasible simulations of cell arrival models can be replaced by feasible simulations of fluid models of comparable accuracy. In a queueing context, it is easy to manipulate fluid buffers. Furthermore, the waiting time concept simply becomes the time it takes to serve (clear) the current buffer. Since fluid models assume a deterministic service rate, these statistics can be readily computed. Typically, though, larger traffic units (say coded frames) are of greater interest than individual cells. Modeling the larger units as discrete traffic and their transport as fluid flow will give us the best of both worlds: we can measure waiting times and enjoy significant savings on simulation computing resources.

Typical fluid models [2, 55] assume that sources are bursty – of the "on-off" type. While in the "off" state, traffic is switched off, whereas in the "on" state traffic arrives deterministically at a constant rate λ. For analytical tractability, the durations of "on" and "off" periods are assumed to be exponentially distributed and mutually independent (that is, they form an alternating renewal process). A Markov model of a set of quantized (fluid) traffic rates is presented in [84].

Fluid traffic models of these types can be analyzed as Markov-modulated constant-rate traffic. The host of generalizations, described above for MMPP, carries over to fluid models as well, including multiple sources and multiple classes of sources.

10.2.4 Autoregressive-type traffic models

Autoregressive-type models define the next variate in the sequence as an explicit function of previous variates (from the same times series or a related one) within a time window stretching from the present into the past. Such models are particularly suitable for modeling VBR coded video – a projected major consumer of bandwidth in emerging high-speed communications networks. The nature of video frames is such that successive frames within a video scene vary visually very little (recall that there are 20-30 frames per second in a high-quality video). Only scene changes (and other visual discontinuities) can cause abrupt changes in frame bit rates (frame sizes). Thus, the sequence of bit rates comprising a video scene may be modeled by an autoregressive scheme, while scene changes can be modeled by some modulating mechanism, such as a Markov chain. However, see Subsections 10.2.5 and 10.2.6 for alternative modeling approaches.

10.2.4.1 Linear autoregressive (AR) processes

The class $AR(p)$ consists of *linear autoregressive models* of order p,

$$X_n = a_0 + \sum_{r=1}^{p} a_r X_{n-r} + \epsilon_n, \quad n > 0, \tag{10.2}$$

where $(X_{-p+1}, \ldots, X_0)$ is a prescribed random vector (usually a multivariate normal vector), the a_r, $0 \leq r \leq p$, are real constants, and the ϵ_n are zero-mean, uncorrelated random variables (white noise), called *residuals*, which are independent of the X_n [5]. In a good model, the residuals ought to be of smaller magnitude than the X_n, in order to "explain" the empirical data.

The recursive form of (2) makes it clear how to generate the next random element in the sequence $\{X_n\}_{n=0}^{\infty}$ from previous ones. This simplicity makes them popular candidates for modeling autocorrelated traffic. A simple AR(2) model was used in [43] to model VBR coded video. More elaborate models can be constructed out of $AR(p)$ models combined with other schemes. For example, in [80], the video bit rate traffic was modeled as a sum $R_n = X_n + Y_n + K_n C_n$, where the first two terms comprise independent AR(1) schemes, and the third term is a product of a simple Markov chain and an independent normal variate from an iid normal sequence. The purpose of having two autoregressive schemes is to achieve a better fit of the empirical autocorrelation function; the third term is designed to capture sample path spikes due to video scene changes.

10.2.4.2 Moving average (MA) processes

The class $MA(q)$ consists of *moving average* models of order q,

$$X_n = \sum_{r=0}^{q} b_r \epsilon_{n-r}, \quad n > 0, \tag{10.3}$$

where the b_r, $0 \leq r \leq q$, are real constants, and the ϵ_n are zero-mean uncorrelated random variables [5]. MA models are autocorrelated time series, since successive variates are defined in terms of common subsets of ϵ_n.

10.2.4.3 Autoregressive moving average (ARMA) processes

Modeling stationary and invertible processes using AR or MA processes often calls for the estimation of a large number of parameters, thereby reducing estimation efficiency. To mitigate this problem, one may try to combine (2) and (3) in a mixed model of the form

$$X_n = a_0 + \sum_{r=1}^{p} a_r X_{n-r} + \sum_{r=0}^{q} b_r \epsilon_{n-r}. \tag{10.4}$$

The class of models defined by (4) is denoted by $\text{ARMA}(p,q)$ and referred to as *autoregressive moving average* models of order (p,q). Clearly, ARMA modeling is more flexible than AR modeling or MA modeling alone.

10.2.4.4 Autoregressive integrated moving average (ARIMA) processes

Related to $\text{ARMA}(p,q)$, is the class $\text{ARIMA}(p,d,q)$ of *autoregressive integrated moving average processes*, obtained by replacing X_n in (4) by the dth differences of the process $\{X_n\}$. ARIMA modeling is more general than ARMA modeling; its scope includes certain types of nonstationary series. The term "integrated" alludes to the fact that an ARMA model is fitted to the differenced data, and that ARMA model is then "integrated" (summed) to yield the target (nonstationary) ARIMA model. For example, an $\text{ARMA}(p,q)$ process may be viewed as an $\text{ARIMA}(p,0,q)$ process and the random walk model can be viewed as an $\text{ARIMA}(0,1,0)$ process.

From a modeling vantage point, ARIMA processes constitute a broad and flexible class of stochastic models whose parameter estimation, forecasting, model identification, and model selection are well-understood (e.g., see [9, 94]). Two points should be kept in mind, though, when considering ARMA and ARIMA processes as models of modern traffic. First, both ARMA and ARIMA processes have autocorrelation functions which decay geometrically in the lag; namely, $\rho(n) \sim r^n$ for some $0 < r < 1$, as $n \to \infty$. In this sense, ARMA and ARIMA processes are inherently short-range dependent models, incapable of parsimoniously capturing the persistence phenomena observed empirically in many modern high-speed networks (see Subsection 10.2.6). Secondly, it has also been observed that the corresponding empirical marginal distributions are typically not Gaussian, but rather, they tend to be skewed to the right [43, 77]. Since the theoretical relationship between skewness/kurtosis of an ARIMA process and the corresponding parameters of its generating white noise process is not yet well-understood, the effect of the marginal distribution mismatch is not clear.

10.2.5 TES traffic models

The need to capture traffic burstiness has motivated the TES (*transform-expand-sample*) modeling approach which strives to simultaneously model both the marginal distribution and autocorrelation function of an empirical record [50, 51, 76], traffic being a special case [77]. The empirical TES methodology assumes that some stationary empirical time series (e.g., traffic measurements over time) are available. It aims to construct a model satisfying the following three fidelity requirements, simultaneously:

1. The model's marginal distribution should match its empirical counterpart (a histogram, in practice).

2.   The model's leading autocorrelations should approximate their empirical counterparts up to a reasonable lag.

3.   The sample paths (histories) generated by simulating the model should "resemble" the empirical time series.

The first two are precise quantitative requirements, whereas the third requirement is a heuristic qualitative one. Nevertheless, the latter is worth adopting, since sample path "resemblance" is informally used by modelers to increase the confidence in the model.

10.2.5.1   TES processes

TES processes constitute a versatile family of stochastic sequences, consisting of two broad classes, called TES^+ and TES^-. The superscript (plus or minus) is a mnemonic reminder of the fact that the TES family gives rise to processes with positive and negative lag-1 autocorrelations, respectively. TES models consist of two stochastic processes in lockstep, called *background* and *foreground* sequences. Background TES sequences have the form

$$U_n^+ = \begin{cases} U_0, & n = 0 \\ \langle U_{n-1}^+ + V_n \rangle, & n > 0 \end{cases} \qquad U_n^- = \begin{cases} U_n^+, & n \text{ even} \\ 1 - U_n^+, & n \text{ odd}. \end{cases} \tag{10.5}$$

Here, U_0 is distributed uniformly on $[0, 1)$; $\{V_n\}_{n=1}^\infty$ is a sequence of iid random variables, independent of U_0, called the *innovation sequence*; and angular brackets denote the *modulo-1 (fractional part) operator* $\langle x \rangle = x - \max\{\text{integer } n: n \leq x\}$. Background sequences play an auxiliary role. The real targets are foreground sequences of the form

$$X_n^+ = D(U_n^+), \quad X_n^- = D(U_n^-), \tag{10.6}$$

where D is a transformation from $[0, 1)$ to the reals, called a *distortion*.

It can be shown that all background sequences are Markovian and stationary; however, it is worth noting that while the transition structure of $\{U_n^+\}$ is stationary, that of $\{U_n^-\}$ is not (it depends on the even or odd time index). More importantly, the marginal distribution of both $\{U_n^+\}$ and $\{U_n^-\}$ is uniform on $[0, 1)$, *regardless* of the probability law of the innovations $\{V_n\}$ [50]. The *inversion method* [21] allows us to transform any background uniform variates to foreground ones with an arbitrary marginal distribution. To illustrate this idea, suppose one has an empirical time series $\{Y_n\}_{n=0}^N$, from which one computes an empirical density $\hat{h}_Y$ and its associated distribution function $\hat{H}_Y$. Then, the random variable $X = \hat{H}_Y^{-1}(U)$ has density $\hat{h}_Y$. Thus, TES foreground sequences can match any empirical distribution.

10.2.5.2   The empirical TES modeling methodology

The empirical TES methodology actually employs a composite two-stage distortion of the form

$$D_{Y,\xi}(x) = \hat{H}_Y^{-1}(S_\xi(x)), \quad x \in [0, 1), \tag{10.7}$$

where $\widehat{H}_Y^{-1}$ is the inverse of the empirical histogram distribution, and S_ξ is a "smoothing" operation, called a *stitching transformation*, parameterized by $0 \le \xi \le 1$, and given by

$$S_\xi(y) = \begin{cases} y/\xi, & 0 \le y < \xi \\ (1-y)/(1-\xi), & \xi \le y < 1. \end{cases} \tag{10.8}$$

For $0 < \xi < 1$, the effect of S_ξ is to render the sample paths of background TES sequences more "continuous-looking." Because stitching transformations preserve uniformity, the inversion method via $\widehat{H}_Y^{-1}$ guarantees that the corresponding foreground sequence would have the prescribed marginal distribution $\widehat{H}_Y$. The empirical TES modeling methodology takes advantage of this fact, which effectively decouples the fitting requirements of the empirical distribution and the empirical autocorrelation function. Since the former is automatically guaranteed by TES theory, one can concentrate on fitting the latter. In practice, fitting is carried out by a heuristic search for pairs, (ξ, f_V), where ξ is a stitching parameter and f_V is an innovation density; the search is declared a success on finding that the corresponding TES sequence gives rise to an autocorrelation function that adequately approximates its empirical counterpart, and whose simulated sample paths bear "adequate resemblance" to their empirical counterparts.

In practice, efficient searches of this kind must rely on software support. TEStool is a visual interactive software environment designed to support TES modeling [34]. TEStool allows the user to read in empirical sample paths and calculate their empirical statistics (histogram, autocorrelation function, and spectral density) in textual and graphical forms. It further provides services to generate and modify TES models and to superimpose the corresponding TES statistics on their empirical counterparts. The search proceeds in an interactive style, guided by visual feedback: each model modification triggers a recalculation and redisplay of the results. TES model autocorrelations and spectral densities are calculated numerically from fast and accurate formulas developed in [50, 51]. This activity is further simplified by restricting the innovation densities f_V to be step functions. Simple densities like that can be readily modified graphically, since steps are visually represented by rectangles, and those can be created, deleted, stretched, and moved easily with the mouse.

Recently, an algorithmic modeling approach has been devised and implemented for TES modeling. The algorithm first carries out a brute-force computation over a subspace of step-function innovation densities and various stitching parameters; recall that the distortion is completely determined by the empirical record and user-supplied histogram parameters. Of those, the algorithm selects the best n combinations of pairs, (f_V, ξ), in the sense that the resulting TES model autocorrelation functions minimize the mean square error with respect to the empirical autocorrelation function. The analyst then selects among those n candidate models the one whose Monte Carlo sample paths bear the "most resemblance" to the empirical record. Experience shows that the TES modeling algorithm produces better and faster results than its heuristic counterpart.

Figure 1 depicts the application of the aforementioned TES modeling algorithm. It displays the final TEStool screen at the end of a TEStool modeling session performed on an empirical record of a (random) sequence of encoded (compressed) video frames [58]. The specific data modeled in Figure 1 was a random sequence of frame bit rates, generated by a VBR video sequence of a football scene,

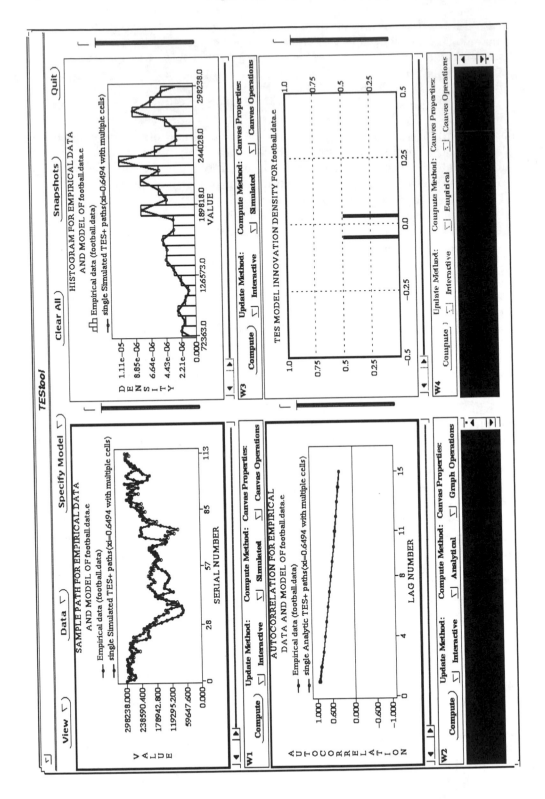

Figure 1. TES model of DCT-compressed VBR video.

whose frames were compressed by a variant of the DCT (discrete cosine transform) [82]; see also [77] for a review of compressed video modeling using the TES methodology. The screen is subdivided into four canvas areas designed to display various types of graphics; the graphical user interface is controlled by buttons and menu selections. Each statistical canvas superimposes an empirical statistic with its TES-model counterpart; a legend at the top of these graphs identifies the constituent curves. The upper-left canvas contains the empirical and model sample paths, the latter being generated by a Monte Carlo simulation; the upper-right canvas contains the corresponding histograms; the lower-left canvas contains the empirical autocorrelation function and its numerically-computed model counterpart; and the lower-right canvas contains a joint specification of a TES sign, stitching parameter, and an innovation density.

Notice that the empirical time series in Figure 1 (lower-left canvas) exhibits very high short-range autocorrelations, attesting to the bursty nature of this type of encoded video. A simple renewal model would not be appropriate, since it cannot capture burstiness due to temporal dependence. In contrast, the TES model displayed in Figure 1 exhibits general good statistical agreement with the empirical data, in accord with the three fidelity requirements stipulated in Subsection 10.2.5; in particular, the histogram fit is very close in the upper-right canvas, as is the autocorrelation fit in the lower-left canvas (the two autocorrelation curves are visually indistinguishable). Such source models can be used to generate synthetic streams of realistic traffic to drive simulations of communications networks.

10.2.6 Self-similar traffic models

Recent traffic studies from working packet networks [7, 24, 33, 59, 97, 75] have revealed new features of packet traffic that have gone unnoticed in the traffic modeling literature, yet seem to have serious implications for designing, engineering, and controlling future high-speed networks. The measured data demonstrate convincingly that packet traffic can be statistically *self-similar* or, more colloquially, *fractal* in nature. Self-similarity means, roughly, that the traffic looks statistically the same over a wide range of time scales, and is related to *long-range dependence* or $1/f$-*noise*. The term "self-similar" was originally coined by Mandelbrot [74], who illustrated that self-similarity is observed in a wide range of physical and mathematical systems and that it is inherent in the study of many irregular or bursty phenomena. This subsection introduces second-order self-similar processes, discusses some of their characteristic features, and exemplifies several stochastic models that can capture self-similarity in a parsimonious manner.

10.2.6.1 Motivation and pictorial evidence

In order to motivate the use of self-similar processes for traffic modeling purposes, we recall the analysis in [59] of high-resolution Ethernet LAN traffic measurements, and the visually-convincing evidence of its self-similar nature (see Figure 2, *ibid.*) Based on 27 consecutive hours of monitored Ethernet traffic, Figure 2 (a)-(e) depicts a plot sequence of time series of packet counts (i.e., number of packets per time unit) for 5 different choices of time units. Starting with a time unit of 100 seconds in Plot (a), each subsequent plot is obtained from the previous one by increasing the time resolution (decreasing the time unit) by a factor of 10 and then focusing on a randomly chosen subinterval (indicated by a darker shade in each plot). The time unit corresponding to the finest time scale is 10 milliseconds in Plot (e); this plot is "jittered" in order to avoid the visually irritating quantization effect associated with such high resolution, that is, a small amount of noise has been added to the actual observed arrival rates. Observe that all plots are visually rather "similar" to each other, so that arrival rates measured over larger time scale (hours, minutes) are quite indistinguishable to the human eye from those measured over smaller time scales (seconds, milliseconds). In particular, no natural length of a "burst" is discernible: at every time scale ranging from milliseconds to minutes and hours, bursts have the same qualitative appearance. This scale-invariant or "self-similar" feature of Ethernet traffic is drastically different from both conventional telephone traffic and from traditional stochastic models of packet traffic. Such models typically give rise to plots of packet counts that are indistinguishable from white noise after aggregating the original time series over a few hundred milliseconds, as illustrated by the plot sequence (a')-(e') in Figure 2; this sequence was obtained by successive aggregations, as in the empirical plot sequence (a)-(e), except that it arose from synthetic traffic generated from a comparable compound Poisson process with the same average packet size and arrival rate as the empirical data. More complicated Markovian arrival processes were observed to give rise to plot sequences, indistinguishable from (a')-(e'). Thus, Figure 2 provides a surprisingly simple method for sharply distinguishing between empirical traffic and standard model-generated traffic, thereby motivating the use of self-similar stochastic processes for traffic modeling purposes.

10.2.6.2 Second-order self-similar processes

Let $X = \{X_t\}_{t=0}^{\infty}$ be a *covariance stationary* (*wide-sense stationary*) stochastic process with mean μ_X, variance σ_X^2, and autocorrelation function $\rho_X(k)$. In particular, X is assumed to have an autocorrelation function of the form

$$\rho_X(k) \approx k^{-\beta} L_1(k), \quad \text{as } k \to \infty, \tag{10.9}$$

where $0 < \beta < 1$ and L_1 is slowly varying at infinity, that is, $\lim_{t \to \infty} L_1(tx)/ L_1(t) = 1$, for all $x > 0$; examples of such slowly varying functions are $L_1(t) = \text{const}$ and $L_1 = \log(t)$. A stochastic process satisfying relation (10.9) is said to exhibit *long-range dependence* [6, 16, 93]. In Mandelbrot's terminology [74], long-range dependence is also referred to as the *Joseph Effect*, in reference to the Old Testament figure who had interpreted Pharaoh's dream of the "seven lean cows and the seven fat cows" to mean the "seven fat years and seven lean years" that ancient Egypt was to experience. Intuitively, this notion captures the persistence (or autocorrelation) phenomena observed in many naturally-occurring empirical time series; these manifest themselves in clusters (runs) of consecutive large (or consecutive small) values. More formally, processes with long-range dependence are characterized by an autocorrelation function that decays hyperbolically in the lag. Moreover, it is easy to see that (10.9) implies $\sum_{k=1}^{\infty} \rho_X(k) = \infty$. This non-summability of the autocorrelations captures the intuition behind long-range dependence, namely, that even though the high-lag autocorrelations are individually small, their cumulative effect is of importance, giving rise to behavior that is markedly different from that of processes with *short-range dependence*; the latter are characterized by geometric decay of the autocorrelation function, that is,

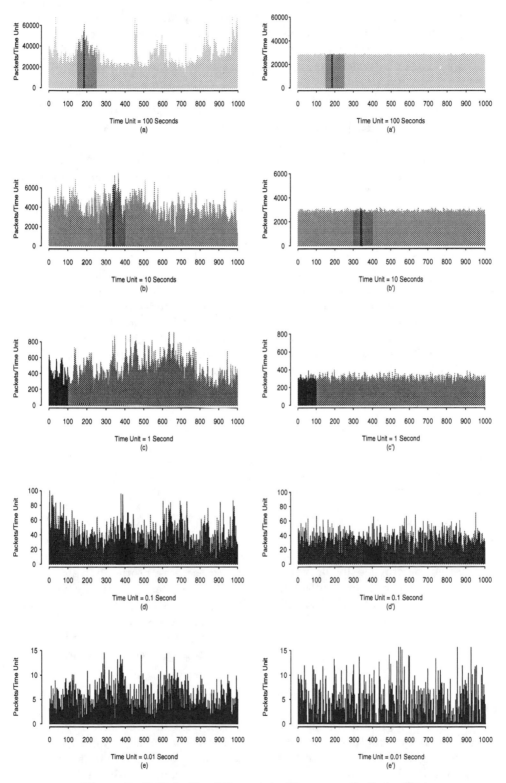

Figure 2. Self-similar Ethernet traffic vs. synthetic traffic.

$\rho_X(k) \approx r^k$, as $k \to \infty$, for some $0 < r < 1$, resulting in a summable autocorrelation function $(0 < \sum_{k=1}^{\infty} \rho_X(k) < \infty)$. In the frequency domain, long-range dependence manifests itself in a spectral density that obeys a power-law near the origin. In fact, under mild regularity conditions on the slowly varying function L_1 in (9), there is long-range dependence in X, if the associated spectral density function, $s_X(\lambda) = \sum_{k=0}^{\infty} \rho_X(k) e^{ik\lambda}$, satisfies

$$s_X(\lambda) \approx \lambda^{-\gamma} L_2(\lambda), \quad \text{as } \lambda \to 0, \tag{10.10}$$

where $0 < \gamma < 1$ and L_2 is slowly varying at 0. Since $s_X(0) = \sum_{k=0}^{\infty} \rho_X(k)$, it follows that long-range dependence is characterized by $s_X(0) = \infty$ (the so-called $1/f$-noise). In contrast, short-range dependence is characterized by $0 \le s_X(0) < \infty$.

For each integer $m \ge 1$, let $X^{(m)} = \{X_k^{(m)}\}_{k=1}^{\infty}$ denote the new covariance stationary time series

$$X_k^{(m)} = \frac{1}{m} \sum_{i=1}^{m} X_{km-m+i} \tag{10.11}$$

(with variance $(\sigma_X^{(m)})^2$ and autocorrelation function $\rho_X^{(m)}$), obtained by averaging the original series, X, over non-overlapping blocks of size m. Self-similarity concepts relate statistical properties of X to their counterparts in $X^{(m)}$, using power laws or invariance, either for all $m \ge 1$, or asymptotically as $m \to \infty$. The process X is called (*exactly*) *self-similar* with self-similarity parameter $H = 1 - \beta/2$, if the stochastic process $\{m^{1-H} X_k^{(m)}\}_{k=1}^{\infty}$ has the same finite-dimensional distributions as X, for all integer $m \ge 1$. X is called (*exactly*) *second-order self-similar* with self-similarity parameter $H = 1 - \beta/2$ if the stochastic process $\{m^{1-H} X_k^{(m)}\}_{k=1}^{\infty}$ has the same variance and autocorrelation function as X, for all integer $m \ge 1$. In terms of the aggregated processes $X^{(m)}$, this implies

$$(\sigma_X^{(m)})^2 = \sigma_X^2 m^{-\beta}, \tag{10.12}$$

$$\rho_X^{(m)}(k) = \rho_X(k) = \tfrac{1}{2} \delta^2(|k|^{2-\beta}), \tag{10.13}$$

where $\delta^2(f(k)) = f(k+1) - 2f(k) + f(k-1)$ is the second central difference operator of a sequence $\{f(k)\}$. Finally, X is called *asymptotically* (*second-order*) *self-similar* with self-similarity parameter $H = 1 - \beta/2$, if for all $k \ge 0$,

$$\rho_X^{(m)}(k) \to \tfrac{1}{2} \delta^2(k^{2-\beta}), \quad \text{as } m \to \infty. \tag{10.14}$$

Thus, an asymptotically self-similar process has the property that for large m, the corresponding aggregated time series, $X^{(m)}$, has a fixed autocorrelation structure that is solely determined by β; moreover, due to the asymptotic equivalence (for large k) of differencing and differentiating, $\rho_X^{(m)}$ agrees asymptotically with the autocorrelation structure of X, given by (10.9). Intuitively, the most striking feature of (exactly or asymptotically) self-similar processes is that their aggregated processes possess a nondegenerate autocorrelation structure, as $m \to \infty$. This behavior is in stark contrast to conventional stochastic models, whose aggregated processes tend to second-order pure noise, i.e.,

$$\rho_X^{(m)}(k) \to 0, \quad \text{as } m \to \infty \tag{10.15}$$

for all $k > 0$.

The importance of self-similar processes derives from the fact that they provide elegant models that capture an important empirical law, commonly referred to as *Hurst's law* or the *Hurst effect*, to be reviewed next. Consider a finite sequence of observations $\{X_k\}_{k=1}^{n}$ with sample mean $\bar{X}(n)$ and sample variance $S^2(n)$, and define $W_k = \sum_{i=1}^{k} X_i - k\bar{X}(n)$. The *rescaled adjusted range statistic* (*R/S statis-*

tic for short) is given by

$$\frac{R(n)}{S(n)} = \frac{\max{(0, W_1, W_2, \ldots, W_n)} - \min{(0, W_1, W_2, \ldots, W_n)}}{S(n)}. \qquad (10.16)$$

Hurst [45] found that many naturally occurring time series appear to be well-modeled by the relation

$$E\left[\frac{R(n)}{S(n)}\right] \approx cn^H, \quad \text{as } n \to \infty, \qquad (10.17)$$

where c is a finite positive constant independent on n, and H has typical values around 0.7 [45]. The parameter H is called the *Hurst parameter*. In contrast, if $\{X_k\}$ is a process with short-range dependence, then

$$E\left[\frac{R(n)}{S(n)}\right] \approx dn^{0.5}, \quad \text{as } n \to \infty, \qquad (10.18)$$

where d is a finite positive constant, independent of n [70]. The discrepancy between (10.17) and (10.18) is generally referred to as the *Hurst effect*.

From a statistical point of view, the most salient feature of self-similar processes is that the variance of the running arithmetic mean decays in the sample size, n, more slowly than the reciprocal of the sample size. More specifically, the variance of the running arithmetic mean of self-similar processes decays like $n^{-\beta}$, for some $0 < \beta < 1$, whereas for processes whose aggregated series converge to second-order pure noise, the same variance decays like n^{-1}. Here, we shall assume, for simplicity, that the slowly varying functions, L_1 in (10.9) and L_2 in (10.10), are asymptotically constant. In fact, it is shown in [16] that a specification of the autocorrelation function satisfying (10.9) (or equivalently, of the spectral density function satisfying (10.10)) coincides with a specification of the sequence $\{(\sigma_X^{(m)})^2\}_{m=1}^{\infty}$, with the property

$$(\sigma_X^{(m)})^2 \approx am^{-\beta}, \quad \text{as } m \to \infty,$$

where a is a finite positive constant independent of m, and $0 < \beta < 1$ is the same as in (10.9). In fact, β is related to the parameter γ in (10.10) by $\beta = 1 - \gamma$. In contrast, for covariance stationary processes whose aggregated series $X^{(m)}$ tend to second-order pure noise (i.e., for which (10.15) holds), the sequence $\{(\sigma_X^{(m)})^2\}_{m=1}^{\infty}$ satisfies

$$(\sigma_X^{(m)})^2 \approx bm^{-1}, \quad \text{as } m \to \infty,$$

where b is a finite positive constant independent of m. The consequences of the slowly-decaying variances, $(\sigma_X^{(m)})^2$, can be disastrous for classical statistical tests and confidence or prediction intervals (see, e.g., [6]), since the usual standard errors (derived for conventional models) are wrong by a factor that tends to infinity in the sample size.

The statistical properties of slowly decaying variances, long-range dependence, and a power-law spectral density are thus seen to be different manifestations of the property that the underlying covariance stationary process, X, is asymptotically or exactly second-order self-similar. Consequently, the problem of testing for self-similarity and estimating it quantitatively can be approached from a number of different angles, utilizing both time domain and frequency domain approaches; these include the time-domain R/S analysis and variance analysis of the aggregated processes, and the frequency-domain periodogram analysis. For details on statistical inference for self-similar processes, see [6, 97] and references therein.

We conclude this subsection by pointing out that for continuous-time processes $X = \{X_t\}_{t \geq 0}$ with zero mean and stationary increments, an alternative definition of self-similarity requires that for all $a > 0$,

$$X_{at} = a^H X_t, \quad t \geq 0, \tag{19}$$

where equality is understood in the sense of equality of the finite-dimensional distributions, and the exponent H is the self-similarity parameter. Nevertheless, we do not use the self-similarity variant (10.19), because the previous definitions, (10.13) and (10.14), have the advantage that they do not obscure the connection with standard time series theory, and they reflect the fact that one is mainly interested in large time scales, m. From a modeling perspective, the crucial point is that both the discrete-time and the continuous-time definitions involve a wide range of time scales. One advantage of (10.19) in the presence of large data sets is that it permits for a quick heuristic estimation of the self-similarity parameter H from simple plots like those in Figure 2. In fact, a naive inference from the successive plots (a)-(e) in that figure (subtracting the sample mean of X and applying simple statistics, such as range and histogram) yields H-values of about 0.8 for the relation (10.19). In contrast, Plots (a')-(e') in that figure reveal pure white noise behavior (i.e., $H = 0.5$) for the synthetic traffic model, which results in an identical but degenerate autocorrelation structure of the $X^{(m)}$ for $m > 100$.

10.2.6.3 Stochastic modeling of self-similar phenomena

Stochastic processes with an autocorrelation structure of the form (10.9) are often viewed as finite approximations of a continuous sum of Gauss-Markov processes; these are often interpreted as suggesting the presence of a multilevel hierarchy of underlying mechanisms which give rise to self-similarity [73]. Some physical examples may be found in references cited by [16]. In general, however, it is very difficult to demonstrate the physical existence of such multilevel hierarchies, or to relate extant mechanisms to self-similar behavior. Consequently, formal mathematical models have been devised to capture self-similarity phenomena, but by and large, these are not amenable to physical interpretation. Two such models, the exactly (second-order) self-similar fractional Gaussian noise process and the asymptotically (second-order) fractional ARIMA process, will be presented next. We also discuss a construction of self-similar models, originally due to Mandelbrot [71] and later extended in [92, 60], which appears to be promising in terms of providing a physical "explanation" for the self-similarity property in high-speed packet traffic (see [97]).

10.2.6.4 Fractional Gaussian noise

A *fractional Gaussian noise* [70] is a stationary Gaussian process, $X = \{X_k\}_{k=1}^{\infty}$, with mean μ_X, variance σ_X^2, and autocorrelation function of the form

$$\rho_X(k) = \frac{1}{2}(\, |k+1|^{2H} - 2\,|k|^{2H} + |k-1|^{2H}), \quad k \geq 1. \tag{10.20}$$

It can be shown that the asymptotic form of (10.20) is $\rho_X(k) \approx H(2H-1)\,|k|^{2H-2}$, as $k \to \infty$ for $0 < H < 1$, and that the attendant aggregated processes, $X^{(m)}$, exhibit long-range dependence in the sense of (10.9) and satisfy (10.13). Thus, fractional Gaussian noise is exactly second-order self-similar with self-similarity parameter H, provided $1/2 < H < 1$. In the case $\mu_X = 0$, a fractional Gaussian noise serves as the increment process of *fractional Brownian motion*, i.e., a continuous-time zero-mean Gaussian process, $B_H = \{B_H(t)\}_{t=0}^\infty$, with $0 < H < 1$ and autocorrelation function $\rho_X(s,t) = 1/2(|s|^{2H} + |t|^{2H} - |t-s|^{2H})$.

Fractional Gaussian noise and fractional Brownian motion have been particularly popular in hydrological modeling (see, e.g., [72]). Despite its rigid autocorrelation structure, fractional Gaussian noise is often a reasonable first approximation of more complex structures due to the fact that certain long-range dependent processes yield fractional Gaussian noise as limits under a special type of central limit theorem. Methods for estimating the three unknown parameters, μ_X, σ_X^2, and H have been developed.

10.2.6.5 Fractional ARIMA(p, d, q) processes

A *fractional ARIMA*(p, d, q) process [37, 44], where p and q are non-negative integers and d is real, is a stochastic sequence, $X = \{X_k\}_{k=1}^\infty$, of the form

$$\Phi[B]\Delta^d(X_k) = \Theta[B]\epsilon_k, \tag{10.21}$$

where $\Phi[B] = 1 - \sum_{i=1}^p \phi_i B^i$, and $\Theta[B] = 1 - \sum_{i=1}^q \theta_i B^i$ are polynomials in the *backward-shift operator* $B(X_k) = X_{k-1}$, $\Delta = 1 - B$ denotes the *differencing operator*, and Δ^d is the *fractional differencing operator* defined by $\Delta^d = (1-B)^d = \sum_{k=0}^\infty \binom{d}{k}(-B)^k$, with $\binom{d}{k}(-1)^k = \frac{\Gamma(-d+k)}{\Gamma(-d)\Gamma(k+1)}$, and the sequence $\{\epsilon_k\}_{k=0}^\infty$ is a white noise process. It is known [37] that for $d \in (-1/2, 1/2)$, X is stationary and invertible, and its autocorrelation function satisfies $\rho_X(k) \approx ak^{2d-1}$ as $k \to \infty$, where a is a finite positive constant independent of k. Moreover, it was shown in [16] that the attendant aggregated time series, $X^{(m)}$, satisfy (10.14) for $0 < d < 1/2$. Thus, (10.9) and (10.14) hold and X is asymptotically second-order self-similar with self-similarity parameter $d + 1/2$, provided $0 < d < 1/2$.

The high-lag autocorrelations of fractional ARIMA(p, d, q) processes are similar to those of the corresponding fractional ARIMA($0, d, 0$) processes. The latter constitute the simplest and most fundamental of the fractionally differenced ARIMA processes and have an infinite-order autoregressive representation, $\Delta^k(X_k) = \epsilon_k$.

The corresponding infinite-order moving average representation, $X_k = \Delta^{-d}(\epsilon_k)$, reveals that ARIMA($0, d, 0$) processes are obtained by subjecting white noise to fractional differencing of order $-d$. Upon setting $H = d + 1/2$, both the fractional Gaussian noise and the ARIMA($0, d, 0$) process have autocorrelations that decay asymptotically as k^{2d-1} (with different constants of proportionality).

One of the main advantages of the ARIMA($0, d, 0$) family over fractional Gaussian noise processes is that the former can be combined with the established class of Box-Jenkins models [9] in a natural way, to yield the family of ARIMA(p, d, q) processes. Fractional ARIMA processes enjoy greater flexible in simultaneous modeling of short-range and long-range behavior than fractional Gaussian noise. The main reason is that a fractional Gaussian noise process is specified by just three parameters, μ, σ^2, and H, which are insufficient to capture a wide range of low-lag autocorrelation structures encountered in practice. ARMA($1, d, 0$) and ARMA($0, d, 1$) models possess much higher flexibility [44].

10.2.6.6 Self-similarity through aggregation

Let $\{I_k\}_{k>0}$ be a sequence of iid integer-valued random variables ("inter-renewal times") with asymptotic tail probabilities obeying the power law

$$P\{I_k \geq t\} \approx t^{-\alpha}h(t), \quad \text{as } t \to \infty, \tag{10.22}$$

where $1 < \alpha < 2$ and h is varying slowly at infinity. For example, the stable (Pareto) distribution with parameter $1 < \alpha < 2$ satisfies the "heavy-tail" condition (10.22). Mandelbrot [74] refers to (10.22) as the *Noah Effect* or the *infinite variance syndrome*, in reference to the Biblical story of Noah and the Deluge ("Big Flood"). Intuitively, this notion captures the phenomenon that empirical time series can fluctuate far away from their mean value, with non-negligible probability. In addition to $\{I_k\}_{k \geq 0}$, let $\{G_k\}_{k \geq 0}$ be an iid sequence ("rewards"), independent of $\{I_k\}$, with $E[G_k] = 0$ and $E[G_k^2] < \infty$. Consider the stationary (delayed) renewal sequence $\{S_k\}_{k \geq 0}$ defined by $S_k = S_0 + \sum_{j=1}^{k} I_j$, $k \geq 1$, with an appropriately chosen S_0. A discrete-time (renewal) reward process, $W = \{W_k\}_{k \geq 1}$, can then be defined by

$$W_k = \sum_{n=1}^{k} G_n 1_{(S_{n-1}, S_n]}(k).$$

Notice that W is stationary in the sense that its finite-dimensional distributions are invariant under time shifts. By aggregating M iid copies, $W^{(1)}, W^{(2)}, \ldots, W^{(M)}$ of W, one obtains the process $W^* = \{W_k^*(M)\}_{k \geq 0}$, given by

$$W_k^*(M) = \begin{cases} 0, & k = 0 \\ \displaystyle\sum_{n=1}^{k} \sum_{m=1}^{M} W_n^{(m)}, & k > 0. \end{cases}$$

It can be shown [71, 92] that for k and M large and $k \ll M$, the process W^* behaves like a fractional Brownian motion. More precisely, the process W^*, properly normalized, converges to the integrated version of fractional Gaussian noise, the notion of convergence being that of finite-dimensional distributions. Thus, the increment process of W^* behaves asymptotically like fractional Gaussian noise.

The ability to produce self-similarity by aggregating an increasing number of iid copies of the rather elementary renewal reward process, W, over increasing time periods relies crucially on the heavy tail behavior, (10.22), of the inter-renewal times I_k. Consequently, the process W can assume the same value, with high probability, over long periods of time. Aggregating a large number of iid copies of W results in an overall sum that approaches a Gaussian distribution; and over long time periods, this procedure introduces significant temporal dependence. These attributes are shared by fractional Brownian motion. A similar construction of asymptotic self-similarity, employing certain AR(1) processes in lieu of the renewal reward process W, may be found in [38].

10.2.6.7 Parsimonious modeling

Since empirical data sets are necessarily finite, it is not possible to determine with certainty whether or not the asymptotic relations (10.9), (10.13), (10.14), etc. hold for data records. However, if sufficient data (empirical or simulated) are available and the underlying process is not self-similar (in the sense that the increasingly aggregated series converge to second-order pure noise as per (10.15)), then the following can be expected to be observed: (*i*) the autocorrelations will eventually decrease exponentially, (*ii*) the spectral density function at the origin will eventually prove to be "continuous," (*iii*) the variances of the aggregated processes will eventually decrease as m^{-1}, and (*iv*) the rescaled adjusted range will eventually increase as $n^{0.5}$, as per (10.18).

In practice, statistical inferences of self-similarity from finite sample sizes are generally problematic. For large data sets, it is often possible to investigate "near" asymptotic behavior of statistics, such as the rescaled adjusted range or the variance of the aggregated processes. Moreover, parsimonious modeling then becomes a necessity, due to the large number of parameters needed to fit a conventional model to a data record which is, in fact, self-similar. For example, modeling long-range dependence via ARMA processes is equivalent to approximating a hyperbolically decaying autocorrelation function by a sum of exponentials. Although this is always possible mathematically, the number of requisite parameters will tend to infinity in the sample size; furthermore, physically meaningful interpretations of the ensuing parameters become increasingly difficult. In contrast, long-range dependence can be parsimoniously modeled via a self-similar process by a single parameter — the Hurst parameter, H. Moreover, self-similar processes capture both short-range and long-range dependence phenomena.

10.2.7 Stochastic intensitities of point processes

Stochastic intensity processes provide an alternative characterization of point processes, which utilizes modern martingale theory [20, 46]. The stochastic intensity formulation is primarily of theoretical, not practical, interest. Since most traffic models are formulated as point processes, a brief overview of stochastic intensities is included here for completeness.

Recall the point process, $N = \{N(t)\}_{t=0}^{\infty}$, as defined in Section 12.2. Intuitively, a stochastic intensity process describes the (random) rate at which points (arrivals) occur in N, given the past history of N and possibly additional information. More precisely, the aforementioned past history is represented by a filtration, $\mathfrak{F} = \{\mathfrak{F}_t\}_{t=0}^{\infty}$, to which N is adapted, i.e., an increasing family of σ-algebras containing the one generated by N, so that the arrival instants, $\{T_n\}_{n=1}^{\infty}$ of N, are $\mathfrak{F}_t$-stopping times.

The setting of the general definition of stochastic intensities is a point process, N, adapted to a filtration, $\mathfrak{F}$. An $\mathfrak{F}$-*intensity* of N is any non-negative, integrable, $\mathfrak{F}$-progressive process, $\lambda = \{\lambda(t)\}_{t=0}^{\infty}$, such that

$$E\left[\int_0^{\infty} C(s)dN(s)\right] = E\left[\int_0^{\infty} C(s)\lambda(s)ds(s)\right], \qquad (10.23)$$

for all $\mathfrak{F}$-predictable processes, $\{C(t)\}_{t=0}^{\infty}$ (see [10] for details as well as the definitions of progressive measurability and predictability of stochastic processes). The stochastic intensity definition in (10.23) is not constructive. Nevertheless, it motivates the term "stochastic intensity," via the "change of variable" from $dN(s)$ to $\lambda(s)ds$. A more transparent motivation is gleaned when a stochastic intensity, λ, is right-continuous and bounded. In this case, a constructive definition of λ is given by the almost sure limit,

$$\lambda(t) = \lim_{h \to 0+} \frac{E\left[N(t+h) - N(t) \mid \mathcal{F}_t\right]}{h}, \quad t \geq 0, \tag{10.24}$$

where the limit is a time derivative of conditional expected point counts [10]. Note that each conditional expectation on the right-hand side of (10.24) is usually a random variable (as is the limit), the exception being Poisson processes, which give rise to deterministic stochastic intensities.

The martingale approach to point processes [10] affords a rigorous mathematical treatment of stochastic intensities, including a martingale-based characterization of point processes and their stochastic intensities. The martingale approach also provides a unified treatment of dynamical point process systems with a martingale-based calculus. A case in point is the elegant martingale-based result, commonly known as the *change-of-intensities* *formula* à la Girsanov, roughly, a change-of-measure result for stochastic intensities akin to a Radon-Nikodym change of the underlying probability measure (*ibid.*).

While the martingale approach to point processes and its calculus has been successfully used to tackle a number of interesting problems in queueing theory, their usefulness in stochastic modeling, statistical inference, and generation of synthetic arrival streams remains largely an open question. On the other hand, intensity-based modeling of point processes has proved successful in the area of survival analysis, largely due to important contributions made by modern martingale theory to statistical inference in the domain of intensity-based point processes.

10.3 TELETRAFFIC METHODS

The traffic models surveyed in Section 10.2 were treated as mathematical objects in their own right. In practice, traffic per se is of interest primarily as an ingredient of server systems, and attention is focused not on traffic statistics, but on queueing statistics such as blocking probabilities and waiting times.

Traditional telephony addressed traffic with a view to engineering, dimensioning, and provisioning the telephone network, mainly for voice services. Queueing-based methods were developed over time and put into use to ensure that a prescribed quality of service (QOS) would be satisfied. For instance, plain old telephone service (POTS) might prescribe a bound on blocking probabilities (known as "grade of service" in traditional telephony), e.g., that at most 1% of dialed calls experience blocking. Even though emerging high-speed communications networks, such as B-ISDN (broadband integrated services digital networks) under ATM (asynchronous transfer mode), are slated to carry new and far more varied traffic than traditional voice and data networks (notably compressed video and images), the methods to be surveyed below will likely be of use in such new communications networks.

The material in Section 10.3 treats traffic within a queueing framework, with a view to practical teletraffic engineering. While the treatment is mathematical, the focus is on explicit and computable formulas and the computational aspects thereof. Some traffic methods for heavy traffic may be found in Subsection 10.3.2. For light-traffic methods, the reader is referred to [8].

10.3.1 Traffic burstiness

A recurrent theme relating to traffic in broadband networks is the traffic "burstiness" exhibited by key services, such as compressed video, file transfer, etc. Intuitively, burstiness is present in a traffic process if the arrival points, $\{T_n\}$, appear to form visual clusters on the time line; equivalently, $\{A_n\}$ tends to give rise to runs of several relatively short interarrival times, surrounded by relatively long ones. The mathematical underpinning of burstiness is more complex. Two main sources of burstiness are due to the shapes of the marginal distribution and autocorrelation function of $\{A_n\}$. For example, burstiness would be facilitated by a bimodal marginal distribution of $\{A_n\}$, or by short-range autocorrelations in $\{A_n\}$. Strong positive autocorrelations are a particularly major cause of burstiness. Since there seems to be no single widely-accepted notion of burstiness, we shall briefly describe some of the commonly-used mathematical measures that attempt to capture it. Engineers tend to employ very simple measures of burstiness which only use first-order traffic statistics, namely, those associated with interarrival distributions. A typical example is the *peak-to-mean ratio* (PMR) of traffic rate — a very crude measure, which also has the shortcoming of dependence on the interval length utilized for rate measurement. Another example is the coefficient of variation of interarrival times, c_A.

Generally, these constitute weak characterizations that only capture very crude aspects of traffic burstiness. More useful measures of burstiness utilize second-order properties of $\{A_n\}$, namely, those associated with its autocorrelation function. Finally, the *Hurst parameter* has been suggested as a measure of burstiness via the concept of self-similarity [59]; see Subsection 10.2.6.

10.3.1.1 The PTC and ISE traffic principles

The two burstiness descriptors, indices of dispersion and peakedness, are motivated, respectively, by two traffic principles: the *Poisson Traffic Comparison* (PTC) principle, and the *Infinite-Server Effect* (ISE) principle. These principles trace their origins to teletraffic analysis and design, and will be explained next.

The PTC principle compares a traffic-related descriptor (statistic), d_X, of a target traffic process, $\{X_n\}$, to the corresponding descriptor, d_E, of a "comparable" Poisson process, $\{E_n\}$, where the E_n are iid exponential random variables. The descriptor statistic is often related to moments of the underlying process. The PTC principle adopts the view that Poisson processes constitute a class of traffic benchmarks; indeed, these classic traffic processes have a long history in teletraffic theory and practice. Furthermore, "comparability" usually means that the benchmark Poisson process is chosen to have the same rate as the target traffic process. The mathematical comparison is most often performed by taking the target-to-benchmark descriptor ratio. In particular, under the PTC principle, a generic burstiness descriptor, b_X, has the form

$$b_X = \frac{d_X}{d_E}. \tag{10.25}$$

Equation (10.25) has the additional appeal that PTC-based descriptors of any Poisson process trivially evaluate to unity, as befits a benchmark. The indices of dispersion descriptors of burstiness, treated in Subsection 10.3.1.2, are motivated by the PTC principle.

The ISE principle adopts the view that to gauge an aspect of a target traffic stream, $\{X_n\}$, one offers that stream to an infinite-server system and observes the stream effect via a suitable statistic, d_X; infinite-server systems are chosen due to their relative tractability. For example, to characterize burstiness, a suitable description would be the equilibrium moments of the number of customers in the system. The PTC principle is often employed implicitly in an ISE context by using a normalized form of the description in such a way that it evaluates to unity for any Poisson process. The ISE principle is also used to construct an approximation to a target traffic stream via a mathematically simpler stream. This is attained by equating equilibrium moments of the number of busy servers induced by the two streams. An example is the *"three moment match"* for creating an interrupted Poisson stream approximation to overflow traffic [57]. The burstiness notion of peakedness, treated in Subsection 10.3.1.3, is subject to the ISE principle.

10.3.1.2 Indices of dispersion

Consider a stationary, non-negative random sequence, $\{X_n\}_{n=0}^{\infty}$, interpreted as the interarrival times of a traffic process. Let the common distribution function of the X_n be denoted by $F_X(x)$; similarly, denote $\lambda_X = 1/E[X_n]$ (traffic rate), $\sigma_X^2 = Var[X_n]$, and $c_X = \lambda_X \sigma_X$. We assume that $0 < \sigma_X < \infty$ and that $\{X_n\}$ is simple, namely, $P\{X_n = 0\} = 0$. The coefficient of variation, c_X, can be used to characterize traffic burstiness, albeit rather weakly, since it utilizes just first-order statistics of $\{X_n\}$ (those pertaining to $F_X(x)$ only). In contrast, the second-order properties are characterized by the autocorrelation function of $\{X_n\}$, given by

$$\rho_X(j) = \frac{E[X_k X_{k+j}] - \lambda_X^{-2}}{\sigma_X^2}, \quad j = 0, 1, \ldots, \qquad (10.26)$$

which measures temporal linear dependence among lagged interarrival times, and will prove to be a key ingredient in the burstiness descriptors to be discussed in the sequel.

The *index of dispersion* for intervals (IDI) [40], associated with $\{X_n\}$, is a burstiness descriptor defined by

$$J_X(n) = \frac{Var[\sum_{j=1}^{n} X_j]}{n \lambda_X^{-2}}. \qquad (10.27)$$

In order to relate the IDI to the PTC principle, denote $v_X(n) = Var[\sum_{j=1}^{n} X_j]$, and note that for a Poisson process with interarrival times, $\{E_n\}$, of mean $1/\lambda_E$, one has $v_E(n) = n\lambda_E^{-2}$. Thus, equation (10.27) is the ratio $J_X(n) = v_X(n)/v_E(n)$, in adherence to the PTC principle. A simple computation provides the explicit relation of $J_X(n)$ to c_X and ρ_X, namely,

$$J_X(n) = c_X^2 \left[1 + 2 \sum_{j=1}^{n-1} \left(1 - \frac{j}{n}\right) \rho_X(j) \right]. \qquad (10.28)$$

Let $\{N(t)\}$ be the "equilibrium" counting process of arrivals in the interval $(0, t]$ (in the sense of stationary increments), so that $E[N(t)] = \lambda_X t$. A related burstiness descriptor is the *index of dispersion for counts* (IDC) [40], associated with $\{X_n\}$, and given by

$$I_X(t) = \frac{Var[N(t)]}{E[N(t)]} = \frac{Var[N(t)]}{\lambda_X t}, \quad t \geq 0. \qquad (10.29)$$

That the IDC adheres to the PTC principle follows from the fact that the mean and variance of a Poisson counting process are equal.

The limiting indices of dispersion, $J_X = \lim_{n \to \infty} J_X(n)$ and $I_X = \lim_{t \to \infty} I_X(t)$, are of particular interest. It can be shown that [40]

$$J_X = I_X = c_X^2 \left[1 + 2 \sum_{j=1}^{\infty} \rho_X(j) \right]. \tag{10.30}$$

The indices of dispersion, $(10.27 - 10.30)$, are more satisfactory descriptors of traffic variability (burstiness) than the coefficient of variation, c_X, since they take into account the autocorrelation function, $\rho_X(j)$, as well. In fact, for renewal traffic, $J_X(n) \equiv J_X = I_X = c_X^2$. Moreover, these indices of dispersion may be usefully applied to the study of the effect of an offered load on a server system [88, 29, 40].

In practical queueing studies, one uses the indices of dispersion, $J_X(n)$ and $I_X(t)$, and especially the simpler common limit $J_X = I_X$, by estimating their values from empirical observations and then fitting a suitable traffic model, such as a Markov-modulated Poisson process (MMPP), which is consistent with the observed index of dispersion [40]. Having thus constructed a model, a queueing system is analyzed with this traffic as the offered load. While this approach has led to some useful results, the index-of-dispersion characterization of traffic burstiness is still fairly weak from a queueing viewpoint. The reason is that indices of dispersion do not consider server systems in their definition. This situation will be remedied in the next section, which will introduce a more powerful traffic burstiness characterization.

10.3.1.3 The peakedness functional

Peakedness is a traffic burstiness descriptor, based on the ISE principle. It gauges the "smoothness" of a traffic stream, X, via its effect (as offered traffic) on an iid infinite-server group, i.e., all service times are mutually independent with common service distribution $B(x)$ (usually the exponential distribution) of mean $1/\mu_B$ (μ_B being the service rate). The *peakedness* descriptor of X is the equilibrium variance-to-mean ratio of $S(t) -$ the number of busy servers at time t. R. Wilkinson [96] had introduced the peakedness concept in teletraffic theory in order to characterize the action of superposition of overflow streams when offered to a secondary trunk group. The object was to compute the grade of service (blocking probability) in the context of his "equivalent random method" (see [48] and Subsection 10.3.5). A simple and more flexible computational method utilizing peakedness for grade-of-service computation was introduced later by W.S. Hayward (see *ibid.* and Subsection 10.3.6). An advantage of peakedness is that it may be used to approximate various statistics in a single-server system, with or without finite buffers, such as waiting times and blocking probabilities [26, 48].

Let $a = \lambda_X/\mu_B$ and note that Little's Law implies $\lim_{t \uparrow \infty} E[S(t)] = a$. The *peakedness functional*, $z_X[B]$, maps the service time distribution, $B(x)$, to the following scalar [96]:

$$z_X[B] = \lim_{t \uparrow \infty} \frac{Var[S(t)]}{E[S(t)]} = \frac{\sigma_S^2}{a}, \tag{10.31}$$

where σ_S^2 is the time-equilibrium variance of the number of busy servers.

For any service distribution, $B(x)$, define $B_0(x) = B(x/\mu_B)$, so that $B_0(x)$ has unit service rate. Define further $B_\mu(x) = B_0(\mu x)$ to be a family of service time distribution functions, parameterized by $\mu > 0$. The *peakedness function* associated with this parametric family is

$$z_{X,B_0}(\mu) = z_X[B_\mu].$$ (10.32)

Let $M_X(t)$ be the expectation function of the traffic process, $\{X_n\}$, that is, the mean number of arrivals in $(0, t]$ when the origin is an arrival point; thus, $M_X(t)$ is the analog of the renewal function in renewal theory. Let $B^c(x)$ designate the complementary service time distribution, that is, $B^c(x) = 1 - B(x)$, and let $\psi_B(x)$ denote the correlation function,

$$\psi_B(t) = \int_0^\infty B^c(u)B^c(t+u)du,$$ (10.33)

of $B^c(x)$. Then one has [26]

$$z_{X,B_0}(\mu) = 1 - \frac{\lambda}{\mu} + 2\mu \int_0^\infty \psi_B(t)dM_X(t).$$ (10.34)

Furthermore, if the density of the expectation function $m_X(t) = dM_X(t)/dt$ exists, then one may also write

$$z_{X,B_0}(\mu) = 1 - \frac{\lambda}{\mu} + 2\mu \int_0^\infty \psi_B(t)m_X(t)dt.$$ (10.35)

An important special case occurs for exponential service time distributions, $B(x) = 1 - e^{-\mu x}$. The notation $z_{X,exp}(\mu)$ and $\psi_{exp}(x)$ will be used for this case. Since

$$\psi_{exp}(y) = \frac{1}{2\mu}e^{-\mu y},$$ (10.36)

it follows that

$$z_{X,exp}(\mu) = 1 - \frac{\lambda x}{\mu} + \int_0^\infty e^{-\mu t}m_X(t)dt.$$ (10.37)

Letting $\widetilde{f}(s) = \int_0^\infty e^{-sx}f(x)dx$ denote the Laplace transform of a function $f(x)$, one has

$$z_{X,exp}(\mu) = 1 - \frac{\lambda_X}{\mu} + \widetilde{m}_X(\mu).$$ (10.38)

Equation (10.38) has an important consequence. It implies that when $\widetilde{m}(s)$ does not depend on μ, then knowledge of $z_{X,exp}(\mu)$ suffices to determine $z_{X,B}(\mu)$ for *any* service distribution $B(x)$. To see that, observe that from $z_{X,exp}(\mu)$, one can determine $\widetilde{m}_X(\mu)$, from which $m_X(t)$ is obtained by inversion. Equation (10.35) can now be used to obtain $z_{X,B}(\mu)$ with the aid of $m_X(t)$.

The transformation theory of peakedness (from one service time distribution to another) may be exhibited in a more satisfactory and illuminating form via the *Mellin transform* [48]. The Mellin transform, $\overline{f}(s)$ of a function $f(x)$, is defined by

$$\overline{f}(s) = \int_0^\infty x^{s-1}f(x)dx,$$

and exists in a strip or half-plane [22] of the complex s-plane. Note that $\psi_B(x) = \psi_{B_0}(\mu x)/\mu$ and $\int_0^\infty \psi_{B_0}(\mu x)dx = 1/(2\mu)$. Introducing the notation, $\zeta_{X,B}(\mu) = z_{X,B_0}(\mu) - 1$ and $h_X(x) = m_X(x) - \lambda_X$, it follows that (10.35) is equivalent to

$$\zeta_{X,B_0}(\mu) = 2\int_0^\infty \psi_{B_0}(\mu x)h_X(x)dx.$$ (10.39)

Applying the Mellin transform to (10.39) results in

$$\overline{\zeta}_{X,B_0}(s) = 2\overline{\psi}_{B_0}(s)\overline{h}_X(1-s), \tag{10.40}$$

which clearly separates the individual effect of the service distribution   (via $\overline{\psi}_{B_0}(s)$) and the traffic stream (via $\overline{h}_X(1-s)$) on $\overline{\zeta}_{X,B_0}(s)$.

Suppose now that it is desired to switch from service distribution F to service distribution G, thereby transforming $z_{X,F_0}(\mu)$ to $z_{X,G_0}(\mu)$. From (10.40), one has

$$\overline{\zeta}_{X,F_0}(s) = 2\overline{\psi}_{F_0}(s)\overline{h}_X(1-s),$$

$$\overline{\zeta}_{X,G_0}(s) = 2\overline{\psi}_{G_0}(s)\overline{h}_X(1-s),$$

where $\overline{\zeta}_{X,F_0}$ and $\overline{\zeta}_{X,G_0}$ are simply related by

$$\frac{\overline{\zeta}_{X,G_0}(s)}{\overline{\zeta}_{X,F_0}(s)} = \frac{\overline{\psi}_{G_0}(s)}{\overline{\psi}_{F_0}(s)}. \tag{10.41}$$

Thus, equation (10.41) permits the calculation of traffic peakedness relative to a new service distribution, based on the corresponding peakedness relative to a given service distribution, provided the Mellin transform exists at least as an extension to a singular integral [65].

Suppose next that $z_{X,F_0}(0)$ is known, and possibly some of its derivatives at the origin, for some service distribution F. From (10.39), one deduces that the nth derivative of ζ_{X,F_0} at the origin is given by

$$\zeta_{X,F_0}^{(n)}(0) = 2\psi_{F_0}^{(n)}(0)\int_0^\infty x^n h(x)dx, \tag{10.42}$$

provided the differentiations are permissible within the integral.   Combining (10.42) with the corresponding formula for $\zeta_{X,G_0}(\mu)$, and denoting

$$R_n = \frac{\psi_{F_0}^{(n)}(0)}{\psi_{G_0}^{(n)}(0)}, \tag{10.43}$$

one obtains

$$\zeta_{G_0}^{(n)}(0) = R_n\zeta_{F_0}^{(n)}(0). \tag{10.44}$$

Equation (10.44) can be used as a transformation formula for the derivatives of the peakedness at $\mu = 0$, analogously to equation (10.41).   For the evaluation of the ratios, R_n in (10.43), observe that for any service distribution, B, one has

$$\psi_{B_0}^{(n)}(0) = \int_0^\infty B_0^c(y)(B_0^c)^{(n)}(y)dy, \tag{10.45}$$

provided the interchange of differentiation and integration is valid.   Applying (10.45) separately to the numerator and denominator of R_n, one obtains the particular cases

$$R_0 = \frac{\int_0^\infty G_0^c(y)^2 dy}{\int_0^\infty F_0^c(y)^2 dy}, \quad R_1(y) \equiv 1. \tag{10.46}$$

From equation (10.46) follows the important peakedness formula

$$z_{X,G_0}(0) = 1 + \frac{\int_0^\infty G_0^c(y)^2 dy}{\int_0^\infty F_0^c 9y)^2 dy} [z_{X,F_0}(0) - 1]. \tag{10.47}$$

Interestingly, (10.46) implies the remarkable conclusion that $z'_{X,F_0}(0)$ is invariant under change of service distribution; of course, this does not apply to the higher derivatives.

Practical applications of the peakedness function $z_B(\mu)$ to queueing problems may be found in [48, 26, 32].

10.3.2 The Erlang B loss function

The Erlang B loss function is used to obtain the loss rate of traffic offered to a finite-server group. It is traditional to express traffic units in *erlang units*, namely, the mean number of arrivals during a time interval whose length is the mean service time.

The Erlang B setting is an M/G/n/n queue. Here, a Poisson arrival stream of rate λ is offered to an iid group of n servers, each with service distribution $B(x)$ of rate μ (i.e., with mean service time $1/\mu$); the offered load is denoted by $a = \lambda/\mu$ and is expressed in erlang units. Let S be the number of busy servers in equilibrium, with probability distribution [56]

$$P\{S = k\} = \frac{a^k/k!}{\sum_{j=0}^n a^j/j!}, \quad k = 0, 1, \ldots.$$

The PASTA property ensures that the distribution of S coincides with its counterpart, embedded at arrival points (see, e.g. [78]). In particular, the probability that all servers are busy is

$$B(n,a) = \frac{a^n/n!}{\sum_{j=0}^n a^j/j!}. \tag{10.48}$$

Equation (10.48) is called the *Erlang B loss function* (Erlang B, for short), and is a function of the server group size and offered load. Thus, the equilibrium mean number of busy servers, called the *carried load*, is $a'(n,a) = a[1 - B(n,a)]$, and the *overflow rate of blocked customers* (in erlangs) is $aB(n,a)$.

Suppose that the servers are numbered 1 to n and the entire arrival stream is offered to server 1, the overflow then offered to server 2 and so on; this discipline is called an *ordered hunt*. Thus, the load carried by server j is

$$\ell(j,a) = a[B(j-1,a) - B(j,a)], \quad 1 \le j \le n. \tag{10.49}$$

When $j = n$, then (10.49) is called the *load on the last server*. Its importance is due to the economic aspects of sizing server groups [56]. A useful identity, relating a, $a'(n,a)$, and $\ell(n,a)$, is

$$a = a'(n,a)\left[1 + \frac{\ell(n,a)}{n - a'(n,a)}\right]. \tag{10.50}$$

The variance of S is

$$\sigma_S^2 = -an + [a + n + 1]a'(n,a) - (a'(n,a))^2 = [1 - \ell(n,a)]a'(n,a).$$

Practical applications of the Erlang B theory require knowledge of sensitivities, approximations, and asymptotic relations for $B(n,a)$, $\ell(n,a)$, etc. To this end, let $\rho = a/n$ be the *offered load per server*, and let $\eta(n,a) = a'(n,a)/n$ be the *carried load per server* (also known as the *efficiency of the server group*). The following are useful identities:

$$\frac{\partial B(n,a)}{\partial a} = \left[\frac{n}{a} - 1 + B(n,a)\right]B(n,a), \tag{10.51}$$

$$\frac{\partial \ell(n,a)}{\partial \rho} = n\left[1 - \frac{\eta(n,a)}{\rho}\right]\frac{n(n\rho - \eta(n,a))(1 - \eta(n,a)) - \eta(n,a)^2}{\eta(n,a)^2}, \tag{10.52}$$

$$\frac{\partial \ell(n,a)}{\partial \eta(n,a)} = \frac{\ell(n,a)}{\eta(n,a)}\frac{n(1 - \eta(n,a))^2 - \eta(n,a) + \ell(n,a)}{(1 - \ell(n,a))(1 - \eta(n,a))}, \tag{10.53}$$

$$\frac{\partial a'(n,a)}{\partial a} = \frac{\sigma_S^2}{a}. \tag{10.54}$$

A numerically convenient formula for calculating $B(n,a)$ is via

$$B(n,a)^{-1} = \sum_{j=0}^{\infty} n^{(j)}a^{-j}, \tag{10.55}$$

where $n^{(j)}$ is the descending factorial, given by

$$n^{(j)} = \begin{cases} 1, & j = 0 \\ n(n-1)\ldots(n-j+1), & j \geq 1. \end{cases}$$

(10.55) is especially useful when $a > n$, since then the series may be conveniently truncated to achieve a given accuracy [47]. Substitution of the identity $a^{-j} = a\int_0^{\infty} e^{-ay}\frac{y^j}{j!}\,dy$, $a > 0$, $j \geq 0$, into (10.55) yields Fortet's integral representation

$$B(n,a)^{-1} = a\int_0^{\infty} e^{-ay}(1+y)^n dy = \int_0^{\infty} e^{-y}\left(1 + \frac{y}{a}\right)^n dy. \tag{10.56}$$

In order to obtain the sensitivity of $B(n,a)$ with respect to n, one extends its representation in (10.56) to complex n and a, by means of the interpolatory analytic function

$$B(x,a)^{-1} = \int_0^{\infty} e^{-y}\left(1 + \frac{y}{a}\right)^x dy, \quad Re[a] > 0. \tag{10.57}$$

Furthermore, this permits the evaluation of the loss function for a non-integral number of servers in practical traffic-related problems. The corresponding extension of equation (10.55) is the asymptotic expansion, as $a \to \infty$ (see [47]),

$$B(x,a)^{-1} \sim \sum_{j=0}^{\infty} x^{(j)}a^{-j}, \quad |\arg a| < \pi. \tag{10.58}$$

Equations (10.50) and (10.52-10.54) may be similarly extended. These extensions are computationally convenient, since the error does not exceed the absolute value of the first neglected term. Furthermore, an integration by parts of (10.57) yields the useful recurrence relation

$$B(x+1,a)^{-1} = \frac{x+1}{a}B(x,a)^{-1} + 1. \tag{10.59}$$

For integral $x = n$, $B(n,a)$ can be conveniently computed via (10.59) from the initial value $B(0,a) = 1$.

The derivative, $\partial B(x,a)/\partial a$, yields the sensitivity with respect to the offered load, a. It may be obtained from (10.57) and simplified to

$$\frac{\partial B(x,a)}{\partial a} = \left[\frac{x}{a} - 1 + B(x,a)\right]B(x,a).$$

Using again (10.57), one has

$$\frac{\partial B(x,a)}{\partial x} = B^2(x,a) \int_0^\infty e^{-y} \left[1 + \frac{y}{a}\right]^x \ln\left(1 + \frac{y}{a}\right) dy. \tag{10.60}$$

The evaluation of (10.57) and (10.60) can be achieved by means of quadrature theory. The basic goal is the evaluation of the integral $I = \int_0^\infty e^{-y} f(y) dy$. It is known that the trapezoidal rule of quadrature is very accurate when the integral has high-order osculation at the endpoints of integration [17]. Using the transformation $w = \ln y$, one has

$$I = \int_{-\infty}^\infty e^{w - e^w} f(e^w) dw. \tag{10.61}$$

For functions of the form $f(y) = O\left(\frac{e^y}{y(\ln y)^2}\right)$, there is sufficient osculation at $-\infty$ and ∞ for the use of the trapezoidal rule. Thus, applying the rule to (10.61) with span $h > 0$,

$$I \simeq h \sum_{j=-\infty}^\infty e^{jh - e^{jh}} f(e^{jh}). \tag{10.62}$$

The quadrature rule (10.62) can be applied to (10.57) and (10.60) to obtain $B(x,a)$ and $\partial B(x,a)/\partial x$, respectively. It should be noted that the requisite computational effort is essentially independent of x and a; this may be contrasted with the corresponding effort associated with (10.58) and (10.59).

An alternate computational method is based on the extension of the Poisson distribution to continuous x via the function $\psi(x,a) = e^{-a} \frac{a^x}{\Gamma(x+1)}$, $x > 1$. From [47], $B(x,a)$ is well approximated by $\psi(x,a)$ when a/x is small, the exact relation being

$$B(x,a)^{-1} = \psi(x,a)^{-1} - \sum_{j=1}^\infty \frac{a^j}{(x+1)\ldots(x+j)}. \tag{10.63}$$

The series is rapidly convergent for $x < -1$ and $a/x \le 1$.

The above computations are accurate but time-consuming. When real-time computations are called for, it is desirable to trade accuracy for time and to construct fast approximate formulas for $\partial B(x,a)/\partial x$, $a'(n,a)$, $\ell(x,a)$, etc.

From (10.57), one deduces that $B(x,a)^{-1}$ and $[aB(x,a)]^{-1}$ are log-convex functions of x and a, respectively, for $a > 0$ and all x [47]; also, $B(x,a)$ is convex in x for $a > 0$, $x > 0$ [52]. The log-convexity of $B(x,a)^{-1}$ or the convexity of $B(x,a)$ implies the inequality $\ell(x,a) \le \eta(x,a)$. Substituting this upper bound into (10.50), which remains valid with n replaced by the continuous variable x, one gets

$$a \le a'(x,a) \left[1 + \frac{a'(x,a)}{x[x - a'(x,a)]}\right],$$

which is a useful first-cut estimate of the offered load, a.

Denote $\alpha = (x+1)/a + B(x,a)$. Then, Jensen's inequality [41] applied to (10.57) implies the lower bound

$$\frac{\partial B(x,a)}{\partial x} \ge -B(x,a) \ln \alpha. \tag{10.64}$$

A simple and useful approximation for $\partial B/\partial x$ may be obtained by the following consideration. The Euler-Maclaurin expansion [89] applied to a functional equation of the form $f(x+1) - f(x) = g(x)$ implies

$$f'(x) \simeq g(x) - \frac{1}{2} g'(x). \tag{10.65}$$

To apply this to obtaining $\partial B/\partial x$, take

$$f(x) = \ln B(x, a),\tag{10.66}$$

whence (10.59) becomes

$$f(x + 1) - f(x) = -\ln \alpha.$$

Relation (10.65) now provides the approximation (see, e.g. [48])

$$B(x, a)^{-1} \frac{\partial B(x, a)}{\partial x} \simeq -\frac{\ln \alpha - 1/(2a\alpha)}{1 - B(x, a)/(2\alpha)}.$$

Interpolation plays a useful role in some applications. For example, $B(x, a)$ may be measured in practice at integral values of x, yet in some calculations the value of $B(x, a)$ is needed at a non-integral value of x. A useful approximation is obtained by applying quadratic interpolation to $f(x)$ in (10.66). Recalling that $\lfloor x \rfloor$ denotes the integral part of x and $\langle x \rangle = x - \lfloor x \rfloor$, its fractional part, one has

$$B(x, a) \simeq B(n, a)^{1 - \langle x \rangle} B(n + 1, a)^{\langle x \rangle} \left[\frac{B(n + 1, a)^2}{B(n, a) B(n + 2, a)} \right]^{\langle x \rangle (1 - \langle x \rangle)/2},\tag{10.67}$$

the worst error of (10.67) occurring at $\langle x \rangle = 0.5$.

An asymptotic formula for $B(x, a)$ (as $x \to \infty$), when a is in the vicinity of x (the *medium traffic condition*), is given by

$$B(x, x + c\sqrt{x}) \sim \sum_{j=0}^{\infty} a_j(c) x^{-\frac{j-1}{2}}, \quad |\arg x| < \frac{\pi}{2}, \; c \text{ real},\tag{10.68}$$

where the coefficients $a_j(c)$ are given by

$$a_j(c) = \int_0^{\infty} e^{-(\frac{1}{2}v^2 + cv)} b_j(v, c)\, dv,$$

with the $b_j(v, c)$ defined by the relation

$$e^{\frac{1}{2}v^2 - v\sqrt{x}} \left[x + \frac{v}{\sqrt{x}} \right]^x (\sqrt{x} + c) = \sum_{j=0}^{\infty} b_j(v, c) x^{-\frac{j-1}{2}}.$$

The first few coefficients are

$$a_0(c) = e^{\frac{1}{2}c^2} \int_c^{\infty} e^{-\frac{1}{2}u^2}\, du,$$

$$a_1(c) = \frac{2}{3} + \frac{1}{3}c^2 - \frac{1}{3}c^3 a_0(c),$$

$$a_2(c) = -\frac{1}{18}c^5 - \frac{7}{36}c^3 + \frac{1}{12}c + \left[\frac{1}{18}c^5 + \frac{1}{4}c^4 + \frac{1}{12} \right] a_0(c).$$

The special case $c = 0$ yields the useful asymptotic relation, as $x \to \infty$,

$$B(x, x)^{-1} \sim \sqrt{\frac{\pi x}{2}} + \frac{2}{3} + \frac{1}{12}\sqrt{\frac{\pi}{2x}}, \quad |\arg x| < \frac{\pi}{2}.\tag{10.69}$$

Equation (10.69) may be used in convenient approximate computations of $a'(n, a)$ and $\ell(n, a)$ for the case of medium traffic. In particular, consider the load $\ell^*(a') = \ell(x, a)$, carried by the virtual server x, as a function of the carried load, $a' = a'(x, a)$. Let $a_0' = a'(x, x)$ denote the carried load at the operating point, (x, x). Then the value of $\ell^*(a')$, resulting from changing the carried load away from the operating point, a_0', can then be approximated by the two-term Taylor expansion of ℓ^* about a_0', namely,

$$\ell^*(a') \simeq \ell^*(a_0') + [a' - a_0']\frac{d\ell^*(a_0')}{da'},$$

where, in view of the relation $\frac{d\ell^*(a')}{da'} = \frac{1}{x}\frac{d\ell(n,a)}{dn}$, one obtains the derivative $\frac{d\ell^*(a')}{da'}$ from (10.53) with the aid of (10.69). A more convenient form of (10.49) for $\ell(x,a)$ is

$$\ell(x,a) = \left[\frac{x}{1 - B(x,a)} - a\right]B(x,a),$$

from which $\ell(x,x)$ is readily obtained, while a more accurate asymptotic version of (10.69), as $x \to \infty$, is

$$B(x,x)^{-1} \sim \sqrt{\frac{\pi x}{2}} + \frac{2}{3} + \frac{1}{12}\sqrt{\frac{\pi}{2x}} - \frac{4}{135x}, \quad |\arg x| < \frac{\pi}{2}.$$

The approximation of $B(x,cx)$, for $c > 1$, applies to the case of *heavy traffic* in large server systems. From [47], one has the asymptotic expansion, as $x \to \infty$,

$$B(x,cx)^{-1} \sim \sum_{k=0}^{\infty} g_k(c)x^{-k}, \quad |\arg x| < \frac{\pi}{2}, c > 1,$$

where the coefficients $g_k(c)$, $k = 0,1,\ldots$, are given by

$$g_k(c) = \left[\frac{c}{c-1}\frac{d}{dc}\right]^k \frac{c}{c-1},$$

implying, in particular,

$$B(x,cx)^{-1} \sim \frac{c}{c-1} - \frac{c}{(c-1)^3}x^{-1} + \frac{2c^2+c}{(c-1)^5}x^{-2}. \tag{10.70}$$

Equation (10.70) may be contrasted with the exact formula (10.63) and the light-traffic Poisson approximation derived from it, as well as to the asymptotic formula (10.58), which approximates heavy traffic with a fixed number of servers.

10.3.3 The Erlang C delay function

The Erlang C setting is an M/M/n/∞ queue. Here, a Poisson arrival stream of rate λ is offered to an infinite FIFO buffer served by n iid servers with exponential service time distribution $B(x) = 1 - e^{-\mu x}$. The offered load is $a = \lambda/\mu$, and the offered load per server is $\rho = a/n < 1$. Since no customers are lost, the carried load is $a'(n,a) = a$.

Let W be the waiting time in the queue (if any) of an arriving customer in equilibrium. The probability, $C(n,a) = P\{W > 0\}$, that an arriving customer finds all servers busy (and, therefore, must wait in line) is given by

$$C(n,a) = \frac{B(n,a)}{1 - \rho[1 - B(n,a)]}. \tag{10.71}$$

Equation (10.71) is called the *Erlang C delay function* (*Erlang C*, for short), and is a function of the server group size and offered load. For the distribution of W, one has

$$P\{W > t\} = C(n,a)e^{-(1-\rho)n\mu t},$$

$$P\{W > t \mid W > 0\} = e^{-(1-\rho)n\mu t},$$

and for its mean and variance,

$$E[W] = \frac{C(n,a)}{[1-\rho]n\mu},$$

$$Var[W] = \frac{1 - [1 - C(n,a)]^2}{([1 - \rho]n\mu)^2}.$$

Let S be the number of customers in the system in equilibrium, and let $P_j = P\{S = j\}$ be the corresponding equilibrium time probabilities. Then,

$$P_j = \begin{cases} n!a^{-n}[1 - \rho]C(n,a), & j = 0 \\[2mm] P_0\dfrac{a^j}{j!}, & 1 \le j \le n - 1 \\[2mm] P_0\dfrac{a^j}{n!n^{j-n}}, & j \ge n \end{cases}$$

and

$$E[S] = P_0\frac{a^n}{n!}\left[a[B(n,a)^{-1} - 1] - n + \frac{1}{[1 - \rho]^2}\right].$$

Finally, let Q be the number of customers in the buffer (excluding customers in service) in equilibrium. Then,

$$P\{Q = j, W > 0\} = [1 - \rho]\rho^j C(n,a),$$

$$P\{Q = j \mid W > 0\} = [1 - \rho]\rho^j.$$

All of the above formulas may be immediately generalized to a continuous number of servers. One need only replace $n!$ by $\Gamma(n+1)$ and use the generalization of $B(n,a)$.

10.3.4 The Engset delay model

The Engset model is a closed system consisting of a finite number, n, of independent sources (customers) and a finite number, s, of iid exponential servers [14]. Each source goes through cycles of service as follows. Initially a source is idle (neither in service nor waiting for service). After an exponentially-distributed time with parameter γ, the source places a service request and is considered busy. All busy sources waiting for a server to become available wait in a FIFO queue and eventually seize a server for an exponentially-distributed service time at rate μ. On service completion the source becomes idle, and the cycle starts over again.

Let $P_j(n,s)$ be the equilibrium probability that j sources are busy. Then,

$$P_j(n,s) = \begin{cases} \dfrac{1}{\sum_{k=0}^{n} \pi_k}, & j = 0 \\[3mm] \pi_j\, P_0(n,s), & j > 0 \end{cases} \tag{10.72}$$

where, letting $\hat{a} = \gamma/\mu$,

$$\pi_j = \begin{cases} \dbinom{n}{j}\hat{a}^j, & 0 \le j \le s \\[3mm] \dfrac{n^{(j)}}{s!s^{j-s}}\,\hat{a}^j, & s+1 \le j \le n. \end{cases} \tag{10.73}$$

Let $\bar{L}$ be the mean number of busy sources and $\bar{I}$ the mean number of idle sources. Then, $\bar{L} = \sum_{j=1}^{n} jP_j(n,s)$ and $\bar{I} = n - \bar{L}$. Since $\hat{a}$ is the offered load per idle source, it follows that the total offered load is $a = \hat{a}\bar{I}$, and the total request rate is $\lambda = a\mu$.

Let $\bar{J}$ be the equilibrium mean sojourn time and $\bar{W}$ the equilibrium mean waiting time. Then, these means are related by $\bar{W} = \bar{J} - 1/\mu$, while from Little's law, $\bar{J} = \bar{L}/\lambda$. Since λ/n is the request rate per source, it follows that $n\lambda^{-1}$ is the mean time between requests. Hence, $\bar{W} + \frac{1}{\mu} + \frac{1}{\gamma} = \frac{n}{\lambda}$.

Let $\Pi_j(n, s)$ be the equilibrium probability that j sources are busy at a request arrival instant. The relation between the (time average) possibilities, P_j, and the corresponding (customer-average) probabilities, Π_j, is given by

$$\Pi_j(n, s) = P_j(n - 1, s).$$

This formulas is an instance of ASTA (Arrivals See Time Averages), provided the arriving customer is excluded (see, eg. [78]). The distribution of the waiting time, W, is given by

$$P\{W > t\} = \Pi_0(n, s) \frac{(n-1)! \, \hat{a}^s}{s!} \left(\frac{\hat{a}}{s}\right)^{n-s-1} e^{s/\hat{a} - \phi(t)} \sum_{j=0}^{n-s-1} \frac{[\phi(t)]^j}{j!},$$

where $\phi(t) = s\mu[t + 1/\gamma]$.

For a single-server Engset model ($s = 1$), equations (10.73) and (10.72) can be simplified using

$$P_0(n, 1) = B(n, 1/\hat{a}), \quad a = 1 - B(n, 1/\hat{a}). \tag{10.74}$$

Furthermore, letting $c = n\gamma$,

$$\bar{J} = n\left[\frac{1}{\lambda} - \frac{1}{c}\right] = n\left[\frac{1/\mu}{1 - B(n, \mu n/c)} - \frac{1}{c}\right]. \tag{10.75}$$

Convenient asymptotic formulas (as $n \to \infty$) for equations (10.74) and (10.75) are derived from the asymptotic results of Subsection 10.3.2 for the special case $s = 1$. These are summarized below for various ranges of μ.

For $\mu > c$,

$$P_0(n, 1) \sim 1 - \frac{c}{\mu} + \frac{c^2}{n\mu(\mu - c)}, \quad \bar{J} \sim \frac{1}{\mu - c} \quad \text{[from (10.70)]}.$$

For $\mu = c$,

$$P_0(n, 1) \sim \sqrt{\frac{2}{\pi n}} - \frac{4}{3\pi n}, \quad \bar{J} \sim \frac{1}{c}\left(\sqrt{\frac{2n}{\pi}} + \frac{2}{3\pi}\right) \quad \text{[from (10.69)]}.$$

For $\mu < c$,

$$P_0(n, 1) \sim e^{-n\mu/c} \frac{(n\mu/c)^n}{n!}, \quad \bar{J} \sim n\left[\frac{1}{\mu} - \frac{1}{c}\right] + \frac{ne^{-n\mu/c}}{\mu} \frac{(n\mu/c)^n}{n!} \quad \text{[from (10.63)]}.$$

10.3.5 The equivalent random method

This section describes Kosten's model and the so-called *equivalent random method*, based on it and due to Wilkinson [96].

In Kosten's model (see Figure 3) a Poisson stream offers a erlangs to a server group of n iid exponential servers, called the *primary group*. The overflow stream, called the *O-stream*, is offered to a second infinite server group of iid servers with the same service distribution as that of the first (finite) group, called the *secondary group*. The equilibrium-offered load of the O stream is denoted by ℓ_O, and is measured in erlangs. The statistics of interest are the equilibrium time variance, σ_S^2, of the number of busy servers in the secondary group, S, and the peakedness,

$z_{O,exp} = \sigma_S^2/\ell_O$.

It can be shown that the O-stream is a renewal traffic process [90]. Consider now the continuous extension, $B(x,a)$, of the ordinary Erlang B function, $B(n,a)$ from (10.48), but with a continuous number of trunks, x, replacing the discrete number of trunks, n. When the O-stream is offered to the secondary server group, the resulting system is a GI/M/∞ queue (*ibid.*)

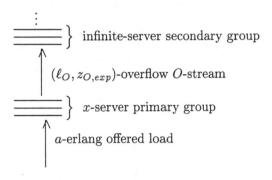

Figure 3. The Kosten traffic model.

The following fundamental formulas,

$$\ell_O = aB(x,a), \tag{10.76}$$

$$z_{O,exp}(\mu) = 1 - \ell_O + \frac{a}{x+1+\ell_O-a}, \tag{10.77}$$

are due to [56]. A direct derivation of these formulas proceeds as follows. Consider a G/G/1/1 queue, in which the arrival stream is a stationary simple point process (that is, arrivals occur singly). Let $N(t)$ be the number of arrivals in $(0,t]$, the origin being an arrival point, and let $M(t) = E[N(t)]$ be the attendant expectation function. The service distribution function is denoted by $B(t)$. Then the equilibrium probability, π, that an arrival finds the server busy is

$$\pi = \frac{\int\limits_0^\infty M(t)\, dB(t)}{1 + \int\limits_0^\infty M(t)\, dB(t)}.$$

To see that, note that in the G/G/1/1 queue, each busy cycle consists of just one arrival being served (the one inaugurating the busy cycle), with the remaining arrivals in that cycle being blocked; furthermore, since 0 is an arrival point, the server becomes busy at time 0. Hence, the numerator is the mean number of blocked arrivals during a service period, while the denominator is the total number of arrivals (served and blocked) during a busy cycle. The ratio is thus the requisite probability. In particular, for exponential service, one has $\pi = \mu\widetilde{M}(\mu)/(1 + \mu\widetilde{M}(\mu))$.

Consider a renewal stream with interarrival time density $a(t)$. Since, in this case, $\mu\widetilde{M}(\mu) = \widetilde{a}(\mu)/(1 - \widetilde{a}(\mu))$, one has

$$\pi = \widetilde{a}(\mu). \tag{10.78}$$

Consider now the primary group in Kosten's model and let γ_j be the probability that the overflow of the jth server finds the $(j+1)$th server busy. Then

$$\gamma_j = \frac{B(j+1,a)}{B(j,a)}. \tag{10.79}$$

The continuous extension of γ_j will be denoted by γ_x. Let $a(t)$ denote the equilibrium density of time between overflows in the O-stream. Then from (10.78) and (10.79),

$$\tilde{a}(\mu) = \frac{B(x+1,a)}{B(x,a)}. \tag{10.80}$$

Now, for any renewal traffic stream, A, one has [15]

$$\tilde{m}_A(\mu) = \frac{\tilde{a}(\mu)}{1 - \tilde{a}(\mu)}. \tag{10.81}$$

Suppose that O is a *renewal* overflow traffic stream in Kosten's setting. Substituting (10.81) into (10.38), and using the relations $\ell_O = \lambda_O/\mu$ (by definition) and $\ell_O = aB(x,a)$ (from (10.76)), yields

$$z_{O,exp}(\mu) = \frac{1}{1 - \tilde{a}(\mu)} - \ell_O. \tag{10.82}$$

Finally, combining (10.59) and (10.80) in (10.82) yields (10.77), with a modicum of algebra.

The Kosten formulas, (10.76) and (10.77), possess a unique solution for (x,a) in terms of ℓ_O and $x_{O,exp}(\mu)$ [52]. The following general traffic problem is of interest. A traffic stream, X, characterized by the offered load and peakedness parameters, $(\ell_X, z_{X,exp})$, is offered to a server group of c iid exponential servers, and the blocking probability is to be determined. This offered traffic is viewed as the overflow stream from a primary group of x servers resulting from a Poisson-offered load of a erlangs, such that equations (10.76) and (10.77) hold [96]. The corresponding parameters, (x,a), are called the *equivalent random parameters*; the "equivalence" is in the sense that the derived (x,a) are "equivalent" to the prescribed $(\ell_X, z_{X,exp})$, and the term "random" refers to a Poisson stream offered load. Figure 4 depicts the scheme.

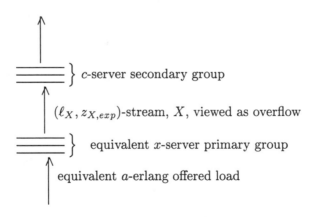

Figure 4. The equivalent random scheme.

The requisite blocking probability, P_b, is given by

$$P_b = \frac{B(x+c,a)}{B(x,a)}. \tag{10.83}$$

Next, view (10.76)-(10.77) as a formal equation system in the equivalent parameters, (x,a), not necessarily in an O-stream context. To obtain (x,a), first solve for a in the equation

$$\ell_X = aB\left(a\,\frac{\ell_X + z_{X,exp}}{\ell_X + z_{X,exp} - 1} - \ell_X - 1, a\right), \tag{10.84}$$

and then compute x by substituting the obtained parameter a from (10.84) into

$$x = a\,\frac{\ell_X + z_{X,exp}}{\ell_X + z_{X,exp} - 1} - \ell_X - 1. \tag{10.85}$$

An initial approximation, (x_0, a_0), readily obtained from (10.84) and (10.85) by expansion in powers of $1/\ell_X$, is given by [81]

$$a_0 = \ell_X z_{X,exp} + 3z_{X,exp}(z_{X,exp} - 1).$$

An accurate computation of a can then be obtained from the recursion

$$a_1 = a_0 - \frac{1}{B(x_0, a_0)}\,\frac{a_0 B(x_0, a_0) - \ell_X}{x_0 + \ell_X + a - a_0 - (x_0 + \ell_X + 1)\ln \alpha_0},$$

$$\alpha_0 = \frac{x_0 + 1}{a_0} + B(x_0, a_0).$$

When a stream characterized by $(\ell_X, z_{X,exp})$ is an O-stream, then x is an integer and the blocking probability P_b, given in (10.83), is exact. Otherwise, a non-integral value of x may be obtained. Nonetheless, (10.83) is used in practice as an approximation of the blocking probability.

When n independent streams, characterized by $(\ell_1, z_1),\ldots,(\ell_n, z_n)$, are superposed, then the superposition stream has parameters (ℓ, z) given by

$$\ell_X = \sum_{i=1}^{n}\ell_i, \quad z_{X,exp} = \frac{\sum_{i=1}^{n}\ell_i z_i}{\sum_{i=1}^{n}\ell_i}.$$

This follows on observing that each $\ell_i z_i$ is the variance of the number of busy servers produced by stream i when offered to an infinite server group of iid exponential servers with the same rate as that of the common server group, serving the superposition traffic stream.

If the target stream is the overflow of the common group, and is in turn offered (cascaded) to yet another group of c servers, then the peakedness of the target stream must be known, in order to compute its blocking probability. The requisite peakedness may be computed from (10.77) by first computing the equivalent random parameters, (x, a), and then considering the target stream to be the overflow from a group of $x + c$ servers with a Poisson-offered load of a erlangs.

Consider the *time congestion statistics* (the limiting fraction of time that all servers are busy) associated with a stream, characterized by $(\ell_X, z_{X,exp})$. The time congestion, P^*, in the cascaded system can be obtained using the results in [11]. For any non-negative integer j, define $\delta_j(n, a)$ by the following recursion on n [49]:

$$\delta_j(n, a) = \begin{cases} 1, & n = 0 \\[2mm] \dfrac{n\delta_j(n-1, a)}{a + j\delta_j(n-1, a)} + 1, & n > 0. \end{cases}$$

Letting $\delta_c(x, a)$ be the continuous extension of $\delta_j(n, a)$, for $j = c$ servers in the server group of the cascaded system, one has

$$P^* = \delta_c(x, a)B(x + c, a). \tag{10.86}$$

Of course, if the stream is not an O-stream, then (10.86) is taken as approximation. In order to compute $\delta_0(x,a)$ when x is not an integer, quadratic interpolation of $\ln\delta_c(x,a)$ is used. Recalling that $\lfloor x\rfloor$ denotes the integral part of x, and $\langle x\rangle$ the fractional part of x, one has

$$\delta_c(x,a) \simeq \delta_c(\lfloor x\rfloor,a)^{1-\langle x\rangle}\delta_c(\lfloor x\rfloor+1,a)^{\langle x\rangle}\left(\frac{\delta_c(\lfloor x\rfloor+1,a)^2}{\delta_c(\lfloor x\rfloor,a)\delta_c(\lfloor x\rfloor+2,a)}\right)^{\langle x\rangle(1-\langle x\rangle)/2}$$

The quantities, $\delta_c(x,a)$, have been shown to obey the inequalities [49]

$$\delta_c(x,a) \geq \frac{1}{2}\left(1+\frac{x-a+1}{c}+\sqrt{\left(1+\frac{x-a+1}{c}\right)^2+4\frac{c(a-1)-x}{c^2}}\right),$$

$$\delta_c(x,a) \leq \frac{1}{2}\left(1+\frac{x-a}{c}+\sqrt{\left(1+\frac{x-a}{c}\right)^2+\frac{4a}{c}}\right),$$

which are useful for real-time computations. In particular, the lower bound is often an excellent approximation.

10.3.6 The Hayward approximation

The equivalent random method is subject to the limitation that the service distribution is assumed to be exponential. This limitation may be removed by use of the so-called Hayward approximation of blocking probabilities for general service distributions.

Consider a c server system of iid servers with service distribution $F(x)$ and rate μ. A traffic stream, X, with parameters (ℓ_X, z_X, F_0) is offered to the server group. The *Hayward approximation* for the blocking probability, P_b, is given by [32]

$$P_b \simeq B\left(\frac{c}{z_{X,F_0}(\mu)}, \frac{a}{z_{X,F_0}(\mu)}\right), \tag{10.87}$$

where the function $B(\cdot,\cdot)$ above is the Erlang B function from (10.48).

Measurements of the peakedness functional are often made on a server whose service distribution, say G, is not in the requisite family, F_0. Thus, it is important to transform the measured information, that is, to transform z_{X,G_0} into z_{X,F_0}. To this end one employs the change-of-service distribution theory surveyed in Subsection 10.3.1.3 in the context of Hayward's approximation (10.87). As in the equivalent random procedure, it is important to calculate the peakedness, z_{Q,F_0}, of the overflow traffic stream, Q; note, though, that Q is generally not an O-stream. A modification of (10.77) yields [32]

$$z_{Q,F_0}(\mu) = z_{X,F_0}(\mu)\left[1-\ell_1+\frac{a_1}{c_1+1+\ell_1+a_1}\right], \tag{10.88}$$

where $c_1 = c/z_{X,F_0}(\mu)$, $a_1 = a/z_{X,F_0}(\mu)$, and $\ell_1 = a_1B(c_1,a_1)$. The peakedness in (10.88) can be further transformed to other service distributions, as described in Subsection 10.3.1.3, and the results used to approximate the blocking probability (10.87).

10.3.7 Waiting times for general traffic

A traffic process with an interarrival time sequence, $A = \{A_n\}$, is termed a *general traffic (GT) stream,* if the only known parameters are the offered load, ℓ_A, and the peakedness, $z_{A,exp}$, with respect to exponential service distributions. Consider a GT stream with a stationary interarrival time sequence, A, of rate λ_A. Suppose that the GT stream is offered to a finite-server group of k iid exponential servers with service rate μ, and that the waiting room is infinite. This queueing model will be denoted by GT/M/k/∞. The objective here is to calculate or approximate the equilibrium distribution, F_W, of the waiting time in queue (excluding service), W.

We begin with renewal traffic for two reasons: First, F_W can then be computed exactly, and second, the method used provides the basis for approximating F_W in non-renewal cases. Accordingly, we start with the GI/M/k/∞ queue. Let $r \in (0, 1)$ be the root of

$$[1 - r]\, z_{A,\,exp}(\mu k(1 - r)) = 1 - \frac{\ell_A}{k}. \tag{10.89}$$

Additionally, the Laplace transform of the renewal density function of A is

$$\tilde{m}_A(s) = z_{A,\,exp}(s) - 1 + \frac{\lambda_A}{s}, \tag{10.90}$$

and the transform of the interarrival time density function, $a(t)$, is

$$\tilde{a}(s) = 1 - \frac{1}{z_{A,\,exp}(s) + \lambda_A/s}. \tag{10.91}$$

From [90], one has

$$F_W^c(t) = \frac{D}{1 - r}\, e^{-\mu k(1 - r)t}, \tag{10.92}$$

$$E[W] = \frac{D}{\mu k(1 - r)^2}, \tag{10.93}$$

where $D^{-1} = \frac{1}{1-r} + \sum_{j=1}^{k} \frac{\binom{k}{j}}{M_j(1 - \tilde{a}(j\mu))} \frac{k(1 - \tilde{a}(j\mu)) - j}{k(1 - r) - j}$, and $M_j = \prod_{i=1}^{j} \tilde{m}_A(i\mu)$.

A significant aspect of (10.89) through (10.93) is that the GI/M/k/∞ queue is analyzed using only ℓ_A and $z_{A,exp}$. Thus, the same analysis could be formally carried out for the GT/M/k/∞ queue by assuming that F_W may be approximated by an exponential distribution of the form (10.92), which is exact for the GI/M/k/∞ queue [26]. The resulting F_W^c from (10.92) and $E[W]$ from (10.93) will then serve as approximations.

If only a single value of $z_{A,exp}(\mu)$ is known, for example, at $\mu = \mu_1$, then the *three moment match* technique [49] may be used to construct an *interrupted Poisson process* (IPP) [57], which in turn provides a usable choice of $z_{A,exp}(\mu)$ with a prescribed value at μ_1. This may now be used in (10.89)-(10.93) to carry out an approximate analysis of the GT/M/k/∞ model.

If at least two values of $z_{A,exp}(\mu)$ are known, then one may assume that the expectation density function is of the form

$$m_A(t) = \lambda_A + Ce^{-\alpha t},$$

for some positive constants C and α, so that for any stationary orderly traffic stream, A,

$$z_{A,\,exp}(\mu) = 1 + \frac{C}{\alpha + \mu}.$$

Thus, practical means are available for estimating queueing delays in a finite-server group of exponential servers, based only on the traffic parameters $(\ell_A, z_{A,exp})$.

10.3.8 Frequency Domain Approach to traffic

The Frequency Domain Approach (FDA) focuses on second-order statistics of offered traffic and their effect on queue response to that traffic [61, 62, 86]; it has been motivated by the need to characterize multimedia traffic in high-speed networks. FDA is distinguished by the fact that it directly utilizes the frequency domain (the traffic spectral functions) and advocates their use as a unified traffic measurement for analyzing and controlling queueing systems with heterogeneous offered traffic.

Analogously to periodic input functions in signal processing, elements of constant, sinusoidal, rectangular pulse, triangle pulse, and their superpositions are used in Li and Hwang [61] to represent various second-order traffic statistics, and to observe their effect on queueing performance. The main finding is that only the low frequencies in the traffic spectrum have a significant effect on queueing statistics. However, this approach has limited applications, since it does not capture the stochastic aspects of a prescribed traffic process. To this end, multistate MMPP traffic processes are used in Sheng and Li [86] to characterize binary sources (on/off traffic), and to study the effect of their second-order statistical properties on queue length and loss rate statistics.

A modeling technique for constructing Markovian traffic processes, which match a prescribed power spectrum, is introduced by Li and Hwang [62]. Let Q be the infinitesimal generator (transition rate matrix) of an N-state Markov chain, which modulates the Poisson rate of the traffic process with rate vector $\vec{\gamma}$. It is shown that the eigenstructure of Q characterizes the effect of such traffic on queueing performance. Assume that Q is diagonalizable with eigenvalues $\vec{\lambda} = (\lambda_0, \lambda_1, ..., \lambda_{N-1})$. The power spectrum of the random (Poisson) rate process can then be expressed by

$$P(\omega) = \sum_{j=0}^{N-1} \frac{1 - 2\psi_j \lambda_j}{\lambda_j^2 + \omega^2}. \tag{10.94}$$

Essentially, each eigenvalue, λ_j, contributes a bell-shaped component in the traffic rate power spectrum. Each bell is described by the average power, ψ_j, the central frequency, $Im[\lambda_j]$, and the half-power bandwidth, $-2Re[\lambda_j]$. Denoting $\vec{\psi} = (\psi_0, \psi_1, ..., \psi_{N-1})$, the traffic power spectrum is characterized by the pair, $(\vec{\psi}, \vec{\lambda})$, and the goal of FDA is to construct $(Q, \vec{\gamma})$ from $(\vec{\psi}, \vec{\lambda})$ — a rather difficult task, involving the so-called *inverse eigenvalue problem*. The technique in Li and Hwang [62] uses a special class of Markov chains, called *circulants*, with infinitesimal generators of the form

$$Q = \begin{pmatrix} a_0 & a_1 & a_2 & \cdots & a_{N-1} \\ a_{N-1} & a_0 & a_1 & \cdots & a_{N-2} \\ a_{N-2} & a_{N-1} & a_0 & \cdots & a_{N-3} \\ \vdots & \vdots & \vdots & \vdots & \vdots \\ a_1 & a_2 & a_3 & \cdots & a_0 \end{pmatrix}, \tag{10.95}$$

where each row is a circular right-shift of the previous one. A circulant-type modulated traffic rate process is then characterized by the pair, $(\vec{a}, \vec{\gamma})$, $\vec{a}$ being the first row of Q. Its power spectrum was shown to have

$$\vec{\lambda} = \sqrt{N}\vec{a}F^*, \quad \vec{\psi} = \frac{1}{N}|\vec{\gamma}F^*|^2, \tag{10.96}$$

where F is a Fourier matrix with $F_{k,j} = \frac{1}{\sqrt{N}}e^{-\frac{2\pi kj}{N}i}$, and $i = \sqrt{-1}$. Then $F^{-1} = F^*$, where F^* is the conjugate transpose of F. More importantly, one can deduce the corresponding pair, $(\vec{a}, \vec{\lambda})$, that gives rise to a prescribed power spectrum characterized by a pair, $(\vec{\lambda}, \vec{\psi})$, by taking the following constrained inverse discrete Fourier transform,

$$\vec{a} = \frac{1}{\sqrt{N}}\vec{\lambda}F \text{ subject to } a_j \geq 0 \text{ for } j > 0 \text{ and } a_0 = -\sum_{j=1}^{N-1}a_j,$$

$$\vec{\gamma} = \sqrt{N}\vec{\beta}F \text{ subject to } \gamma_j \geq 0, \text{ for all } j \text{ with } \beta_j = \sqrt{\psi_j}e^{i\theta j}, \tag{10.97}$$

where $\vec{\theta} = (\theta_0, \theta_1, \ldots, \theta_{N-1})$ is a phase vector. Hence, $\vec{a}$ depends only on $\vec{\lambda}$, and $\vec{\gamma}$ only on $(\vec{\psi}, \vec{\theta})$. It should be noted that while $\vec{\theta}$ is unrelated to the power spectrum, it influences the steady-state distribution and higher-order traffic statistics [62]. The numerical analyses in [62, 85] are based on the Folding Algorithm, which is a fast computational method for the analysis of a finite quasi-birth-death process with level-dependent transitions [98].

A significant advantage of using circulants in FDA is the ability to identify the impact of first-order (marginal distribution), second-order (power spectrum or autocorrelation function), and higher-order traffic statistics (such as the bispectrum and trispectrum) on queue length and loss-rate statistics. Interestingly, only second-order traffic statistics appear to have a significant effect on queueing performance, with the low frequencies exerting the most effect. This observation has important potential applications to traffic measurement, especially since first-order and second-order statistics are much easier to measure than higher-order ones. FDA has been applied, in this vein, to traffic rate control [63], to link capacity allocation and network control [64], and to the design and analysis of buffer congestion control [85].

10.3.9 Effective bandwidth of traffic

Most communications services are subject to performance constraints designed to guarantee a minimal quality of service (QOS). For example, the delivery of a video frame to a monitor must be both timely and not overly lossy. Consider a general traffic stream offered to a deterministic server, and assume that some prescribed parametrized performance constraints are required to hold. The *effective bandwidth* of the traffic stream corresponds to the minimal deterministic service rate, required to meet these constraints. Queueing-oriented performance constraints include bounds on such statistics as queueing delay quantiles or averages, server utilization, overflow probabilities, etc. [53]. This approach has the advantage of focusing directly on the relevant performance criteria rather than on the statistics of the offered traffic.

To illustrate this viewpoint, suppose that a traffic stream shares a link with buffering capacity B and bandwidth c. Suppose further that the QOS constraint is that the overflow probability satisfy

$$P\{W > B\} \leq e^{-\delta B}, \tag{10.98}$$

where W denotes the stationary workload in the buffer and δ is a QOS parameter. In this example, the corresponding effective bandwidth of the traffic stream, $\alpha(\delta)$, is the minimal deterministic service guaranteeing (10.98). Operationally, if the goal is to ensure that the overflow probability incurred by a traffic stream offered to a deterministic server is small (say, 10^{-9} for a particular δ), then a link bandwidth of at least $\alpha(\delta)$ should be allocated to carry that traffic. As $\delta \geq 0$ increases, the overflow constraint becomes more stringent and the effective bandwidth increases from the mean arrival rate to the peak rate. Viewed this way, the effective bandwidth concept serves as a compromise between two alternative bandwidth allocation schemes, representing a pessimistic and an optimistic outlook. The strict one allocates bandwidth based on stream peak rate, seeking to eliminate losses, whereas the lenient one allocates bandwidth based on stream average rate, merely seeking to guarantee stability.

Generally, effective bandwidths are difficult to compute, as their value depends both on traffic stream statistics as well as possible interactions with cross traffic traversing the same system. Occasionally, a direct calculation of the effective bandwidth is feasible; a case in point are Markov fluid sources, where the overflow probability can be computed explicitly as a function of the service rate via spectral expansions [27, 35, 39].

An alternative approach is to consider asymptotically large buffers B and find approximate effective bandwidths for the streams, see [12, 19, 54, 95]. For example, consider an unbounded buffer shaped by a heterogeneous mix of multiplexed streams, of which n_j are of type $j \in J$, where J is the set of traffic types. For service rate c, it can be shown that

$$\sum_{j \in J} n_j \alpha_j(\delta) \leq c \Leftrightarrow \lim_{B \to \infty} \frac{1}{B} \log P\{W > B\} \leq -\delta,$$

where $\alpha_j(\delta)$ is the *effective bandwidth* of stream j and depends only on the statistics of stream j, say, in terms of the interarrival times, $\{A_i^j\}_{i=1}^{\infty}$. This approximation can be established for a relatively large class of arrival streams which are stationary, ergodic and have weak dependencies, such as Markov or mixing processes. For example, a traffic stream with slotted arrivals has an effective bandwidth, given by

$$\alpha_j(\delta) = \frac{\Lambda_j(\delta)}{\delta}, \text{ where } \Lambda_j(\delta) = \lim_{n \to \infty} \frac{1}{n} \log E[\exp(\delta \sum_{i=1}^{n} A_i^j)],$$

which can be computed for many standard traffic models, including Markov fluid and MMPP.

The notion of effective bandwidth provides a useful tool for studying resource requirements of telecommunications services and the impact of different management schemes on network performance. A study of the impact of filtering, traffic discard, and traffic shaping on the effective bandwidth of traffic streams may be found in [18]. Extensions of these results to networks of queues are proposed in [19].

10.4 DISCUSSION AND OPEN PROBLEMS

From a practical point of view, stochastic models of traffic streams are considered relevant to network traffic engineering and performance analysis, to the extent that they are able to predict system performance measures to a reasonable degree

of accuracy; additionally, a practitioner's confidence in a traffic model is greatly enhanced if the model can capture visual features of the empirical traffic, in addition
to approximating its statistics. The fundamental systems, of which traffic is a key
ingredient, are queueing systems. While traditional analytical models of traffic
were often devised and selected for the analytical tractability they induced in the
corresponding queueing systems, this selection criterion is largely absent from
recent traffic models. In particular, queueing systems with offered traffic consisting of autoregressive-type processes (Subsection 10.2.4), TES processes (Subsection
10.2.5), and self-similar processes (Subsection 10.2.6) are difficult to solve
analytically. Consequently, these are only used to drive Monte Carlo simulation
models. On the other hand, some fluid models (Subsection 10.2.3) are analytically
tractable, but only subject to considerable restrictions. Similarly, FDA models
(Subsection 10.3.8) and effective bandwidth models (Subsection 10.3.9) are largely
analytically tractable, but at the cost of simple models, and restrictive modeling
assumptions. Thus, the most significant traffic research problem is to solve
analytically induced queueing systems; or in the absence of a satisfactory solution,
to devise approximate traffic models that lead to analytically tractable queueing
systems.

Traditional traffic models (Subsections 10.2.1-10.2.4) have served admirably in
advancing traffic engineering and understanding performance issues, primarily in
traditional telephony. The advent of modern high-speed communicatons networks
(e.g., B-ISDN under ATM) is ushering in a dramatically heterogeneous traffic mix.
The inherent burstiness of several important B-ISDN services (mainly compressed
video and file transfer) is bringing to the fore some serious modeling inadequacies
of traditional models, particularly in regard to temporal dependence, both short-
range and long-range. This realization has brought about renewed interest in
traffic modeling and has driven the development of new models utilizing non-tradi-
tional paradigms, such as TES processes and self-similar processes.

The importance of empirical second-order traffic statistics, such as indices of
dispersion (Subsection 10.3.1.2) and peakedness (Subsection 10.3.1.3), has already
been recognized in traditional traffic engineering, and again in recent work on traffic models, including TES, self-similar, and FDA models. Of those, only the TES
modeling paradigm addresses itself directly to capturing both the empirical marginal distribution (histogram) and empirical autocorrelation function, simultaneously, and additionally calls for qualitative "resemblance" of sample paths. The availability of TES modeling software, incorporating both heuristic and algorithmic fitting procedures, renders the TES methodology a practical paradigm for constructing traffic models directly from empirical data, but the absence of corresponding analytically tractable queueing models currently restricts it to Monte Carlo
simulations. An important traffic research problem, motivated by the TES approach, is to find when first-order and second-order statistics are insufficient for
traffic characterization, and whether using higher-order spectral functions (such as
the bispectrum and trispectrum) can remedy such characterization problems.

Initially, the validity and efficacy of traffic models for high-speed networks was
difficult to assess due to the paucity or unavailability of empirical data, but more
recently, increasing volumes of traffic measurements from working high-speed networks (e.g., CCS/SS7 at 56 kbps, ISDN at 1.5 Mbps, Ethernet at 10 Mbps) have
been made available to researchers. In particular, the traffic studies mentioned in
Subsection 10.2.6 suggest that certain Ethernet traffic has distinguishing features,
often referred to as *fractal properties*; these typically pertain to notions such as

slowly decaying variances, long-range dependence, $1/f$-noise, or self-similarity. Statisticians are now aware that ignoring long-range dependence can have drastic consequences for many statistical methods [6]. However, traffic engineers and network managers will only be convinced of the practical relevance of fractal traffic models by direct arguments, concerning the impact of fractal properties on network performance. Thus, fractal traffic (including stochastic modeling, statistical inference, synthetic traffic generation, and queueing analysis) opens a new and challenging area of mathematical research, where few results are available, and what little is known (e.g., [33, 59]) is mostly preliminary in nature and illustrates the difficulties presented by these models.

Recent results reported in [23, 79] demonstrate that performance measures of queueing systems with fractal traffic can differ drastically from those predicted by corresponding systems with traditional traffic models (see also [30, 59]). Traditional models are way too optimistic as compared to bursty TES traffic [66] and self-similar models. A case in point is the tail behavior of the steady-state queue-length distribution $P\{Q > x\}$ (x large) in a single-server, infinite-capacity queue [23, 79]. For Markovian offered traffic, these tail probabilities are known to be *asymptotically exponential* [1, 36], i.e.,

$$P\{Q > x\} \sim e^{-\eta x}, \text{ as } x \to \infty, \tag{10.99}$$

where $\eta > 0$ is called the *asymptotic decay rate*. Recall that relation (10.99) underlies the concept of effective bandwidth, where admission control or service capacity allocation are based on tail probabilities of select random variables (e.g., equation (10.98)). In contrast to (10.99), traffic streams with long-range dependence (in particular, fractional Brownian motion-based models) give rise to corresponding *Weibull-like* asymptotic tail probabilities, i.e.

$$P\{Q > x\} \sim e^{-\gamma x^{\beta}}, \text{ as } x \to \infty, \tag{10.100}$$

where γ is a constant, and $\beta = 2 - 2H \in (0, 1)$ [23, 79]. Equations (10.99) and (10.100) have very different asymptotic characteristics, the former providing relatively optimistic predictions as compared to the latter; see also [33, 79] for additional queueing examples. An open problem of considerable interest is whether other traffic models tend to give rise to conservative performance measures, as compared to their empirical counterparts.

If second-order or fractal properties (e.g., the Joseph Effect and Noah Effect) lead to fundamentally distinct system behavior, then the ability of tractable traffic models to capture such properties and their impact (say, on queueing statistics) become of interest. For example, while non-fractal models (such as Markov-based ones) have inherently short-range dependence, it is nevertheless known that adding parameters can lead to models with "approximate" fractal features, e.g., Markovian models with "supplemental" states, and TES-based or FDA-based models with low frequency components of "enhanced" magnitude, particularly the DC component. A judicious choice of a traffic model could lead to tractable queueing models capable of approximating their intractable counterparts. However, one can expect to give up model parsimony, and this approach may work for some performance aspects and fail for others.

Overall, the variety of traffic classes introduced by emerging high-speed communications networks will doubtlessly prove to be a rich source of research, both theoretical and applied. One can fully expect traffic engineers to call upon statisticians and probabilists to develop new theoretical and applied tools to assist in traffic handling at every stage of a communications systems life cycle, including network design, analysis, and operation.

ACKNOWLEDGEMENTS

We are grateful to San-Qi Li for the material on the frequency domain approach, and to Gustavo de Veciana for the material on effective bandwidths. We also thank Robert Cooper, Gustavo de Veciana, and Robert Maier for reading and commenting on the manuscript. Special thanks are due to Tomasz Rolski for a careful reading of the paper and for numerous useful suggestions.

BIBLIOGRAPHY

[1]   Abate, J., Choudhury, G.L. and Whitt, W., Asymptotics for steady-state tail probabilities in structured Markov chains, *Stoch. Models* **10** (1994), 99-143.

[2]   Anick, D., Mitra, D. and Sondhi, M.M., Stochastic theory of a data-handling system with multiple sources, *The Bell System Tech. J.* **61**:8 (1982), 1871-1894.

[3]   Asmussen, S., *Applied Probability and Queues*, Wiley, New York 1987.

[4]   Asmussen, S. and Koole, G., Marked point processes as limits of Markovian arrival streams, *J. Appl. Prob.* **30** (1993), 365-372.

[5]   Bendat, J.S. and Piersol, A.G., *Random Data*, Wiley-Interscience, New York 1986.

[6]   Beran, J., Statistical methods for data with long-range dependence, *Statistical Science* **7**:4 (1992), 404-427.

[7]   Beran, J., Sherman, R., Taqqu, M.S. and Willinger, W., Long-range dependence in variable-bit-rate video traffic, *IEEE Trans. on Commun.* **43** (1995), 1566-1579.

[8]   Błaszczyszyn, B., Rolski, T. and Schmidt, V., Light-traffic approximations in queues and related stochastic models, In: *Advances in Queueing: Theory, Methods, and Applications* (ed. by J.H. Dshalalow), CRC Press, Boca Raton, Florida (1995), 379-406.

[9]   Box, G.E.P. and Jenkins, G.M., *Time Series Analysis: Forecasting and Control*, Prentice Hall 1976.

[10]   Brémaud, P., *Point Processes and Queues: Martingale Dynamics*, Springer Verlag 1981.

[11]   Brockmeyer, E., The simple overflow problem in the theory of telephone traffic, *Teleteknik* **5** (1954), 361-374.

[12]   Chang, C.S., *Stability, Queue Length and Delay, Part II: Stochastic Queueing Networks*, IBM Research Report No. **RC 17709**, T.J. Watson Research Center, Yorktown Heights, New York 1992.

[13]   Çinlar, E., *Introduction to Stochastic Processes*, Prentice-Hall, Englewood Cliffs 1975.

[14]   Cooper, R.B., *Introduction to Queueing Theory*, North Holland, New York 1981.

[15]   Cox, D.R., *Renewal Theory*, Methuen and Co., London 1962.

[16]   Cox, D.R., Long-range dependence: A review, In: *Statistics: An Appraisal*

(ed. by H.A. David and H.T. David), The Iowa State University Press, Ames, Iowa (1984), 55-74.

[17] Davis, P.J. and Rabinowitz, P., *Methods of Numerical Integration*, Academic Press, New York 1975.

[18] de Veciana, G. and Walrand, J., Effective bandwidths: Call admission, traffic policing and filtering for ATM networks, *Queueing Systems* **20** (1994), 37-59.

[19] de Veciana, G., Courcoubetis, C. and Walrand, J., Decoupling bandwidths for networks: A decomposition approach to resource management for networks, Memorandum **UCB/ERL M93/50** (1993), Berkeley.

[20] Dellacherie, C. and Meyer, P.-A., *Probabilities and Potential B*, North Holland 1982.

[21] Devroye, L., *Non-Uniform Random Variate Generation*, Springer-Verlag, New York 1986.

[22] Doetsch, G., *Theorie und Anwendung der Laplace-Transformation*, Dover Publications, New York 1943.

[23] Duffield, N.G. and O'Connell, N., Large deviations and overflow probabilities for the general single-server queue, with applications, *Dublin Institute for Adv. Studies* (1993), preprint.

[24] Duffy, D.E., McIntosh, A.A., Rosenstein, M. and Willinger, W., Statistical analysis of CCSN/SS7 traffic data from working CCS subnetworks, *IEEE J. Select. Areas Commun.* **12**:3 (1994), 544-551.

[25] Eckberg, A.E., Generalized peakedness of teletraffic processes, *Proc. 10th International Teletraffic Congress*, Montreal, Canada 1983.

[26] Eckberg, A.E., Approximations for bursty (and smoothed) arrival delays based on generalized peakedness, *Proc. 11th International Teletraffic Congress*, Kyoto, Japan 1985.

[27] Elwalid, A.I. and Mitra, D., Effective bandwidth of general Markovian traffic sources and admission control of high speed networks, *IEEE/ACM Trans. Networking* **1**:3 (1993), 329-343.

[28] Fendick, K. and Whitt, W., Measurements and approximations to describe the offered traffic and predict the average workload in a single-server queue, *Proc. of the IEEE* **77** (1989), 171-194.

[29] Fendick, K.W., Saksena, V.R. and Whitt, W., Dependence in packet queues, *IEEE Trans. on Comm.* **37** (1989), 1173-1183.

[30] Fowler, H.J. and Leland, W.E., Local area network traffic characteristics, with implications for broadband network congestion management, *IEEE J. Select. Areas Commun.* **9**:7 (1991), 1139-1149.

[31] Franken, P., Koenig, D., Arndt, U. and Schmidt, V., *Queues and Point Processes*, Akademie-Verlag 1981.

[32] Fredericks, A.A., Congestion in blocking systems − A simple approximation technique, *Bell System Tech. J.* **59**:6 (1980), 805-827.

[33] Garrett, M.W. and Willinger, W., Analysis, modeling and generation of self-similar VBR video traffic, *Proc. of the ACM Sigcomm '94*, London, UK (1994), 269-280.

[34] Geist, D. and Melamed, B., TEStool: An environment for visual interactive modeling of autocorrelated traffic, *Proc. of the 1992 Intern. Conf. on Commun.*, Chicago, Illinois (1992), 1285-1289.

[35] Gibbens, R.J. and Hunt, P.J., Effective bandwidths for the multi-type UAS channel, *Queueing Systems* **9** (1991), 17-28.

[36] Glynn, P.W. and Whitt, W., Logarithmic asymptotics for steady-state tail probabilities in a single-server queue, In: *Studies in Applied Probab., Essays in honor of Lajos Takács* (ed. by J. Galambos and J. Gani), Applied Probability Trust, Sheffield, England (1994), 131-156. Also in *J. Appl. Prob.* special volume **31A** (1994).

[37] Granger, C.W.J. and Joyeux, R., An introduction to long-memory time series models and fractional differencing, *Time Series Anal.* **1** (1980), 15-29.

[38] Granger, C.W.J., Long memory relationships and the aggregation of dynamic models, *J. Econometrics* **14** (1980), 227-238.

[39] Guérin, R., Ahmadi, H. and Naghshineh, M., Equivalent capacity and its application to bandwidth allocation in high-speed networks, *IEEE JSAC* **9** (1991), 968-981.

[40] Gusella, R., Characterizing the variability of arrival processes with indexes of dispersion, *IEEE J. on Selected Areas in Commun.* **9**:2 (1991), 203-211.

[41] Hardy, G.H., Littlewood, J.E. and Polya, G., *Inequalities*, Cambridge University Press 1959.

[42] Heffes, H. and Lucantoni, D.M., A Markov modulated characterization of packetized voice and data and related statistical multiplexer performance, *IEEE J. on Selected Areas in Commun.* **SAC-4** (1986), 856-868.

[43] Heyman, D., Tabatabai, A. and Lakshman, T.V., Statistical analysis and simulation study of video teletraffic in ATM networks, *IEEE Trans. Circuits and Systems for Video Technology* **2** (1992), 49-59.

[44] Hosking, J.R.M., Fractional differencing, *Biometrika* **68** (1981), 165-176.

[45] Hurst, H.E., Long-term storage capacity of reservoirs, *Trans. Amer. Soc. Civil Eng.* **116** (1951), 770-799.

[46] Jacod, J., *Calcul Stochastique et Problème de Martingales*, Lecture Notes in Math. **714**, Springer-Verlag 1979.

[47] Jagerman, D.L., Some properties of the Erlang loss function, *Bell System Tech. J.* **53**:3 (1974), 525-551.

[48] Jagerman, D.L., Methods in traffic calculations, *Bell System Tech. J.* **63**:7 (1984), 1283-1310.

[49] Jagerman, D.L., Laplace transform inequalities with application to queueing, *Bell System Tech. J.* **64**:7 (1985), 1755-1764.

[50] Jagerman, D.L. and Melamed, B., The transition and autocorrelation structure of TES processes - part I: General theory, *Stoch. Models* **8**:2 (1992), 193-219.

[51] Jagerman, D.L. and Melamed, B., The transition and autocorrelation structure of TES processes - part II: Special cases, *Stoch. Models* **8**:3 (1992), 499-527.

[52] Jagers, A.A. and Van Doorn, E.A., On the continued Erlang loss function, *Oper. Res. Letters* **5**:1 (1986), 43-46.

[53] Kelly, F.P., Effective bandwidths at multi-class queues, *Queueing Systems* **9** (1991), 5-15.

[54] Kesidis, G., Walrand, J. and Chang, C.S., Effective bandwidths for multi-class Markov fluids and other ATM sources, *IEEE/ACM Trans. Networking* **1**:4 (1993), 424-428.

[55] Kobayashi, H. and Ren, Q., A mathematical theory for transient analysis of communications networks, *IEICE Trans. on Commun.* **E75-B**:12 (1992), 1266-1276.

[56] Kosten, L., *Stochastic Theory of Service Systems*, Pergamon Press 1973.

[57] Kuczura, A., The interrupted Poisson process as an overflow process, *Bell System Tech. J.* **52**:3 (1973), 437-448.

[58] Lee, D.S., Melamed, B., Reibman, A. and Sengupta, B., TES modeling for analysis of a video multiplexer, *Perf. Evaluation* **16** (1992), 21-34.

[59] Leland, W.E., Taqqu, M.S., Willinger, W. and Wilson, D.V., On the self-similar nature of Ethernet traffic (extended version), *IEEE/ACM Trans. on Networking* **2**:1 (1994), 1-15.

[60] Levy, J.B. and Taqqu, M.S., On renewal processes having stable inter-renewal intervals and stable rewards, *Ann. Sc. Math. Quebec* **11** (1987), 95-110.

[61] Li, S.Q. and Hwang, C.L., Queue response to input correlation functions: Discrete spectral analysis, *IEEE/ACM Trans. on Networking* **1**:5 (1993), 522-533.

[62] Li, S.Q. and Hwang, C.L., Queue response to input correlation functions: Continuous spectral analysis, *IEEE/ACM Trans. on Networking* **1**:6 (1993), 678-692.

[63] Li, S.Q. and Chong, S., Fundamental limits of input rate control in high-speed network, *Proc. IEEE Infocom '93*, San Francisco, CA (1993), 662-671.

[64] Li, S.Q., Chong, S. and Hwang, C.L., Link capacity allocation and network control by filtered input rate in high speed networks, *IEEE/ACM Trans. on Networking* **3**:1 (1995), 10-25.

[65] Lighthill, M.J., *Introduction to Fourier Analysis and Generalized Functions*, Cambridge University Press 1958.

[66] Livny, M., Melamed, B. and Tsiolis, A.K., The impact of autocorrelation on queueing systems, *Management Science* **39** (1993), 322-339.

[67] Lucantoni, D.M., Meier-Hellstern, K.S. and Neuts, M.F., A single-server queue with server vacations and a class of non-renewal arrival processes, *Adv. Appl. Prob.* **22** (1990), 676-705.

[68] Larson, H.J. and Shubert, B.O., *Probabilistic Models in Engineering Sciences*, Wiley, New York 1979.

[69] Mandelbrot, B.B., Self-similar error clusters in communication systems and the concept of conditional stationarity, *IEEE Trans. Commun. Tech.* **COM-13** (1965), 71-90.

[70] Mandelbrot, B.B. and Van Ness, J.W., Fractional Brownian motions, fractional noises and applications, *SIAM Review* **10** (1968), 422-437.

[71] Mandelbrot, B.B., Long-run linearity, locally Gaussian processes, H-spectra and infinite variances, *Intern. Econom. Review* **10** (1969), 82-113.

[72] Mandelbrot, B.B. and Wallis, J.R., Some long-run properties of geophysical records, *Water Resources Research* **5** (1969), 321-340.

[73] Mandelbrot, B.B. and Taqqu, M.S., Robust R/S analysis of long run serial correlation, *Proc. 42nd Session ISI* **XLVIII**:2 (1979), 69-99.

[74] Mandelbrot, B.B. *The Fractal Geometry of Nature*, Freeman, New York 1983.

[75] Meier-Hellstern, K., Wirth, P.E., Yan, Y.-L. and Hoeflin, D.A., Traffic models for ISDN data users: Office automation application, *Proc. 13th Intern. Teletraffic Congress*, Copenhagen, Denmark (1991), 167-172.

[76] Melamed, B., An overview of TES processes and modeling methodology, In: *Perf. Evaluation of Computer and Commun. Systems* (ed. by L. Donatiello and R. Nelson), Springer-Verlag Lecture Notes in Computer Science (1993),

359-393.

[77]   Melamed, B. and Sengupta, B., TES modeling of video traffic, *IEICE Trans. on Commun.* **E75-B**:12 (1992), 1292-1300.

[78]   Melamed, B. and Yao, D.D., The ASTA property, In: *Advances in Queueing: Theory, Methods, and Open Problems* (ed. by J.H. Dshalalow), CRC Press, Boca Raton, Florida (1995), 195-224.

[79]   Norros, I., A storage model with self-similar input, *Queueing Systems* **16** (1994), 387-396.

[80]   Ramamurthy, G. and Sengupta, B., Modeling and analysis of a variable bit rate video multiplexer, *Proc. of INFOCOM '92*, Florence, Italy (1992), 817-827.

[81]   Rapp, L.Y., Planning of junction networks in a multi-exchange area − part I, *Ericson Technics* **20** (1962), 22-130.

[82]   Reibman, A.R., DCT-based embedded coding for packet video, *Signal Processing: Image Communication* **3** (1991), 231-237.

[83]   Schassberger, R.A., Investigation of queueing and related systems with the phase method, *Proc. 7th Intern. Teletraffic Congress* (1973).

[84]   Sen, P., Maglaris, B., Rikli, N.-E. and Anastossiou, D., Models for packet switching of variable-bit-rate video sources, *IEEE J. on Selected Areas in Communications* **7**:5 (1989), 865-869.

[85]   Sheng, H.D. and Li, S.Q., Spectral analysis of packet loss rate at statistical multiplexer for multimedia services, *IEEE/ACM Trans. on Networking* **2**:1 (1994), 53-65.

[86]   Sheng, H.D. and Li, S.Q., Second order effect of binary sources on characteristics of queue and loss rate, *IEEE Trans. on Commun.* **42**:3 (1994), 1162-1173.

[87]   Sigman, K., *Stationary Marked Point Processes*, Chapman and Hall, New York 1995.

[88]   Sriram, K. and Whitt, W., Characterizing superposition arrival processes in packet multiplexers for voice and data, *IEEE J. Select. Areas in Commun.* **4**:6 (1986), 833-846.

[89]   Steffenson, J.F., *Interpolation*, Chelsea, New York 1950.

[90]   Takács, L., *Introduction to the Theory of Queues*, Oxford University Press, New York 1962.

[91]   Takine, T., Sengupta, B. and Hasegawa, T., An analysis of a discrete-time queue for broadband ISDN with priorities among traffic classes, *IEEE Trans. on Commun.* **42**:2/3/4 (1994), 1837-1845.

[92]   Taqqu, M.S. and Levy, J.B., Using renewal processes to generate long-range dependence and high variability, In: *Dependence in Probability and Statistics* (ed. by E. Eberlein and M.S. Taqqu), Progress in Prob. and Stat., Birkhauser, Boston **11** (1986), 73-89.

[93]   Taqqu, M.S., Self-similar processes, *Encyclopedia of Statistical Sciences* **8** (ed. by S. Kotz and N. Johnson), Wiley, New York 1987.

[94]   Wei, W.W.S., *Time Series Analysis: Univariate and Multivariate Methods*, Addison-Wesley 1990.

[95]   Whitt, W., Tail probabilities with statistical multplexing and effective bandwidths in multi-class queues, *Telecommun. Systems* **2** (1993), 71-107.

[96]   Wilkinson, R., Theories for toll traffic engineering in the U.S.A., *Bell System Tech. J.* **35**:2 (1956), 421-514.

[97]   Willinger, W., Taqqu, M.S., Leland, W.E. and Wilson, D.V., Self-similarity

in high-speed packet traffic: Analysis and modeling of Ethernet traffic measurements, *Statistical Science* **10** (1995), 67-85.

[98]   Ye, J. and Li, S.Q., Folding algorithm: A computational method for finite QBD processes with level-dependent transitions, *IEEE Trans. on Commun.* **42**:2 (1994), 625-639.

Chapter 11
Fluid models for single buffer systems[1]

V.G. Kulkarni

ABSTRACT This chapter considers a stochastic fluid model of a buffer content process $\{X(t), t \geq 0\}$ that depends upon an external environment process $\{Z(t), t \geq 0\}$ as follows: whenever the environment is in state z the X process changes state at rate $\eta(z)$. The X process is restricted to stay in $[0, B]$, where $B \leq \infty$. The aim is to study the steady state distribution of the bivariate process $\{(X(t), Z(t)), t \geq 0\}$. Three main cases are considered: the environment is (*i*) a continuous time Markov chain (CTMC), (*ii*) a CTMC and white noise, and (*iii*) an Ornstein-Uhlenbeck process. Spectral representations are obtained for the steady state distributions. Finally, an extension to state dependent drift is considered, where the rate of change of the X process depends on both the X and Z processes. The chapter ends with some interesting open problems in this area.

CONTENTS

11.1 INTRODUCTION

A stochastic fluid flow system is an input-output system where the input is modeled as a continuous fluid that enters and leaves a storage device, called a buffer, according to randomly varying rates. Such models are motivated as approximations to discrete queueing models of manufacturing systems, high-speed data networks, etc. They have also been used in transportation systems, theory of dams, queueing theory, etc.

The main purpose of this chapter is to provide a general framework for a variety of fluid models that have been studied in the literature. This is accomplished by introducing a stochastic process to model the external random environment, which modulates the input and output rates of the fluid to the buffer. We then proceed to classify the models according to the stochastic nature of the random environment.

---

[1]This research was partially supported by NSF Grant No. NCR-9406823.

The fluid models where the external environment is a continuous time Markov chain have been used in data communication networks. For example, Anick et al. [3], Kosten [20, 21], and Kosten and Vrieze [22] treat a single buffer that receives input data from several independent sources, each source switching between on and off states according to a two-state CTMC. (This is referred to as the AMS model.) The data is removed from the buffer at a fixed rate. Mitra [28, 29] considers a generalization of this model with multiple input sources and output channels, both subject to on-off switching. This work continues in Elwalid and Mitra [7] and Stern and Elwalid [37].

More recently, Asmussen [2] and Kulkarni and Karandikar [26] have studied fluid models where the environment process is a CTMC and white noise. These models are useful in communication networks when the effect of jitter in the channels needs to be accounted for. These are also called *second order models*, since they introduce the variance terms explicitly in the model.

A special limiting case of the AMS model leads to fluid models where the external environment is an Ornstein-Uhlenbeck (OU) process. Such models are studied by Simonian [35], Simonian and Virtamo [36], and Kulkarni and Rolski [27].

Although we are motivated by telecommunication applications, fluid models have been used earlier in transportation systems to model the flow of vehicles at a traffic intersection (see Newell [32]) and in dam theory (see Moran [30]). In queueing theory, the work content process can be thought of as a fluid model where the fluid arrives in instantaneous quantities and leaves at a continuous rate (see Gaver and Miller [10] and Prabhu [33]). In a recent paper, Chen and Yao [6] consider a single source model with general on and off times. This can be thought of as a fluid model with a simple two-state semi-Markovian environment process. Kella and Whitt [16] study a stochastic fluid model with Levy input.

The rest of the chapter is organized on the basis of the stochastic nature of the external environment. General stability results are stated in Section 11.2. In Section 11.3 we consider the case where the environment is a CTMC and summarize the important results. Section 11.4 describes several applications of the model of Section 11.3 to high-speed networks. It includes applications to a leaky-bucket control scheme, the statistical multiplexing problem, and the admission control policies using effective bandwidth concepts. Relevant literature is cited at appropriate places.

Section 11.5 deals with a bivariate environment process: one component is a CTMC and the other component is white noise. In Section 11.6 we consider an Ornstein-Uhlenbeck process as a random environment. In Section 11.7 we consider the case where the input and output rates depend on the buffer content as well as the external environment. This case is useful in modeling the buffer-sharing schemes in high-speed networks. The chapter concludes with a brief discussion of the open problems in the area.

11.2 THE MODEL

In this section we describe a general model of fluid entering and leaving a single buffer system. The input and output rates of the fluid depend on an an external environment as follows: Let $Z(t)$ be the state of the environment at time t and $X(t)$ the amount of fluid in the buffer at time t. Let $\dot{\eta}(Z(t))$ be the net input rate (entry rate $-$ exit rate) at time t. $\eta(\cdot)$ is called the *drift function*. (In Section

11.7 we shall consider the case where the input and output rates depend on $X(t)$ as well   as $Z(t)$, i.e., the drift function is given by $\eta(Z(t), X(t))$.)

When the buffer capacity is infinite, the dynamics of the buffer content process $X = \{X(t), t \geq 0\}$ is given by

$$\frac{dX(t)}{dt} = \begin{cases} \eta(Z(t)) & \text{if } X(t) > 0, \\ (\eta(Z(t)))^+ & \text{if } X(t) = 0, \end{cases} \tag{11.1}$$

where $(x)^+ = \max(x, 0)$. The special form at $X(t) = 0$ ensures that the X process does not become negative. When the buffer capacity is finite, say B, the dynamics is given by

$$\frac{dX(t)}{dt} = \begin{cases} (\eta(Z(t)))^+ & \text{if } X(t) = 0, \\ \eta(Z(t)) & \text{if } 0 < X(t) < B, \\ (\eta(Z(t)))^- & \text{if } X(t) = B, \end{cases} \tag{11.2}$$

where $(x)^- = \min(0, x)$. The form at $X(t) = B$ prevents the buffer content from exceeding B.

Definition 11.1. The X process is called a *fluid input-output process (or a fluid process, for short) driven by the Z process.*

In this chapter we shall assume the buffer capacity to be infinite unless otherwise mentioned. To solve the equation (11.1) with the initial condition $X(0) = x$, we write

$$Y(t) = x + \int_0^t \eta(Z(s))ds. \tag{11.3}$$

Integrating both sides of equation (11.1) and manipulating the result, we get (see Prabhu [33])

$$X(t) = Y(t) - \int_0^t (\eta(Z(s)))^- 1_{\{X(s) = 0\}}ds. \tag{11.4}$$

Using the standard argument of queueing theory, we get

$$X(t) = Y(t) - \inf_{0 \leq u \leq t} (0, Y(u)) \tag{11.5}$$

$$= \sup_{0 \leq u \leq t} (Y(t), \int_u^t \eta(Z(s))ds. \tag{11.6}$$

The following theorem gives the result about the existence of the stationary distribution for the buffer-content process. (See Borovkov [4].)

Theorem 11.2. *Suppose $\{Z(t), t \geq 0\}$ is stationary and ergodic with*

$$E(\eta(Z(t))) < 0. \tag{11.7}$$

Then,

$$X^* = \sup_{u \leq 0} \int_u^0 \eta(Z(s))ds \tag{11.8}$$

is an a.s. finite random variable and

$$\lim_{t \to \infty} P(X(t) > x) = P(X^* > x), \quad x \geq 0. \tag{11.9}$$

In the following sections we consider fluid processes driven by various stochastic processes Z.

11.3 DRIVING PROCESS: CTMC

In this section we study the fluid process driven by a CTMC. Let $\{Z(t), t \geq 0\}$ be an irreducible CTMC on state space $S = \{1, 2, \ldots, M\}$ and infinitesimal generator matrix $Q = [q_{ij}]$. The drift function is given by $\eta(i) = d(i)$, $i \in S$; $d(i)$ is called the *drift in state i*. Let

$$\pi_{ij}(t) = P(Z(t) = j \mid Z(0) = i), \quad i, j \in S, \tag{11.10}$$

$$\text{and } \pi_j = \lim_{t \to \infty} P(Z(t) = j \mid Z(0) = i), \quad i, j \in S. \tag{11.11}$$

It is well known (see Ross [34], Kulkarni [23]) that $\Pi(t) = [\pi_{ij}(t)]$ satisfies the following equations:

$$\frac{d\Pi(t)}{dt} = \Pi(t)Q, \quad \Pi(0) = I. \tag{11.12}$$

Furthermore, $\pi = (\pi_1, \pi_2, \ldots, \pi_M)$ is given by the unique solution to

$$\pi Q = 0, \tag{11.13}$$

$$\sum_{i \in S} \pi_i = 1. \tag{11.14}$$

Transient behavior. It is clear that $(X, Z) = \{(X(t), Z(t)), t \geq 0\}$ is a bivariate Markov process. For $0 \leq x, y < \infty$, and $i, j \in S$, let

$$F(t, x, j; y, i) = P(X(t) \leq x, Z(t) = j \mid X(0) = y, Z(0) = i). \tag{11.15}$$

The next theorem gives the forward differential equations satisfied by the transition probabilities $\{F(t, x, j; y, i)\}$. First we need the following notation:

$$F(t, x; y) = [F(t, x, j; y, i)]_{i, j \in S}, \tag{11.16}$$

$$D = diag(d(1), d(2), \ldots, d(M)). \tag{11.17}$$

We call D the *drift matrix*.

Theorem 11.3. *The transition probabilities* $\{F(t, x, j; y, i)\}$ *satisfy the equations*

$$\frac{\partial F(t, x; y)}{\partial t} + \frac{\partial F(t, x; y)}{\partial x}D = F(t, x; y)Q, \tag{11.18}$$

with the boundary conditions

$$F(t, 0, j; y, i) = 0 \quad \text{if } d(j) > 0. \tag{11.19}$$

If the buffer capacity is finite, the additional boundary conditions are

$$F(t, B, j; y, i) = \pi_{ij}(t) \quad \text{if } d(j) < 0. \tag{11.20}$$

The solution of the differential equations of Theorem 11.3 is a complicated task. Hence, we turn our attention to the steady-state behavior.

Limiting behavior. For the finite capacity buffer, the (X, Z) process is always stable. In the case of the infinite capacity buffer, from Theorem 11.3, it is clear that the buffer content process is stable, i.e., it has a limiting distribution if

$$d = \sum_{i \in S} \pi_i d(i) < 0. \tag{11.21}$$

The above condition makes intuitive sense, since d is the net input rate to the buffer in steady state.

We assume from now on that the above stability condition holds when the buffer capacity is infinite, and study the distribution of (X, Z). Define

$$F(x, j) = \lim_{t \to \infty} F(t, x, j; y, i), \quad 0 \le x < \infty, i \in S. \tag{11.22}$$

Note that we have implicitly assumed that the limiting distribution is independent of the initial state. Let

$$F(x) = [F(x, 1), F(x, 2), ..., F(x, M)]. \tag{11.23}$$

The next theorem gives the equations satisfied by $F(x)$.

Theorem 11.4. *$F(x)$ satisfies*

$$\frac{dF(x)}{dx}D = F(x)Q, \tag{11.24}$$

with the boundary conditions

$$F(0, j) = 0 \quad if\ d(j) > 0. \tag{11.25}$$

When the buffer capacity is finite, the additional boundary conditions are:

$$F(B, j) = \pi_j \quad if\ d(j) < 0. \tag{11.26}$$

Proof. See Mitra [28, 29].

Now we develop a spectral representation of $F(x)$. Towards this end we try

$$F(x) = e^{\lambda x}\phi, \tag{11.27}$$

where λ is a scalar and ϕ is an M-dimensional row vector. Substituting in Equation (11.24) we get

$$\lambda e^{\lambda x}\phi D = e^{\lambda x}\phi Q. \tag{11.28}$$

This yields

$$\phi(\lambda D - Q) = 0. \tag{11.29}$$

Now, a non-zero vector ϕ satisfying equation (11.29) exists if

$$det(\lambda D - Q) = 0. \tag{11.30}$$

The next theorem discusses the solutions (called the eigenvalues) λ to the above equation. (See Mitra [29].) We need the following notation:

$$S_+ = \{i \in S : d(i) > 0\}, \tag{11.31}$$

$$S_0 = \{i \in S : d(i) = 0\}, \tag{11.32}$$

$$S_- = \{i \in S : d(i) < 0\}, \tag{11.33}$$

$$M_+ = |S_+|, \tag{11.34}$$

$$M_0 = |S_0|, \tag{11.35}$$

$$M_- = |S_-|. \tag{11.36}$$

Theorem 11.5. *Equation (11.30) has $M_+ + M_-$ solutions (counting multiplicities) $\{\lambda_i, i = 1, 2, ..., M_+ + M_-\}$. When $d < 0$, exactly M_+ have negative real parts, 1 is zero, and $M_- - 1$ have positive real parts.*

Now consider the infinite buffer case with $d < 0$. Number the λ_i's as follows:

$$Re(\lambda_1) \le Re(\lambda_2) \le ... \le Re(\lambda_{M_+}) <$$
$$Re(\lambda_{M_+ + 1}) = 0 < Re(\lambda_{M_+ + 2}) \le ... \le Re(\lambda_{M_+ + M_-}). \tag{11.37}$$

We assume that all the eigenvalues $\{\lambda_i, i = 1, 2, ..., M_+ + M_-\}$ are distinct. Let $\phi_i = (\phi_{i1}, \phi_{i2}, ..., \phi_{iM})$ be the eigenvector corresponding to the eigenvalue λ_i such that the pair (λ_i, ϕ_i) satisfies equation (11.29). (Note: π is the eigenvector corresponding to eigenvalue 0.) Then the general solution $F(x)$ to equation (11.24) is given by

$$F(x) = \sum_{i=1}^{M_+ + M_-} a_i e^{\lambda_i x} \phi_i, \tag{11.38}$$

where $\{a_i, i = 1, 2, \ldots, M_+ + M_-\}$ are scalar unknowns to be determined from the appropriate boundary conditions. The following theorem shows how this can be done. (See Mitra [29].)

Theorem 11.6. (i) *Infinite capacity buffer with $d < 0$. The a_j's are given by the solution to*

$$a_j = 0 \qquad if \ Re(\lambda_j) > 0, \tag{11.39}$$

$$a_{M_+ + 1} = 1, \tag{11.40}$$

$$\sum_{i=1}^{M_+ + 1} a_i \phi_{ij} = 0 \qquad if \ j \in S_+. \tag{11.41}$$

(ii) *Buffer with finite capacity B. The a_j's are given by the solution to*

$$\sum_{i=1}^{M_- + M_+} a_i \phi_{ij} = 0 \qquad if \ j \in S_+, \tag{11.42}$$

$$\sum_{i=1}^{M_- + M_+} a_i \phi_{ij} e^{\lambda_i B} = \pi_j \qquad if \ j \in S_-. \tag{11.43}$$

The condition in equation (11.39) arises because $F(x)$ is a bounded function of x. Note that in both cases there are as many linear equations as there are unknown a_j's. This completes the spectral representation that we set out to obtain. We illustrate an example.

Example 11.7. Suppose the input is generated by an *on-off* source. Such a source stays *on* for an $exp(\alpha)$ amount of time and stays *off* for an $exp(\beta)$ amount of time. It generates fluid at rate R when it is on and does not produce any fluid when it is off. The fluid is removed at a constant rate c from the buffer.

In this case the environment process Z is a two-state CTMC on state space $S = \{1 = \text{on}, 2 = \text{off}\}$ with the following generator matrix:

$$Q = \begin{bmatrix} -\alpha & \alpha \\ \beta & -\beta \end{bmatrix}. \tag{11.44}$$

The drift matrix is given by

$$D = \begin{bmatrix} R - c & 0 \\ 0 & -c \end{bmatrix}. \tag{11.45}$$

Assume that the buffer capacity is infinite and that $R\beta/(\alpha + \beta) < c$ for stability. The two eigenvalues are $\lambda_1 = 0$ and $\lambda_2 = \lambda = \beta/c - \alpha/(R - c) < 0$. The final solution is given by

$$F(x, 1) = \frac{\beta}{(\alpha + \beta)}(1 - e^{\lambda x}), \tag{11.46}$$

$$F(x, 2) = \frac{\alpha}{(\alpha + \beta)} - \frac{\beta(R - c)}{c(\alpha + \beta)} e^{\lambda x}. \tag{11.47}$$

11.4 APPLICATIONS

In this section we consider several applications of the fluid model of the previous section to high-speed telecommunication networks.

11.4.1 Congestion control

A basic preventative congestion-control strategy used in high-speed networks is called the *leaky bucket* mechanism (see Elwalid and Mitra [7], Gün and Guérin [12], Guérin et al. [13], Buttó et al. [5]), which operates as follows:

Tokens enter a token pool of size M at rate γ. Each token gives permission for transmission of a single bit of information, i.e., γ is in bits/sec and M is in bits. The user generating data behaves like an on-off source, as described in Example 11.7 above. If a token is waiting in the token pool, an arriving bit of data removes it from the token pool and enters the network. If no tokens are in the pool, the incoming data waits in the data buffer of size B. When the data buffer is full, the arriving data is lost. Similarly, if the token pool is full, the arriving tokens are lost.

Now, let $Y(t)$ be the amount of data in the data buffer at time t, $Z(t)$ the state of the source at time t, and $W(t)$ the amount of tokens in the token pool at time t. The logic of the leaky bucket implies that the data buffer and the token pool cannot be simultaneously nonempty, i.e., $W(t)Y(t) = 0$ for all t. Now define

$$X(t) = Y(t) - W(t) + M. \tag{11.48}$$

It can be seen that $\{X(t), t \geq 0\}$ is a fluid process on $[0, M + B]$ driven by a two-state CTMS, as given in Example 11.7, with drifts $d(1) = \gamma$ and $d(2) = \gamma - R$. Then the limiting distribution of $X(t)$ can be written down by using the results in Example 11.1. The limiting distributions of $Y(t)$ and $W(t)$ can then be obtained as follows:

$$P(Y(t) = 0) = P(X(t) \leq M), \tag{11.49}$$

$$P(Y(t) > x) = P(X(t) > x + M) \quad \text{for } 0 \leq x < B, \tag{11.50}$$

$$P(Y(t) = B) = P(X(t) = M + B), \tag{11.51}$$

$$P(W(t) = 0) = P(X(t) \geq M), \tag{11.52}$$

$$P(W(t) > x) = P(X(t) < M - x) \quad \text{for } 0 \leq x < M, \tag{11.53}$$

$$P(W(t) = M) = P(X(t) = 0). \tag{11.54}$$

11.4.2 Multiplexing in high-speed networks

The two-state source described in Example 11.7 is a special case of a Markov modulated fluid source (MMFS, for short). An MMFS is described by two parameters (Q, r), where Q is the generator matrix of a CTMC on state space S and $r = [r(i)]_{i \in S}$ is a vector. When the CTMC is in state i the source produces traffic at rate $r(i)$. In high-speed networks several such sources of fluid traffic are multiplexed onto a single buffer, i.e., the output from several such sources is superimposed to form a single input stream to the buffer. Such a situation can be modeled by simply constructing a large environment process Z that keeps track of the state of each source. However, the size of the state space of the composite process undergoes a combinatorial explosion, and it makes the computation infeasible. In this section we discuss how to exploit the structure of the composite process to make the computation easier. The results here are based on Stern and Elwalid [37].

Consider the situation where K MMFS's are multiplexed onto a single infinite buffer. The kth source has parameters (Q_k, r_k). Let $Z_k(t)$ be the state of the kth source at time t and assume that $\{Z_k(t), t \geq 0\}$ is an irreducible CTMC on state-space $S_k = \{1, 2, \ldots, N_k\}$. The fluid is removed from the buffer at rate c.

Let $X(t)$ be the amount of fluid in the buffer at time t. Then it can be seen that $\{X(t), t \geq 0\}$ is a fluid process driven by the CTMC $\{Z(t) = (Z_1(t), Z_2(t), \ldots, Z_K(t)), t \geq 0\}$. The generator matrix of the Z process is given by

$$Q = Q_1 \oplus Q_2 \oplus \ldots \oplus Q_K, \tag{11.55}$$

where $\oplus$ represents Kronecker sum. The drift matrix is given by

$$D = R_1 \oplus R_2 \oplus \ldots \oplus R_K - cI, \tag{11.56}$$

where

$$R_k = diag(r_k), \quad \text{for } 1 \leq k \leq K. \tag{11.57}$$

Notice that the Q and D matrices are of size $\prod_{k=1}^K N_k$, which is a combinatorially large number. R_k is an N_k by N_k diagonal matrix whose iith element is $r_k(i)$. Fortunately, the problem of computing the (eigenvalue, eigenvector) pairs for these matrices can be reduced to K coupled (eigenvalue, eigenvector) problems involving smaller matrices Q_k and D_k, as explained below.

For $1 \leq k \leq K$, define

$$A_k(\lambda) = R_k - \frac{1}{\lambda} Q_k. \tag{11.58}$$

The main result is given in the next theorem.

Theorem 11.8. *A pair (λ, ϕ) satisfies equation (11.29) if and only if the following equations hold:*

$$g_k(\lambda)\phi_k = \phi_k A_k(\lambda), \tag{11.59}$$

$$\sum_{k=1}^K g_k(\lambda) = c, \tag{11.60}$$

$$and \quad \phi = \phi_1 \otimes \phi_2 \otimes \ldots \otimes \phi_K. \tag{11.61}$$

Example 11.9. Consider the multiplexing of K identical and independent on-off sources as described in Example 11.7. From Anick et al. [3] we get

$$F(x, i) = \sum_{n=0}^m \phi_{ni} exp(-\lambda_n x), \tag{11.62}$$

where $m = K - \lceil \frac{c}{R} \rceil$ ($\lceil x \rceil$ is the largest integer less than or equal to x) and λ_n, $n = 0, 1, \ldots, m$ are the positive roots of the following $K + 1$ quadratic equations:

$$A_n \lambda^2 + B_n \lambda + C_n = 0, \quad n = 0, 1, \ldots, K, \tag{11.63}$$

where

$$A_n = R^2 (\frac{K}{2} - n)^2 - (\frac{KR}{2} - c)^2, \tag{11.64}$$

$$B_n = 2R(\alpha - \beta)(\frac{K}{2} - n)^2 - K(\alpha + \beta)(\frac{KR}{2} - c), \tag{11.65}$$

$$C_n = -(\alpha + \beta)^2 \{(\frac{K}{2})^2 - (\frac{K}{2} - n)^2\}. \tag{11.66}$$

Furthermore, ϕ_n is the eigenvector corresponding to λ_n such that pair (ϕ_n, λ_n) satisfies equation (11.29) and can be computed easily by using the above theorem.

11.4.3 Effective bandwidths

Now suppose that the network provides assurance that the incoming data will be dropped with a probability that is bounded above by a given number ϵ. Typically, $\epsilon \approx 10^{-8}$. This Quality of Service (QOS) criterion can be mathematically expressed as

$$G(B) = \lim_{t \to \infty} P(X(t) \geq B) < \epsilon. \tag{11.67}$$

The following theorem gives (see Elwalid and Mitra [8]) a simple yet powerful result in the asymptotic region

$$B \to \infty, \epsilon \to 0, \text{ such that } \frac{\log \epsilon}{B} \to z \in (-\infty, 0]. \tag{11.68}$$

Theorem 11.10. *In the asymptotic region in equation (11.68), the QOS criterion (11.67) is satisfied if*

$$\sum_{k=1}^{K} g_k(z) < c, \tag{11.69}$$

and it is violated if

$$\sum_{k=1}^{K} g_k(z) > c. \tag{11.70}$$

(Note that the case $\sum_{k=1}^{K} g_k(z) = c$ is left as indeterminate in the above theorem. In this case, the QOS criterion may or may not be satisfied.) The quantity $g_k(z)$ is called the *effective bandwidth* (or *effective capacity*) of the kth source, as it depends upon the QOS parameter z and other source parameters (Q_k, r_k). If the sum of the effective bandwidths of the sources is less then the channel capacity, the QOS criterion is satisfied for all the multiplexed sources. This simple additive structure provides a very useful call admission criterion. Elwalid and Mitra [8] study important properties of the effective bandwidths. The concept of effective bandwidths has its roots in the theory of large deviations and it has appeared in many other contexts. See Gibbens and Hunt [11], Guérin et al. [13], Kelly [17], Kesidis et al. [18], etc.

11.5 DRIVING PROCESS: CTMC + WHITE NOISE

The fluid process studied in the previous section has piecewise deterministic sample-paths. In practice, the input and output rates depend deterministically on an external environment, but in addition, there is a small random component, called jitter, that introduces further randomness. We model this situation by a fluid process driven by a composite process $\{(Z(t), W(t)), t \geq 0\}$ where the Z component is a CTMC, as described in Section 11.3 (with state space $S = \{1, 2, \ldots, M\}$ and generator Q), and $\{W(t), t \geq 0\}$ is a standard white noise process. (See Karlin and Taylor [15].) We consider the following drift function:

$$\eta(Z(t), W(t)) = d(Z(t)) + \sigma^2(Z(t))W(t). \tag{11.71}$$

One way to interpret equations (11.1) and (11.2) is the following Itô stochastic differential equation (see Harrison [14] and Karlin and Taylor [15]):

$$dX(t) = d(Z(t)) + \sigma^2(Z(t))dB(t), \tag{11.72}$$

where $\{B(t), t \geq 0\}$ is the standard Brownian motion. The boundary behavior at 0 (and at B, if required) needs to be studied carefully. This fluid model is studied by Asmussen [2] and Kulkarni and Karandikar [26]. Kulkarni and Karandikar [26] study the spectral representation of the steady state distribution of the (X, Z) process, while Asmussen [2] studies the (X, Z) process via change of measure techniques. Here we concentrate on the steady-state results of Kulkarni and Karandikar [26].

When the buffer is finite, the fluid process is always stable. When it is infinite, the stability condition remains the same as in equation (11.21). Assume that the process is stable and let $F(x, j)$, $j \in S$, and $F(x)$ be as defined in equations (11.22) and (11.23). The equations satisfied by $F(x)$ are given in the next theorem. We need the following notation:

$$\Sigma = \frac{1}{2} \begin{bmatrix} \sigma^2(1) & 0 & \cdots & 0 \\ 0 & \sigma^2(2) & \cdots & 0 \\ 0 & 0 & \ddots & 0 \\ 0 & 0 & \cdots & \sigma^2(M) \end{bmatrix}. \tag{11.73}$$

$$S_+ = \{i \in S : \sigma^2(i) > 0\}, \tag{11.74}$$

$$S_{0+} = \{i \in S : \sigma^2(i) = 0, d(i) > 0\}, \tag{11.75}$$

$$S_{00} = \{i \in S : \sigma^2(i) = 0, d(i) = 0\}, \tag{11.76}$$

$$S_{0-} = \{i \in S : \sigma^2(i) = 0, d(i) < 0\}. \tag{11.77}$$

Theorem 11.11. *F(x) satisfies*

$$\frac{d^2 F(x)}{dx^2}\Sigma - \frac{dF(x)}{dx}D + F(x)Q = 0 \tag{11.78}$$

with the following boundary conditions:

$$F(0, i) = 0, \quad \text{for } i \in S_+ \cup S_{0+}. \tag{11.79}$$

If the buffer capacity, B, is finite, it satisfies the following additional boundary conditions:

$$F(B, i) = \pi(i), \quad \text{for } i \in S_+ \cup S_{0-}. \tag{11.80}$$

As in the previous section, we derive a spectral representation for $F(x)$. Assume that $F(x)$ is as given in equation (11.27), and substitute it into equation (11.78). We see that (λ, ϕ) is a valid (eigenvalue, eigenvector) combination if

$$det(\lambda^2\Sigma - \lambda D + Q) = 0, \tag{11.81}$$

$$\phi(\lambda^2\Sigma - \lambda D + Q) = 0. \tag{11.82}$$

The next theorem describes the nature of the solutions (λ, ϕ) to the equations (11.81) and (11.82). We use the following notation: $M_+ = |S_+|$, $M_{0+} = |S_{0+}|$, $M_{00} = |S_{00}|$, and $M_{0-} = |S_{0-}|$.

Theorem 11.12. *Equation (11.82) has $2M_+ + M_{0+} + M_{0-}$ solutions (counting multiplicities). When $d < 0$, exactly $M_+ + M_{0-} - 1$ have positive real parts, 1 is zero, and $M_+ + M_{0+}$ have negative real parts.*

Now assume that all the eigenvalues $\{\lambda_i, \; i = 1, 2, \ldots, 2M_+ + M_{0+} + M_{0-}\}$

are distinct and arranged in ascending order of their real parts. Let ϕ_i be the eigen-vector that satisfies equation (11.82) for $\lambda = \lambda_i$. Then, the general solution $F(x)$ to equation (11.78) is given by

$$F(x) = \sum_{i=1}^{2M_+ + M_{0+} + M_{0-}} a_i e^{\lambda_i x} \phi_i, \qquad (11.83)$$

where $\{a_i, i = 1, 2, \ldots, 2M_+ M_{0+} + M_{0-}\}$ are scalar unknowns to be determined from the appropriate boundary conditions. The following theorem is analogous to Theorem 11.6.

Theorem 11.13. (*i*) *Infinite capacity buffer with $d < 0$. The a_j's are given by the solution to*

$$a_j = 0 \quad if \ Re(\lambda_j) > 0, \qquad (11.84)$$

$$a_j = 1 \quad if \ \lambda_j = 0, \qquad (11.85)$$

$$\sum_{i=1}^{M_+ + M_{0+}} a_i \phi_{ij} = 0 \quad if \ j \in S_+ \cup S_{0+}. \qquad (11.86)$$

(*ii*) *Buffer with finite capacity B. The a_j's are given by the solution to*

$$\sum_{i=1}^{2M_+ + M_{0+} + M_{0-}} a_i \phi_{ij} = 0 \quad if \ j \in S_+ \cup S_{0+}, \qquad (11.87)$$

$$\sum_{i=1}^{2M_+ + M_{0+} + M_{0-}} a_i \phi_{ij} e^{\lambda_i B} = \pi_j \quad if \ j \in S_+ \cup S_{0-}. \qquad (11.88)$$

Example 11.14. Consider the extreme case of $M = 1$. Thus, the Z process does not change state and has the generator $Q = [0]$. Let $\sigma^2(1) = \sigma^2$ and $d(1) = d$. Then, the $\{X(t), t \geq 0\}$ process reduces to a standard Brownian motion on $[0, B]$, with reflection at 0 and B. Equation (11.81) becomes

$$\tfrac{1}{2}\sigma^2 \lambda^2 - d\lambda = 0.$$

Hence we get $\lambda_1 = 0$, $\lambda_2 = 2d/\sigma^2$. If $B = \infty$ and $d < 0$ we get

$$F(x, 1) = 1 - exp\{\tfrac{2d}{\sigma^2}x\} \quad \text{for } x \geq 0.$$

If B is finite we get

$$F(x, 1) = \frac{1 - exp\{\tfrac{2d}{\sigma^2}x\}}{1 - exp\{\tfrac{2d}{\sigma^2}B\}} \quad \text{for } 0 \leq x \leq B.$$

These results match with known distributions. See Harrison [14] and Karlin and Taylor [15].

It can easily be seen that the results of this section reduce to those of the previous section if we set $\sigma^2(i) = 0$ for all $i \in S$.

11.6 DRIVING PROCESS: ORNSTEIN-UHLENBECK PROCESS

Consider the multiplexing of K on-off sources of Example 11.9. Suppose $R = R(K)$ goes to zero and $c = c(K)$ goes to ∞ as $K \to \infty$ in such a way that

$$R(K)\sqrt{Kf(1-f)} \to r, \qquad (11.89)$$

$$c(K) - KfR(K) \to c, \tag{11.90}$$

where $f = \alpha/(\alpha + \beta)$. Under this asymptotic behavior, the fluid process of Example 11.9 converges to the fluid process driven by an Ornstein-Uhlenbeck process with drift parameter $-(\alpha + \beta)z$ and variance parameter $2(\alpha + \beta)$. The drift function for this limiting fluid process is given by

$$\eta(z) = rz - c. \tag{11.91}$$

(See Kulkarni and Rolski [27], Simonian [35], Simonian and Virtamo [36].)

This motivates the study of a fluid process driven by a general Ornstein-Uhlenbeck process. Thus, we assume that $\{Z(t), t \geq 0\}$ is an Ornstein-Uhlenbeck process, i.e., it is a diffusion process on $(-\infty, \infty)$ with drift parameter $\mu(b - z)$ and variance parameter σ^2, where μ and σ^2 are non-negative constants. (See Karlin and Taylor [15].) We consider the drift function given in equation (11.91) with $r = 1$, without loss of generality. Now define the following transformed processes:

$$X'(t) = \frac{\mu X(t/\mu)}{\sigma/\sqrt{2\mu}}, \tag{11.92}$$

$$Z'(t) = \frac{Z(t/\mu) - b}{\sigma/\sqrt{2\mu}}. \tag{11.93}$$

Then the transformed process $\{Z'(t), t \geq 0\}$ is an Ornstein-Uhlenbeck process with drift $-z$ and variance parameter 2. The process $\{X'(t), t \geq 0\}$ is a fluid process driven by Z' with the following drift function:

$$\eta(z') = z' - \gamma, \tag{11.94}$$

where $\gamma = (c - b)/(\sigma/\sqrt{2\mu})$. From now on we omit the primes for clarity and consider this normalized (X, Z) process with a single parameter γ.

Next we study the stationary distribution of the bivariate process (X, Z). Theorem 11.2 implies that the process is stable if $\gamma > 0$. We assume this to be the case from now on. Now, in steady-state, the X process has a mass at zero whenever the Z process is below γ. Hence the bivariate process has an absolutely continuous density $f(x, z)$ on $S = \{(x, z) \mid x > 0, \ -\infty < z < \infty\} \cup \{(x, z) \mid x = 0, z > \gamma\}$, and an absolutely continuous density $f_0(z)$ on $S_0 = \{z < \gamma\}$. The next theorem gives the equations satisfied by these densities:

Theorem 11.15. *The densities $f(x, z)$ and $f_0(z)$ satisfy the following equations:*

$$\frac{\partial^2 f(x, z)}{\partial z^2} + \frac{\partial}{\partial z}(zf(x, z)) = (z - \gamma)\frac{\partial f(x, z)}{\partial x}, \quad (x, z) \in S, \tag{11.95}$$

$$\frac{d^2 f_0(z)}{dz^2} + \frac{d}{dz}(zf_0(z)) = (z - \gamma)f(0, z), \quad z \in S_0, \tag{11.96}$$

where $f(0, z) = \lim_{x \to 0} f(x, z)$.

The solution to the above equations is given in the next theorem. First we need the following notation:

$$\omega_k = \tfrac{1}{2}(\sqrt{\gamma^2 + 4k} + \gamma), \quad k \geq 0, \tag{11.97}$$

$$H_k(z) = (-1)^k exp(z^2/2)\frac{d^k}{dz^k}exp(-z^2/2), \quad k \geq 0, \tag{11.98}$$

$$g_k(z) = exp(-\omega_k^2/2)exp(-(z-\omega_k)^2/2)H_k(\frac{z-2\omega_k}{\sqrt{2}}), \quad k \geq 0. \qquad (11.99)$$

The $H_k(z)$ functions defined above are the standard Hermite polynomials. (See Andrews [1].) With this notation we have:

Theorem 11.16. (Knessl and Morrison [19]) *The densities $f(x,z)$ and $f_0(z)$ are given by:*

$$f(x,z) = \frac{1}{\sqrt{2\pi}} \sum_{k=0}^{\infty} a_k\omega_k exp(-\omega_k x)g_k(z), \quad (x,z) \in S, \qquad (11.100)$$

$$f_0(z) = \frac{1}{\sqrt{2\pi}} exp(-z^2/2) - \frac{1}{\sqrt{2\pi}} \sum_{k=0}^{\infty} a_k g_k(z), \quad z \in S_0. \qquad (11.101)$$

The constants a_k, $k \geq 0$ are given by

$$a_0 = exp(\gamma\zeta(1/2))\prod_{m=1}^{\infty} \frac{\omega_m}{\omega_m - \gamma}exp(-\gamma/m), \qquad (11.102)$$

$$a_k = 2^{-k/2}\frac{\gamma}{\omega_k - \gamma}exp((\zeta(1/2) - 1/\sqrt{k})\omega_k + (k\psi - 1)/2) \cdot$$

$$\prod_{m=1, m \neq k}^{\infty} \frac{\omega_m}{\omega_m - \omega_k}exp(-\omega_k/\sqrt{m} - k/(2m)), \quad k \geq 1, \qquad (11.103)$$

where ψ is Euler's constant and $\zeta(\cdot)$ is Riemann zeta function.

One consequence of the above theorem is the following asymptotic result when x is large:

$$\lim_{t\to\infty} P(X(t) > x) \sim e^{-\gamma x}. \qquad (11.104)$$

This result has been improved in Kulkarni and Rolski [27] who prove the following bound for all $x \geq 0$, using change of measure techniques:

$$\lim_{t\to\infty} P(X(t) > x) \leq e^{-\gamma x}exp(-\gamma^2/2). \qquad (11.105)$$

11.7 STATE-DEPENDENT DRIFT

Here we consider a further generalization of the basic model of Section 11.2: when the external environment is z and the buffer content is x, the net input to the buffer is given by $\eta(z,x)$. There is no general theory for such a case. The stability condition for the infinite buffer case can be intuitively seen to be the following:

$$\limsup_{x\to\infty} E(\eta(Z,x)) < 0, \qquad (11.106)$$

where Z has the steady state distribution of $\{Z(t), t \geq 0\}$. We describe below one case for which explicit results are available.

The buffer capacity is B. The given J thresholds are $0 = B_0 < B_1 < B_2 < \ldots < B_J = B$. The environment process is a CTMC on state-space $S = \{1,2,\ldots, M\}$ with rate matrix Q and steady state distribution π. The drift function is a step function of x as follows:

$$\eta(i,x) = d(i,j) \quad \text{for } B_{j-1} \leq x < B_j, \ 1 \leq j \leq J. \qquad (11.107)$$

If B is finite, the system is always stable. If B is infinite, the stability condition is:

$$\sum_{i=1}^{M} \pi_i d(i, J) < 0. \tag{11.108}$$

We assume this to hold if $B = \infty$ and study the steady state distribution of the $\{(X(t), Z(t)), t \geq 0\}$ process. The results here are based on Elwalid and Mitra [9].

Let $F(x, i)$ and $F(x)$ be as defined by equations (11.22) and (11.23). For $1 \leq j \leq J$, we use the notation

$$F^j(x, i) = F(x, i) \quad \text{for } B_{j-1} < x < B_j, i \in S, \tag{11.109}$$

$$F^j(x) = F(x) \quad \text{for } B_{j-1} < x < B_j, \tag{11.110}$$

$$D^j = diag(d(1, j), d(2, j), \ldots, d(M, j)), \tag{11.111}$$

$$S_+^j = \{i \in S: d(i, j) > 0\}, \tag{11.112}$$

$$S_0^j = \{i \in S: d(i, j) = 0\}, \tag{11.113}$$

$$S_-^j = \{i \in S: d(i, j) < 0\}. \tag{11.114}$$

From the results of Section 11.3, we get the following theorem.

Theorem 11.17. $\{F^j(x), 1 \leq j \leq J\}$ *satisfy the following equations:*

$$\frac{dF^j(x)}{dx} D^j = F^j(x)Q, \tag{11.115}$$

with the following boundary conditions:

$$F^1(0, i) = 0 \quad \text{if } i \in S_+^1, \tag{11.116}$$

$$F^j(B_j-, i) = F^{j+1}(B_j+, i) \text{ if } i \in S_-^j \cap S_-^{j+1}, 1 \leq j \leq J-1$$

$$\text{or} \quad \text{if } i \in S_+^j \cap S_+^{j+1}, 1 \leq j \leq J-1. \tag{11.117}$$

If the buffer capacity is finite, the additional boundary conditions are

$$F^J(B_i-, i) = \pi_i \quad \text{if } i \in S_-^J. \tag{11.118}$$

Proof. See Elwalid and Mitra [9].

We follow the methodology of Section 11.3 to obtain the spectral representation for $F(x)$ given in the next theorem.

Theorem 11.18. *Let* $\{(\lambda_i^j, \phi_i^j), 1 \leq i \leq M, 1 \leq j \leq J\}$ *be the (eigenvalue, eigenvector) pairs for the following generalized eigenvalue problems:*

$$\phi(\lambda D^j - Q) = 0 \quad 1 \leq j \leq J. \tag{11.119}$$

Then, $F^j(x)$ *has the following spectral representation:*

$$F^j(x) = \sum_{i=1}^{M} a_i^j \phi_i^j e^{\lambda_i^j x} \quad 1 \leq j \leq J, \tag{11.120}$$

where the scalars $[a_i^j, 1 \leq i \leq M, 1 \leq j \leq J]$ *are chosen to satisfy the boundary conditions in equations (11.116-11.118). If the buffer capacity is infinite the conditions generated by equation (11.118) are replaced by the following:*

$$a_i^J = 0 \quad \text{if } Re(\lambda_i^J) > 0, \tag{11.121}$$

$$a_i^J = 1 \quad \text{if } Re(\lambda_i^J) = 0. \tag{11.122}$$

Equation (11.117) says that the bivariate process $\{(X(t), Z(t)), t \geq 0\}$ has no mass at (B_j, i) if the drift (in state i) on both sides of B_j has the same sign. Otherwise, there may be a positive mass at (B_j, i). This makes intuitive sense.

Note that there are as many equations as there are unknown a_i^j's. Hence the above theorem gives a complete solution to the steady-state distribution of the bivariate process.

When $\eta(z, x)$ is not a step function in x, one can approximate it by a step function in x and use the above results. Hence we have an approximate numerical procedure for solving the general problem when the external environment is a CTMC.

Example 11.19. Consider an on-off source (see Example 11.7). Suppose it produces two types of fluid (at rates R^1 and R^2) when it is on. The type 2 fluid is always accepted in the buffer if there is space for it. The type 1 fluid is accepted only if the buffer content is less than a given threshold $0 < B_1 < B$. (Thus, the type 2 fluid will suffer fewer losses than the type 1 fluid and hence will have a better QOS.) The buffer is emptied at a fixed rate c.

This situation fits into the model analyzed above with a two-state CTMC as an external environment and a two-step drift function (i.e., $M = 2$, $J = 2$). The Q matrix is given in Example 11.7. The two drift matrices are given by

$$D^1 = \begin{bmatrix} R^1 + R^2 - c & 0 \\ 0 & -c \end{bmatrix}, \tag{11.123}$$

$$D^2 = \begin{bmatrix} R^2 - c & 0 \\ 0 & -c \end{bmatrix}. \tag{11.124}$$

For a solution (in the case of an infinite buffer), we refer the reader to Kulkarni et al. [25].

As discussed in Subsection 11.4.3, one can develop the concepts of effective band-widths for multiclass traffic using a shared buffer approach. The models developed in the current section have been found useful in the area of multiplexing multipriority traffic. Some work in this direction is in Kulkarni et al. [24, 25].

11.8 FURTHER WORK

11.8.1 Other driving processes

One possible extension is to consider a semi-Markov process as a driving process and extend the results of Section 11.3 to this case. However, the work of Chen and Yao [6] suggests that the analysis is going to be rather hard.

Another possibility is to extend the results of Section 11.6 to the case where the driving process is a bivariate process $\{(Z_1(t), Z_2(t)), t \geq 0\}$, with $Z_1(t)$ being a CTMC and $Z_2(t)$ an Ornstein-Uhlenbeck process. The drift function is the same as in Section 11.6. Such a driving process is motivated by the multiplexing problems where a large number of small sources (giving rise to the Ornstein-Uhlenbeck component) are multiplexed along with a small number of large sources (giving rise to the CTMC component) onto a single buffer. The solution promises to be extremely complicated.

11.8.2 State dependent drifts

The results of Section 11.7 can be extended to other driving processes. For example, the driving process can be the CTMC + White Noise as in Section 11.5 or it can be the Ornstein-Uhlenbeck process of Section 11.6. These models are motivated by the buffer-sharing models as explained in Example 11.19.

Work is currently in progress on a process that satisfies the following stochastic differential equation:

$$dX(t) = d(Z(t), X(t)) + \sigma^2(Z(t), X(t))dB(t), \qquad (11.125)$$

where $\{Z(t), t \geq 0\}$ is a CTMC. As in Section 11.7, we first concentrate on the case where $d(z, x)$ and $\sigma^2(z, x)$ are step functions of x.

11.8.3 Multiclass fluid models

This subsection is motivated by the desire to extend the fluid models to the multiclass case along the same line as in the multipriority queues. The simplest case is to assume that there are K classes. When the external environment is in state z, the fluid of class k arrives at rate $R(z, k)$. The buffer is emptied at a maximum rate of c. Let $X_k(t)$ be the amount of fluid of class k in the buffer at time t and define $X(t) = (X_1(t), X_2(t), \ldots, X_K(t))$. The aim is to study the limiting distribution of $\{(X(t), Z(t)), t \geq 0\}$. Of course we need to specify how the K classes are treated. A simple case is the Full-Service-Static-Priority discipline, under which the highest priority fluid that is in the buffer is always served first at the maximum possible rate.

Zang [38] has attempted to solve the problem via transform techniques. The joint distribution of $X(t)$ is rather messy. Narayanan [31] has developed the transforms of the marginal steady-state distributions of $X_k(t)$. Note that $Y_k(t) = \sum_{r=1}^{k} X_r(t)$ is a standard fluid model driven by $Z(t)$, assuming class 1 is the highest priority and class K is the lowest priority class. Hence, the steady state expected values of X_k are readily available.

BIBLIOGRAPHY

[1]   Andrews, L.C., *Special Functions for Engineers and Applied Mathematicians*, 2nd ed., McGraw-Hill, New York 1992.

[2]   Asmussen, S., Stationary distribution for fluid models and Markov-modulated reflected Brownian motion, *Stoch. Models* (1995), (to appear).

[3]   Anick, D., Mitra, D. and Sondhi, M.M., Stochastic theory of a data-handling system with multiple sources, *The Bell System Tech. Journal* **61** (1982), 1871-1894.

[4]   Borovkov, A.A., *Stochastic Processes in Queueing Theory*, Springer-Verlag, New York 1976.

[5]   Buttó, M., Cavallero, E., and Tonietti, A., Effectiveness of the leaky bucket policing mechanism in ATM networks, *IEEE J. Sel. Areas Commun. SAC-***9** (1991), 335-342.

[6]   Chen, H. and Yao, D.D., A fluid model for systems with random disruptions, *Opns. Res.* **40** Supp. 2 (1992), S239-S247.

[7]   Elwalid, A.I. and Mitra, D., Analysis and design of rate based congestion

control of high speed networks I: Stochastic fluid models, access regulation, *Queueing Sys.* **9** (1991), 29-64.

[8]   Elwalid, A.I. and Mitra, D., Effective bandwidth of general Markovian sources and admission control of high speed networks, *IEEE/ACM Trans. on Networking* **1** (1993), 329-343.

[9]   Elwalid, A.I. and Mitra, D., Fluid models for the analysis and design of statistical multiplexing with loss priorities on multiple classes of bursty traffic, *INFOCOM'92* (1992), 415-425.

[10]   Gaver, D.P. and Miller, R.G., Limiting distributions for some storage problems, In: *Studies in Appl. Prob. and Mgmt. Science*, ed. by K. Arrow, S. Karlin, and H. Scarf, Stanford Univ. Press 1962.

[11]   Gibbens, R.J. and Hunt, P.J., Effective bandwidths for multi-type UAS channel, *Queueing Sys.* **9** (1991), 17-28.

[12]   Gün, L. and Guérin, R., Bandwidth management and congestion control framework for the broadband network architecture, *Computer Networks & ISDN Systems* **26** (1993), 61-78.

[13]   Guérin, R., Ahmadi, H., and Naghshineh, M., Equivalent capacity and its application to bandwidth allocation in high-speed networks, *IEEE J. Sel. Areas Commun.* SAC-**9** (1991), 968-981.

[14]   Harrison, J.M., *Brownian Motion and Stochastic Flow Systems*, Wiley, New York 1985.

[15]   Karlin, S. and Taylor, S., *A Second Course in Stochastic Processes*, Academic Press, New York 1981.

[16]   Kella, O. and Whitt, W., A tandem fluid network with Levy input, *Technical Report* **90-16**, Yale University, CT (1990).

[17]   Kelly, F.P., Effective bandwidths at multi-type queues, *Queueing Sys.* **9** (1991), 17-28.

[18]   Kesidis, G., Walrand, J, and Chang, C.-S., Effective bandwidths for multiclass Markov fluid and other ATM sources, *IEEE/ACM Trans. on Networking* **1** (1993), 424-428.

[19]   Knessl, C. and Morrison, J.A., Heavy traffic analysis of a data-handling system with many sources, *SIAM J. Appl. Math.* **51** (1991), 187-213.

[20]   Kosten, L., Stochastic theory of a multi-entry buffer, part I, Delft Progress Report (1974), 10-18.

[21]   Kosten, L., Stochastic theory of a multi-entry buffer, part 2, Delft Progress Report (1974), 44-55.

[22]   Kosten, L. and Vrieze, O.J., Stochastic theory of a multi-entry buffer, part 3, Delft Progress Report (1975), 103-115.

[23]   Kulkarni, V.G., *Modeling and Analysis of Stochastic Systems*, Chapman and Hall, New York 1995.

[24]   Kulkarni, V.G., Chimento, P.F., and Gün, L., Effective bandwidth vectors for multiclass traffic multiplexed in a partitioned buffer, *Tech. Report* **TR 94-5**, Dept. of Opns. Res., Univ. of North Carolina, Chapel Hill, NC (1994).

[25]   Kulkarni, V.G., Gün, L., and Chimento, P.F., Effective bandwidth vector for two priority ATM traffic, *IEEE INFOCOM'94* (1994), 1056-1063.

[26]   Kulkarni, V.G. and Karandikar, R.L., Second-order fluid flow models: reflected Brownian motion in a random environment, *Opns. Res.*, (to appear).

[27]   Kulkarni, V.G. and Rolski, T., Fluid model driven by an Ornstein-Uhlenbeck process, *Prob. in Eng. and Info. Sciences* (to appear).

[28] Mitra, D., Stochastic fluid models, *Performance* 87, Elsevier Science Publishers, North Holland 1988.

[29] Mitra, D., Stochastic theory of fluid models of multiple failure-susceptible producers and consumers coupled by a buffer, *Adv. Appl. Prob.* 20 (1988), 646-676.

[30] Moran, P.A.P., A probability theory of a dam with a continuous release, *Quart. J. Math.* 2 (1956), 130-137.

[31] Narayanan, A., Stochastic fluid flow models for polling systems and multi-class queues, *Ph.D. Thesis*, Dept. of Opns. Res., Univ. of North Carolina, Chapel Hill, NC 1993.

[32] Newell, G.F., *Applications of Queueing Theory*, Chapman and Hall, London 1982.

[33] Prabhu, N.U., *Stochastic Storage Processes: Queues, Insurance Risk, and Dams*, Springer-Verlag, New York 1980.

[34] Ross, S.M., *Stochastic Processes*, Wiley, New York 1983.

[35] Simonian, A., Stationary analysis of a fluid queue with input rate varying as an Ornstein-Uhlenbeck process, *SIAM J. Appl. Math.* 51 (1991), 828-842.

[36] Simonian, A. and Virtamo, J., Transient and stationary distributions for fluid queues and input processes with a density, *SIAM J. Appl. Math.* 51 (1991), 1732-1739.

[37] Stern, T.E. and Elwalid, A.I., Analysis of separable Markov-modulated rate models for information handling systems, *Adv. Appl. Prob.* 23 (1991), 105-139.

[38] Zang, J., Performance study of Markov modulated fluid flow models with priority traffic, *IEEE INFOCOM'93* (1993), 10-17.

Part IV

Applied Techniques and Statistical Inference in Queueing Models

Chapter 12
Computational methods in queueing

Henk C. Tijms

ABSTRACT This paper presents a survey of computational methods for basic queueing models including models with batch Poisson input. Both exact and approximate methods for the state and waiting-time probabilities are discussed.

CONTENTS

12.1 INTRODUCTION

Computational analysis of queueing systems is more than getting numerical answers. Its essence is to find probabilistic ideas which make the computations transparent and natural. However, once an algorithm has been developed according to these guidelines, one should always verify that it works in practice. Queueing problems have not been solved satisfactorily until numerical answers can be obtained with sufficient ease and accuracy.

This chapter restricts itself to basic queueing models with only one service node and an infinite-capacity waiting room. We will discuss methods for the computation of queue-size and waiting-time probabilities in equilibrium. The emphasis is on the following approaches:

- Markov chains
- Regenerative processes
- Discrete Fast Fourier Transform
- Asymptotic expansions
- Interpolation

These methods are of course not the only computational tools for queueing models. Other techniques, such as the root-finding method and the matrix-geometric method, are not discussed here, but can be found in the books of Chaudhry and Templeton [1] and Neuts [2]. The discussion in this chapter relies heavily on Tijms [5].

This chapter is organized as follows. In Section 12.2 we describe the general model and introduce some notation. The various computational approaches are outlined in Section 12.3.

12.2 MODEL AND NOTATION

The general model we will consider is the batch-arrival $GI^X/G/c$ queue. This queueing system is abbreviated as $GI/G/c$ queue when the customers arrive singly (i.e., the batch size $X = 1$). In the $GI^X/G/c$ queueing system, batches of customers arrive according to a renewal process. Each batch consists of a random number of customers that is distributed according to a generic variable X. The service times of the customers are independent random variables that are distributed according to a generic variable S and that are independent of the arrival process. There are c identical servers available; each server can handle only one customer at a time. The queue discipline specifying which customer is to be served next is first-come-first-served. Let

$A(t) =$ the probability distribution function of the interarrival time between two
      batches
$B(x) =$ the probability distribution function of the service time of a customer
$\{\beta_s, s \geq 1\} =$ the probability distribution of the batch size X
$\lambda =$ the long-run average arrival rate of batches.

It is assumed that the server utilization ρ defined by $\rho = \dfrac{\lambda E(X)E(S)}{c}$ is less than 1.

Define the following quantities:

$L_q =$ the long-run average number of customers waiting in queue (excluding cus-
      tomers in service)
$W_q =$ the long-run average delay in queue per customer (excluding service time)
$p_j =$ the long-run fraction of time that j customers are present in the system
      $(j = 0, 1, \ldots)$
$W_q(x) =$ the long-run fraction of customers whose delay in queue is no more

      than x $(x \geq 0)$.

These quantities are well-defined under weak regularity conditions in addition to the assumption of $\rho < 1$.

12.3 COMPUTATIONAL APPROACHES

In this section we outline various computational approaches mentioned in the introduction. We use specific queueing models to illustrate the ideas of the various methods. A more detailed discussion can be found in Chapter 4 of Tijms [5].

12.3.1 Markov chain methods

The embedded Markov chain approach using a discrete-time Markov chain and the continuous-time Markov chain approach using the flow-rate technique are well-known and can be found in any textbook. It is perhaps less well-known that the practical applicability of these approaches can be greatly enhanced by using the result that the distribution of any positive random variable can be arbitrarily closely approximated by a mixture of Erlangian distributions with the same scale parameters, see e.g., Tijms [5]. Such a mixture has the probability density

$$\sum_{i=1}^{\infty} q_i \mu^i \frac{t^{i-1}}{(i-1)!} e^{-\mu t},$$

where $q_i \geq 0$ $(i = 1, 2, \ldots)$ and $\sum_{i=1}^{\infty} q_i = 1$. For a random variable with this density we have the useful interpretation that, with probability q_i, the random variable is distributed as the sum of i independent and exponentially distributed phases each having the same mean $1/\mu$. This interpretation together with the memoryless property of the exponential distribution enables us to set up a Markov chain analysis when the interarrival time and/or service time have phase-type distributions of the above form. The state description in the Markov chain analysis is further facilitated by the fact that the exponential phases have the *same* means.

Let us use the GI/G/1 queue as a vehicle to illustrate the usefulness of the Markov chain approach combined with the phase-method. Suppose that the service-time density $b(t)$ is given by

$$b(t) = \sum_{i=1}^{m} q_i \mu^i \frac{t^{i-1}}{(i-1)!} e^{-\mu t}, \quad t \geq 0. \tag{12.1}$$

In other words, each service time consists, with probability q_i, of i independent exponential phases which have to be consecutively processed. A very simple and powerful method can now be given to compute the waiting-time probabilities $W_q(x)$. To do so, define the embedded process $\{X_n, n \geq 1\}$ by

X_n = the number of uncompleted service phases present just before the nth arrival.

The process $\{X_n\}$ is a discrete-time Markov chain that is ergodic. Denote its equilibrium probabilities by z_j ($= \lim_{n \to \infty} (1/n) \sum_{k=1}^{n} P\{X_k = j\}$). Suppose for the moment that we have computed the z_j's. Then $W_q(x)$ follows by noting that the conditional waiting time of a customer finding upon arrival j uncompleted phases present has an Erlang-j distribution. Thus,

$$1 - W_q(x) = \sum_{j=0}^{\infty} z_j \sum_{k=0}^{j-1} e^{-\mu x} \frac{(\mu x)^k}{k!}, \quad x \geq 0.$$

A faster convergent series representation is

$$1 - W_q(x) = \sum_{k=0}^{\infty} e^{-\mu x} \frac{(\mu x)^k}{k!} (1 - \sum_{j=0}^{k} z_j), \quad x \geq 0.$$

Next we discuss how to compute the z_j's. We shall give a generally applicable method that is much more efficient than the usual approach of truncating the infinite set of equilibrium equations. The probabilities z_j are the unique solution to the linear equations

$$z_j = \sum_{i=0}^{\infty} p_{ij} z_i, \quad j = 0, 1, \ldots, \tag{12.2}$$

together with the normalizing equation $\sum_{j=0}^{\infty} z_j = 1$, where the p_{ij}'s are the one-step transition probabilities of the Markov chain $\{X_n\}$. The p_{ij}'s are given by

$$p_{ij} = \sum_{k=1}^{m} q_k \int_{0}^{\infty} e^{-\mu t} \frac{(\mu t)^{i+k-j}}{(i+k-j)!} dA(t) \quad \text{for } 1 \leq j \leq i+m$$

and $p_{i0} = 1 - \sum_{j=1}^{i+m} p_{ij}$, $i \geq 0$. Solving numerically the infinite system of linear equations (12.2) is less simple than it looks. A finite system of linear equations resulting from a brute-force truncation will become very large when the server utilization ρ gets close to 1. However, we can do much better in view of the asymptotic expansion

$$\frac{z_j}{z_{j-1}} \approx \eta \quad \text{for } j \text{ large enough} \tag{12.3}$$

for some constant η. Such asymptotic expansions are prevalent in queueing systems. It has been shown in Tijms [5] that the constant η is the unique solution to the equation

$$\sum_{i=1}^{m} q_i \eta^{m-i} \int_0^\infty e^{-\mu(1-\eta)t} dA(t) - \eta^m = 0$$

on the interval $(0,1)$. It is standard fare in numerical analysis to compute a root of a nonlinear function in a single variable. Once the constant η has been computed, we can reduce the infinite system of linear equations (12.2) to a finite system of linear equations by replacing z_j by $z_N \tau^{j-N}$ for $j > N$, where N is an appropriately chosen integer such that $z_j/z_{j-1} \approx \tau$ holds with sufficient accuracy when $j > N$. Thus we solve the finite system of linear equations

$$z_j = \sum_{k=0}^{N} a_{jk} z_k, \quad j = 0, 1, \ldots, N-1,$$

$$\sum_{j=0}^{N-1} z_j + \frac{z_N}{1-\eta} = 1,$$

where $a_{jk} = p_{kj}$ for $0 \le k \le N-1$ and $a_{jN} = \sum_{i=N}^{\infty} \eta^{i-N} p_{ij}$. How large the value of N should be chosen, has to be determined experimentally and depends, of course, on the required accuracy in the state probabilities. Empirical investigations show that remarkably small values of N are usually good enough for practical purposes. This is true for any value of ρ, including values very close to 1. An appropriate value of N is typically in the range of $1-200$ so that standard codes for solving linear equations can be routinely used.

12.3.2 Regenerative approach

The regenerative approach is a powerful technique to derive *recursion* schemes for calculating the state probabilities in a wide variety of queueing models of the M/G/1 type. This approach uses basic results from the theory of regenerative processes and a simple up- and downcrossing argument.

The regenerative approach is explained via the batch-arrival $M^X/G/1$ queue. Let us define a cycle as the time interval between two consecutive arrivals of batches finding the system empty. Also define the random variables:

T = the length of one cycle
T_j = the cumulative amount of time that j customers are present during one cycle $(j = 0, 1, \ldots)$
N_k = the number of service completion epochs in one cycle at which k customers are left behind $(k = 0, 1, \ldots)$.

By a basic result in the theory of regenerative processes, the time-average state probabilities p_j are given by

$$p_j = \frac{E(T_j)}{E(T)}, \quad j = 0, 1, \ldots \tag{12.4}$$

Next we derive two relations between $E(T_j)$ and $E(N_k)$. The first relation is based on the up- and downcrossing argument that during each cycle the number of state transitions out of the set of states $\{k+1, k+2, \ldots\}$ must be equal to the number of state transitions into this set. The expected number of transitions out of the set $\{k+1, k+2, \ldots\}$ per cycle equals $E(N_k)$ by definition. Using a general result based on the PASTA property, i.e., *Poisson arrivals see time averages* (see Corollary 1.7.2 in Tijms [5]), it follows that the expected number of state transitions into the set $\{k+1, k+2, \ldots\}$ per cycle is given by $\lambda \sum_{i=0}^{k} E(T_i) \sum_{s > k-i} \beta_s$. Thus, we obtain the fundamental relation

$$E(N_k) = \lambda \sum_{i=0}^{k} E(T_i) \sum_{s > k-i} \beta_s, \quad k = 0, 1, \ldots. \tag{12.5}$$

The second relation between $E(T_j)$ and $E(N_k)$ is obtained by splitting the cycle into disjoint intervals via the service completion epochs and calculating $E(T_j)$ as the sum of the contributions of the disjoint intervals. In doing so, we need again the memoryless property of the Poisson arrival process. Define now the quantity A_{kj} by

A_{kj} = the expected amount of time that j customers are present during one service time, starting with k customers present ($j \geq k \geq 0$).

Then, we obtain as a second relation

$$E(T_j) = \sum_{s=1}^{j} \beta_s A_{sj} + \sum_{k=1}^{j} E(N_k) A_{kj}, \quad j = 1, 2, \ldots. \tag{12.6}$$

Defining for fixed $t > 0$ and $j = 0, 1, \ldots$, the compound Poisson probability

$r_j(t) = P\{$in total j customers arrive during a time interval of length $t\}$,

it follows that $A_{kj} = \int_0^\infty r_{j-k}(t)\{1 - B(t)\}dt$. Inserting (12.5) into (12.6), dividing both sides of the resulting equation by $E(T)$, and using (12.4) with $p_0 = 1/[\lambda E(T)]$, we obtain the recursive relation

$$p_j = \lambda p_0 \sum_{s=1}^{j} \beta_s a_{j-s} + \lambda \sum_{k=1}^{j} \left(\sum_{i=0}^{k} p_i \sum_{s > k-i} \beta_s \right) a_{j-k}, \quad j = 1, 2, \ldots, \tag{12.7}$$

where the constants a_n are given by

$$a_n = \int_0^\infty r_n(t)\{1 - B(t)\}dt, \quad n = 0, 1, \ldots.$$

Except for deterministic service, this representation of a_n is not very useful for computational purposes. If the service-time density $b(t)$ is given by (12.1), we can give a recursive scheme for the a_n's by using the fact that a_n can be interpreted as the expected amount of time that $n+1$ customers are present during a service time starting with one customer present. We then have $a_n = \sum_{i=1}^{m} q_i \alpha_n^{(i)}$, where the numbers $\alpha_n^{(i)}$ are recursively computed from

$$\alpha_0^{(i)} = \frac{1}{\lambda + \mu} + \frac{\mu}{\lambda + \mu} \alpha_0^{(i-1)}, \quad 1 \leq i \leq r,$$

$$\alpha_n^{(i)} = \frac{\lambda}{\lambda + \mu} \sum_{j=1}^{n} \beta_j \alpha_{n-j}^{(i)} + \frac{\mu}{\lambda + \mu} \alpha_n^{(i-1)}, \quad n \geq 1 \text{ and } 1 \leq i \leq r,$$

with $\alpha_n^{(0)} = 0$ by convention. Although for Erlangian services the recursive scheme (12.7) is a practically useful algorithm, a computationally faster discrete Fast

Fourier Transform method needs to be applied to the generating function of the p_j's; see also the next subsection.

The regenerative approach discussed above is a versatile approach that can be applied to many variants of the M/G/1 queue, including single-server queues with state-dependent Markovian input and/or state-dependent service times; see Tijms [5].

12.3.3 Discrete Fast Fourier Transform

In many queueing systems, it is possible to derive an explicit expression for the generating function of the state probabilities. Then the discrete Fast Fourier Transform (FFT) method provides an extremely powerful computational tool to recover numerically the state probabilities from the explicit expression for the generating function. To illustrate this consider the $M^X/G/1$ queue. Using the relation (12.7), it can be shown that the generating function $P(z) = \sum_{j=0}^{\infty} p_j z^j$ allows the explicit expression

$$P(z) = (1-\rho)\frac{1 - \lambda A(z)\{1 - \beta(z)\}}{1 - \lambda A(z)\{1 - \beta(z)\}/(1-z)}, \tag{12.8}$$

where

$$\beta(z) = \sum_{j=1}^{\infty} \beta_j z^j \text{ and } A(z) = \int_0^{\infty} e^{-\lambda\{1-\beta(z)\}t}\{1 - B(t)\}dt.$$

Note that the integral representation of $A(z)$ can be reduced to a simple algebraic expression for most service-time distributions of practical interest.

The FFT method proceeds now as follows, to recover the p_j's from the right-hand side of (12.8). An integer N ($= 2^m$) is chosen so that the tail probability $\sum_{j \geq N} p_j$ is negligibly small and hence, $P(z) \approx \sum_{j=0}^{N-1} p_j z^j$. Then the complex numbers

$$f_k = P(e^{2\pi ik/N}), \quad k = 0, 1, \ldots, N-1$$

are calculated using the explicit expression for $P(z)$. Here, i is the complex number with $i^2 = -1$. The FFT method recovers the p_j's from the f_k's by using the inversion formula

$$p_j \approx \frac{1}{N}\sum_{k=0}^{N-1} f_k e^{-2\pi ijk/N}, \quad j = 0, 1, \ldots, N-1.$$

The breakthrough in the discrete Fourier analysis is that the multiplications in the above inversion formula can be done in an extremely fast way using a recursive procedure. A total of only $N\log_2(N)$ multiplications is only required. Details can be found in the book of Press et al. [3]; see also Tijms [5].

The FFT method requires that an *explicit* expression is available for the generating function of the state probabilities. In some queueing systems the generating function is an explicit expression except for a few unknown constants. For example, as shown in Tijms [5], the generating function $P(z) = \sum_{j=0}^{\infty} p_j z^j$ for the $M^X/D/c$ queue with a deterministic service time D is given by

$$P(z) = \frac{\sum_{j=0}^{c-1} p_j(z^j - z^c)}{1 - z^c e^{\lambda D\{1 - \beta(z)\}}}. \tag{12.9}$$

This is an explicit expression except for the unknowns $p_0, \ldots, p_{c-1}$. A possible approach is now to take an initial guess for $p_0, \ldots, p_{c-1}$ and to apply the FFT method. This yields estimates for the p_j's, including new estimates for $p_0, \ldots,$

p_{c-1}. The FFT method is now repeatedly applied until the estimates for $p_0,\ldots,$ p_{c-1} have sufficiently converged. To speed up this naive iterative procedure, it is important that after each iteration the estimates for the p_j's are normalized in order to satisfy the equation

$$\sum_{j=0}^{c-1} jp_j + c \sum_{j=c}^{\infty} p_j = c\rho.$$

This equation expresses that the average number of busy servers is $c\rho$. Another approach is to determine first $p_0,\ldots,p_{c-1}$ by using a root-finding method and then to apply the FFT method. Since $P(z)$ is regular for $|z| \leq 1$, the c complex roots of the denominator of (12.9) must also be roots of the numerator of (12.9), yielding c linear equations for $p_0,\ldots,p_{c-1}$; cf. Chaudhry and Templeton [1]. The roots can be found by solving $c\ln(z) + \lambda D\{1 - \beta(z)\} = 2\pi ik$, by Newton-Raphson, for appropriately chosen integers k.

12.3.4 Asymptotic expansions

In many cases the generating function of the state probabilities p_j enables us to obtain an asymptotic expansion for p_j for j sufficiently large. Such an asymptotic expansion is extremely useful for computational purposes. We need the following result. Suppose that $\{p_j, j \geq 0\}$ is a discrete probability distribution whose generating function $P(z) = \sum_{j=0}^{\infty} p_j z^j$, $|z| \leq 1$, can be represented as

$$P(z) = N(z)/D(z), \tag{12.10}$$

where it is assumed that $N(z)$ and $D(z)$ have no common zeros. In many cases such a representation naturally arises, where the denominator $D(z)$ is a "nice" function involving no unknown parameters. This finding is illustrated by the expressions (12.8) and (12.9). Let us now assume that the functions $N(z)$ and $D(z)$ can be analytically extended outside the unit circle $|z| \leq 1$. For example, this is the case for the right-hand side of (12.9) when we assume that the power series $\beta(z) = \sum_{j=1}^{\infty} \beta_j z^j$ has a convergence radius larger than 1. In addition to the assumption that $N(z)$ and $D(z)$ are analytic functions whose domains of definition can be extended to a region $|z| < R$ in the complex plane with $R > 1$, we assume that the following conditions are satisfied:

C1.   The equation $D(z) = 0$ has a real root τ in the interval $(1, R)$.
C2.   The function $D(z)$ has no zeros in the annulus $1 < |z| < \tau$ of the complex plane.
C3.   The zero $z = \tau$ of $D(z)$ is of multiplicity 1 and is the only zero of $D(z)$ on the boundary $|z| = \tau$.

Then it can be shown (see e.g., Appendix C in Tijms [5]) that

$$p_j \approx \sigma \tau^{-j} \text{ for } j \text{ large enough}, \tag{12.11}$$

where the amplitude factor σ is given by

$$\sigma = -\frac{1}{\tau} \frac{N(\tau)}{D'(\tau)}.$$

It is a very useful empirical finding that in practical applications the asymptotic expansion (12.11) applies for relatively small values of j. In practice, the computation of the root τ of the equation $D(\tau) = 0$ offers no difficulties, since the function $D(z)$ is typically a nice function involving no unknown parameters. However, the amplitude factor σ is in general difficult to obtain. Fortunately, it is often sufficient to know only the decay factor τ. By (12.11), we have $p_j/p_{j+1} \approx \tau$ for j large enough and this result is often sufficient for computational purposes. To illustrate this, consider the recursive scheme (12.7). The recursive calculations can be halted as soon as p_j/p_{j+1} has been sufficiently converged to τ. The number τ can be computed on beforehand from $D(\tau) = 0$. The computation of the root of a nonlinear function of a single variable is a standard fare in numerical analysis. Another demonstration of the usefulness of the result $p_j/p_{j+1} \approx \tau$ for j large enough is given in Subsection 12.3.1, when solving the infinite set of equilibrium equations.

The asymptotic expansion for the state probabilities is not only useful to compute the p_j's, but it can sometimes also be used to obtain a computationally useful asymptotic expansion for the waiting-time probabilities $W_q(x)$. To illustrate this, consider the $M^X/M/c$ queue. Then,

$$1 - W_q(x) = \frac{1}{\rho} \sum_{j=c}^{\infty} p_{j+1} \sum_{k=0}^{j-c} e^{-c\mu x} \frac{(c\mu x)^k}{k!}, \quad x \ge 0,$$

where the service rate $\mu = 1/E(S)$. The generating function of the p_j's is given by

$$P(z) = \frac{(1/c)\sum_{i=0}^{c-1}(c-i)p_i z^i}{1 - \lambda z\{1 - \beta(z)\}/\{c\mu(1-z)\}}.$$

Assuming that $\beta(z) = \sum_{j=1}^{\infty}\beta_j z^j$ has a convergence radius $R > 1$, we obtain from the asymptotic expansion of the p_j's that

$$1 - W_q(x) \approx \frac{\sigma\tau^{-c}}{\tau - 1}e^{-c\mu(1 - 1/\tau)x} \quad \text{for } x \text{ large enough}, \tag{12.12}$$

where τ is the unique solution of the equation

$$\lambda\tau[1 - \beta(\tau)] = c\mu(1 - \tau)$$

on the interval $(1, R)$, and the constant σ is given by

$$\sigma = \frac{(\tau-1)\sum_{i=0}^{c-1}(c-i)p_i\tau^i/c}{1 - \lambda\tau^2\beta'(\tau)/(c\mu)}.$$

The derivation of the above results can be found in Tijms [5]. Incidentally, for the $M^X/M/c$ queue the unknowns $p_0, ..., p_{c-1}$ can be directly computed by applying first the recursive scheme

$$i\mu p_i = \sum_{k=0}^{i-1} p_k\lambda \sum_{s \ge i-k}\beta_s, \quad i = 1, ..., c-1,$$

and then using the normalizing equation $\sum_{i=0}^{c-1}ip_i + c(1 - \sum_{i=0}^{c-1}p_i) = c\rho$. The recursive scheme is initialized with $\bar{p}_0 = 1$ and gives the values $\bar{p}_1, ..., \bar{p}_{c-1}$. The values $\bar{p}_i$ differ from the desired values p_i by a multiplicative constant.

Similarly, we have for the $M^X/D/c$ queue with a deterministic service time D that

$$1 - W_q(x) \approx \frac{\sigma\{\beta(\tau) - 1\}}{(\tau - 1)^2\tau^{c-1}E(X)}e^{-\lambda\{\beta(\tau) - 1\}x} \quad \text{for } x \text{ large enough}, \tag{12.13}$$

where τ is the unique root of the equation

$$\tau^c e^{\lambda D\{1 - \beta(\tau)\}} - 1 = 0$$

on the interval $(1, R)$ and the constant σ is given by

$$\sigma = [c - \lambda D\tau \beta'(\tau)]^{-1} \sum_{j=0}^{c-1} p_j (\tau^j - \tau^c).$$

In the next section it will be shown how asymptotic expansions (12.12) and (12.13) for the special cases of the $M^X/M/c$ queue and the $M^X/D/c$ queue can be used to approximate the waiting-time probabilities in the general $M^X/G/c$ queue.

12.3.5 Interpolation

For most queueing models it is not possible to give exact algorithms that are computationally feasible. In those situations one has to resort to approximate methods for calculating measures of system performance. Useful approximations to complex queueing systems are often obtained by using exact results for simpler, but related queueing systems. In this subsection it will be demonstrated that linear interpolation with respect to the squared coefficient of variation of the service time, may be a powerful tool.

The motivation of the interpolation approach is provided by the well-known Pollaczek-Khintchine formula for the average delay in queue per customer in the standard M/G/1 queue. The explicit expression for W_q in the M/G/1 queue allows for the representation

$$W_q = (1 - c_s^2)W_q(\text{det}) + c_s^2 W_q(\exp), \tag{12.14}$$

where

$$c_s^2 = \frac{\sigma^2(S)}{E^2(S)}$$

denotes the squared coefficient of variation of the service time S, and $W_q(\text{det})$ and $W_q(\exp)$ denote the respective values of W_q in the M/D/1 queue and the M/M/1 queue with the same mean service time $E(S)$. Representation (12.14) can be used as a two-moment approximation in more complex queueing systems. For example, consider the multi-server M/G/c queue. A simple algorithm for the exact computation of the average waiting time can in general not be given, but the interpolation formula (12.14) yields an excellent approximation to W_q, provided that c_s^2 is not too large (say, $0 \le c_s^2 \le 2$). An explicit expression for $W_q(\exp)$ in the M/M/c queue is given by

$$W_q(\exp) = \frac{(c\rho)^c}{c!c\mu(1-\rho)} \left[\sum_{k=0}^{c-1} \frac{(c\rho)^k}{k!} + \frac{(c\rho)^c}{c!(1-\rho)} \right]^{-1},$$

with $\mu = 1/E(S)$. The average waiting time $W_q(\text{det})$ in the M/D/c queue is not as easy to compute as $W_q(\exp)$, but a very good approximation for $W_q(\text{det})$ is provided by the Cosmetatos approximation

$$W_q^{app}(\text{det}) = \frac{1}{2}\left[1 + (1-\rho)(c-1)\frac{(\sqrt{4+5c}-2)}{16c\rho}\right]W_q(\exp).$$

An interesting question is whether an interpolation formula like (12.14) can also be used to approximate the waiting-time probabilities $W_q(x)$ in the $M^X/G/c$ queue. The answer to this question is negative, but fortunately it turns out that the interpolation formula can be used when we consider the waiting-time percentiles rather than the waiting-time probabilities. The waiting-time percentile $\xi(p)$ is defined by

$$W_q(\xi(p)) = p \quad \text{for } W_q(0) < p < 1.$$

Numerical computations yield that the percentiles $\xi(p)$ in the $M^X/G/c$ queue allow the two-moment approximation

$$\xi_{app}(p) = (1 - c_s^2)\xi_{det}(p) + c_s^2\xi_{exp}(p), \qquad (12.15)$$

provided that c_s^2 is not too large (say, $0 \leq c_s^2 \leq 2$). Here, $\xi_{det}(p)$ and $\xi_{exp}(p)$ denote the corresponding percentiles in the $M^X/D/c$ queue and the $M^X/M/c$ queue. In practice, one is typically interested in $\xi(p)$ for p close to 1. Then the percentiles $\xi_{exp}(p)$ and $\xi_{det}(p)$ can be calculated by using the asymptotic expansions (12.12) and (12.13) for $W_q(x)$. The reader is referred to Tijms [5] for more detailed discussion.

A glance at the tables in the queueing tablebook by Seelen et al. [4] shows that the interpolation formulas (12.14) and (12.15) can also be used to approximate performance measures in the GI/G/c queue by the corresponding measures in the special cases of the GI/D/c queue and the GI/M/c queue. Relatively simple algorithms that lead to exact solutions for GI/D/c and GI/M/c queues are discussed in Tijms [5].

12.4 OPEN PROBLEMS

The above discussion was restricted to queueing models with an infinite-capacity waiting room. However, in many practical applications, one has to deal with finite-buffer queues. A design problem of practical importance is how to choose the buffer size such that the overflow probability does not exceed some predefined value. To solve this problem, it is important to have an effective method to compute (approximately) the state probabilities in a finite-buffer queue for various values of the buffer size. These computations could be greatly facilitated when the state probabilities in the finite-capacity model could be related to those in the corresponding infinite-capacity model. Some partial results on proportionality relations between the state probabilities in the finite-capacity and infinite-capacity models have been recently obtained; see Chapter 4 in Tijms [5]. However, much research on this topic remains to be done. Similar questions are still open for (approximate) relations between the workload distributions in finite-buffer queues and the corresponding infinite-buffer queues.

BIBLIOGRAPHY

[1]   Chaudhry, M.L. and Templeton, J.G.C., *A First Course in Bulk Queues*, Wiley, New York 1983.

[2]   Neuts, M.F., *Matrix-Geometric Solutions in Stochastic Models - An Algorithmic Approach*, The John Hopkins University Press, Baltimore 1981.

[3]   Press, W.H., Flannery, B.P., Teukolsky, S.A. and Vetterling, W.T., *Numerical Recipes*, Cambridge University Press, Cambridge 1986.

[4]   Seelen, L.P., Tijms, H.C. and Van Hoorn, M.H., *Tables for Multi-Server Queues*, North-Holland, Amsterdam 1985.

[5]   Tijms, H.C., *Stochastic Models: An Algorithmic Approach*, Wiley, New York 1994.

Chapter 13

Statistical analysis of queueing systems[1]

U. Narayan Bhat, Gregory K. Miller, and S. Subba Rao

ABSTRACT This paper provides an overview of the literature on the statistical analysis of queueing systems. Topics discussed include: model identification, parameter estimation using the maximum likelihood, method of moments and Bayesian frameworks, a discussion of covariance structure and autocorrelation in queueing systems, estimation from simulation experiments, hypothesis testing, and other related aspects. The bibliography, fairly exhaustive, should provide the reader with a source of articles that comprise the core of work done up to the present time.

CONTENTS

13.1 INTRODUCTION

Statistical analysis is an integral part of formulating a mathematical model for a real system. A model is not of much use unless it is related with the system through empirical data analyses, parameter estimation, and tests of relevant hypotheses. However, in queueing theory, statistical analyses have taken a back seat due to two major reasons. Unlike in the study of time series, the input process of queues is fully described and statistical analyses of various elements of the input process can be carried out through established procedures, thus making the inference studies of the underlying process less urgent. From a global view, stochastic processes underlying queueing systems are special cases of more general stochastic processes, and the general theory of inference on stochastic processes has

---

[1]This paper is a revised and updated version of the paper with the same title published in *Queueing Systems* 1 (1987), 217-247.

---

made major strides over the last three decades (see Basawa and Prakasa Rao [8], Jacobsen [73], Prabhu and Basawa [100], Karr [81], and Basawa and Prabhu [7]). Obviously, these theories are applicable in queueing theory as well.

Despite these considerations, queueing systems present special problems not usually confronted in general statistical investigations and provide much more specific structure so that one can go beyond the general theory on stochastic processes. In this survey, we provide an overview of this subject area, with particular emphasis on topics not covered by books and earlier survey articles.

The paper is arranged in eleven sections. Section 13.2 deals with statistical problems arising from model identification. In Sections 13.3, 13.4, and 13.5 we survey estimation problems encountered in queueing systems. Section 13.4 pays special attention to the relatively new and fruitful area of Bayes estimation in queues. Section 13.6 considers problems dealing with covariance and autocorrelation structure, incorporating the estimation of a process mean. In Section 13.7 we explore estimation using two special forms of data: cross-sectional and transitional. Performance measures and their estimation are described in Section 13.8, while Section 13.9 examines estimation via simulation of queueing systems. Hypothesis tests and sequential analysis are the focus of Section 13.10. The final Section 13.11 identifies some related topics and prospects for future work. We use standard Kendall's [82] notation in representing different queueing systems, with any abandonment thus being noted.

Earlier articles giving overviews of statistical analysis in queueing theory emphasizing various aspects are the following: Cox [29], Harris [65], Reynolds [115], Eschenbach [40], Gross and Harris [61, Section 6.6], and the earlier version of this work by Bhat and Rao [16]. Also, some of the techniques discussed in Cox and Lewis [30] and Lewis [92] are pertinent to the statistical analysis of queueing systems.

13.2 IDENTIFICATION OF MODELS

In the formulation of a queueing model, one starts with the identification of its elements and their properties. The system structure is easily determined. What remains is the determination of the form and properties of the input and service processes. Four major steps are essential in this analysis: (i) collection of data, (ii) tests for stationarity in time, (iii) tests for independence, and (iv) distribution selection and/or estimation.

13.2.1 Collection of data

The required data form depends largely on the proposed model and the nature of results sought. For instance, in an M/M/1 queue, the traffic intensity can be estimated as the ratio of the estimates of the arrival and service rates. Alternatively, noting that the traffic intensity provides the utilization factor for the system, we may use the empirical utilization factor as its estimate. Some of the pitfalls of this approach are indicated by Cox [29], who notes that if ρ is the traffic intensity, the efficiency of this approach is given by $1 - \rho$. Also see the discussion by Burke following Cox's article in [29] regarding the bias resulting from estimating the load factor in an M/M/s loss system as (average number of customers in systems)/(1-probability of loss), and also see Descloux [37].

An additional problem relates to the sampling plan of the investigation. How long should the system be observed - for a specific amount of time or until a specified number of events have occurred? If the arrival process is Poisson, Birnbaum [18] has shown that the second alternative is better in determining the sample size than the first one. But when nothing is known regarding the processes, no such statements can be made and the efficiency of different schemes should be considered in individual cases. Another aspect of the sampling plan is the mode of observation; for discussions of what are known as the snap reading method and systematic sampling, the reader is referred to Cox [29], and page 86 of Cox [28], respectively.

13.2.2 Tests for stationarity

A comprehensive treatment of tests for stationarity has been given by Cox and Lewis [30]. In addition to the treatment of data on the occurrence of events as a time series and the determination of second-order properties of the counting processes, they consider statistical problems related to renewal processes and provide tests of significance in some general, as well as some specific cases. Lewis [92] updates this study and considers topics such as trend analysis of nonhomogeneous Poisson processes.

In many queueing systems (such as airport and telephone traffic), the nonstationarity of the arrival process leads to a periodic behavior. Furthermore, even though the process is nonstationary when the entire period is considered, it might be possible to consider it as a piecewise stationary process in which stationarity periods can be identified (e.g., a rush hour). Under such circumstances, a procedure that can be used to test the stationarity of the process, as well as to identify stationary periods, is the Mann-Whitney-Wilcoxon test (see, for example, Bradley [22], Conover [27], or Randles and Wolfe [104]), appropriately modified to handle ties in ranks (Putter [107], Mielke [95]). The data for the test can be obtained by considering two adjacent time intervals $(0, t_1]$ and $(t_1, t_2]$ and observing the number of arrivals during such intervals for several time periods.   Let $X_1, X_2, \ldots, X_n$ be the number of arrivals during the first interval for n periods, and let $Y_1, Y_2, \ldots, Y_m$ be the number of arrivals during the second interval for m periods (usually $m = n$). If F and G represent the distributions of the X's and Y's, respectively, then the hypothesis to be tested is $F = G$ against the alternative $F \neq G$, for which the Mann-Whitney-Wilcoxon statistic can be used. Using this test, successive stationary periods can be delineated and the system can be studied in detail within such periods (see Moore [97], who gives an algorithm for the procedure).

To analyze cyclic trends of the type discussed above, we may also use the periodogram method described by Lewis [92] for the specific case of a non-homogeneous Poisson process. Another test in the framework of the non-homogeneous Poisson process is proposed by Joseph et al. [80]. They consider the output of an $M/G/\infty$-queue where G is assumed to be known. Along with the assumption on the empty queue at the start of observation, statistical inference about the input intensity is then considered. The authors give a test for the hypothesis of a constant input intensity versus an increasing input intensity. Also discussed is a comparison between the input of two independent Poisson processes, all where only the output processes are observed.

13.2.3 Tests for independence

While formulating queueing models, several assumptions of independence are made regarding its elements. Thus, most of the models assume that interarrival times and service times are independent sequences of i.i.d. random variables. If there are reasons to make such assumptions, statistical tests can be used for verification. Some of the tests that can be used to verify independence of a sequence of observations are tests for serial independence in point processes, described in Lewis [92], and various tests for trend analysis and renewal processes, given by Cox and Lewis [30]. To verify the assumption of independence between interarrival and service times, nonparametric tests seem appropriate. Spearman's rho and Kendall's tau (Bradley [22], Conover [27], Randles and Wolfe [109]) tests for the correlation between two sequences of random variables are used, whereas Cramer-von Mises type statistics (see Koziol and Nemec [85] and references cited therein) test for bivariate independence directly from the definition of independence applied to random variables.

Tests for the dependence structure can also be carried out on the output process such as the number of customers in the system. Then a check for Markovian dependence can be made using well-known tests for Markov chains (see Bhat [13], Chapter 5 and references therein).

13.2.4 Distribution selection

The next step in the model identification process is the determination of the best model for arrival and service processes. The distribution selection problem is a standard one and based on the nature of data and availability of analytical models distributions. For a comprehensive discussion of this problem, readers are referred to Gross and Harris [61]. It is advisable to start with simple distributions such as the Poisson, since analysis under such assumptions is considerably similar. After all, a mathematical model is essentially an approximation of a real process. The simpler the model is, the easier it is to analyze and to extract information from it. Thus, the selection of a distribution should be made with regard to the tradeoff between the advantages of the sophistication of the model and our ability to derive useful information from it.

In infinite server queues, which can also be considered as systems with random translations of stochastic point processes, the estimation problem arising in many situations includes the identification of a service distribution. Often, the problem is compounded because only partial information is available. For a discussion of some problems in these and other specific situations we refer the reader to Brown [23], George and Agrawal [50], Phelan and Prabhu [105], and Ramalhoto [108].

13.3 PARAMETER ESTIMATION: THE MAXIMUM LIKELIHOOD METHOD

Estimation problems in queueing theory come in different types, some of which are (*i*) parameter estimation based on the maximum likelihood principle, (*ii*) the method of moments and nonparametric methods, (*iii*) Bayesian estimation, and (*iv*) process mean value estimation based on autocorrelation and second-order properties of an underlying stochastic process. Clarke [26], Beneš [9], Wolff [134], and Cox [29] have explored the parameter estimation problem via the maximum likelihood method for the M/M/s type queues. Harris [65] has extended their work to the $M/E_2/1$ queue. This method has been used for parameter estimation in embedded Markov chain models by Goyal and Harris [59] and Harishchandra and Rao [64]. Basawa and Prabhu [6] derived moment estimates as well as maximum likelihood estimates for general queues over a random time horizon. Halfin [62] and Heyde [67] elevate the discussion of moment estimators for the mean by considering a wider class of linear estimators. McGrath et al. [94], together with Armero [2] and Thiruvaiyaru and Basawa [129], have discussed Bayesian methods as applied to estimation in queueing systems. A good review of autocorrelation functions in queues was made by Reynolds [115]. Notable papers using this approach for process mean value estimation are Daley [33, 34], Blomqvist [19, 20], Gafarian and Ancker [46], Reynolds [114], and Aigner [1]. Throughout the remainder of this section we review maximum likelihood estimates (m.l.e.). In the upcoming sections we describe the methods (*ii*), (*iii*), and (*iv*) mentioned above.

13.3.1 Parameter estimation in Markovian systems

For estimation of parameters in Markovian systems, two papers by Moran [98, 99] are of historical significance. In the first paper he obtains the m.l.e. of the sum $\lambda + \mu$ of birth (λ) and death (μ) rates and in the second he derives the m.l.e. for the ratio $\lambda/(\lambda + \mu)$ in a simple birth and death process. In the first problem, he uses the fact that the inter-event time in a birth and death process is exponential and in the second one he uses the random walk property, in which the walk takes a positive jump with probability $\lambda/(\lambda + \mu)$ and a negative jump with probability $\mu/(\lambda + \mu)$. The two estimates determine individual estimates for λ and μ.

A landmark paper in parameter estimation in queues is by Clarke [26], who derives the m.l.e.'s for the arrival rate λ and the service rate μ in the M/M/1 queue. Let the system be observed for a length of time t such that the time spent in a busy state is a preassigned value t_b. Let n_a, n_s, t_e represent the number of arrivals, number of service completions, and the time spent in the empty state, respectively, during $[0, t]$. Furthermore, let n_0 be the initial queue length. Also assume that the system is in the steady state. The likelihood function can be written as

$$L(\lambda, \mu) = \left(\frac{\lambda}{\mu}\right)^{n_0}\left(1 - \frac{\lambda}{\mu}\right)\lambda^{n_a}\mu^{n_s}e^{-(\lambda+\mu)t_b}e^{-\lambda t_e}, \tag{13.1}$$

and the m.l.e.'s of λ and μ are found from the equations

$$\widehat{\lambda} = (\widehat{\mu} - \widehat{\lambda})(n_a + n_0 - \widehat{\lambda}t) \text{ and } \widehat{\lambda} = (\widehat{\lambda} - \widehat{\mu})(n_s - n_0 - \widehat{\mu}t_b). \tag{13.2}$$

Estimating $\widehat{\mu}$ from the second equation gives a quadratic in $\widehat{\lambda}$. Of the two solutions, any negative solution is rejected, and for the remaining values of $\widehat{\lambda}$, corresponding $\widehat{\mu}$ is obtained. Furthermore, any pair $(\widehat{\lambda}, \widehat{\mu})$ would be rejected for which $\widehat{\mu} \leq 0$ or $\widehat{\lambda}/\widehat{\mu} \geq 1$. If both solutions are valid, then the solution which maximizes the likelihood function is chosen.

For large $n_s - n_0$ Clarke gives a simple approximation for $\widehat{\lambda}$ and $\widehat{\mu}$ as

$$\widehat{\lambda} \cong (n_a + n_0)/t, \quad \widehat{\mu} \cong (n_s - n_0)/t_b. \tag{13.3}$$

The consistency of $\widehat{\lambda}$ and $\widehat{\mu}$ has been examined by Samaan and Tracy [119], who could establish only a weak consistency for $\widehat{\lambda}$. If we ignore the initial queue size, the estimates of λ and μ are, respectively, n_a/t and n_s/t_b. A possibility to render it depends on whether we have observations from a very long realization or we observe a number of independent and fairly short realizations making up a sample. In the latter case, as Cox [29] points out using Fisher's information measure, the information provided by the initial state could be remarkably high. A practical consequence is that in a given situation, it may often be advantageous to split the observations into a number of independent periods while observing the initial state in each section.

The above analysis can be extended to most Markovian queueing systems. In particular, for those queues satisfying the generalized birth-death process, the conditional likelihood function is of the form

$$e^{-\sum(\lambda_i+\mu_i)t_i}\prod\lambda_i^{n_{a_i}}\mu_i^{n_{s_i}},\tag{13.4}$$

where λ_i,μ_i are the rates of arrival and service completions in state i, n_{a_i} and n_{s_i} are the numbers of arrival and service completions in state i, and t_i is the total time spent in state i during the observation interval $(0,t]$. For a finite state birth-death queue, ignoring the impact of the initial queue size, the m.l.e.'s of λ_i and μ_i are given by

$$\widehat{\lambda}_i = n_{a_i}/t_i \ \ (0 \le i \le M-1), \ \ \widehat{\mu}_i = n_{s_i}/t_i \ \ (1 \le i \le M).\tag{13.5}$$

The above results and similar estimates for parameters in M/M/s, M/M/∞, and machine interference problems have been given by Wolff [134], where many details are provided. Parameter estimates in M/M/2 queues with heterogeneous servers have been considered by Dave and Shah [36]. For an extension of these methods to a simple Markovian queueing network, commonly known as the Jackson network, see Thiruvaiyaru et al. [130], where joint asymptotic normality of the estimators is also established.

To see the effect of neglecting the initial queue length on the estimates one can compare (13.5) and (13.3). Estimates can also be different if the observational procedures are different. For example, for the M/M/∞ model, let us assume that during an interval $[0,t]$, the following are observed:

n_0: number of calls at the start of the period,
n_a: number of calls arriving during the period,
n_d: number of calls terminating during the period,
$\bar{n}$: average number of calls during the period.

Beneš [10] has shown that n_0, n_a, n_d, and $\bar{n}$ are sufficient statistics to estimate the arrival rate λ and the service rate μ. These estimates are

$$\widehat{\lambda} = \tfrac{1}{t}(n_a+n_d) - \widehat{\mu}\bar{n}$$

$$\widehat{\mu} = \frac{1}{\bar{n}\,t}\left\{ n_d - n_0 - \bar{n} + \left[(n_d - n_0 - \bar{n})^2 + 4\bar{n}(n_a+n_d)\right]^{1/2}\right\}.\tag{13.6}$$

For the same model, Wolff [134] gives estimates based on the counts up and down out of a particular state. Beneš uses the number of arrivals and services in the observation period.

The means, variances, and correlation coefficients of these estimators, namely $\widehat{\lambda}$ and $\widehat{\mu}$ above, are also given by Beneš [10].

As pointed out by Cox [29], confidence intervals for λ, μ, and ρ in an M/M/1 system can be obtained by observing that $2\widehat{\lambda}(t_e + t_b)$ can be treated as a chi-square variate with $2n_a$ degrees of freedom and $2\widehat{\mu}t_b$ as a chi-square variate with $2n_s$ degrees of freedom, and thus forming the ratio to yield a variate having an F distribution. For details, readers are referred to Lilliefors [93] and Gross and Harris [61]. For an extension of these results to the M/M/2 queue see Jain and Templeton [78].

A consequence of estimating parameters is a notable discrepancy between the state distributions for the model with estimated parameters and for the actual system with known parameters. An estimated traffic intensity based on estimated arrival and service rates can be greater than or equal to one, even when the actual system traffic intensity is less than one. To alleviate this problem, Schruben and Kulkarni [120] suggest to bound the traffic intensity from above by 1, while estimating arrival and service rates, proposed by Clarke [26], as a possible solution.

13.3.2 Estimation in non-Markovian systems

Regarding the general problem in non-Markovian systems, Cox [29] observes that the maximum likelihood estimates of the arrival and service distributions can be determined for the following, more general class of queueing systems: (*i*) arrivals occur as a point process with a specified probabilistic structure except for unknown parameters; (*ii*) each customer is associated with a service time that is a random variable independent of the arrival pattern; (*iii*) given the arrival epochs and service times, the entire process is either uniquely determined or has a distribution independent of the unknown parameters.

Under these conditions, the likelihood function will be the product of likelihood functions for arrival patterns and for the service times, the time spent in service, say x_1, by the very last customer, and the probability for the initial number of customers. The parameters can be estimated, at least numerically, by maximizing the likelihood function once a plausible functional form has been chosen for the interarrival time and service time densities. For an illustration of this approach, see Gross and Harris [61], page 386, where m.l.e.'s for parameters in the $M/E_2/1$ queue are considered.

13.3.3 The GI/G/1 system–estimation of arrival and service time parameters

Basawa and Prabhu [6] obtain the m.l.e.'s of parameters of the arrival and service time distributions with continuous densities $f(u; \theta)$ and $g(v; \phi)$, respectively. The sampling scheme is to observe the queue until the first n customers have departed from the system and the service times of these n customers, say $(v_1, v_2, \ldots, v_n)$. Let the nth departure epoch be D_n and observe the interarrival times of all customers who arrive during $(0, D_n]$, obtaining the interarrival sequence $(u_1, u_2, \ldots, u_{N_A})$, where $N_A = N_A(D_n) = \max\{k : u_1 + u_2 + \ldots + u_k \le D_n\}$. Under this sampling scheme, the likelihood function is

$$L_n(f, g) = \left\{ \prod_{i=1}^{N_A} f(u_j; \theta) \right\} \left\{ \prod_{i=1}^{n} g(v_j; \phi) \right\} [1 - F(x_n; \theta)], \qquad (13.7)$$

where

$$x_n = x_n(D_n) = D_n - \sum_{j=1}^{N_A} u_j.$$

Since the factor $[1 - F(x_n; \theta)]$ causes difficulty in obtaining simple estimates, consider the alternative approximate likelihood function obtained by dropping the last term in (13.7):

$$L_n^a(f,g) = \left\{ \prod_{i=1}^{N_A} f(u_j;\theta) \right\} \left\{ \prod_{i=1}^{n} g(v_j;\phi) \right\}. \tag{13.8}$$

If $\widehat{\theta}_n^a, \widehat{\phi}_n^a$ are the m.l.e.'s of θ and ϕ based on $L_n^a(f,g)$, they are solutions of the equations

$$\sum_{j=1}^{N_A} \frac{\partial}{\partial\theta} \log f(u_j;\theta) = 0, \quad \sum_{j=1}^{n} \frac{\partial}{\partial\phi} \log g(v_j;\phi) = 0. \tag{13.9}$$

They prove that $\widehat{\theta}_n^a, \widehat{\phi}_n^a$ are consistent estimators of θ and ϕ and that

$$\begin{bmatrix} \sqrt{n}(\widehat{\theta}_n^a - \theta) \\ \sqrt{n}(\widehat{\theta}_n^a - \phi) \end{bmatrix} \xrightarrow{D} N_2 \left\{ \begin{pmatrix} 0 \\ 0 \end{pmatrix}, \begin{pmatrix} \sigma_\theta^2/\eta & 0 \\ 0 & \sigma_\phi^2 \end{pmatrix} \right\}, \tag{13.10}$$

where N_2 represents a bivariate normal density with

$$\sigma_\theta^2 = \left[E\left(\frac{\partial}{\partial\theta}\log f\right)^2 \right]^{-1}, \quad \sigma_\phi^2 = \left[E\left(\frac{\partial}{\partial\phi}\log g\right)^2 \right]^{-1}, \tag{13.11}$$

$\eta = \max(1,\rho)$, and ρ being the traffic intensity.

Let $\widehat{\theta}_n$ and $\widehat{\phi}_n$ be the estimators based on the full likelihood function (13.7). It is seen that $\widehat{\phi}_n = \widehat{\phi}_n^a$, and $\widehat{\theta}_n$ differs from $\widehat{\theta}_n^a$, but it can be shown that $\widehat{\theta}_n$ and $\widehat{\theta}_n^a$ have the same limiting distributions whenever

$$\frac{1}{\sqrt{n}} \frac{\partial}{\partial\theta} \log(1 - F(x_n;\theta)) \xrightarrow{P} 0. \tag{13.12}$$

This condition is satisfied for Erlangian arrivals. For large samples, estimators of θ and ϕ can be determined from (13.9) at least numerically, if not in closed form. Using (13.10), confidence intervals for θ and ϕ can also be constructed. From a practical point of view, it is significant to note that the limit properties of these statistics are obtained without the assumption on the existence of equilibrium. Basawa and Prabhu also consider m.l.e.'s for arrival and service rates in the M/M/1 queue based on a sample function observed over a fixed interval $(0,t]$, as done by Wolff [134], and obtain limit distributions of the m.l.e.'s without any restrictions on ρ. For a numerical comparison of estimates under full and approximate likelihood functions, see Jain [76].

In a subsequent paper, Basawa and Prabhu [7] have provided a unified framework for the estimation problem described above where the observation period is $(0,T]$, with a suitable stopping time T. Four different stopping rules are considered. It is shown that the limit distribution does not depend on the particular stopping rule if a random normalization is used. They assume that the interarrival and service time distributions belong to the class of non-negative exponential families. Basawa and Prabhu also derive similar results using a generalized linear model for interarrival and service time distributions.

13.3.4 Estimation using embedded Markov chains

Embedded Markov chains facilitate the study of the underlying stochastic process in a queueing system such as M/G/1 as a Markov chain. Let Q_n be the number of customers in the system immediately after the nth departure in an M/G/1 queue. It is well known that $\{Q_n, n = 0, 1, 2, \ldots\}$ is a Markov chain. Let p_{ij} be the transition probability $P(Q_{n+1} = j \mid Q_n = i)$. Consider the problem of estimating parameters associated with the arrival and service processes. Suppose the process is observed until the total number of departures reaches a fixed value N. Let the queue length (i.e., the number of customers in the system) be observed at every departure epoch. Now for the random variables $(Q_0, Q_1, \ldots, Q_N)$, the likelihood function can be expressed as

$$L = P(Q_0 = i_0) \prod_{k=1}^{N} P(Q_k = i_k \mid Q_{k-1} = i_{k-1}). \tag{13.13}$$

Depending on the form of the service time distribution, an explicit expression for the likelihood function can be written down and maximized in the usual manner to determine maximum likelihood estimates. The same general formulation holds when the service times are state dependent. Goyal and Harris [59] consider two such systems: (i) service times are exponential but with different means when the queue size is 1 and when it is > 1, (ii) service times are exponential with means linearly dependent on the number of customers in the system ($\mu_n = nu$). They derive m.l.e.'s or utilization factors (arrival rate/service rate) in the case of these two systems when the effect of the initial queue length can or cannot be ignored. Depending on the complexity of likelihood functions to be maximized, some equations will have to be solved using numerical approximation methods.

Goyal and Harris [59] also consider a second sampling method in which, in addition to the queue length at departure epochs, service and idle periods are recorded. As one would expect, the derivation gets more cumbersome. Jain [74] considers a system in which service times are Erlangian with different means when the queue size is 1 and when it is > 1. Jain also illustrates how Billingsley's [17] results can be used to determine the precision of the derived estimates and how Fisher's information matrices can be used to determine simultaneous confidence regions.

Another approach to maximum likelihood estimation using embedded Markov chains is to observe only the number of arrivals during successive service periods. In particular, when the arrivals are Poisson and the service times are Erlangian, the number of arrivals during service intervals form a sequence of i.i.d. random variables with a probability distribution consisting of only two parameters, the shape parameter k and the traffic intensity ρ. Thus, if X_n denotes the number of arrivals during the service of the $(n+1)$th customer in the $M/E_k/1$ queue, then X_n has the negative binomial distribution given by

$$P(X_n = x) = f(x, \rho) = \binom{x+k-1}{x} \left(\frac{\rho}{\rho+k} \right)^x \left(\frac{k}{\rho+k} \right)^k, \quad x = 0, 1, 2, \ldots. \tag{13.14}$$

Suppose the system is observed only at departure epochs, and let $x_1, x_2, \ldots, x_n$ be the number of arrivals during the first n service times, respectively. The likelihood function for this sample is then

$$L(x_1, x_2, \ldots, x_n; \rho) = \prod_{i=1}^{n} \binom{x_i + k - 1}{x_i} \left(\frac{\rho}{\rho+k} \right)^{x_i} \left(\frac{k}{\rho+k} \right)^k. \tag{13.15}$$

The maximum likelihood estimate of ρ is found to be $\hat{\rho} = \sum x_i / n$. This estimator is unbiased and consistent, since $E(\hat{\rho}) = \rho$ and $\text{Var}(\hat{\rho}) = \rho(\rho+k)/(kn)$. Furthermore, it turns out that $\hat{\rho}$ is also the minimum variance bound (MVB) estimator and therefore the uniformly minimum variance unbiased estimator (UMVUE) of ρ. It can be shown that the probability distribution of X belongs to the one-parameter exponential family and hence $T = \sum x_i$ is a sufficient statistic for ρ. Finally, for large values of n,

$$\frac{1}{\sigma}\sqrt{n}(\hat{\rho} - \rho) \xrightarrow{D} N(0,1), \tag{13.16}$$

where

$$\sigma^2 = \left[E\left(\frac{\partial}{\partial\rho}\log f(x,\rho)\right)^2 \right]^{-1} = \frac{\rho(\rho+k)}{k}. \tag{13.17}$$

Using (13.16, 13.17), large sample confidence intervals can be computed. These results are due to Harishchandra and Rao [64]. A numerical comparison of confidence intervals for ρ based on the asymptotic normality of the maximum likelihood estimate and on the chi-square distribution method proposed by Lilliefors [93] indicates that the former method yields a shorter interval based on the same sample size (Jain [75]).

Unfortunately, similar approaches for the $E_k/M/1$ queue do not work, since in order to obtain an embedded Markov chain the random variables $\{X_n\}$ are defined as potential services during an interarrival period and they are not observable during idle periods.

13.4 PARAMETER ESTIMATION: THE METHOD OF MOMENTS

When it is not possible to observe the system completely, several interesting estimation problems arise. One such situation occurs when we observe the output process of the M/G/1 queue and we wish to estimate the mean interarrival and mean service times. If the process is in equilibrium then, as pointed out by Cox [29], the arrival rate can be estimated with full asymptotic efficiency, since it should nearly equal the departure rate over a long time period. If the service time distribution is also exponential, no inference is possible about the mean service time, since in that case, the limiting distribution of the output is the same as that of the input, i.e., Poisson (Burke [24] and Reich [113]). On the other hand, if the service time distribution is other than exponential, an estimation of the mean service time is feasible. This is facilitated by the relation (Gross and Harris [61], p. 387)

$$C(t) = \frac{\lambda}{\mu}B(t) + \left(1 - \frac{\lambda}{\mu}\right)\int_0^t B(t-x)\lambda e^{-\lambda x}dx, \tag{13.18}$$

where $C(t)$ and $B(t)$ are the distributions of inter-departure and service times, respectively. In particular, if the service time is a constant ($= \mu^{-1}$), it can be shown that its estimate is given by the minimum observed inter-departure time.

When the service time distribution is other than exponential or deterministic, the method of moments can be used. From (13.18), the Laplace-Stieltjes transforms (LST's) are found to be

$$C^*(x) = \frac{\{1 + s/\mu\}B^*(s)}{(1 + s/\lambda)}, \tag{13.19}$$

where $C^*(s)$ and $B^*(s)$ are, respectively, the LST of the inter-departure and service times. If β_r and γ_r are the cummulants of service and inter-departure times, expanding (13.19) in powers of s and taking logarithms yields

$$\gamma_1 = \frac{1}{\lambda}, \quad \gamma_2 = \beta_2 - \beta_1^2 + \frac{1}{\lambda^2}, \quad \gamma_3 = \beta_3 - 2\beta_1^3 + \frac{2}{\lambda^3}, \text{ etc.} \tag{13.20}$$

If we assume a particular form for $B(x)$, we can have as many equations (13.20) as there are parameters in $B(x)$. Consequently, we can estimate these parameters by equating them to the observed moments of the inter-departure times. However, there are some problems in using this method. This has to do with the fact that a dependence of successive inter-departure times and autocorrelation need to be taken into consideration when calculating moments of the inter-departure times from observed data. We may either test for the absence of correlation by computing the autocorrelation coefficient or see that data are spread sufficiently far apart to make sure that the sample is approximately random (see Harris [65] and Cox [29]).

When observations of the waiting times are available, the estimation of the arrival rate and the service time parameters for the M/G/1 queue can be made by using the Pollaczek-Khintchine formula. For further details, once again, see Cox [29] and Gross and Harris [61], p. 388.

For the GI/G/1 queue, Basawa and Prabhu [6] propose the following moment estimates for the means a and b of the arrival and service time distributions:

$$\widehat{a}_n = \left(\frac{1}{N_A}\right)\sum_{j=1}^{N_A} u_j, \quad \widehat{b}_n = \frac{1}{n}\sum_{j=1}^{n} v_j. \tag{13.21}$$

The sampling scheme and the quantities u_j, v_j, N_A, and n are the same as described earlier in connection with the maximum likelihood method. It should be observed that while $\widehat{b}_n$ is the usual sample mean, $\widehat{a}_n$ is based on a random number of observations. They show that $\widehat{a}_n$ and $\widehat{b}_n$ are consistent estimators for a and b and further, that

$$\begin{bmatrix} \sqrt{n}(\widehat{a}_n - a) \\ \sqrt{n}(\widehat{b}_n - b) \end{bmatrix} \xrightarrow{D} N_2 \left\{ \begin{pmatrix} 0 \\ 0 \end{pmatrix}, \begin{pmatrix} \sigma_1^2/\eta & 0 \\ 0 & \sigma_2^2 \end{pmatrix} \right\}, \tag{13.22}$$

where N_2 is the bivariate normal density, σ_1^2 and σ_2^2 are, respectively, variances of arrival and service times, and $\eta = \max\{1, \rho\}$. As observed earlier, these properties of the estimates do not require the imposition of steady state conditions. Furthermore, the estimates are "natural" estimates and are simple. However, if it is required to find estimates of either the arrival or service time parameters, it is not clear whether this simplicity can still be maintained.

13.5 BAYES ESTIMATION

As an alternative to the estimation procedures outlined above, some techniques using Bayesian frameworks have been developed. An early paper on the subject is by Muddapur [100], who derives the Bayesian estimates of λ and μ in the M/M/1 and M/M/∞ models using the same likelihood function given by Clarke [26]. In conjunction with Clarke's likelihood function, Muddapur uses both a natural conjugate prior joint density for λ and μ, as well as a product of Gamma distributions, as a prior to produce two sets of Bayes estimates for parameters in the M/M/1 queue. Muddapur also follows Clarke in deriving estimates in the M/M/∞ case by assuming that the initial queue size has a Poisson distribution.

Many years later, the M/M/1 queue was considered by Armero [2]. Here, Armero obtains the posterior distribution of ρ and the posterior predictive distributions of the waiting time in the queue, and the number of customers in the system given independent samples of arrival and service times. Let $z_1 = \{x_1,...,x_n\}$ be a random sample of service times from an exponential distribution with parameter θ.

Likewise, $z_2 = \{y_1, \ldots, y_m\}$ are random interarrival times from an exponential distribution with parameter λ. Then the joint likelihood of θ and λ is $p(z_1, z_2 \mid \theta, \lambda) = \theta^n \lambda^m \exp(-\theta r_1 - \lambda r_2)$ where $r_1 = \sum x_i$ and $r_2 = \sum y_i$. Assuming a prior on λ and θ of independent Gamma distributions with parameters (a_0, b_0) and (α_0, β_0), respectively, Armero derives the posterior distribution for λ and θ as $p(\theta, \lambda \mid z_1, z_2) = Ga(\theta \mid a_n, b_n) Ga(\lambda \mid \alpha_m, \beta_m)$ where $Ga(\cdot \mid a, b)$ is a Gamma distribution with parameters a and b and $a_n = a_0 + n$, $b_n = b_0 + r_1$, and $\alpha_m = \alpha_0 + m$, $\beta_m = \beta_0 + r_2$. Using the relation between the Gamma and F distribution, the posterior distribution of ρ is:

$$\rho \sim \frac{\alpha_m b_n}{a_n \beta_m} F(2\alpha_m, 2\alpha_n), \tag{13.23}$$

where $F(a, b)$ is an F distribution with parameters a and b.

Recently, Armero and Bayarri [4] have extended this work in the M/M/1 case by deriving posterior predictive distributions of the number of customers in the system and queue, along with posterior predictive distributions for the waiting times in the system and in the queue. In addition, Armero and Bayarri find posterior distributions for the length of busy and idle periods in the queue. All of the posterior predictive distributions given are expressed in terms of the Gauss hypergeometric function and most turn out to have no moments, due to the form of the assigned prior distribution of ρ whose right hand tail does not approach 0 as ρ goes to 1. This seemingly undesirable property does not surface when other choices of priors are used.

McGrath et al. [94] and McGrath and Singpurwalla [95] present a *subjective* Bayesian approach to inference in queueing; the first of the two papers being concerned with modeling aspects. As a setup, let $\theta = (\lambda, \mu)$ and let θ have prior distribution $\pi(\theta \mid w)$, with w being a vector of "hyperparameters", say $w = (\alpha_1, \alpha_2, \beta_1, \beta_2, \gamma)$ with $\alpha_i, \beta_i (i = 1, 2) > 0$, and γ representing the dependence between λ and μ.

First, suppose $\gamma = 0$ and describe the independence between λ and μ as the product of two Erlang densities with means α_i / β_i and variances α_i / β_i^2 for $i = 1, 2$ respectively. Thus,

$$\pi(\theta \mid w) = \exp(-\lambda \beta_1 - \mu \beta_2) \lambda^{\alpha_1 - 1} \mu^{\alpha_2 - 1} \prod_{i=1}^{2} \frac{\beta_i^{\alpha_i}}{\Gamma(\alpha_i)}, \tag{13.24}$$

which by relationship of the Erlang distribution with the F distribution, produces the following density at ρ:

$$\pi(\rho \mid w) = \frac{\Gamma(\alpha_1 + \alpha_2)}{\Gamma(\alpha_1)\Gamma(\alpha_2)} \left(\frac{\beta_1}{\beta_2}\right)^{\alpha_1} \frac{\rho^{\alpha_1 - 1}}{\left(1 + \left(\frac{\beta_1}{\beta_2}\right)\rho\right)^{\alpha_1 + \alpha_2}} \quad \text{for } \rho \geq 0. \tag{13.25}$$

In the case of dependence, the authors suggest the use of the Bivariate Lognormal Distribution (BVL) as a prior. This implies that the prior distribution for ρ is also a BVL.

The effects of uncertainty in parameters on the assumed distributions can then be analyzed. For the M/M/1 queue, the interarrival and service time distributions are both exponential and independent denoted $a(t \mid \lambda) = \lambda e^{-\lambda t}$ and $s(t \mid \mu) = \mu e^{-\mu t}$. The assignment of (13.24) as a prior leads to a Pareto density at t for any single interarrival time.

Similarly, $s(t \mid \alpha_2, \beta_2)$, the density at time t for any single service time, is also

Pareto. In this case, it is shown that $\{T_i\}$ and $\{S_i\}$, the sequences of interarrival and service times, are each exchangeable sequences. Thus, the setup in which uncertainty is involved in the parameters leads to the consideration of queues with exchangeable interarrival and service times. The authors also discuss effects of the uncertainty in parameters on traditional measures of performance.

In the second paper, McGrath and Singpurwalla isolate discussion to inference procedures. Let $\mathcal{H}$ be some background information and let $\mathcal{S}$ denote any additional new information that is collected, such as data. McGrath and Singpurwalla state that the concept of $\mathcal{H}$ is important, since in the subjective Bayesian framework, all probabilities are conditional on $\mathcal{H}$. Also, let $\mathcal{D}$ be the design of the experiment and $\mathcal{E}$ (see example below) be the narrative specification of an experiment. The quantities that are known before the experiment is running or specified in advance are a part of $\mathcal{D}$, whereas the quantities that are made available to us after the experiment is performed are in $\mathcal{S}$. The authors are concerned with those members of $\mathcal{S}$ that are "informative" versus those that are "noninformative" and thus will not be included in the likelihood function, i.e., if $\mathcal{S} = (w_1, w_2)$, then w_1 is noninformative with respect to θ if w_1 is independent of w_2 given θ. Several types of experiments are considered for the M/M/1 queue such as: $\mathcal{E}_1$ = Observe the queue for a fixed duration of time, beginning at time t_0. Record the interarrival and service times only; further assume that the initial system size is unobservable. The experiment $\mathcal{E}_1$ is paired with its corresponding $\mathcal{D}_1$ and $\mathcal{S}_1$ to yield a likelihood for θ. Then the "noninformative" parts are omitted and an application of Bayes rule (using prior (13.24)) leads to the posterior density at θ being proportional to (also see Armero [2])

$$\lambda^{\alpha_1 + m - 1} \left[\exp\left\{ -\lambda\left(\beta_1 + \sum_{i=0}^{m} a_i \right) \right\} \right] \mu^{\alpha_2 + n - 1} \left[\exp\left\{ -\mu\left(\beta_2 + \sum_{j=1}^{n} s_j \right) \right\} \right],$$

$$(13.26)$$

where $\{a_0, a_1, \ldots, a_m\}$ are interarrival times and $\{s_1, \ldots, s_n\}$ are service times. This posterior density is the product of Erlang densities and thus the choice of conjugate (Erlang) prior leads to a convenient way to update statements of uncertainty regarding θ. Four other separate cases of experiments involving the M/M/1 queue are explored in a similar fashion.

In contrast, Thiruaiyaru and Basawa [124] take an *empirical* Bayes approach for estimating parameters of various queueing systems. Rather than stating a single prior distribution for a set of parameters, the empirical Bayes approach allows for a specification of a *family* of prior distributions whose parameters (the super- or hyper-parameters) take on values in a specified set. Then the prior distribution is estimated by using data to make assessments of these superparameters.

Concerning the M/M/1 queue, suppose that N queues are observable (independent) in which the arrival times for the first n customers, denoted U_{ik} for $i = 1, \ldots, n$, and the service times of the first m customers, denoted V_{jk} for $j = 1, \ldots, m$, are recorded for $k = 1, \ldots, N$. Given the arrival rates λ_k and service rates μ_k, we have the following distributions for $U_k = \{U_{ik}, i = 1, \ldots, n\}$ and $V_k = \{V_{jk}, j = 1, \ldots, m\}$:

$$f_{U_k}(u_k \mid \lambda_k) = \lambda_k^n \exp\left\{ -\lambda_k \sum_{i=1}^{n} u_{ik} \right\}, \quad f_{V_k}(v_k \mid \mu_k) = \mu_k^m \exp\left\{ -\mu_k \sum_{j=1}^{m} v_{jk} \right\}.$$

$$(13.27)$$

As priors, assign $\{\lambda_1,\ldots,\lambda_N\}$ to have i.i.d. Gamma (α_1,β_1) distribution and assign $\{\mu_1,\ldots,\mu_N\}$ to have i.i.d. Gamma (α_2,β_2) distribution. Let $X_k = (U_k, V_k)$, $\theta_k = (\lambda_k,\mu_k)$, and denote the traffic intensity of the kth queue by ρ_k. The aim is to estimate ρ_{N+1} on the basis of $(X_1, X_2,\ldots,X_{N+1})$. Suppressing $(N+1)$ and assuming that (α_1,β_1) and (α_2,β_2) are known, the authors arrive at the following Bayes estimator for ρ (w.r.t. squared error loss):

$$\hat{\rho}^B = \frac{(n+\alpha_1)\left(\sum\limits_{j=1}^{m} V_j + \beta_2\right)}{(m+\alpha_2-1)\left(\sum\limits_{i=1}^{n} U_i + \beta_1\right)}, \tag{13.28}$$

and the corresponding *empirical Bayes* (*EB*) estimator being the same as above with α_i and β_i replaced by $\hat{\alpha}_i$ and $\hat{\beta}_i$, respectively, for $i = 1,2$, where $\hat{\alpha}_i$ and $\hat{\beta}_i$ are the one step maximum likelihood estimators of α_i and β_i. A discussion of how to obtain these one step MLE's is given by the authors. Asymptotic properties such as consistency and rate of convergence to the Bayes estimator are also given for the empirical Bayes estimator.

The above analysis uses the joint distribution of X and θ to assess properties of ρ in the empirical Bayes context. Using the conditional distribution of X given θ, a "frequentist" approach can be implemented. Let $(X_1,\theta_n)\ldots(X_N,\theta_N)$ be the data from the first N queues. Suppose that N and the number of arrivals and service completions in these N queues are fixed, and the number of arrivals n_c and service completions m_c in the current $(N+1)$th queue go to infinity, such that $T_c = n_c + m_c \to \infty$ with $m_c/T_c \to d$ and $n_c/T_c \to (1-d)$ for some d in $(0,1)$. Then as $T_c \to \infty$,

$$\sqrt{T_c}(\hat{\rho}^{EB} - \rho),\ \sqrt{T_c}(\hat{\rho}_B - \rho)\ \text{and}\ \sqrt{T_c}(\hat{\rho}^{MLE} - \rho), \tag{13.29}$$

all have the same asymptotic normal distribution: $N(0,(1-d)^2\rho^2 + d)$.

Singpurwalla [124] gives a strong critique of some of Thiruvaiyaru and Basawa's approaches in a discussion following the article, and Thiruvaiyaru and Basawa respond in neat order in the Rejoinder [129].

13.6 COVARIANCE STRUCTURE, AUTOCORRELATION, AND MEAN VALUE FUNCTION

In the previous sections, we have identified procedures to estimate various parameters of the arrival and service processes based on random samples of observations on a queueing system. These estimates in turn provide estimates of the queue characteristics such as the mean queue length $E[X(t)]$ and mean waiting time $E[W_n]$. However, it is possible to estimate these quantities directly from sample observations. Thus, consider the $X(t)$ process, which is assumed to be stationary in the wide sense. To estimate the process mean value $\mu = E[X(t)]$, one can observe $X(t)$ over a fixed time interval $(0, T]$ and construct a suitable sample mean. Two obvious choices for estimating μ are (see Reynolds [115])

$$\hat{\mu}_1 = \sum_{r=1}^{n} X(rh)/n \ \text{and} \ \hat{\mu}_2 = \frac{1}{T}\int_0^T X(t)dt, \ \text{where } nh = T. \tag{13.30}$$

It can be seen that $\hat{\mu}_2$ is the limiting form of $\hat{\mu}_1$ as $n \to \infty$ and $h \to 0$. Both $\hat{\mu}_1$ and $\hat{\mu}_2$ are unbiased for μ. To assess their precision as well as compare them with other estimators for μ, we need to evaluate their standard errors. These are given by

$$\text{Var}(\widehat{\mu}_1) = \frac{\sigma^2}{n}\left[1 + \frac{2}{n}\sum_{j=1}^{n-1}(n-j)\rho(jh)\right] \text{ and } \text{Var}(\widehat{\mu}_2) = \frac{2\sigma^2}{T}\int_0^T\left(1 - \frac{x}{T}\right)\rho(x)dx, \quad (13.31)$$

where $\sigma^2 = \text{Var}\{X(t)\}$, $\rho(x)$ is the autocorrelation function of the $X(t)$ process defined by $\rho(x) = \text{cov}[X(t), X(t+h)]/\sigma^2$, and $\rho(jh)$ are similarly defined for the discrete process $X(rh)$ sampled from the $X(t)$ process. From the above variance expressions, it is obvious that the sampling errors of $\widehat{\mu}_1$ and $\widehat{\mu}_2$ (and other such estimators) cannot be evaluated without knowledge of the covariance structure of $X(t)$. Furthermore, it is noted that for large n,

$$\text{Var}(\widehat{\mu}_1) \cong \frac{\sigma^2}{n}\left[1 + 2\sum_{j=1}^{\infty}\rho(jh)\right]. \quad (13.32)$$

These results have motivated the study of the covariance structure in queues and of the asymptotic behavior of $\sum_{j=1}^{\infty}\rho(jh)$. For a good review of the work in the area, see Reynolds [115].

To obtain confidence intervals for the process mean in the M/M/1 system, Gebhard [51] uses (13.32) with the result

$$\text{Var}(\widehat{\mu}_1) \cong A(\widehat{\mu}_1)\frac{\sigma^2}{n}, \quad \text{where } A(\widehat{\mu}_1) = 1 + 2\rho(1+\rho)/(1-\rho)^2. \quad (13.33)$$

By applying the central limit theorem for dependent variables, Gebhard shows that the distribution of $\widehat{\mu}_1$ is asymptotically normal with mean $E[X(t)]$ and the variance given by (13.33). These results lead to the construction of confidence intervals.

As seen above, a standard estimation procedure when estimating the mean value of a process (e.g., the mean number of busy servers, probability of delay, and mean queue length) is to average the measurements collected from data given each one equal weight. Although this approach is attractive due to its simplicity, it does not produce the best estimator in the sense of having the minimal variance. In fact, Descloux [38] shows that if equal weights are used and observations are taken over a fixed time interval, and are equally spaced, then increasing the number of measurement points can sometimes increase the variance of the estimator. Thus, down-weighting some observations due to the covariance structure is a possibility. Halfin [62] presents methods to accomplish this; he finds the minimum variance linear estimator for the expected value of a stationary stochastic process when the covariance function of the process is a sum of decaying exponentials. This condition is often met in Markovian queueing models with a limited waiting space.

To introduce Halfin's method, let $X(t)$ be a stationary stochastic process, such as those described above, with expected value m and covariance function ρ having

form $\rho(x) = \sum_{i=1}^{K}a_ie^{-r_i|x|}$ where $a_i > 0$, $i = 1, 2, \ldots, K$, $0 < r_1 < r_2 < \ldots < r_K$. Thus, $E[X(t)] = m$, and $\text{cov}(X(t), X(s)) = \rho(t-s)$. Next, let the process $X(t)$ be observed in the fixed interval $[0, T]$, and let $\mu(t)$ be a function of bounded variation for $t \leq T$ such that $\mu(t) = 0$ for $t < 0$ and $\mu(T) \neq 0$. If $\mu(T) = 1$, then μ is normalized. Let $Z = Z(X(t), T, \mu) = (1/\mu(T))\int_0^T X(t)\mu(dt)$. Since $E[X(t)] = m$ for all t, we have $E[Z] = (1/\mu(T))\int_0^T E[X(t)]\mu(dt) = m$. Thus for each μ, Z is an unbiased linear estimator for the mean of $X(t)$. The most common form of μ when the process is continually observed is $\mu_u(x) = x/T$, corresponding to uniform weights. Halfin goes on to give the variance of Z as

$$\text{Var}(Z) = (1/\mu(T))^2 \int_0^T \int_0^T \text{cov}(X(t), X(s))\mu(dt)\mu(ds). \quad (13.34)$$

If V represents the set of all right continuous functions with bounded variation on $[0, T]$, then the problem of finding the optimal weight function for the observations is expressed as

$$\min_{\mu \in V} \int_0^T \int_0^T \rho(t - s)\mu(dt)\mu(ds), \quad \text{subject to } \mu(T) = 1. \quad (13.35)$$

Noting the similarity of this equation with one that is well known in detection theory, Halfin constructs the best linear estimators by examining the unique solution of the integral equation $\int_0^T \rho(t - s)\mu_0(ds) = g(t) \ \ 0 \le t \le T$, where $g(t)$ is a constant.

A substantial extension to Halfin's results has been given by Heyde [67] by relaxing the condition of decaying exponential assumptions on the covariance function to a broader class of stochastic processes. Heyde shows that in this class of queueing processes, the method of moments estimator has the same asymptotic variance as the minimum variance unbiased estimator.

We now proceed to address four separate ideas that are tied to covariance structure and autocorrelation functions of processes associated with queues. This section concludes with a brief discussion of some other issues related to autocovariance/autocorrelation calculations in queueing processes.

13.6.1 Sample size determination

A use of the above results is in determination of the minimum sample size in simulation runs consistent with some required precision. For a detailed discussion of estimation in simulation of queueing processes, see Section 13.9. It is clear that a larger sample will be necessary to estimate the process mean when the observations are dependent. In particular, when the observations are independent, the required sample size is proportional to σ^2, while when the observations are serially correlated, the required sample size n_c is proportional to $\sigma(1 + 2\sum \rho(jh))$.

The key to sample size determination is the evaluation of $\sum \rho(jh)$. For simple queues, this may not pose much of a problem. If we consider the GI/M/1 queue, the equilibrium distribution is geometric. Using this fact and the property of stochastic monotonicity in Markov chains for this system, Daley [33] shows that as $\rho \to 1$, $(1 - \rho)^2 \sum \rho(jh) \to 1 + \alpha_2/2\alpha_1^2$, where α_1 and α_2 are the first and second moments of the arrival time distribution. Combining this result with (13.33), it can be concluded that the coefficient of variation of $\hat{\mu}_1$ is constant for ρ near 1 when $n \cong k(1 - \rho)^{-2}$ for some constant k. A similar result should hold for the GI/M/s queue as well.

13.6.2 Time slicing and event sequencing

We have considered the estimator $\hat{\mu}_1$ when the $X(t)$ process is observed only at times $rh \ (r = 1, 2, \ldots, n; nh = T)$. This sampling scheme is referred to as *time-slicing* by Gafarian and Ancker [46]. In order to consider $\hat{\mu}_2$, or Halfin's methods, we must observe the process continuously during $[0, T]$, and the sampling method is then called *event sequencing*. In the latter case, $\hat{\mu}_2$ can be approximated by $\hat{m}_2 = \frac{1}{T}\sum_{i=1}^{N} S_i$ where N is the number of customers arriving in $[0, T]$, and S_i is the service time of the ith customer. For small values of ρ in the M/M/1 case, it can be shown that for large T, $\mathrm{var}(\hat{m}_2) \cong 2\rho^2/(\lambda T)$ (see Beneš [11]).

Comparisons of the efficiencies of $\hat{\mu}_1$ and $\hat{\mu}_2$ require a more detailed investigation, which has been done by Gafarian and Ancker and later on by Reynolds [114], for the M/M/c type of queues. Recall that the efficiency of $\hat{\mu}_1$ relative to $\hat{\mu}_2$ is defined by $E_h = \mathrm{Var}(\hat{\mu}_2)/\mathrm{Var}(\hat{\mu}_1)$. For a class of processes for which $\rho(h) = e^{-ch}$ $(c > 0)$,

$$E_h = \frac{2(1 - e^{-ch})^2(nch - 1 + e^{-nch})}{(ch)^2[n(1 - e^{-2ch}) - 2e^{-ch}(1 - e^{-nch})]}. \tag{13.36}$$

The constant c is a measure of how rapidly the correlation between two samples, h time units apart, decreases as h increases. By graphing E_h against ch and n, Gafarian and Ancker conclude that for any reasonable efficiency $E_h > .90$, the sampling interval h must be less than the constant $1/c$, preferably by a substantial margin. Notice, as alluded to in Halfin [62], that the M/M/1, M/M/∞, and the M/M/c loss system satisfy condition $\rho(h) = e^{-ch}$. Qualitatively similar results have been obtained by Reynolds [114] for finite Markov queues, wherein the eigenvalue structure is exploited in exhibiting $\rho(h)$ and E_h. He concludes that $E_h < 1$ for all $h > 0$ and, therefore, that event sequencing is always more efficient than time slicing. However, from a practical point of view it is easier to handle time slicing rather than event sequencing. Since $E_h \to 1$ as $h \to 0$, a value of h can be determined to insure that E_h is as close to 1 as required, in which case time slicing can be used with only this acceptable loss of efficiency.

It is worthy to note that Halfin gives the variance of the best linear estimator of the mean of the $X(t)$ process, so that efficiencies comparing μ_0 with $\hat{\mu}_1$ and $\hat{\mu}_2$ are theoretically possible. Halfin also gives a discussion on the asymptotic properties of this variance and how it compares with the asymptotic variance of other estimators.

In the process of time slicing, the question arises whether the process should be sliced evenly or at non-uniform intervals. For a number of problems, including the M/M/∞ queue, Gafarian and Ancker have shown that it is not optimal to space the sample points evenly unless $n = 2$. In the case of the queue M/D/∞, however, Grassmann [60] comes to a different conclusion. He shows that the optimal way to estimate the number of customers in the queue M/D/∞ is by sampling at times 0, $s, 2s, \ldots, ks$, where s is the constant service time and $T = ks$ is the sampling interval.

13.6.3 Estimation of waiting time

Consider the sequence $\{W_r\}$, where W_r is the waiting time of the rth customer. We shall assume the queue discipline to be first-come-first-served. The mean waiting time $\mu_w = E\{W_r\}$ is estimated by $\hat{\mu}_w$ (Blomqvist [19]), where $\hat{\mu}_2 = \frac{1}{n}\sum_{r=1}^{n} W_r$. We have

$$E(\hat{\mu}_w) = \mu_w, \quad \mathrm{Var}\{\hat{\mu}_w\} = \frac{\sigma_w^2}{n}\left[1 + \frac{2}{n}\sum_{j=1}^{n-1}(n-j)\rho_j\right], \tag{13.37}$$

where $\rho_j = \gamma_j/\gamma_0$, $\gamma_j = \mathrm{Cov}(W_r, W_{r+j})$, and $\sigma_w^2 = \mathrm{Var}\{W_n\}$.

In the case of the M/G/1 queue, Blomqvist has shown that for large n,

$$\text{Var}(\widehat{\mu}_w) \cong \frac{1}{n}\left(2\sum_{r=0}^{\infty}\gamma_r - \gamma_0\right). \tag{13.38}$$

The sum $\sum \gamma_r$ is evaluated in terms of the derivatives of the LST of the waiting time distribution. Blomqvist has demonstrated numerically the effect of autocorrelation in the case of the $M/E_k/1$ queue by tabulating $n\,\text{Var}\,(\widehat{\mu}_w)/\,\text{Var}\,(W_n)$, i.e., the factor by which the variance of the sample mean of uncorrelated observations should be multiplied to allow for the effect of the autocorrelation.

A later paper by Blomqvist [21] gives heavy traffic results for the covariance functions in the GI/G/1 queue and considers the problem of estimating $P(W > w)$ under steady state conditions. Further details of the behavior of the correlation structures for the waiting time process can be found in Daley [33] and Craven [32].

13.6.4 Direct estimates versus m.l.e.

Let $W_n^S = W_n + S_n$ be the time spent by the customer in the system. A direct estimator for $\mu_w^S = E(W)$, based on the $X(t)$ process, was proposed by Jenkins [79] as

$$\widehat{\mu}_w^S = \frac{\int_0^T X(t)dt}{A(T)} = \frac{I}{A}, \tag{13.39}$$

where $A(T)$ is the number of arrivals in $(0, T]$. An approximation to $\text{Var}(\widehat{\mu}_w^S)$ is given by Jenkins in the form

$$\text{Var}\left(\frac{I}{A}\right) \cong \left[\frac{E(I)}{E(A)}\right]^2\left[\frac{\text{Var}(I)}{E^2(I)} + \frac{\text{Var}(A)}{E^2(A)} - \frac{2\text{Cov}(I, A)}{E(I)E(A)}\right]. \tag{13.40}$$

For large T, evaluation of these relevant quantities in the M/M/1 case leads to

$$\text{Var}(\widehat{\mu}_w^S) \cong \rho^2(1 + \rho)^2 \big/ \{\lambda^3(1 - \rho)^4 T\}. \tag{13.41}$$

On the other hand, the m.l.e. of μ_w^S is $\widehat{\mu}_w^{*S} = (\widehat{\mu} - \widehat{\lambda})^{-1}$, where $\widehat{\mu}$ and $\widehat{\lambda}$ are the m.l.e.'s of μ and λ. This has the variance

$$\text{Var}(\widehat{\mu}_w^{*S}) \cong \rho^2(1 + \rho^2) \big/ \{\lambda^3(1 - \rho)^4 T\}. \tag{13.42}$$

Comparing (13.41) to (13.42), we find that the efficiency of the direct estimator $\widehat{\mu}_w^S$ relative to $\widehat{\mu}_w^{*S}$ is $E = (1 + \rho^2)/(1 + \rho)^2$, which varies between 1 and 1/2 as ρ varies between 0 and 1. This shows that, except for the case of a low traffic intensity, there is a marked loss of efficiency in estimation using only the waiting time of individual customers.

13.6.5 Optimal prediction

A natural extension of the covariance structure of the output process is the use of their properties in prediction. For the GI/M/1 queue, Stanford et al. [125] provide an algorithm to obtain the optimal mean square predictor $E[Y_n \mid \widetilde{Q}_k]$, where Y_n can be Q_n (the embedded queue length at the nth arrival epoch), W_n (the waiting time of the nth customer), or S_n (the system time for the nth customer), and $\widetilde{Q}_k$ $= (Q_0, Q_1, ..., Q_n)$. The mean squared errors of the predictors are obtained through a bounding approach. These results have been extended to the GI/M/s system by Woodside et al. [135]. In a companion paper to their previous one, Pagurek et al. [103] consider similar problems pertaining to the queue M/G/1.

13.7 ESTIMATION FROM SPECIAL TYPES OF DATA

Sometimes it is possible to observe a number of identical queueing systems simultaneously and obtain *cross-sectional* data which the estimates are to be based on. Other times it is barely possible to even observe a single queue, save the service completion and commencement times. This latter type of data is termed *transactional*. This section examines estimation techniques that are available in these two special cases of data collection.

13.7.1 Cross-sectional data

Previous sections have examined parameter estimation from single sampling realizations or time series data. However, if cross-sectional data are available, then one can use existing proportionality relations such as $L = \lambda W$, leading to ratio and least squares estimation methods that may provide statistics with useful properties. An interesting paper on parameter estimation using this approach in an M/M/1 setting is Aigner [1]. Suppose that the following observations are made: $u = $ time between successive arrivals, $v = $ service times, $n = $ number of customers in the system, $q = $ number of customers in the queue, $z = $ waiting time in the system, and $x = $ waiting time in the queue.

Then for the queue M/M/1, the following hold:

$$E(n) = \lambda E(z), \ E(x) = \rho E(z), \ E(q) = \lambda E(x), \ E(q) = \rho E(n), \ E(n) = \mu E(x),$$

where λ and μ are the arrival and service parameters, respectively.

A "direct" ratio (R) estimator for λ, based on a sample of size N, is given by $\widehat{\lambda}_{nz}^{R} = \bar{n}/\bar{z}$, where $\bar{n}$ and $\bar{z}$ are, respectively, the sample means of observations on n and z. We find that $\widehat{\lambda}_{nz}^{R}$ is asymptotically unbiased under random sampling assumptions and its asymptotic variance being $\text{Var}(\widehat{\lambda}_{nz}^{R}) = \frac{\lambda^2}{N}\frac{1-\rho}{\rho}$. Moreover, in this case $\widehat{\lambda}_{nz}^{R}$ is also the maximum likelihood estimator $\widehat{\lambda}_{nz}^{ML}$.

The least squares (LS) estimator for λ, proposed by Aigner, is generated by a homogeneous regression of n on z. This estimator is then given by $\widehat{\lambda}_{nz}^{LS} = \sum_{i=1}^{N} n_i z_i \Big/ \sum_{i=1}^{N} z_i^2$, where (n_i, z_i), $i = 1, 2, \dots, N$, denote the paired observations of n and z. In general, even though the LS estimator could be biased and inconsistent, for the M/M/1 queue it is found that LS estimators are consistent. The large sample variance of $\widehat{\lambda}_{nz}^{LS}$ is obtained as $\text{Var}(\widehat{\lambda}_{nz}^{LS}) = \frac{3}{2}\left[\frac{\lambda^2}{N}\frac{1-\rho}{\rho}\right]$.

Another estimate of λ with the same asymptotic variance as $\widehat{\lambda}_{nz}^{R}$ is $\widehat{\lambda}_{qx} = \bar{q}/\bar{x}$.

Comparing estimators $\widehat{\lambda}_{nz}^{R}$, $\widehat{\lambda}_{nz}^{LS}$, $\widehat{\lambda}_{qx}^{R}$ for λ with the m.l.e. $\widehat{\lambda}_{u}^{ML} = 1/u$, Aigner observes that when the traffic intensity is low, $\widehat{\lambda}_{u}^{ML}$ is most efficient, if $\rho > .5$, $\widehat{\lambda}_{nz}^{ML,R}$ and $\widehat{\lambda}_{qx}^{R}$ are generally more efficient than $\widehat{\lambda}_{u}^{ML}$, and the least-squares estimator $\widehat{\lambda}_{nz}^{LS}$ is always less efficient than the alternatives.

Similar arguments for estimators of μ lead to the following observation: The best estimator (with the least asymptotic variance for all values of ρ) of μ is given by $\widehat{\mu}_{nz}^{ML} = \frac{1+\bar{n}}{\bar{z}}$ with asymptotic variance $(\mu^2/N)(1-\rho)$. The major drawback of this estimator is that it requires observations on waiting times that are generally expensive to collect.

Finally, based on asymptotic variance, except where ρ is small ($< .4$), the best estimator of ρ is the m.l.e. and the LS estimator $\hat{\rho}_n^{ML,LS} = \frac{\bar{n}}{1+\bar{n}}$, which uses only the number of customers in the system. This estimator has asymptotic variance $\rho(1-\rho)^2/N$. When ρ is small, the estimator $\hat{\rho}_q^{ML} \simeq \frac{1+\bar{q}}{2+\bar{q}}$ has a smaller asymptotic variance given by $\frac{\rho^2}{N}\left[\frac{\rho+1-\rho^2}{1+(1-\rho)^2}\right]$.

13.7.2 Transactional data

Recently, an attempt at estimating quantities associated with queues by only observing times of service commencement and service completion has been undertaken. Larson [86] calls such data *transactional*, such as the transactions that an Automatic Teller Machine might record at a bank. Larson and others have developed algorithms for estimating mean waiting times and queue lengths when only transactional data are available. Larson has attached the name Queue Inference Engine (QIE) to his routines.

Consider a homogeneous Poisson arrival process with rate λ. Suppose that over a fixed interval $[0,T]$ N Poisson events occur. Let the N ordered arrival times be $0 \leq X_1 \leq X_2 \leq \ldots \leq X_N \leq T$. Larson shows that service completion times within a congestion period impose a set of inequalities on the arrival times of other customers who wait in the queue. These inequalities in turn produce conditioning information in the context of order statistics, which is used to deduce the queue behavior. Larson's results follow from a calculation of the *a priori* probability that arrivals during a congestion period obey the established time orderings of the observed departure times.

To illustrate, consider a congestion period that lasts from time $t = 0$ to time $t = t_{N+1}$ and define $\boldsymbol{t} = (t_1, t_2, \ldots, t_N)$, where t_i is the departure epoch of the ith customer served, and $O(\boldsymbol{t}) = \{X_1 \leq t_1, X_2 \leq t_2, \ldots, X_N \leq t_N\}$, where X_i is the arrival time of the ith customer to enter the queue during the congestion period. Furthermore, let $N(t) = $ number of arrivals to system in $(0,t]$, $t \leq t_N$. In order to present the algorithms, it becomes necessary to calculate $Pr\{O(\boldsymbol{t}) \,|\, N(t_N) = N\}$. Two recursive procedures for calculating this value are given, which depend on the following probabilities:

$$\Psi_k(\boldsymbol{t},T) \equiv Pr\{X_1 \leq t_1, X_2 \leq t_2, \ldots, X_k \leq t_k \,|\, N(T) = k\},$$

where $k \leq N, T \geq t_k$, and $\Psi_0 = 1$ and

$$\alpha_{ki}(\boldsymbol{t}) = Pr\{X_1 \leq t_1, X_2 \leq t_2, \ldots, X_i \leq t_i, \ldots, X_k \leq t_i \,|\, N(t_N) = k\}$$

for $k \geq i$.

$\Psi_k(\boldsymbol{t},T)$ is the *a priori* conditional probability that the arrival times in $[0,T]$ satisfy the first k departure time inequalities given k arrivals in $[0,T]$, and $\alpha_{ki}(\boldsymbol{t})$ is the *a priori* conditional probability that all arrivals in $[0,t_N]$ occur before the ith departure and obey the first $(i-1)$ other departure time inequalities, given exactly k arrivals in $[0,t_N]$. We then can write $Pr\{O(\boldsymbol{t}) \,|\, N(t_N) = N\} = \Psi_N(\boldsymbol{t},t_N) = \alpha_{NN}(\boldsymbol{t})$. The following two lemmas are given that can be used to recursively calculate $\Psi_k(\boldsymbol{t},T)$ and $\alpha_{ki}(\boldsymbol{t})$:

Lemma 13.1. $\Psi_k(\boldsymbol{t},T) =$

$$\sum_{j=1}^{k}\binom{k}{k-j+1}\left(\frac{t_1}{T}\right)^{k-j+1}\left(\frac{T-t_1}{T}\right)^{j-1}\Psi_{j-1}(t-t_1,T-t_1).\qquad(13.43)$$

Lemma 13.2.

$$\alpha_{ki}(t)=\sum_{j=0}^{k-i+1}\binom{k}{j}\alpha_{(k-j)(i-1)}(t)\left(\frac{t_i-t_{i-1}}{t_N}\right)^{j},\quad k\geq i.$$

Also necessary for deducing queueing statistics is an analysis of the event $X_k\leq t_i$. Thus, define

$$\beta_{ki}(t)=Pr\{X_k\leq t_i\,|\,O(t),N(t_N)=N\}.\qquad(13.44)$$

Similar to that above, Larson gives two recursive ways in which he calculates the $\beta_{ki}(t)$'s, depending on the choice of implementing either Lemma 13.1 or Lemma 13.2 (using the Ψ's or the α's). From this, a matrix $\beta(t)\equiv(\beta_{ki}(t))$ can be calculated. It is shown that the number of computations necessary in order to complete the matrix $\beta(t)$ is $O(N^5)$. (See Addendum for a reduction of this order.) Notice that when computing the values of $\alpha_{ki}(t)$ one is filling out a lower triangular matrix $A(t)\equiv(\alpha_{ki}(t))$. Using the elements of the matrices $A(t)$ and $\beta(t)$, Larson presents recursive methods to compute the following:

1) the expected cumulative number of arrivals to the system up to time t, given $O(t)$ and $N(t_N)=N$;
2) the average queue length over a congestion period of length T;
3) the expected time spent in queue by customers during a congestion period;
4) the distribution of queue length upon arrival of a random customer during a congestion period.

As a follow up to Larson [86], Bertsimas and Servi [12] propose an $O(n^3)$ algorithm that yields estimation of the transient queue length during a busy period as well as delay of each customer on the busy period, all from transactional data. The algorithm is generalized for a time varying input Poisson process and then finally to the case of general stationary interarrival times. An on line algorithm is presented that allows the user to update current estimates for queue length after each departure.

The above works of Larson and Bertsimas and Servi were implemented with optimistic results in the setting of a communication network by Gawlick [49]. Gawlick uses the algorithms to estimate the queue lengths for a transmission line purely from transmission completion times of packets of information. Gawlick concludes that the QIE's accuracy is noteworthy in light of the strong possibility that the Poisson input assumption is violated for his real data set. Gawlick also provides evidence to suggest that the QIE's accuracy depends on the coefficient of variation of the interarrival times of the transmitted packets.

Daley and Servi [35] use taboo probabilities in Markov Chains in order to construct a simpler $O(n^3)$ algorithm for busy periods with n customers in order to calculate the queue length distribution from transactional data assuming Poisson input. An $O(n^2\log n)$ approximate algorithm is presented that is verified to be within the prescribed accuracy of the $O(n^3)$ algorithm. When the arrival rate is assumed to be known, the algorithm can be extended to Erlangian arrivals and finite buffers as well as handle the real time updating problem, as discussed in Bertsimas and Servi [12]. For the case of Erlangian arrivals, Daley and Servi's approach yields a faster algorithm by avoiding the numerical integration that is pivotal in the work of Bertisimas and Servi [12].

13.8 ESTIMATION RELATED TO PERFORMANCE MEASURES

In the case of queueing models, convenience of observation plays a major role. For example, as observed by Neal and Kuczura [101], traffic parameters in the Bell System were generally determined by obtaining the following three measurements during a period of time $(0, t)$: (i) $A(t)$, the number of calls, (ii) $O(t)$, the number of unsuccessful calls (overflow), and (iii) $L_d(t)$, an estimate of usage based on discrete samples. Using these functions, one can determine

$$\text{call congestion} = O(t)/A(t); \quad \text{load } \widehat{\alpha} = \frac{L_d(t)/n}{1 - O(t)/A(t)}. \quad (13.45)$$

Intuitively, we may identify call congestion as representing the probability of a loss and load $\widehat{\alpha}$ as representing the effective traffic intensity. These measures can be obtained from analytical models by determining arrival and service rates. However, it is convenient to get the quantities directly, as shown above, instead of using the detailed information on the arrival and service processes. When such measurements are made, accuracy of the result as measured by the standard error of the sample function is of interest. This aspect has received considerable attention from several investigators; see Kosten et al. [84], Syski [127], Beneš [11], Riordan [116], Descloux [31], and Neal and Kuczura [101]. In particular, Descloux considers the joint distribution of three processes in an M/M/s loss system: $N(t) =$ number of busy channels at time t and $A(t)$ and $O(t)$ defined above. He derives $\text{cov}[A(t), O(t)]$ and the variance of the call congestion $O(t)/A(t)$. Comparing call congestion with the proportion of time when all servers are busy (time congestion), Descloux concludes that the standard deviation of call congestion is larger than that of the time congestion when the offered load is approximately less than the number of channels and the inequality is reversed otherwise.

In practice, Little's law $L = \lambda W$ is a convenient formula for indirect estimation of L from W, or vice versa. Glynn and Whitt [55] use central limit theorem versions of Little's law to investigate the asymptotic efficiency of such estimators. In particular, they show that an indirect estimator for L using the natural estimator for W plus the known arrival rate λ is more efficient than a direct estimator for L, provided that the interarrival and waiting times are negatively correlated. Let $\{A_n, n \geq 1\}$ be the sequence of arrival epochs and $\{W_n, n \geq 1\}$ be the waiting times of arriving customers. The basic central limit theorem used by Glynn and Whitt states that there exist constants λ and ω with $\lambda > 0$ and $\omega < \infty$ and covariance matrix C such that

$$\sqrt{n} \left(A_n - n\lambda^{-1}, \sum_{k=1}^{n} W_k - n\omega \right) \xrightarrow{D} N(0, C). \quad (13.46)$$

In their terminology, a natural estimator is the simple average, the direct estimator of a function is the one obtained using the natural estimator, and the indirect estimator results when one of the parameters is known.

The mean sojourn time (time spent by the customer in the system) is an important performance measure for queueing systems in practice. The direct estimate involves collecting the time instants of arrivals and departures of customers. A simple estimate of the mean sojourn time in the M/M/1 queue, proposed by Toyoizumi [131], is obtained by observing the number of arrivals and departures over a certain period of time only at "regular" (i.e., of equal length) intervals. Toyoizumi compares this customer count estimate (CCE) with the direct estimate (DE) and a test-customer estimate (TCE) obtained by inserting test customers at regular intervals and measuring their sojourn times. Let $A(t)$ be the number of arrivals during $(0, t]$ and $D(t)$ be the number of departures during the same period. Observe $A(t)$ and $D(t)$ at "regular" time points $\{t_i, \ i = 1, 2, \ldots\}$ with $t_{i+1} - t_i = \Delta$, $i = 1, 2, \ldots$; here we obtain the sequences $A_i(\Delta)$ and $D_i(\Delta)$, $\{i = 1, 2, \ldots\}$ of arrivals and departures and let $L(t)$ be the number of customers in the system at time t. Let $L_i(\Delta) = A_i(\Delta) - D_i(\Delta)$. For the average number $\bar{L}$ of customers in the system we have

$$\bar{L} \equiv \frac{1}{n} \sum_{i=1}^{n} L_i(\Delta) \equiv \frac{1}{n} \left(\sum_{i=1}^{n} A_i(\Delta) - \sum_{i=1}^{n} D_i(\Delta) \right). \tag{13.47}$$

Furthermore, $\bar{\lambda} \equiv \frac{1}{t_n} A(t_n)$. By Little's formula, an estimate of the mean sojourn time is

$$W_c \equiv \frac{\bar{L}}{\bar{\lambda}} = \frac{\sum\limits_{i=1}^{n} (A_i(\Delta) - D_i(\Delta))}{A(t_n)}. \tag{13.48}$$

Since all three estimates, CCE, DE, and TCE, are unbiased, Toyoizumi compares the variances of these estimates and concludes that based on the efficiency of these estimates represented by the measurement period, CCE is a convenient method for estimating the sojourn time.

Taking averages is a common method in practice to estimate a performance measure such as waiting time. The question then arises whether to obtain the time average or the customer average. For instance, consider (see Glynn et al. [56]) the workload $U(t)$ (virtual waiting time) in the M/M/1 queue. Let $N(t)$ be the Poisson arrival process with associated arrival times $\{T_n, n \geq 1\}$. Clearly, $\{U(T_n -), n \geq 1\}$ is the sequence of waiting times just before arrivals. Define the time average

$$V(t) = \frac{1}{t} \int_0^t U(s)\,ds, \ t > 0,$$

and the customer average

$$W(t) = \frac{1}{N(t)} \sum_{k=1}^{N(t)} U(T_k -), \ N(t) > 0.$$

It is well known that when the traffic intensity $\rho < 1$, both of the above converge with probability one to $\rho/(1 - \rho)$ as $t \to \infty$. It can also be shown that $\sigma_W^2 - \sigma_V^2 = \frac{(2 - \rho)(1 + \rho)}{(1 - \rho)^3} > 0$, where σ_W^2 and σ_V^2 are respective asymptotic variances. This result indicates that the time average is always asymptotically more efficient for the M/M/1 system. Glynn et al. [56] address this question more generally by establishing a joint central limit theorem for customer and time averages. Under appropriate regularity conditions, for $V(t) \to v$ and $W(t) \to w$ a.s. as $t \to \infty$,

$$\sqrt{t}\{(W(t) - V(t)) - (w - v)\} \xrightarrow{D} N(0, \sigma_d^2), \tag{13.49}$$

where $\sigma_d^2 = \sigma_w^2 + \sigma_v^2 - 2\sigma_{vw}$, with σ_{vw} being the covariance term.

The probability of long delay in the queue is a performance measure considered by Gaver and Jacobs [48] in the M/G/1 system. When the traffic intensity ρ is less than 1, the virtual waiting time $W(t)$ in this system has the moment generating function

$$E[e^{sw}] = \lim_{t \to \infty} E[e^{sW(t)}] = \frac{1 - \rho}{1 - \rho A(s)}, \tag{13.50}$$

where $A(s)$ is the moment generating function of distribution H. If $A(s)$ exists for $s < s_0$, $s_0 > 0$, then there exists a smallest real zero $s = \kappa > 0$ of the denominator of (13.50). Using κ we can write

$$P\{W > w\} \sim D(\kappa)e^{-\kappa w}, w \to \infty. \tag{13.51}$$

The distribution H can be interpreted as the distribution of the forward recurrence time in a renewal process made up of service times. Following up on this interpretation, we can show that κ is a positive solution of the equation

$$\frac{\lambda}{\theta}[\phi(\theta) - 1] = 1, \tag{13.52}$$

where $\phi(\theta)$ is the moment generating function of the service time distribution. A nonparametric estimate of κ, say $\hat{\kappa}$, is obtained via the Taylor series expansion of (13.52).

Gaver and Jacobs provide asymptotic results for the distribution of $\hat{\kappa}$ as the sample size $n \to \infty$ and show that $\hat{\kappa}$ is a consistent estimator of κ, and if $E[\exp(2\kappa X)] < \infty$, then the distribution of $\hat{\kappa} - \kappa$ is asymptotically normal (X is the random variable representing service time).

13.9 ESTIMATION FROM SIMULATION OF QUEUEING PROCESSES

Whenever the results cannot be obtained analytically, often one must resort to simulation in order to study the properties of a particular process. Simulation also provides a vehicle for examining analytical results in the context of a practical situation. See the book by Fishman [44] for an introduction. Much work has been done in the area of simulating stochastic processes; here we examine some techniques that have been applied to queueing systems.

Much of the literature on simulation in queueing revolves around the concept of *regeneration*. The idea is that there exist points in time where the process probabilistically restarts itself. In other words, the realization of the process between any two regeneration points is a replica of the process between any other two regeneration points in a probabilistic sense. Regeneration leads to the independence of certain events that occur between the regeneration times; this in turn leads to classical statistical analysis on the resulting i.i.d. random variables. For an introduction to regeneration, see Crane and Lemoine [31] or Shedler [123].

In this section we will concentrate on issues that need to be considered before the simulation is carried out, as well as problems of estimation of mean values, higher moments, and quantiles. The section concludes with a brief discussion of estimation in simulated networks.

13.9.1 Simulation run length considerations

It is useful to have an idea *a priori* of how long a simulation needs to be performed in order to achieve prescribed accuracy of estimates. Equally important is the decision to either conduct one long simulation run or to perform many shorter runs. The first of these questions is addressed in Whitt [132], where the author provides heuristic formulas that estimate the necessary run lengths. These formulas are based on heavy traffic limits and are associated with diffusion approximations for queues. The goal is to provide formulas that will aid in designing the experiment before any simulation data have been collected. Whitt points out that simulation length depending on the value of the traffic intensity ρ and heavy traffic queue length limits helps make a decision how the run length should change as ρ increases toward its stability limit. Statistical precision of estimates is measured by the relative standard error, $\bar{r}$, defined as the ratio of the simulation standard error to the simulation estimate of the mean. Whitt states that in order to achieve uniform relative standard error over all choices of ρ, the simulation length should be proportional to $1/(1-\rho)^2$ arrivals. Suggestions are made on how to handle the variability of the arrival and service processes which, along with ρ, have an impact on necessary simulation length.

Law and Kelton [90] consider sequential procedures used to determine the run length of a simulation when the goal is to estimate the steady state mean of, say, the mean delay in the M/M/1 queue.

In a simulation experiment one often must decide between extending the observation period from $[0,T]$ to $[0,mT]$ or carrying out m independent simulations of the system on $[0,T]$, the objective being the reduction of the standard error of an estimate. For the case where replication is deemed superior to extension, see Gafarian and Ancker [46]. In this connection, we also cite the work of Cox [29], based on Fisher's information measure.

The question of one long run versus several shorter runs is also discussed in Whitt [133] where he uses the concept of estimation efficiency, defined as the reciprocal of the product of the mean squared error and the total simulation run length. In some special cases, Whitt reaches definitive conclusions as to which strategy is most advantageous. However, a general guideline is not provided (see also the references cited in Whitt [133]).

13.9.2 Estimation of means

The estimation of the mean of a queueing process in the steady state using simulation is one of the significant problems considered in the literature. Fishman [42] discusses simulation of a multiserver queue in which twelve population descriptors can be estimated from three summary statistics. Let $\{t_i \quad i = 0,1,...\}$ be a sample sequence of times spent in states $i = 0,1,...$, where state i denotes that there are i jobs in or waiting for service, and define

$$a_1 = \sum_{i=0}^{m-1} t_i, \; a_2 = \sum_{i=m}^{\infty} t_i, \text{ and } a_3 = \sum_{i=m}^{\infty} it_i. \tag{13.53}$$

It is important to realize that these statistics translate as: $a_1 = $ time during which at least one server is idle, $a_2 = $ time during which all servers are busy, and $a_3 = $ time spent in the system when all servers are busy. From these three statistics, the following estimates are obtained:

1. sample mean number of jobs in the queue $= \hat{h}_1 = (a_3 - ma_2)/(a_1 + a_2)$;
2. sample mean number of jobs in the queue, given that all servers are busy, $= \hat{h}_2 = a_3/(a_2 - m)$;
3. sample probability of a delay upon arrival $= \hat{h}_3 = a_2/(a_1 + a_2)$,

where m is the number of servers in the system. Fishman states that the other nine population descriptors can be regarded as linear combinations of these three statistics. Some of these additional descriptors are the sample mean number of jobs in the system, sample mean waiting time, sample probability of an idle server, and the sample mean busy period. Through appropriate transformations, point and interval estimates for all twelve characteristics are given; furthermore, a discussion of simultaneous confidence intervals for these descriptors is included.

Much work on the estimation of the mean delay in queues has been done using simulation, including Law [88], who compares estimation of the mean delay in the M/M/1 queue by replication of simulations with estimation via batch means. Law shows empirically that batch means are to be preferred to replication by using the criteria of probability of coverage and the half length of the intervals. The idea of replication is as follows: Suppose that k independent simulation runs are made of the process, each being of length m. Let $\bar{X}_i(m)$ be the sample mean of the m observations in the ith run. Then, as a point estimator of μ (the mean of the process) use $\bar{\bar{X}}(k,m) = \sum_{i=1}^{k} \bar{X}_i(m)/k$. Let $\mu(m) = E[\bar{X}_i(m)]$. The replication method leads to the following confidence interval for μ:

$$\bar{\bar{X}}(k,m) \pm t_{k-1,1-\alpha/2}\sqrt{\hat{\sigma}^2[\bar{\bar{X}}(k,m)]}, \tag{13.54}$$

where

$$\hat{\sigma}^2[\bar{\bar{X}}(k,m)] = \sum_{i=1}^{k}[\bar{X}_i(m) - \bar{\bar{X}}(k,m)]^2/k(k-1). \tag{13.55}$$

Two possible sources of error using this interval are: 1) the possibility that $\mu(m) \neq \mu$, and 2) the possible nonnormality of the $\bar{X}_i(m)$'s. Law discusses the effects of making these errors when using this interval.

An approach using batch means is based on simulation of one long run of the process and then breaking the run up into k adjacent in time batches of length m and calculating means within each batch. Letting $\bar{X}_i(m)$ be the mean of the ith batch, once again use $\bar{\bar{X}}(k,m)$ as the point estimator. If m is sufficiently large, then these sample means will be nearly uncorrelated and so we have essentially the same setting as before, and thus we are led to a confidence interval having the same form as that above. Possible sources for error when using this technique are: 1) correlation structure between the $\bar{X}_i(m)$'s, 2) the $\bar{X}_i(m)$'s are not identically distributed with mean μ, and 3) the nonnormality of the $\bar{X}_i(m)$'s. By using both techniques in the simulation of an M/M/1 queue it is found that for a specified half length of an interval, batch means have greater coverage than replication, or for specified coverage, batch means have a shorter half length.

In a separate paper, Law [85] considers the M/G/1 queue and addresses estimation of the mean delay, denoted d, along with the average queue length Q, the mean waiting time w in the system, the average number of customers L in the system, and the average amount of work V in a simulated system. He shows that it is more efficient (in terms of asymptotic variance) to estimate all of the above *from the estimate of d* rather than do it directly. Conventionally, one might estimate d by performing a simulation run of length n customers and then calculating $\tilde{d}_1 = \sum_{i=1}^{n} D_i/n$, where D_i is the delay of the ith customer. Likewise, it is natural to estimate Q, by running a simulation for t time units, and use $\tilde{Q}_1 = \frac{1}{t}\int_0^t N_q(s)ds$, where $N_q(s)$ is the queue length at time s.

Law uses regeneration to construct alternative estimators to these natural ones, and ultimately suggests that in the case of the M/G/1 queues, one can obtain an estimator that is more efficient than $\tilde{d}_1$ for estimating d. He then suggests that it is more efficient to estimate d, Q, w, L, and V indirectly from this estimator rather than to estimate them directly.

Law and Kelton [90] survey four methods of confidence interval construction for the mean delay in the M/M/1 queue; two of which are regenerative and the other two nonregenerative. Their results indicate that of the two regenerative methods considered, the method by Fishman [43] performs better under the criteria of coverage whose cost requires a longer run than other regenerative methods on average. The nonregenerative method of choice is given to be the one by Law and Carson [89]. For a new method of estimating the mean μ of a queueing process that incorporates the skewness of the underlying distribution, see Choobineh and Ballard [25]. They use the concept of lower and upper partial variance defined, respectively, as $PV_L(x) = \int_{-\infty}^{\mu} (x - \mu)^2 dF(x)$ and $PV_U = \sigma^2 - PV_L(x)$ and approximate $PV_L(x)$ by $APV_L(x) = (1 - P)\sigma^2$, where $P = Pr\{x \leq h\}$. Their confidence interval for μ takes the form

$$\left(\bar{X} - t_{n-1,1-\alpha/2}[2(1 - \hat{P})S^2/n]^{1/2}, \; \bar{X} + t_{n-1,1-\alpha/2}[2\hat{P}S^2/n]^{1/2} \right), \qquad (13.56)$$

where $\hat{P} = \sum_{i=1}^{n} I_i/n$ and $I_i = 1(= 0)$ if $X_i - \bar{X} \leq 0$ (> 0), and S^2 is the sample variance. For a more complete development, see Choobineh and Ballard [25] as well as references therein.

13.9.3 Estimation of other moments and quantiles

Fishman [41] proposes a method for estimating the variances of some sample performance measures using simulation techniques. He also describes ways to remove bias from these sample measures. Define t_i as the time spent in state i in a run length $t = \sum_{i=0}^{\infty} t_i$, then let $g_j = \sum_{i=0}^{\infty} c_{ij}t_i$. By choosing appropriate c_{ij} one can define system performance measures such as the sample mean queue length or the sample mean waiting time. If a simulation is started with an arrival to the empty system and it runs for J epochs (an epoch being the time between regeneration points), then let $y_{j1} = \sum_{i=0}^{\infty} c_{i1}t_{ij}$ and $y_1 = (1/J)\sum_{j=1}^{J} y_{j1}$ so that $g_1 = Jy_1$ is a linear combination of i.i.d. random variables. If y_1 has mean μ_1 and variance σ_1^2 and the covariance between y_1 and y_k is σ_{k1}, then the distribution of $J^{1/2}(y_1 - \mu_1)/\sigma_1$ converges to a standard normal as J increases. Estimators of the form y_1/y_k are given, along with their variance (all up to order $1/J$), as

$$E(y_1/y_k) = (\mu_1/\mu_k)(1 + \theta_1)/J$$
$$\mathrm{var}(y_1/y_k) = (\mu_1/\mu_k)^2\theta_2/J \qquad (13.57)$$
$$\text{with } \theta_1 = \sigma_k^2/\mu_k^2 - \sigma_{k1}/(\mu_k\mu_1)$$
$$\text{and } \theta_2 = \sigma_k^2/\mu_k^2 - 2\sigma_{k1}/(\mu_k\mu_1) + \sigma_1^2/\mu_1^2.$$

Fishman provides an example of the above techniques applied to the M/M/1 queue to show the behavior of the above estimates.

Glynn and Iglehart [54] develop a methodology in order to estimate the variance of the waiting time process in the M/M/1 queue using regenerative simulation. More generally, let X (taking values in a set S) be a regenerative process with regeneration times $T(\cdot)$. Define a continuous time process $X(\cdot)$ by $X(t) = X_{[t]}$. Under general conditions there exists a distribution π such that

$$r_t(f) = \frac{1}{t} \int_0^t f(X(s))ds \rightarrow \int_S f(y)\pi(dy) \equiv r(f) \qquad (13.58)$$

almost surely for a broad class of functions $f : S \to \mathbb{R}$. The interpretation of $r(f)$ is then as a steady state mean of $f(X(\cdot))$. Also to be estimated are the variations of $f(X(\cdot))$ around its steady state limit. Let $f_c(\cdot) = f(\cdot) - r(f)$ and set $v(f) = r(f_c^2)$. Then, $v(f)$ is the steady state variance of $f(X(\cdot))$. Generally, let $\mu_m(f) = r(f_c^m)$; then, $\mu_m(f)$ is the mth steady state central moment of $f(\cdot)$. It is $\mu_m(f)$ for which Glynn and Iglehart develop estimation methodology. They provide limit theorems, including consistency and central limit type results, and develop confidence intervals for the central moments. Also given are numerical results for the case of steady state variance, which include a simulation of the M/M/1 queue with $\rho = 0.5$, in order to estimate the variance of the waiting time distribution.

An alternative approach to further understanding the behavior of estimates is to look at the behavior of the gradient of performance measures. Shalmon and Rubinstein [122] explore this problem by considering the gradient of moments associated with the waiting time in the M/G/1 queue. Estimators explored are those obtained through finite difference approximation methods, likelihood ratio transformations, and from infinitesimal perturbations. Several heavy traffic formulas show that for the gradient of the mean waiting time with respect to arrival or service rates, the likelihood ratio estimator reduces the length of the simulation required by the finite difference estimator by an order of magnitude. A major point is the derivation of analytical formulas for the variance of regenerative estimators for the M/G/1 queue, which include those for mean values and arbitrarily higher moments of the waiting time as well as those for the likelihood ratio gradient estimator of the moments of the waiting time. Shalmon and Rubinstein's appendix lists the formulas that are obtained which remain valid for an arbitrary traffic intensity as well as approximation formulas for heavy traffic.

Often the goal of simulation of a queueing process is to obtain information about the tails of an underlying distribution rather than the mean or variance. To address this issue, the estimation of quantiles becomes necessary. Seila [121] proposes a method for estimation of the quantiles of simulated distributions associated with stochastic processes in the M/M/1 queue. He compares his method of quantile estimation with that of Iglehart [70].

Let $\{X(t)\}$ denote a discrete valued regenerative process with regeneration times $\{T_i\}$. Define $N_i = T_i - T_{i-1}$ as the length of the ith *cycle* of the process. Suppose that the queueing process, such as the one in the M/M/1 system, is simulated, and waiting time data in the form of X_i for $i = 1, \ldots, T_n - 1$ is collected. For $i = 1, \ldots, n$, define $Y_i(x)$ to be the number of observations in the ith cycle that are less than or equal to x and let $Z_i(x) = Y_i(x) - N_i F(x)$. $\{Z_i(x)\}$ constitutes an i.i.d. sequence with mean 0 and variance denoted $\sigma^2(x)$. Then, the sample c.d.f., based on n cycles of the data, is

$$\widehat{F}_n(x) = T_n^{-1} \sum_{i=1}^{n} Y_i(x) = \frac{\sum_{i=1}^{n} Y_i(x)}{\sum_{i=1}^{n} N_i}. \tag{13.59}$$

$\widehat{F}_n(x)$ is the proportion of observations in n cycles that are less than or equal to x. Thus it follows that the sample pth-quantile of the data based on n cycles is

$$\widehat{Q}_n = \inf\{x : \widehat{F}_n(x) \geq p\} = X_{([(T_n - 1)p] + 1)}, \tag{13.60}$$

where $X_{(i)}$ denotes the ith order statistic and the brackets indicate the "greatest integer" function. (Note $\{X(t)\}$ converges in distribution to a random variable X

with c.d.f. $F(x)$ having pth quantile defined by $q = \inf\{x\colon F(x) \geq p\}$.) Iglehart develops the asymptotic properties of $\widehat{Q}_n$ to estimate quantiles by linear interpolation of $\widehat{F}_n(x)$ between the points in a grid specified by the user.

Seila's batch quantile method of estimating q is as follows: Suppose that we have a simulation run of length n consisting of m cycles separated in r batches, $n = mr$. The observations in the ith batch are denoted by $X_{T_{(i-1)m}}, \ldots, X_{T_{im}-1}$; the corresponding order statistics being $X^i_{(1)} \leq X^i_{(2)} \leq \ldots \leq X^i_{(M_i)}$, where $M_i = T_{im} - T_{(i-1)m}$ is the number of observations in the ith batch. Let $\widehat{Q}_{m,i} = X^i_{([M_i p + 1])}$ designate the sample pth quantile from the ith batch. Note that $\{\widehat{Q}_{m,i}\}$ is an i.i.d sequence (due to regeneration) and in fact is asymptotically normal with mean q (see Iglehart). Define $\bar{Q}_n$ to be the sample mean of $\{\widehat{Q}_{m,i}\}$ for $i = 1, 2, \ldots, r$. Then,

$$\lim_{m \to \infty} P \left\{ \frac{\sqrt{mr}\,(\bar{Q}_{m,r} - q)}{\sigma(q)/\mu f(q)} \leq y \right\} = \Phi(y), \tag{13.61}$$

where $\Phi(y)$ is the normal c.d.f. Thus, for a fixed r and large m, the batch quantile estimator $\bar{Q}_n$ has the same asymptotic distribution as $\widehat{Q}_n$ (see Iglehart). Furthermore, by the classical central limit theorem, $\bar{Q}_n$ is asymptotically normal for fixed m and large r. One drawback is that $\bar{Q}_n$ is generally biased, since sample quantiles are usually biased for finite samples. Empirical studies have shown this to be a problem for the M/M/1 queue. In order to reduce the bias, Seila [121] proposes the use of the "two fold jackknife" in order to arrive at a final estimator.

Hsu and Nelson [68] use the idea of control variate estimation to estimate the delay of the nth customer that arrives at the M/M/1 system (denote this by Y). Control variate estimation is a variance reduction technique that estimates a characteristic of a random variable Y by using knowledge of a random variable X, which is simultaneously observable and statistically dependent on Y. Our interest is focused on an upper quantile of the delay distribution. For the M/M/1 queue, the control variate X is the difference between the sum of service times of the first $n + h - 1$ customers (there are h in the system initially) and the sum of the interarrival times of the first n customers. Through the use of the Erlang distribution, the 95th percentile of the distribution of X is available, the goal being to estimate the 95th percentile of the distribution of Y. Using simulation techniques, four estimators (three of which the authors propose) are compared. Methods for confidence interval estimation are also given. Hsu and Nelson's recommendation as the best general purpose quantile estimator is an estimator that is obtained after inverting a likelihood ratio test.

13.9.4 Estimation in simulated networks

Iglehart and Shedler [71] investigate the statistical efficiency of estimation techniques for passage times in closed queueing networks, a passage time being loosely defined as the time it takes to traverse a specified part of a network. Two types of simulation, the "marked job" and the "decomposition" methods, are compared. They present a method for calculating variance constants that are used in central limit theorems to produce confidence intervals for certain characteristics. The conclusion drawn is that the "decomposition" method is superior in the sense that it produces tighter confidence intervals. Attention is focused on networks in which the service times are exponentially distributed. In a follow up article (Iglehart and Shedler [72]) they incorporate a general service time distribution in order to compare the "marked job" method to a third method, the "labeled job" strategy. Again using central limit theorem arguments they show that confidence intervals constructed for the expected value of a general function of the limiting passage time using labeled jobs rather than marked jobs is more statistically efficient.

Further information regarding estimation in queueing networks can be found in the references contained in Iglehart and Shedler [71, 72]. Another source of information is the book by Shedler [123].

13.10 HYPOTHESIS TESTING

Hypothesis testing in queueing systems still remains a vast unexplored area. A hypothesis testing problem arises when we are required to make inferences about parameters of arrivals and service time distributions or measures such as the traffic intensity, as well as forms of distributions, based upon sampled data from the system. These inferences may in turn lead to control procedures. As seen in the previous sections, the sampled data can be drawn from a queueing process such as queue length, waiting time, or the output process. In this section, we explore these aspects of parameter testing in some detail.

13.10.1 Significance tests for arrival and service parameters in the M/M/1 system

For the M/M/1 queue, significance tests for λ and μ can be based on the chi-squared distribution. As noted in Subsection 13.3.1, $2\widehat{\lambda}t = 2\widehat{\lambda}(t_e + t_b)$ and $2\widehat{\mu}t_b$ are chi-squared variates with $2n_a$ and $2n_s$ degrees of freedom, respectively, and therefore a test for $\rho = \lambda/\mu$ can be based on the F distribution. This procedure assumes that the state of the system (empty or busy) is under observation throughout the interval $(0, t)$.

As proposed by Cox [29], a simple test for the proportion of the total idle time over the time axis in the M/M/1 system, which is also the probability that the waiting time is zero, is obtained by observing that under the null hypothesis that the system is M/M/1, the number n_{ae} of arrivals at the empty system has a binomial distribution with parameters n_a and t_e/t.

13.10.2 A test for exponential service time using waiting time data

Sometimes a full observation of a system may not be possible. Suppose that only observations of waiting times $W_1, W_2, \ldots, W_n$ of the first n successive customers are available. We assume that the interarrival times have exponential distribution, and we wish to test the hypothesis that the service times are exponential, which can be stated as $H_0: G = M$ in the M/G/1 queue (Thiagarajan and Harris [128]).

The main difficulty arises from the fact that $\{W_i\}$ are correlated. Assume that none of W_i's are zero. Then we have the well-known relationship $W_{n+1} = W_n + Y_n$, where $Y_n = v_n - u_n$, with u_n being the nth interarrival time and v_n the nth service time. Here, $\{u_n\}$ and $\{v_n\}$ are assumed to be i.i.d. random variables, and therefore $\{Y_n\}$ are i.i.d. as well. Under $H_0: G = M$, the conditional densities of Y, given $Y > 0$ and $Y < 0$, are, respectively,

$$g(y \mid Y > 0) = \mu e^{-\mu y} \ (y > 0), \text{ and } g(y \mid Y < 0) = \lambda e^{\lambda y}. \qquad (13.62)$$

This suggests that the test $G = M$ be stated as follows: Split the data into two groups, one consisting of positive numbers and another consisting of negative numbers. Test for exponentiality separately, using the test proposed by Gnedenko (see Gnedenko et al. [51]). For details and tests for cases when there are zero waiting times, see Thiagarajan and Harris [128].

13.10.3 A uniformly most powerful test for ρ in the $M/E_k/1$ system

Using the assumption that the number of customers X_n arriving during the nth service period in the $M/E_k/1$ queue form a sequence of i.i.d. random variables with a negative binomial distribution, Harishchandra and Rao [64] developed a likelihood ratio test for ρ based on the sample $x = (x_1, x_2, \ldots, x_n)$. By the Neyman-Pearson lemma, a uniformly most powerful test of size α for $H_0 : \rho = \rho_0$ against $H_a : \rho > \rho_0$ is given by

$$\phi(x) = \begin{cases} 1 & \text{if } \sum x_i > c \\ \gamma(x) & \text{if } \sum x_i = c \\ 0 & \text{if } \sum x_i > c, \end{cases} \qquad (13.63)$$

where c and γ are determined such that $\alpha = P[\sum x_i > c \mid \rho_0] + \gamma P[\sum x_i = c \mid \rho_0]$ and $\varphi(x)$ is the probability of rejecting H_0. Note that this is a randomized test. The power function of the test is given by

$$\beta(\rho) = P(\sum x_i > c \mid \rho) + \gamma P(\sum x_i = c \mid \rho).$$

13.10.4 Sequential parameter control

When operating a queueing system, monitoring and controlling the parameters of the system are essential to ensure that the system performance is up to design standards, and in order to respond to exigencies of the environment. The parameter control problem, in effect, involves a problem of testing the hypothesis $H_0 : \theta = \theta^0$, where θ^0 is the desired vector of parameters, against a suitable alternative. If the hypothesis is not rejected at a chosen level of significance, we conclude that the system parameters have not changed, while the rejection of the hypothesis is indicative of the fact that the system has changed from the desired state. Once detection of change is achieved, an appropriate control action can be taken.

Bhat and Rao [15] consider the problem of controlling the traffic intensity in the $M/G/1$ and $GI/M/1$ queues. In an analogy with statistical quality control, the technique used is such that the queueing system is left undisturbed as long as it satisfies certain conditions, but when the conditions are violated, it is readjusted so as to make it consistent with the original objectives. For an ideal control technique, the type I and type II errors should be under control.

Let $t_0, t_1, \ldots$ be a discrete set of epochs at which the system is observed and Q_n be the number of customers observed at epoch t_n, $n = 0, 1, \ldots$. The technique is based on the sample function $\{Q_n\}$. In general, the parameter control technique consists of two phases and two sets of control limits for $\{Q_n\}$. The first phase (a warning phase) indicates the time at which the sample function gets out of the region covered by the upper and lower control limits, say c_u and c_ℓ; the second phase (the testing phase) is intended to see whether the process returns to the control region within a specified amount of time and involves two limits, say d_u and d_ℓ. For the M/G/1 and GI/M/1 queues, if we select $\{t_n\}$ to be the epochs such that $\{Q_n\}$ is an embedded Markov chain, the first set of limits are approximately determined using the equilibrium distribution of $\{Q_n\}$. Let $Q^* = Q_\infty$ and α_u and α_ℓ be two specified probabilities. Then c_u and c_ℓ are integers such that

$$c_u = \min\{k \mid P(Q^* \geq k) \leq \alpha_u\}, \quad c_\ell = \max\{k \mid P(Q^* \leq k) \leq \alpha_\ell\}. \quad (13.64)$$

It is somewhat harder to determine the second set of limits (d_u and d_ℓ) as this involves first passage distributions. Bhat and Rao have shown that in the M/G/1 case, if β_u and β_ℓ are two specified probabilities and $c_\ell = 0$, then

$$d_u = \min\{n \mid P(T_\ell > n) \leq \beta_u\}, \quad d_\ell = \min\{n \mid k_0^n \leq \beta_\ell\}, \quad (13.65)$$

where T_ℓ is the length of a busy period initiated by a single customer and k_0 is the probability that no customers arrive during a service phase. Note that c_u, c_ℓ, d_u, and d_ℓ are determined under $H_0 : \rho = \rho_0$.

Once these limits are obtained, the control technique may be described for given values of α_u, α_ℓ, β_u, β_ℓ as follows:

(i)   starting with an initial queue length i and traffic intensity ρ_0, leave the system alone as long as Q_n lies between c_u and c_ℓ, or when it goes out of these limits if it returns within bounds before d_u and d_ℓ transitions, respectively;

(ii)   if the queue length does not return within bounds between d_u or d_ℓ consecutive transitions as the case may be, conclude that the traffic intensity has changed from ρ_0 and reset the system to bring the traffic intensity back to the level ρ_0;

(iii) repeat (i) and (ii) using the last state of the system as the initial state.

Tables for c_u, c_ℓ, d_u, and d_ℓ are provided by Bhat and Rao [15] for the M/E_k/1 queue for some selected values of k. The second phase of the sequential parameter control technique requires the knowledge of first passage distributions and a great deal of numerical work. An attempt to alleviate this problem in the system M/G/1 has been made by Bhat [14] using a modified distribution free procedure based on a censored sample of service periods in which no customer arrivals occur. Such periods can be thought of as Bernoulli trials and tests based on the number of service periods with no arrivals can be easily constructed. As one expects, the relative efficiency of this procedure falls as $\rho \to 1$.

13.10.5 Sequential probability ratio tests

When the difference between parameter values under the null and alternative hypothesis is large, a sequential test has the advantage of using a considerably smaller sample size. With this objective, Rao et al. [111] have developed a procedure for testing the hypothesis $H_0 : \rho = \rho_0$ versus $H_1 : \rho = \rho_1$ using Wald's Sequential Probability Ratio Test (SPRT) for the systems M/G/1 and GI/M/s in which the queue length processes have embedded Markov chains $\{Q_n\}$. Let the transition probabili-ties of the chain be $p_{ij}(\rho)$ and let n_{ij} be the number of transitions $i \to j$ in $\{Q_n\}$ up to and including the nth transition. Then, the likelihood ratio for the SPRT is

$$L_n = \prod_{i,j} p_{ij}^{n_{ij}}(\rho_1) \Big/ \prod_{i,j} p_{ij}^{n_{ij}}(\rho_0). \quad (13.66)$$

Let $A = (1 - \beta)/\alpha$ and $B = \beta/(1 - \alpha)$, where α and β are the probabilities of Type I and Type II errors, respectively. The SPRT procedure is: accept H_1 if $L_n \geq A$, accept H_0 if $L_n \leq B$ and observe the next queue length Q_{n+1} and compute L_{n+1}, then repeat the procedure if $B < L_n < A$. The mechanics of applying the test are easier if logarithms are used. For the case of the systems M/M/1, M/E_k/1, E_k/M/s, M/M/s/s, and the machine interference problem, the logarithm of (13.66) takes the form $\log L_n = an + \sum_{i,j} n_{ij} c_{ij}$, where a and c_{ij} are constants depending upon ρ_0, ρ_1, and the transition probabilities of the embedded Markov chain.

Even though Rao et al. [111] used finite Markov chain results in deriving the SPRT, Rao and Bhat [110] have subsequently shown that for the type of Markov chains used in the analysis of queues, the sequential probability ratio test terminates in a finite number of steps with probability one, simplifying some of the computational aspects of the test procedure.

When the state space of $\{Q_n\}$ is finite, the operating characteristics (OC) curve for the SPRT can be obtained as

$$L(\rho) \cong \frac{A^{t_0(\rho)} - 1}{A^{t_0(\rho)} - B^{t_0(\rho)}} \quad \text{if } t_0(\rho) \neq 0,$$

$$\cong \frac{\log A}{\log A - \log B} \quad \text{if } t_0(\rho) = 0,$$

(13.67)

where $t_0(\rho)$ is the non-zero real root of the equation $\lambda_0(t) = 1$. Here, $\lambda_0(t)$ is the largest real positive latent root of the matrix

$$P(t) = \left\{ p_{ij}(\rho) \left[\frac{p_{ij}(\rho_1)}{p_{ij}(\rho_0)} \right]^t \right\}.$$

(13.68)

Note that $\lambda_0(t)$ is a function of ρ. The average sample number (ASN) is also proved by the authors. A limitation of the procedure adopted in the determination of OC and ASN functions is that the state space must be restricted to a finite number for the associated Markov chain. Furthermore, the computation of ASN and OC functions is quite tedious.

Clearly, sequential probability ratio tests can be developed for an i.i.d. sequence $\{X_n\}$ of the number of customers arriving during service periods in M/G/1 queues. Jain and Templeton [77] have used this technique in deriving a SPRT in the case of M/E_k/1 queues. As noted in Bhat [14], however, the analogous sequence $\{X_n\}$ in GI/M/1 queues cannot be used to develop a SPRT since X_n is the potential (therefore not always observable) number of service completions during an interarrival period in this system.

13.10.6 Large sample tests

Drawing upon the asymptotic tests for Markov processes derived by Billingsley [17], Wolff [134] gives a number of tests for the parameters in the M/M/1, M/M/∞, M/M/s loss, and M/M/s systems, where the arrival and service rates are λ_n and μ_n when the number of customers in the system is n. For the M/M/s loss system, his approach is illustrated as follows. Suppose the queue is fully observed. Let u_j be the number of upward and d_j the number of downward transitions from state j, let γ_j be the total time spent in state j, and n the total number of observed transitions during $(0, T]$. The null hypothesis is stated as $H_0: \theta = \theta^0$, where $\theta^0 = (\lambda_0^0, \lambda_1^0, \ldots, \lambda_{s-1}^0, \mu_1^0, \mu_2^0, \ldots, \mu_s^0)$. Then the likelihood ratio statistics can be obtained as

$$\chi^2 = \sum_{j=0}^{s-1} u_j \log(u_j / \lambda_j^0 \gamma_j) + \sum_{j=1}^{s} d_j \log(d_j / \mu_j^0 \gamma_j) + \sum_{j=0}^{s} (\lambda_j^0 + \mu_j^0) - n, \qquad (13.69)$$

which under the null hypothesis is asymptotically distributed as χ^2 with $2s$ degrees of freedom.

When the system is not Markovian but there are embedded Markov chains identified in them, large sample tests can also be constructed using i.i.d. sequences of random variables responsible for the Markovian structure. Thus, in the embedded Markov chain characterization of the $M/G/1$ queue, if X_n is the number of customers arriving during the nth service period, then $\{X_n\}$ is an i.i.d. sequence with the traffic intensity $E\{X_n\} = \rho$. For large n, we may consider $Z_n = \dfrac{\overline{X}_n - \rho_0}{s/\sqrt{n}}$ as being approximately standard normal for testing the hypothesis that $\rho = \rho_0$, where s is the standard deviation of the sample. In general, this is a distribution-free test. When the form of the service distribution is known, better tests can be constructed. In particular, for the $M/E_k/1$ queue, Harishchandra and Rao [64] use $Z_n = (\overline{X}_n - \rho_0)/\sqrt{\dfrac{\rho_0(\rho_0 + k)}{kn}}$ and compare the power of this test with the power of the exact UMP test given by equation (13.63) for $n = 30$. Even for moderate sample sizes, the approximation by the large sample test appears to be good.

For testing $\rho = \rho_0$ in the $M/G/1$ queue using the i.i.d. sequence $\{X_n\}$, additional large sample tests can be developed; see Basawa and Bhat [5]. For more information and some asymptotic normality results in $GI/G/1$ queues that lead to large sample tests, see also Basawa and Prabhu [6].

Rao and Harishchandra [112] propose a large sample test based on a normal approximation for ρ. Both single and multiple server models with general interarrival and service time distributions are considered. First consider a single server model, where the interarrival times are i.i.d. with distribution function $F(x)$ having mean μ_x and variance σ_x^2. The service times are i.i.d. with distribution $G(y)$ having mean μ_y and variance σ_y^2. Thus, the traffic intensity is $\rho = \mu_x/\mu_y$. Assume that the first customer arrives at time $t = 0$, and that we observe the system until the nth departure, which occurs at time τ_n. Service times are denoted by $\{y_i\}$ and interarrival times by $\{x_i\}$. All interarrival times of customers arriving in $(0, \tau_n]$ are observed, and so we get to see $x_1, x_2, \ldots, x_m$, where $m = \max\{j : x_1 + x_2 + \ldots + x_j \le \tau_n\}$. The hypothesis to be tested is $H_0 : \rho = \rho_0$ versus $H_1 : \rho \ne \rho_0$. Let $\varphi(t) = \log t$ and $\delta = \varphi(\mu_y) - \varphi(\mu_x) = \log \rho$.

Following Basawa and Prabhu [6], Rao and Harishchandra give lemmas that allow the construction of the following limit theorems:

$$\sqrt{m}[\varphi(\overline{x}_m) - \varphi(\mu_x)] \xrightarrow{L} N(0, \sigma_x^2/\beta\mu_x^2),$$

$$\sqrt{n}[\varphi(\overline{y}_n) - \varphi(\mu_y)] \xrightarrow{L} N(0, \sigma_y^2/\mu_y^2), \qquad (13.70)$$

where $\overline{x}_m = \sum_{i=1}^{m} x_i/m$ and $\overline{y}_n = \sum_{j=1}^{n} y_j/n$ and $\beta = \max(1, \rho)$. This leads to $T_{mn} = \dfrac{mn\beta\delta_{mn}}{m\beta\eta_y^2 + n\eta_x^2}$ having a normal distribution with mean $(mn\beta\delta/m\beta\eta_y^2 + n\eta_x^2)$ and variance $(mn\beta/m\beta\eta_y^2 + n\eta_x^2)$ for large m and n, where $\delta_{mn} = \varphi(\bar{y}_n) - \varphi(\bar{x}_m)$ and η_x

$= \sigma_x/\mu_x$ and $\eta_y = \sigma_y/\mu_y$. Therefore, $Z = \left(\dfrac{mn\beta}{m\beta\eta_y^2 + n\eta_x^2}\right)^{1/2}(\delta_{mn} - \delta)$ has an as-

ymptotic standard normal distribution. Thus, for testing $H_0: \rho = \rho_0$ versus H_1: $\rho \neq \rho_0$ in terms of $H_0': \delta = \delta_0$ versus $H_1': \delta \neq \delta_0$, the statistic Z can be used. The possible unknowns β, η_x, and η_y can be replaced by consistent estimates and a similar limit theorem stated. The above easily generalizes to s servers.

Along with varied results on estimators and their large sample properties, Basawa and Prabhu [7] present two tests and their limiting null distributions. Suppose that k independent GI/G/1 queues are observed in the interval $(0, T_i)$, where T_i is a stopping time and $i = 1, 2, \ldots k$. The sample data collected include $\{(A_i(T_i), u_{ij}), i = 1, 2, \ldots, k, j = 1, 2, \ldots, A_i(T)$, and $(D_i(T_i), v_{il}), l = 1, 2, \ldots, D_i(T_i)\}$, where $A_i(T_i)$ is the number of arrivals for the ith queue, and $D_i(T_i)$ is the corresponding number of departures, and also u_{ij} is the ith interarrival time, v_{il} being the service time of the lth customer. The densities of u_{ij} and v_{il} are assumed to be members of the exponential family, namely:

$$f_i(u_{ij}; \theta_i) = a_{1i}(u_{ij})\exp\{\theta_i h_{1i}(u_{ij}) - k_{1i}(\theta_i)\}$$

$$g_i(v_{il}; \phi_i) = a_{2i}(v_{il})\exp\{\phi_i h_{2i}(v_{il}) - k_{2i}(\phi_i)\}. \tag{13.71}$$

The log-likelihood function in this case is

$$\ln L(\theta, \phi) = \sum_{i=1}^{k}\left[\left\{\theta_i \sum_{j=1}^{A_i} h_{1i}(u_{ij}) - A_i k_{1i}(\theta_i)\right\} + \left\{\phi_i \sum_{l=1}^{D_i} h_{2i}(v_{il}) - D_i k_{2i}(\phi_i)\right\}\right]. \tag{13.72}$$

Let $\mu_{1i}(\theta) = k_{1i}'(\theta) = E_\theta(h_1(u_{ij}))$ and $\mu_{2i}(\phi) = k_{2i}'(\phi) = E_\phi(h_2(v_{il}))$. Since μ_1 and μ_2 are positive, the authors considered the log-linear model $\ln \mu_{1i}(\theta) = \sum_{j=1}^{p} a_{ij}\alpha_j$ and $\ln \mu_{2i}(\phi) = \sum_{j=1}^{q} b_{ij}\beta_j$, where (a_{ij}) and (b_{ij}) are known covariates and the vectors $\alpha^T = (\alpha_1, \alpha_2, \ldots, \alpha_p)$ and $\beta^T = (\beta_1, \beta_2, \ldots, \beta_1)$ are the unknown parameters of interest.

Under this setup, Basawa and Prabhu consider a test of fit, corresponding to log linearity of the interarrival time and service time means within the exponential family, i.e.,

$$H: \ln \mu_{1i}(\theta) = \sum_{j=1}^{p} a_{ij}\alpha_j, \quad \ln \mu_{2i}(\phi) = \sum_{j=1}^{q} b_{ij}\beta_j. \tag{13.73}$$

The second test considered by the authors is one of homogeneity in the M/M/1 setting. Both tests result in likelihood ratio test statistics that have limiting chi-squared distributions under the null hypotheses.

13.10.7 Miscellaneous tests

Continuing the discussion on Bayesian estimation, Armero [3] develops a test for stationarity in the M/M/c setting. From an inferential point of view, the choice of $H_0: \rho < 1$ versus $H_a: \rho \geq 1$ is based on the posterior probabilities

$$P(H_0 \mid z) = P(\rho < 1 \mid z), \text{ and } P(H_a \mid z) = P(\rho \geq 1 \mid z) = 1 - P(\rho < 1 \mid z), \tag{13.74}$$

where $z = \{r_1, r_2\}$.

Consider this as a decision problem with actions a_0 and a_1 corresponding to acceptance of $H_0(H_a)$ and a loss function $L(a_i, \rho)$ $i = 0, 1$, which is the loss when the value of the traffic intensity is ρ and the statistician takes action a_i. From the above posterior distribution of ρ, the expected loss of action a_i is defined as $\ell^*(a_i) = E_{\rho \mid z}[L(a_i, \rho)]$, $i = 0, 1$, and the Bayes action taken is that which makes the posterior expected loss the smallest one.

Armero also discusses the incorporation of a vague prior information including an illustration of results that stem from the use of Jeffrey's method to derive non-informative priors.

13.11 OTHER RELATED TOPICS AND FUTURE PROSPECTS

In this section we briefly mention some related topics that, in our opinion, indicate the type of problems that are significant in the future development of the subject area.

(1) Applied queueing theory relies heavily on results from Markovian systems. The robustness of such results is therefore a significant factor. Queueing theorists have tackled heavily non-Markovian queues primarily by indirect methods using approximate systems. When simpler systems are used to approximate much more complex ones, a validation is essential. At the present time, simulation and validation through some simple cases seem to be the more prevalent methods available for this purpose. In this case of experimentation, more sensitive analysis is needed. A wider use of statistical techniques related to point and interval estimation needs to be made. A study of system robustness using approximation systems does not confront the problem at its roots. What is needed is information on the impact of changing distribution assumptions for system elements.

(2) One of the major problems in gathering necessary data from queueing systems for statistical analysis is the inability to obtain complete information, either due to the system structure or prohibitive costs. An example is the test for the exponentiality of the service time in the M/G/1 queue, using waiting time data, as given by Thiagarajan and Harris [128]. The sequential parameter control technique presented in Section 13.10 is aimed at using observations only on the queue length process. It is our belief that if analysis techniques are to be useful to the practitioner, they should make use of only observations that are easy to collect.

Another example is given by Nedelman [102] in modeling multiple malaria infections as a time dependent infinite server queueing model. The queue is only partially observable: it can be ascertained only as empty and non-empty and any continuous observation of the queue is impossible. The parameters are then estimated by maximizing the likelihood numerically.

(3) In queueing theory, most of the resulting random variables are non-normal, while most of the standard tests are based on normal variates. Therefore, an area that needs exploration is the use of nonparametric tests. We may use simple test procedures such as the Kolmogorov-Smirnov test to see whether a queueing situation can in fact be represented by a specific queueing model with known properties. When steady state distributions are characterized, this procedure can be easily applied. However, for more complex systems and hypotheses, we need to develop simple tests.

(4) An area of application of queueing theory that has emerged during the

past few years is manufacturing systems. These models are normally queueing network oriented, but with special characteristics appropriate in the manufacturing context. An example is the research by Duenyas and Hopp [39] in estimating the variance of output from a closed queueing network of single server exponential queues. Since exact computations become cumbersome, using renewal theory concepts, approximations are introduced and their accuracy is examined through simulation. As manufacturing system applications of queueing theory become more sophisticated, the solutions for deeper and increasingly difficult inference problems will be needed.

(5) The lack of time homogeneity presents major problems for the application of standard techniques of estimation. An interesting study of such a system is given by Hantler and Rosberg [63], who discuss the problem of optimal estimation for an M/M/c queue with time varying parameters. First, they derive minimum variance unbiased estimators of the arrival rate, mean service time, and the traffic intensity in a time homogeneous queue, and these results are then used to derive the minimal mean square error linear estimators of the parameters at any moment in a time varying queue. Using a discrete time scale for convenience, the parameters λ_k, $1/\mu_k$, and $\rho_k = \lambda_k/\mu_k$ for interval k are first estimated. The second step in estimating the time varying parameters is the use of the Kalman-Bucy filter. At any moment, the Kalman-Bucy filter computes the instantaneous parameter estimator as a linear function of the previous estimator and the current observation. The weights of these two values are then optimally selected from the use of the covariance function of the time varying parameter process. Also see Patterson [104], who derives regression estimates of the intensity function of a non-homogeneous Poisson process numerically by using an $M(t)/G/\infty$ service system.

(6) A relatively new subject in the estimation of parameters in queueing systems is the use of perturbation analysis in the construction of estimators. A good paper on the subject is by Suri and Zazanis [126]. Before this paper appeared, most of the statistical properties of perturbation analysis were validated only through experimentation. Suri and Zazanis consider, for the M/G/1 queue, the sensitivity of the mean system time of a customer for a parameter of the arrival or service distribution. They show that the steady state value of the estimate of this sensitivity derived by perturbation analysis is unbiased and they present an algorithm that is implemented on a single sample path of the process, which gives asymptotically unbiased and strongly consistent estimates. No previous knowledge of perturbation analysis is assumed, so the article serves as a sort of introduction. This paper has been widely cited in various perturbation analysis research articles as applied to the inference in queueing systems, and so we refer the reader desiring more information to Rubinstein and Szidarovszky [118], Glasserman [52, 53], Heidelberger et al. [66], Zazanis [136], L'Ecuyer [91], Hu [69], Konstantopoulos and Zazanis [83], and Fu and Hu [45], just to name a few.

As queueing theory finds new application areas, new problems emerge. For example, during recent years, queueing problems in computer communications systems have been a major area of research in queueing theory. These are mainly network-related systems and consequently, statistical analysis of queueing networks has become a necessity (see Gawlick [49]). Special types of queueing systems abound in applications. Estimation techniques relative to such systems will have to be specialized as well. For an illustration of this, see Rubin and Robson [117].

Finally, we may also mention that new perspectives on statistical inference are also influencing research on inference on queueing systems. The subjective Bayesian approach to the theory of queues initiated by McGrath et al. [94], just to name one, is in this spirit. In any case, statistical analysis of queueing systems is an area that brings together the practitioner and theoretician with a common purpose and it is our belief that this should continue in future research on inference in queues.

BIBLIOGRAPHY

[1]　Aigner, D.J., Parameter estimation from cross-sectional observations on an elementary queueing system, *Oper. Res.* **22** (1974), 422-428.

[2]　Armero, C., Bayesian analysis of M/M/1/∞/FIFO queues, In: *Bayesian Statistics* 2 (ed. by J.M. Bernardo, M.H. DeGroot, D.V. Lindley, and A.F.M. Smith) (1985), 613-617.

[3]　Armero, C., Bayesian inference in Markovian queues, *Queueing Sys.* **15** (1994), 419-426.

[4]　Armero, C. and Bayarri, M.J., Bayesian prediction in M/M/1 queues, *Queueing Sys.* **15** (1994), 401-417.

[5]　Basawa, I.V. and Bhat, U.N., Some large sample tests for traffic intensity in M/G/1 queues, (1995), to appear.

[6]　Basawa, I.V. and Prabhu, N.U., Estimation in single server queues, *Nav. Res. Log. Quart.* **28** (1981), 475-487.

[7]　Basawa, I.V. and Prabhu, N.U., Large sample inference from single server queues, *Queueing Sys.* **3** (1988), 289-304.

[8]　Basawa, I.V. and Prabhu, N.U., *Statistical Inference in Stochastic Processes*, Special Issue of *J. of Stat. Plann. Infer.* **39**:2 (1994).

[9]　Basawa, I.V. and Prakasa Rao, B.L.S., *Statistical Inference for Statistical Processes*, Academic Press, New York 1980.

[10]　Beneš, V.E., A sufficient set of statistics for a simple telephone exchange model, *Bell Sys. Tech. J.* **36** (1957), 939-964.

[11]　Beneš, V.E., The covariance function of a simple trunk group with applications to traffic measurement, *Bell Sys. Tech. J.* **40** (1961), 117-148.

[12]　Bertsimas, D.J. and Servi, L.D., Deducing queuing from transactional data: the queue inference engine revisited, *Oper. Res.* **40**:S2 (1992), S217-S228.

[13]　Bhat, U.N., *Elements of Applied Stochastic Processes*, 2nd edition, Wiley, New York 1984.

[14]　Bhat, U.N., A sequential technique for the control of traffic intensity in Markovian queues, *Ann. Oper. Res.* **8** (1987), 151-164.

[15]　Bhat, U.N. and Rao, S.S., A statistical technique for the control of traffic intensity in queueing systems M/G/1 and GI/M/1, *Oper. Res.* **20** (1972), 955-966.

[16]　Bhat, U.N. and Rao, S.S., Statistical analysis of queueing systems, *Queueing Sys.* **1** (1987), 217-247.

[17]　Billingsley, P., *Statistical Inference for Markov Processes*, University of Chicago Press, Chicago 1961.

[18]　Birnbaum, A., Statistical methods for Poisson processes and exponential populations, *J. Amer. Stat. Assoc.* **49** (1954), 254-266.

[19]　Blomqvist, N., The covariance function of the M/G/1 queueing system, *Skandinavisk Aktuarietidskrift* **50** (1967), 157-174.

[20] Blomqvist, N., Estimation of waiting time parameters in the GI/G/1 queueing systems, Part I: General results, *Skandinavisk Aktuarietidskrift* **51** (1968), 178-197.

[21] Blomqvist, N., Estimation of waiting time parameters in the GI/G/1 queueing systems, Part II: Heavy traffic approximations, *Skandinavisk Aktuarietidskrift* **52** (1969), 125-136.

[22] Bradley, J.V., *Distribution-Free Statistical Tests*, Prentice-Hall, Englewood Cliffs, NJ 1968.

[23] Brown, M., An M/G/∞ estimation problem, *Ann. Math. Stats.* **41** (1970), 651-654.

[24] Burke, P.J., The output of a queueing system, *Oper. Res.* **4** (1956), 699-704.

[25] Choobineh, F. and Ballard, J.L., A method of confidence interval construction for simulated output analysis, *Oper. Res. Letters* **8** (1989), 265-270.

[26] Clarke, A.B., Maximum likelihood estimates in a simple queue, *Ann. Math. Stats.* **28** (1957), 1036-1040.

[27] Conover, W.J., *Practical Nonparametric Statistics*, Wiley, New York 1971.

[28] Cox, D.R., *Renewal Theory*, Methuen and Co., London 1962.

[29] Cox, D.R., Some problems of statistical analysis connected with congestion, In: *Proc. of the Symp. on Congestion Theory* (ed. by W.L. Smith and W.B. Wilkinson), University of North Carolina at Chapel Hill, NC 1965.

[30] Cox, D.R. and P.A. Lewis, *The Statistical Analysis of Series of Events*, Methuen and Co., London 1966.

[31] Crane, M.A. and Lemoine, A.J., *An Introduction to Regenerative Method for Simulation Analysis*, Lecture Notes in Control and Information Sciences **5**, Springer-Verlag, New York 1977.

[32] Craven, B.D., Asymptotic correlation in a queue, *J. Appl. Prob.* **6** (1977), 573-583.

[33] Daley, D.J., The serial correlation coefficients of waiting times in a stationary single server queue, *J. Austral. Math. Soc.* **8** (1968), 683-699.

[34] Daley, D.J., Monte Carlo estimation of mean queue size in a stationary GI/M/1 queue, *Oper. Res.* **16** (1968), 1002-1005.

[35] Daley, D.J. and Servi, L.D., Exploiting Markov Chains to infer queue length from transactional data, *J. Appl. Prob.* **29** (1992), 713-732.

[36] Dave, U. and Shah, Y.K., Maximum likelihood estimates in the M/M/2 queue with heterogeneous servers, *J. Opnl. Res.* **31** (1980), 423-426.

[37] Descloux, A., On the accuracy of loss estimates, *Bell Sys. Tech. J.* **44** (1965), 1139-1164.

[38] Descloux, A., Some properties of the variance of the switch-count load, *Bell Sys. Tech. J.* **55** (1976), 59-88.

[39] Duenyas, I. and Hopp, W.J., Estimating variance of output from cyclic exponential queueing systems, *Queueing Sys.* **7** (1990), 337-354.

[40] Eschenbach, W., Statistical inference for queueing models, *Stat.* **15** (1984), 451-462.

[41] Fishman, G.S., Statistical analysis for queueing simulations, *Mgt. Sci.* **20** (1973), 363-369.

[42] Fishman, G.S., Estimation in multiserver queueing simulations, *Oper. Res.* **22** (1974), 72-78.

[43] Fishman, G.S., Achieving specific accuracy in simulation output analysis, *Comm. Assoc. Comput. Mach.* **20** (1977), 310-315.

[44] Fishman, G.S., *Principles of Discrete Event Simulation*, Wiley, New York 1978.

[45] Fu, M.C. and Hu, J., Derivative sample path estimators for the GI/G/m queue, *Mgt. Sci.* **39** (1993), 359-383.

[46] Gafarian, A.V. and Ancker, C.J., Mean value estimation from digital computer simulations, *Oper. Res.* **14** (1966), 25-44.

[47] Gaver, D.P. and Jacobs, P.A., On inference and transient response for M/G/1 models, In: *Teletraffic Analysis and Computer Performance Evaluation* (ed. by O.J. Boxma, J.W. Cohen and H.C. Tijms), Elsevier Science, North Holland (1986), 163-170.

[48] Gaver, D.P. and Jacobs, P.A., Nonparametric estimation of a long delay in the M/G/1 queue, *J. Roy. Stat. Soc.* **B50** (1988), 392-402.

[49] Gawlik, R., Estimating disperse network queues: the queue inference engine, *Computer Comm. Rev.* **20** (1990), 111-118.

[50] George, L.L. and Agrawal, A.C., Estimation of a hidden service distribution of an $M/G/\infty$ system, *Nav. Res. Log. Quart.* **20** (1973), 549-555.

[51] Gebhard, R.F., A limiting distribution of an estimator of mean queue length, *Oper. Res.* **11** (1963), 1000-1003.

[52] Glasserman, P., Infinitesimal perturbation analysis of a birth and death process, *Oper. Res. Letters* **7** (1988), 43-49.

[53] Glasserman, P., The limiting value of derivative estimators based on perturbation analysis, *Stoch. Mod.* **6** (1990), 229-257.

[54] Glynn, P.W. and Iglehart, D.L., Estimation of steady-state central moments by the regenerative method of simulation, *Oper. Res. Letters* **5** (1986), 271-276.

[55] Glynn, P.W. and Whitt, W., Indirect estimation via $L = \lambda W$, *Oper. Res.* **37** (1989), 82-103.

[56] Glynn, P.W., Melamed, B. and Whitt, W., Estimating customer and time averages, *Oper. Res.* **41** (1993), 400-408.

[57] Gnedenko, B.V., Belyayev, Y.K. and Solovyev, A.D., *Mathematical Methods of Reliability Theory*, Academic Press, New York 1969.

[58] Goyal, T.L., Some confidence intervals for queues with state-dependent Erlang service, *Sys. Cyber. in Mgt.* (SCIMA) **4** (1975), 19-30.

[59] Goyal, T.L. and Harris, C.M., Maximum likelihood estimation for queues with state dependent service, *Sankhya A* **34** (1972), 65-80.

[60] Grassmann, W.K., The optimal estimation of the expected number in a $M/D/\infty$ queueing system, *Oper. Res.* **29** (1981), 1208-1211.

[61] Gross, D. and Harris, C.M., *Fundamentals of Queuing Theory*, 2nd edition, Wiley, New York 1985.

[62] Halfin, S., Linear estimators for a class of stationary queueing processes, *Oper. Res.* **30** (1982), 515-529.

[63] Hantler, S.L. and Rosberg, Z., Optimal estimation for an M/M/c queue with time varying parameters, *Stoch. Mod.* **5** (1989), 295-313.

[64] Harishchandra, K. and Rao, S.S., Statistical inference about the traffic intensity parameter of $M/E_k/1$ and $E_k/M/1$ queues, Report, Indian Inst. of Mgt., Bangalore, India 1984.

[65] Harris, C.M., Some new results in the statistical analysis of queues, In: *Proc. of a Conf. on Math. Meth. in Queueing Theory* (ed. by A.B. Clarke), Lecture Notes in Economics and Math. Sys., Springer-Verlag **98** (1974), 157-183.

[66]   Heidelberger, P., Cao, X., Zazanis, M.A. and Suri, R., Convergence properties of infinitesimal perturbation analysis estimates, *Mgt. Sci.* **34** (1988), 1281-1302.

[67]   Heyde, C.C., Asymptotic efficiency results for the method of moments with application to estimation for queueing processes, *Queueing Theory and Appl.*, in: *Queueing Theory Appl.* (eds. O.J. Boxma and R. Syski) 405-412, North-Holland, Amsterdam 1988.

[68]   Hsu, J.C. and Nelson, B.L., Control variates for quantile estimation, *Mgt. Sci.* **36** (1990), 835-851.

[69]   Hu, J.Q., Consistency of infinitesimal perturbation analysis estimators with rates, *Queueing Sys.* **8** (1991), 265-278.

[70]   Iglehart, D.L., Simulating stable stochastic systems, VI: quantile estimation, *J. Assoc. for Comp. Mach. (JACM)* **23** (1976), 347-360.

[71]   Iglehart, D.L. and Shedler, G.S., Regenerative simulation of response times in networks of queues: statistical efficiency, *Acta Info.* **15** (1981), 347-363.

[72]   Iglehart, D.L. and Shedler, G.S., Statistical efficiency of regenerative simulation methods for networks of queues, *Adv. Appl. Prob.* **15** (1983), 183-197.

[73]   Jacobsen, M., *Statistical Analysis of Counting Processes*, Springer-Verlag, New York 1982.

[74]   Jain, S., Estimation of $M/E_r/1$ queueing systems,   *Comm. in Statistics: Theory and Methods* **A20** (1991), 1871-1879.

[75]   Jain, S., Comparison of confidence intervals of traffic intensity for $M/E_k/1$ queueing systems, *Statistische Hefte* **32** (1991), 167-174.

[76]   Jain, S., Relative efficiency of a parameter for a M/G/1 queueing system based on reduced and full likelihood functions, *Comm. in Statistics: Computation and Simulation* **B21** (1992), 597-606.

[77]   Jain, S. and Templeton, J.G.C., Problem of statistical inference to control traffic intensity, *Seq. Anal.* **8** (1989), 135-146.

[78]   Jain, S. and Templeton, J.G.C., Confidence interval for M/M/2 queue with heterogeneous servers, *Oper. Res. Letters* **10** (1991), 99-101.

[79]   Jenkins, J.H., The relative efficiency of direct and maximum likelihood estimates of mean waiting time in the simple queue M/M/1, *J. Appl. Prob.* **9** (1972), 396-403.

[80]   Joseph, L., Wolfson, D.B. and Wolfson, C., Is multiple sclerosis an infectious disease?   Inference about an input process based on the output, *Biometrics* **46** (1990), 337-349.

[81]   Karr, A.F., *Point Processes and their Statistical Inference*, 2nd edition, Marcel Dekker, New York 1991.

[82]   Kendall, D.G., Stochastic processes occurring the theory of queues and their analysis by the method of imbedded Markov chains, *Ann. Math. Stat.* **24** (1953), 338-354.

[83]   Konstantopoulos, P. and Zazanis, M.A., Sensitivity analysis for stationary and ergodic queues, *Adv. Appl. Prob.* **24** (1992), 738-750.

[84]   Kosten, L., Manning, J.R. and Garwood, F., On the accuracy of measurements of probabilities of loss in telephone systems, *J. Roy. Stat. Soc.* **B11** (1949), 54-67.

[85]   Koziol, J.A. and Nemac, A.F., On a Cramer-von Mises type statistic for testing bivariate independence, *Canad. J. Stats.* **7** (1979), 43-52.

[86]   Larson, R., The queue inference engine: deducing queue statistics from transactional data, *Mgt. Sci.* **36** (1990), 586-601. Addendum (1991), *ibid.*

37, 1062.

[87]   Law, A.M., Efficient estimators for simulated queueing systems, *Mgt. Sci.* **22** (1975), 30-41.

[88]   Law, A.M., Confidence intervals in discrete event simulation: a comparison of replication and batch means, *Nav. Res. Log. Quart.* **24** (1977), 667-678.

[89]   Law, A.M. and Carson, J.S., A sequential procedure for determining the length of a steady-state simulation, *Oper. Res.* **27** (1979), 1011-1025.

[90]   Law, A.M. and Kelton, W.D., Confidence intervals for steady state simulations II: A survey of sequential procedures, *Mgt. Sci.* **28** (1982), 550-562.

[91]   L'Ecuyer, P., A unified view of the IPA, SF, and LR gradient estimation techniques, *Mgt. Sci.* **36** (1990), 1364-1383.

[92]   Lewis, P.A.W., Recent results in the statistical analysis of univariate point processes, In: *Stochastic Point Processes* (ed. by P.A.W. Lewis), Wiley (1972), 1-54.

[93]   Lilliefors, H.W., Some confidence intervals for queues, *Oper. Res.* **14** (1966), 723-727.

[94]   McGrath, M.F., Gross, D. and Singpurwalla, N.D., A subjective Bayesian approach to the theory of queues, Part I - Modeling, *Queueing Sys.* **1** (1987), 317-333.

[95]   McGrath, M.F. and Singpurwalla, N.D., A subjective Bayesian approach to the theory of queues, Part II - Inference and information in M/M/1 queues, *Queueing Sys.* **1** (1987), 335-353.

[96]   Mielke, Jr., P.W., A note on some squared rank tests with existing ties, *Technometrics* **9** (1967), 312-314.

[97]   Moore, S.C., Approximate techniques for non-stationary queues, Ph.D. Dissertation, Southern Methodist University, Dallas, TX 1972.

[98]   Moran, P.A.P., Estimation methods for evolutive processes, *J. Roy. Stat. Soc.* **B13** (1951), 141-146.

[99]   Moran, P.A.P., The estimation of the parameters of a birth and death process, *J. Roy. Stat. Soc.* **B15** (1953), 241-245.

[100] Muddapur, M.V., Bayesian estimates of parameters in some queueing models, *Ann. Inst. Stat. Math.* **24** (1972), 327-331.

[101] Neal, S.R. and Kuczura, A., A theory of traffic-measurement errors for loss systems with renewal input, In: *Proc. of a Conf. on Math. Meth. in Queueing Theory*, (ed. by A.B. Clarke), Lecture Notes in Economics and Math. Sys., Springer-Verlag **98** (1974), 199-229.

[102] Nedelman, J., Estimation for a model of multiple malaria infections, *Biometrics* **41** (1985), 447-453.

[103] Pagurek, B., Stanford, D.A. and Woodside, C.M., Optimal prediction of times and queue lengths in the M/G/1 queue, *J. Opnl. Res. Soc.* **39** (1988), 585-593.

[104] Patterson, R.L., Regression estimates of inputs to an M(t)/G/∞ service system, *Appl. Math. Comp.* **24** (1987), 47-63.

[105] Phelan, M.J. and Prabhu, N.U., Estimation from an infinite server queueing system with two demands, in: *Queueing Theory Appl.* (eds. O.J. Boxma and R. Syski) 429-441, North-Holland, Amsterdam 1988.

[106] Prabhu, N.U. and Basawa, I.V., *Statistical Inference in Stochastic Processes*, Marcel Dekker, New York 1991.

[107] Putter, J., The treatment of ties in some nonparametric tests, *Ann. Math. Stats.* **26** (1955), 368-386.

[108] Ramalhoto, M.F., Some statistical problems in random translations of stochastic point processes, *Ann. Oper. Res.* **8** (1987), 229-242.

[109] Randles, R.H. and Wolfe, D.A., *Introduction to the Theory of Nonparametric Statistics*, Wiley, New York 1979.

[110] Rao, S.S. and Bhat, U.N., A sequential test for a denumerable Markov chain and an application to queues, *J. Math. Phy. Sci.* **25** (1991), 521-527.

[111] Rao, S.S., Bhat, U.N. and Harishchandra, K., Control of traffic intensity in a queue - A method based on SPRT, *Opsearch* **21** (1984), 63-80.

[112] Rao, S.S. and Harishchandra, K., A large sample test for the traffic intensity in the GI/G/s queue, *Nav. Res. Log. Quart.* **33** (1986), 545-550.

[113] Reich, E., Waiting times when queues are in tandem, *Ann. Math. Stats.* **28** (1957), 768-773.

[114] Reynolds, J.F., Asymptotic properties of mean length estimators for the finite Markov queue, *Oper. Res.* **20** (1972), 52-57.

[115] Reynolds, J.F., The covariance structure of queues and related processes: A survey of recent work, *Adv. Appl. Prob.* **7** (1975), 383-415.

[116] Riordan, J., *Stochastic Service Systems*, Wiley, New York 1962.

[117] Rubin, G. and Robson, D.S., A single server queue with random arrivals and balking: Confidence interval estimation, *Queueing Sys.* **7** (1990), 283-306.

[118] Rubinstein, R.Y. and Szidarovszky, F., Convergence of perturbation analysis estimates for discontinuous sample functions: A general approach, *Adv. Appl. Prob.* **20** (1988), 59-78.

[119] Samaan, J.E. and Tracy, D.S., Properties of some estimators for the simple queue M/M/1, In: *Proc. of the Third Symp. in Oper. Res.: Methods of Oper. Res.*, University of Mannheim (1978), 377-387.

[120] Schruben, L. and Kulkarni, R., Some consequences of estimating parameters for the M/M/1 queue, *Oper. Res. Letters* **1** (1982), 75-78.

[121] Seila, A.F., A batching approach to quantile estimation in regenerative simulations, *Mgt. Sci.* **28** (1982), 573-581.

[122] Shalmon, M. and Rubinstein, R., Error analysis for regenerative queueing estimators with special reference to gradient estimators via likelihood ratio, *Ann. Oper. Res.* **36** (1992), 383-396.

[123] Shedler, G.S., *Regeneration and Networks of Queues*, Springer Verlag, New York 1987.

[124] Singpurwalla, N., Discussion of Thiruvaiyaru and Basawa's "Empirical Bayes estimation for queueing systems and networks", *Queueing Sys.* **11** (1992), 203-206.

[125] Stanford, D.A., Pagurek, B. and Woodside, C.W., Optimal prediction of times and queue lengths in the GI/M/1 queue, *Oper. Res.* **31** (1983), 322-337.

[126] Suri, R. and Zazanis, M.A., Perturbation analysis gives strongly consistent sensitivity estimates for the M/G/1 queue, *Mgt. Sci.* **34** (1988), 39-64.

[127] Syski, R., *Introduction to Congestion Theory in Telephone Systems*, Oliver and Boyd, London 1960.

[128] Thiagarajan, T.R. and Harris, C.M., Statistical tests for exponential service from M/G/1 waiting-time data, *Nav. Res. Log. Quart.* **26** (1979), 511-520.

[129] Thiruvaiyaru, D. and Basawa, I.V., Empirical Bayes estimation for queueing systems and networks, *Queueing Sys.* **11** (1992), 179-202, Rejoinder, *ibid.*, 207-210.

[130] Thiruvaiyaru, D., Basawa, I.V. and Bhat, U.N., Estimation for a class of simple queueing networks, *Queueing Sys.* **9** (1991), 301-312.

[131] Toyoizumi, H., Evaluating mean sojourn time estimates for the M/M/1 queue, *Comp. Math. Appl.* **24** (1992), 7-15.

[132] Whitt, W., Planning queueing simulations, *Mgt. Sci.* **35** (1989), 1341-1366.

[133] Whitt, W., The efficiency of one long run versus independent replications in steady-state simulation, *Mgt. Sci.* **37** (1991), 645-666.

[134] Wolff, R.W., Problems of statistical inference for birth-and-death queueing models, *Oper. Res.* **13** (1965), 343-357.

[135] Woodside, C.M., Stanford, D.A. and Pagurek, B., Optimal prediction of queue lengths and delays in GI/M/m multiserver queues, *Oper. Res.* **32** (1984), 809-817.

[136] Zazanis, M.A., Infinitesimal perturbation analysis estimates for moments of the system time of an M/M/1 queue, *Oper. Res.* **38** (1990), 364-369.

Chapter 14

Perturbation analysis for control and optimization of queueing systems: An overview and the state of the art[1]

Yu-Chi Ho and Christos G. Cassandras

ABSTRACT Perturbation Analysis (PA) has evolved over the past ten years as a methodology for estimating performance gradients with respect to parameters of Discrete Event Dynamic Systems (DEDS's), and particularly, queueing systems, from information extracted from a single observed sample path. This paper is a self-contained tutorial on the state of the art in this field. We develop the principles of Infinitesimal PA (IPA) in terms of evaluating derivatives of sample functions and using them as unbiased estimators of performance derivatives for certain classes of queueing systems. We then present extensions of IPA, including an overview of recent results. Finally, we provide a broader view of PA as a means for global performance response surface estimation for both continuous and discrete parameters.

CONTENTS

14.1 INTRODUCTION

The area of Perturbation Analysis (PA) of stochastic Discrete Event Dynamic Systems (DEDS's), and especially, queueing systems, is motivated by the need to develop efficient methodologies for control and optimization problems for which no closed-form solutions or effective analytical techniques are available. The main tenet of PA is that a great deal of information is contained in the sample paths of such systems, beyond the usual statistics collected, such as the means and variances of various output variables. When simulating a queueing system, for example, instead of looking at the discrete event simulation process simply as a special case of statistical analysis of experiments (i.e., a "black box" with input parame-

---

[1]This work has been supported in part by the National Science Foundation under grants ECS-8801912, ECD-90-44673, and EID-92-12122; by Army contracts DAAL-03-91-G-0194 and 03-92-G-0115; by the Naval Research Laboratory under contract N000014-91-J-2025; and by a grant from United Technologies.

---

ters and final output results), we can take advantage of our knowledge regarding the dynamics of this system and extract additional useful information from a single experiment; such information includes performance gradients, higher-order derivatives, and sensitivities with respect to various parameters of interest. In a broader sense, PA has to do with the problem of enhancing the efficiency of discrete-event simulation or obtaining more information related to performance evaluation for the same computing budget.

It is also important to note that the principles of PA apply to any sample path. In particular, PA can be applied to a sample path not generated through simulation, but as a result of a sequence of events and state transitions observed on actually operating DEDS's, such as a manufacturing system or a computer network. The implication is that PA is also a technique suitable for on-line applications. As performance sensitivity information is obtained in real time, it can be incorporated into a variety of control schemes, ultimately intended to optimize system performance. Examples may be found in [8, 40, 46]. Finally, it is worth pointing out that many of the developments relevant to PA have been based on the observation that DEDS's share many conceptual commonalities with continuous-variable dynamic systems (CVDS's) governed by differential equations. Many of the successes in the optimization and control of differential-equation-based dynamic systems can be transplanted to the DEDS's domain. In fact, the analog to PA in CVDS's is simply the familiar idea of linearization and of variational differential equations.

This chapter is intended to be a tutorial on PA techniques used for sample-path-based gradient estimation in queueing systems, including an overview of recent developments. The issue of gradient estimation based on sample-path data has attracted considerable attention over the past decade and includes techniques other than PA, such as the Likelihood Ratio methodology (see [24, 42, 43]). In this chapter, however, we will limit ourselves to PA and overview material, which can also be found in recent books by Glasserman [19], Ho and Cao [35], and Cassandras [11], and an overview paper by Ho [36]. The chapter is composed of five major sections. After introducing our basic modeling framework, notation, and terminology in Section 14.2, we present in Section 14.3 the basic principles of Infinitesimal PA (IPA) and its application to gradient estimation based on a single sample path. Section 4 is concerned with extensions of IPA, including a general approach based on the notion of "smoothing" and known as Smoothed PA (SPA). In Section 14.5 we briefly discuss extensions of PA as means for global performance response surface estimation for both continuous and discrete parameters. This leads us to new developments aiming at the parallel generation of multiple sample paths corresponding to different parameter values, some of which may be used in new parallel processing computer architectures.

14.2 MODELING FRAMEWORK AND NOTATION

We begin by introducing the modeling framework we will adopt for queueing systems and the corresponding notation. Let X be a countable *state space*, and E a countable *event set*. For any state $x \in X$, let $\Gamma(x) \subseteq E$ be the set of *feasible* (or *enabled*) events that can occur at that state; this reflects the fact that it is not always physically possible for some events to occur. Given a state $x \in X$ and an event $i \in \Gamma(x)$, we define a *state transition function*, $f: X \times E \to X$; this function is undefined for $i \notin \Gamma(x)$, since i is not a feasible event at that state. Note that this state transition function may be replaced by *transition probabilities* $p(x'; x, i)$ representing the probability that a transition from state x to state x' occurs whenever event i takes place. We also assume that an initial state $x_0 \in X$ is specified. One can easily see that the five-tuple (E, X, Γ, f, x_0) is simply a state automaton.

A *timed* state automaton $(E, X, \Gamma, f, x_0, V)$ is obtained when the model above is endowed with a *clock structure*, $V = \{\mathbf{v}_i, i \in E\}$, defined as follows. We associate with every event $i \in E$ a real-valued clock sequence $\mathbf{v}_i = \{V_{i,1}, V_{i,2}, \ldots\}$, which we view as an input to the system; here, $V_{i,k}$ is the kth *lifetime* of event i. Associated with any event i that happens to be feasible at some state is a *clock value* (or *residual lifetime*) y_i. Whenever event i becomes enabled for the kth time, we set its clock value y_i to $V_{i,k}$. The clock for this event then ticks down to zero, at which point the event is said to occur (note the difference between the time when an event is *enabled*, i.e., its clock starts ticking down from $V_{i,k}$, and the time when this event *occurs*, i.e., its clock reaches zero). Thus, for any state x, there is a set of associated clock values, one for each event $i \in \Gamma(x)$. The *triggering event* e' is the event which occurs next at that state, defined as the feasible event with the smallest clock value:

$$e' = arg \min_{i \in \Gamma(x)} \{y_i\}. \tag{14.1}$$

Once this event is determined, the next state, x', is specified by $x' = f(x, e')$ (or the probabilistic mechanism specified by $p(x'; x, i)$ defined above). The amount of time spent at state x defines the *interevent time* (between the event that caused a transition into x and the triggering event e'):

$$y^* = \min_{i \in \Gamma(x)} \{y_i\}. \tag{14.2}$$

Thus, time is updated through

$$t' = t + y^*. \tag{14.3}$$

In addition, clock values are updated through $y_i' = y_i - y^*$, i.e., we use the residual lifetime of event i as the new clock value of i in the new state x'. This is true for all events i which remain feasible in x', except for two cases: (a) If $i = e'$, the clock value is set to a new lifetime obtained from the next available element in the event's clock sequence $\mathbf{v}_i$ (since this event is enabled afresh), or (b) If $i \notin \Gamma(x)$ but $i \in \Gamma(x')$, again the clock value is set to a new lifetime obtained from the next available element in the event's clock sequence (since this event was not feasible in x, but becomes feasible in x'). In this way, every enabled event becomes a triggering event sooner or later as its clock ticks down to zero.

Given the initial state $x_0 \in X$, the feasible event set $\Gamma(x_0)$ is known. Therefore, the process defined above starts with clock values given by $y_i = V_{i,1}$ for all $i \in \Gamma(x_0)$ and its operation results in a timed event sequence $\{e_k, t_k\}$, $k = 1, 2, \ldots$, or equivalently, a sequence $\{x_k, t_k\}$, $k = 1, 2, \ldots$, where x_k is the kth state entered due to an event occurring at time t_k. This defines a *trajectory* or *sample path* of this system. Note that the clock structure V is completely independent of the state transition function and can be constructed in advance for any system with a given event set E.

In a general stochastic setting, the lifetime sequences $\mathbf{v}_i = \{V_{i,1}, V_{i,2}, \ldots\}$, $i \in E$, consist of random elements. Thus, the last step in our framework is to define a *stochastic time automaton* $(E, X, \Gamma, f, x_0, G)$ by replacing the clock structure V by a set of probability distribution functions $G = \{G_i, i \in E\}$. In this case, $\mathbf{v}_i = \{V_{i,1}, V_{i,2}, \ldots\}$ is a random sequence; for simplicity, we assume for each event

$i \in E$ that the lifetimes $V_{i,1}, V_{i,2}, \ldots$ are independent and identically distributed (i.i.d.) random variables with a common distribution G_i. Thus, to generate a sample path of the system, we obtain a sample from G_i whenever a lifetime for an event i is needed. The state sequence generated through this mechanism is a stochastic process known as a *Generalized Semi-Markov Process* (GSMP) (for details see the books by Glasserman [19], Ho and Cao [35], and Cassandras [11]). The procedure we have described also corresponds to the standard way of generating sample paths of queueing systems through discrete event simulation (e.g., [39]). In this case, lifetimes for each event $i \in E$ are provided through an appropriate random variate generation mechanism driven by some pseudo-random number generator.

To help fix the ideas above, consider a simple single-server queueing system with random arrival and service times operating in First-Come-First-Served (FCFS) fashion. The state space is $X = \{0, 1, 2, \ldots\}$, representing the possible number of customers waiting in the queue, including a customer in service. The event set is $E = \{a, d\}$, where a denotes an "arrival" and d denotes a "departure". The feasible event set of any $x > 0$ is $\Gamma(x) = \{a, d\} = E$, while $\Gamma(0) = \{a\}$, since no departure is feasible when the queue is empty. The clock structure is specified by two distributions, G_a and G_d, characterizing the sequences of interarrival times and service times, respectively. The state transition function $x' = f(x, e')$ is simply

$$
x' = \begin{cases} x + 1 & \text{if } e' = a \\ x - 1 & \text{if } e' = d \text{ and } x > 0. \end{cases}
$$

Assuming the initial state $x = 0$, the first triggering event is an arrival, which causes a state transition to $x' = 1$. By subsequently drawing interarrival and service time samples from the distributions G_a and G_d, respectively, a trajectory or sample path of this system can be completely determined.

Mathematically, we often denote a sample path simply as (θ, ξ), where θ is a vector of parameters characterizing the state transition function and/or the clock structure G, and ξ represents all random occurrences in the queueing system of interest. In particular, we let the underlying sample space be $[0, 1]^\infty$, and ξ a sequence of independent random variables uniformly distributed on $[0, 1]$. In the example above, θ obviously consists of the parameters characterizing the interarrival and service time distributions, and ξ is the underlying sequence of random numbers from which all lifetimes in the clock structure are obtained, i.e., all samples of interarrival and service times. Given a sample path (θ, ξ) we can evaluate its *sample performance* $L(\theta, \xi)$ via a statistical experiment, i.e., a discrete-event simulation run or data collected from an actual system in operation. This, then, serves as an estimate of the actual *performance measure* of interest, which we consider to be the expectation $J(\theta) \equiv E[L(\theta, \xi)]$. In Perturbation Analysis (PA), we are often interested not only in $J(\theta)$, but also in $J(\theta + \Delta\theta)$. We use the adjectives *nominal* and *perturbed* to qualify the performance measures $J(\theta)$ and $J(\theta + \Delta\theta)$, as well as the corresponding sample paths (θ, ξ) and $(\theta + \Delta\theta, \xi)$.

14.3 INFINITESIMAL PERTURBATION ANALYSIS (IPA)

We begin with a brief review of the basic idea of IPA in Subsection 14.3.1. In Subsections 14.3.2-14.3.4, we present a formal development of IPA as a method for estimating gradients of performance measures of queueing systems with respect to parameters of their event lifetime distributions.

14.3.1 A short history and the basic idea of IPA

The methodology now known as IPA was a problem-driven innovation. In 1977, the first author was presented with an interesting consulting problem [28]. The FIAT Motor Company in Turin, Italy, had installed a production monitoring system on one of their automobile engine production lines that could be modeled as a simple serial queueing network with finite queue (buffer) capacity between servers (machines). The automatic line monitoring system recorded service initiations, completions, idlings, and blockings of various machining stations, as well as the movement of the engine parts among them; in short, a complete operating history of the underlying queueing system. A tremendous amount of production information was being generated. The following questions were asked: "Besides the standard statistical information, such as downtime, throughput, and utilization, that were being generated by the monitoring system from the information collected, could the same information be used further for control purposes? In particular, we (FIAT) are interested in whether or not the buffer spaces between machines are optimally distributed for maximal throughput, given a limited budget for buffer spaces." The attempt to answer the latter question ([35], p. 20) led to the following three generalizable ideas that have formed the foundations of IPA since its inception and early development in the work of Ho and Cassandras [29], Ho and Cao [31], and Ho et al. [30]:

(i) A change in some parameter θ (e.g., increasing the size of the buffer space by 1) can *generate* perturbations in the timing events in the sample path of a DEDS.

(ii) Perturbations in the timing of one event (e.g., termination of a service period of a machine) can be *propagated* to another event (e.g., termination of a service period at a downstream machine, effected via the termination of an idle period at that machine).

(iii) Since many performance measures of a DEDS depend on the timing of events over its sample paths, perturbations in the timing of events will induce perturbations in a sample performance function $L(\theta, \xi)$.

Observations (i)-(iii) suggest a method for calculating efficiently the perturbed sample performance $L(\theta + \Delta\theta, \xi)$ or the derivative $dL/d\theta$ from the nominal sample path (θ, ξ) alone, since all three facts only require information directly observable on (θ, ξ). However, we must keep in mind that what we are actually interested is $J(\theta) = E[L(\theta, \xi)]$, as defined in Section 14.2, and hence the derivative $dJ/d\theta$. The crucial question then arises: when can $dL/d\theta$ be used as a "good" estimate of $dJ/d\theta$? By "good" we normally mean properties of an estimator, such as unbiasedness and consistency. We further discuss this issue in the next section. Before proceeding, we point out that M. Bello, in a 1977 M.I.T. MS thesis, also had a version of the basic idea of PA as applied to the M/D/1 queue (see [2] and [51]).

14.3.2 The interchangeability issue

In a narrow sense, IPA is a technique for the *efficient* computation of the n-dimensional gradient vector of performance measure $J(\theta) = E[L(\theta, \xi)]$ of a queueing system with respect to the n parameters constituting vector θ, using *only one* statistical experiment of the system. This is in contrast to the traditional method of making $n + 1$ experiments (one for the nominal system under θ and n additional ones for each parameter perturbed) and taking differences to approximate the gradient vector, i.e., for a scalar θ:

$$\frac{dJ}{d\theta} \approx \frac{E[L(\theta + \Delta\theta, \xi)] - L(\theta, \xi)]}{\Delta\theta} = \frac{\frac{1}{N_i}\sum_{i=1}^{N} L(\theta + \Delta\theta, \xi_i) - \frac{1}{N_i}\sum_{i=1}^{N} L(\theta, \xi_i)}{\Delta\theta}. \tag{14.4}$$

This equation presents a numerically difficult task, since we are dividing the difference of two nearly equal random quantities by a small number and suffer the effects of the twin evils, of noise and nonlinearity: making $\Delta\theta$ small to better approximate the derivative of a nonlinear function forces us to divide two very small quantities where the numerator becomes dominated by the noise. Instead, IPA proposes to calculate directly the sample derivative $dL(\theta, \xi)/d\theta$ using information from the nominal trajectory (θ, ξ) alone. The basic idea is as follows: if the perturbations imposed on the trajectory (θ, ξ) are sufficiently small, then we can assume that the event sequence of the perturbed trajectory $(\theta + \Delta\theta, \xi)$ remains unchanged with respect to the nominal one. In this case, the derivative $dL(\theta, \xi)/d\theta$ can be calculated rather easily, as will be seen in Subsection 14.3.3. However, if the sample derivative $dL(\theta, \xi)/d\theta$ that we obtain is to serve as a good estimate of $dJ/d\theta$, then we require that this estimate be *unbiased*, i.e.,

$$E\left[\frac{dL(\theta, \xi)}{d\theta}\right] = \frac{dE[L(\theta, \xi)]}{d\theta} \equiv \frac{dJ}{d\theta}. \tag{14.5}$$

The validity of this equation is what we refer to as the "interchangeability issue" (see also [5]), since it involves interchanging the expectation (E) and differential operator $(d/d\theta)$. We are interested in the right-hand side of (14.5), but IPA provides the left-hand-side. In nontechnical terms, this question translates into "How can one extract an information from a sample path observed under one value of the system parameter, θ, about a sample path under a different value, $\theta' = \theta + \Delta\theta$? Would not the two sample paths behave entirely dissimilarly sooner or later?"

Returning momentarily to the dilemma presented by equation (4), i.e., the division of two small numbers, note that there is yet another approach to it. Suppose we can efficiently compute the difference $E[L(\theta + \Delta\theta, \xi)] - E[L(\theta, \xi)]$ as a quantity proportional to $\Delta\theta$. Then $dJ/d\theta$ in (14.4) may become numerically well-behaved, since no division by $\Delta\theta$ is required. That such an efficient computation is possible for many DEDS's should not be too surprising. Because of the event-driven dynamics, there is a great deal of duplication in the computation of $L(\theta + \Delta\theta, \xi)$ and that of $L(\theta, \xi)$. One should, therefore, be able to derive $L(\theta + \Delta\theta, \xi)$ from $L(\theta, \xi)$ with only minimal additional work. When this approach is used, it totally bypasses the interchangeability issue. Ramifications of this approach, both analytical and computational, will be further discussed in Section 14.4.

It should be clear at this point that the study of IPA involves two components: (a) The derivation of a sample function derivative $dL(\theta, \xi)/d\theta$ from information observed on a nominal sample path under θ alone, and (b) the interchangeability issue, i.e., whether $dL(\theta, \xi)/d\theta$ is an unbiased estimator of the performance measure derivative $dJ/d\theta$. We examine each of these components in the next two sections.

14.3.3 IPA sample function derivatives

In this section, we consider any queueing system or, more generally, any DEDS that falls within the modeling framework of Section 14.2. Let θ be a parameter (for simplicity a scalar) that characterizes one or more lifetime distributions $G_i(t; \theta)$ of events $i \in E$; in particular, θ does *not* affect the state transition mechanism. Given a sample function $L(\theta, \xi)$, we are interested in deriving an expression for the derivative $dL/d\theta$. If we accomplish this goal, we will then study the conditions under which $dL/d\theta$ can be used as an unbiased estimator of $dJ/d\theta$.

We begin by making one simplifying assumption regarding the models to be considered. In particular, we will assume that once an event i is activated, i.e., $i \in \Gamma(x)$ for some state x, it cannot be deactivated. In other words, once a clock for some event starts running down, it cannot be interrupted. This is known as the *non-interruption condition* (see [19]). In a simulation setting, this means that a scheduled event cannot be deferred or canceled. For most systems encountered in practice, this is not a serious limitation. To keep notation consistent, we will henceforth reserve Greek letters (usually α, β) to index events, and reserve Roman index letters (such as i, j, k, n, m) for counting event occurrences. We also define T_k to be the kth event occurrence time, $k = 1, 2, \ldots$, of an event in a sample path of our system (regardless of its type), and $T_{\alpha, n}$ to be the nth occurrence time of an event of type α. We can now immediately make the following simple observation: when event α takes place at time $T_{\alpha, n}$, it must have been activated at some point in time $T_{\beta, m} < T_{\alpha, n}$ upon the occurrence of some event β (note that it is possible that $\beta = \alpha$). In turn, it is also true that event β at $T_{\beta, m}$ must have been activated by some event γ at time $T_{\gamma, k} < T_{\beta, m}$, and so on, all the way back to time $T_0 = 0$. Recalling that $V_{\alpha, k}$ denotes the kth lifetime of event α, it follows that since (a) α was activated at time $T_{\beta, m}$, (b) α cannot be deactivated until it occurs (by the non-interruption assumption above), and (c) α finally occurs at time $T_{\alpha, n}$, we have $V_{\alpha, n} = T_{\alpha, n} - T_{\beta, m}$. Similarly, we have $V_{\beta, m} = T_{\beta, m} - T_{\gamma k}$, and so on. Therefore, we can always write

$$T_{\alpha, n} = V_{\beta_1, k_1} + \ldots + V_{\beta_s, k_s} \tag{14.6}$$

for some s. Clearly, a similar expression can be written for any event time T_k, since $T_k = T_{\alpha, n}$ for some $\alpha \in E$ and $n = 1, 2, \ldots$. The expression in (14.6) can be rewritten in a more convenient form by introducing *triggering indicators*, i.e., functions $\eta(\alpha, n; \beta, m)$ taking values in $\{0, 1\}$ and defined by:

$\eta(\alpha, n; \beta, m) = 1$,   if the nth occurrence of event α is triggered by the mth occurrence of β

$\eta(\alpha, n; \alpha, n) = 1$,   for all $\alpha \in E$, $n = 1, 2, \ldots$

$\eta(\alpha, n; \beta', m') = 1$,   if $\eta(\alpha, n; \beta, m) = 1$ and $\eta(\beta, m; \beta', m') = 1$ for some $\beta \in E$

$\eta(\alpha, n; \beta, m) = 0$,   otherwise.

Then:

$$T_{\alpha, n} = \sum_{\beta, m} V_{\beta, m} \eta(\alpha, n; \beta, m). \tag{14.7}$$

The sequence of (β, m) pairs that leads to $T_{\alpha, n}$ with $\eta(\alpha, n; \beta, m) = 1$ defines the *triggering sequence* of (α, n).

Event time derivatives. Since the parameter θ affects one or more of the event lifetime distributions $G_\alpha(x, \theta)$, a change in θ generally affects all event lifetimes in the sequence $\{V_{\alpha, 1}, V_{\alpha, 2}, \ldots\}$. Therefore, we view lifetimes as functions $V_{\alpha, k}(\theta)$ of

θ. Moreover, let us adopt the standard inverse transform technique (e.g., [39]) for generating variates from a distribution $G_\alpha(t;\theta)$. In particular, if $\{U_1, U_2, \ldots\}$ is a sequence of random numbers uniformly distributed over $[0,1]$, then a variate $V_{\alpha,k}$ generated by this method satisfies $U_k = G_\alpha(V_{\alpha,k};\theta)$ or, equivalently, $V_{\alpha,k} = G_\alpha^{-1}(U_k;\theta)$, since G_α is non-decreasing. Omitting details (see Ch. 9 of [11]), one can then obtain, through elementary probabilistic calculations, the following expression for the derivative $dV_{\alpha,k}(\theta)/d\theta$:

$$\frac{dV_{\alpha,k}}{d\theta} = -\frac{[\partial G_\alpha(t;\theta)/\partial\theta]_{(V_{\alpha,k},\theta)}}{[\partial G_\alpha(t;\theta)/\partial t]_{(V_{\alpha,k},\theta)}}, \tag{14.8}$$

where the notation $[\cdot]_{(V_{\alpha,k},\theta)}$ means that the term in brackets is evaluated at $t = V_{\alpha,k}$ (the observed event lifetime along the nominal sample path) and the given value of the parameter θ. Thus, through (14.8), the derivative of an event lifetime is obtained from the knowledge of the corresponding distribution and the observed lifetime along the sample path. It is worth pointing out that if θ is known to be either a scale or a location parameter of $G_\alpha(t;\theta)$, then the knowledge of the distribution itself is not required in (14.8). This leads to some robustness properties of IPA further discussed in [10].

Returning to (14.7), since $T_{\alpha,n}$ is expressed as a sum of event lifetimes, it is natural to expect that the derivative of $T_{\alpha,n}$ can be obtained in terms of the derivatives of these event lifetimes, under certain conditions. In particular, let us make the following assumptions:

(A1) For all $\alpha \in E$, $G_\alpha(t;\theta)$ is continuous in θ, and $G_\alpha(0;\theta) = 0$.

(A2) For all $\alpha \in E$ and $k = 1, 2, \ldots, V_{\alpha,k}(\theta)$ is almost surely continuously differentiable in θ.

Under these conditions, we can guarantee that the event sequence leading to time $T_{\alpha,n}$ (assumed finite) remains unchanged for sufficiently small perturbations of θ. As a result, the sequence of triggering indicators in (14.5) also remains unchanged. Therefore, from (14.7), the event time derivatives $dT_{\alpha,n}(\theta)/d\theta$ are given by

$$\frac{dT_{\alpha,n}}{d\theta} = \sum_{\beta,m} \frac{dV_{\beta,m}}{d\theta}\eta(\alpha,n;\beta,m), \tag{14.9}$$

where the interchange of summation and differentiation is permitted since the sum above is finite. This expression captures the fact that event time perturbations are *generated* by the effect of θ on event lifetimes $V_{\beta,m}$, and subsequently *propagate* from (β,m) to (α,n) through the events contained in the triggering sequence of (α,n).

Example 14.1. *IPA for the GI/G/1 queueing system.* To illustrate the precise way in which (14.9) can be used along an observed sample path of some system, let us consider a typical sample path of a GI/G/1 queueing system, as shown in Figure 1 below. In this case, there are two events, a (arrivals) and d (departures). Focusing on the departure event at $T_9 = T_{d,4}$, note that the 4th departure was activated at time $T_7 = T_{a,4}$ (when the second busy period starts), i.e., $\eta(d,4;a,4) = 1$. The 4th arrival, in turn, was activated at $T_4 = T_{a,3}$, the previous arrival, which in turn was activated at $T_2 = T_{a,2}$. Finally, the second arrival was activated at $T_1 = T_{a,1}$, and the first arrival was activated at time zero. We therefore get:

$$T_{d,4} = V_{d,4} + V_{a,4} + V_{a,3} + V_{a,2} + V_{a,1}.$$

Thus, the triggering sequence of $(d,4)$ is $\{(a,1),(a,2),(a,3),(a,4)\}$.

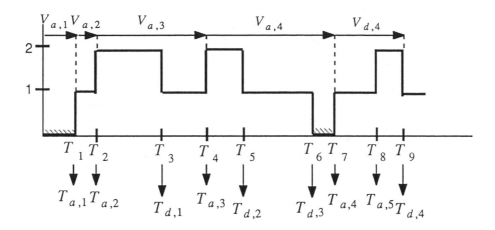

Figure 1. Sample path of a GI/G/1 system.

Now, suppose the parameter θ is the mean of the service time distribution $G_d(t;\theta)$. Thus, all service times (departure event lifetimes) $V_{d,k}$, $k = 1,2,\ldots$, are affected by perturbations in θ, whereas interarrival times are not (i.e., $dV_{a,k}/d\theta = 0$ for all $k = 1,2,\ldots$). Therefore,

$$\frac{dT_{d,4}}{d\theta} = \frac{dV_{d,4}}{d\theta} + \frac{dV_{a,4}}{d\theta} + \frac{dV_{a,3}}{d\theta} + \frac{dV_{a,2}}{d\theta} + \frac{dV_{a,1}}{d\theta} = \frac{dV_{d,4}}{d\theta}.$$

As another example, the derivative $dT_{d,3}(\theta)/d\theta$ is given by

$$\frac{dT_{d,3}}{d\theta} = \frac{dV_{d,3}}{d\theta} + \frac{dV_{d,2}}{d\theta} + \frac{dV_{d,1}}{d\theta} + \frac{dV_{a,1}}{d\theta} = \frac{dV_{d,3}}{d\theta} + \frac{dV_{d2}}{d\theta} + \frac{dV_{d,1}}{d\theta}.$$

Here, unlike the case of $dT_{d,4}(\theta)/d\theta$, three successive perturbations are accumulated, since the first three customers all belong to the same busy period. Note that the presence of $(a,4)$ in the triggering sequence of $(d,4)$ has the effect of "resetting" the perturbations accumulated by the 3rd departure.

To complete the discussion of event time derivatives, we present below a general-purpose algorithm for evaluating such derivatives along an observed sample path. Let us define a "perturbation accumulator" Δ_α for every event $\alpha \in E$. Based on (14.9), an accumulator Δ_α is updated at event occurrences in two ways:

1. It is incremented by $dV_\alpha/d\theta$ whenever an event α occurs.
2. It becomes dependent on an accumulator Δ_β whenever an event β (possibly $\beta = \alpha$) occurs that activates an event α.

Note that because of the non-interruption condition, the addition of $dV_\alpha/d\theta$ to Δ_α can be implemented either at the time of the actual occurrence of α or at the time of its activation. In the algorithm below, it is assumed that the system starts out at some given state x_0. No particular stopping condition is specified, since this may vary depending on the problem of interest (e.g., stop after a desired total number of event occurrences, stop after a desired number of event type α occurrences, etc.)

Algorithm for evaluating event time derivatives

1. INITIALIZATION:

If event α is feasible at x_0: $\Delta_\alpha := dV_{\alpha,1}/d\theta$,
else, for all other $\alpha \in E$: $\Delta_\alpha := 0$

2.   WHENEVER EVENT β IS OBSERVED:

If event α is activated by β with new lifetime V_α:

2.1.   Compute $dV_\alpha/d\theta$ through (14.8)

2.2. $\Delta_\alpha := \Delta_\beta + dV_\alpha/d\theta$.

It is easy to verify that the application of this algorithm to our GI/G/1 example above yields the same expression for $dT_{d,4}/d\theta$ as the one derived from first principles. Assuming the queue is initially empty, we set $\Delta_d = 0$ (because d is not feasible at this state) and $\Delta_a = 0$ (which is always true, since interarrival times are independent of θ). Then, the accumulator Δ_d, which keeps track of $dT_{d,k}/d\theta$ for all $k = 1, 2, \ldots$, is updated in one of two ways:

(a) when d occurs, a new d event is activated as long as d remains feasible in the new state, i.e., as long as $x > 1$ (if $x = 1$, the queue becomes empty and d is no longer feasible). Thus,

$$\Delta_d := \Delta_d + dV_d/d\theta. \tag{14.10}$$

(b) when a occurs, a new d is activated only if $x = 0$, i.e., a new busy period starts. In this case,

$$\Delta_d := \Delta_a + dV_d/d\theta = dV_d/d\theta. \tag{14.11}$$

Note that (14.8) corresponds to idea (i) in Subsection 14.3.1 pertaining to the generation of perturbations due to changes in θ, whereas (14.9) corresponds to idea (ii) pertaining to propagation of perturbations. The discussion that follows, in particular equations (14.15)-(14.17) below, corresponds to idea (iii) regarding the way in which perturbations in θ are transformed into perturbations in a sample function $L(\theta)$.

We next proceed with the derivation of sample function derivatives. In many cases of interest, a sample performance function $L(\theta)$ can be expressed in terms of event times $T_{\alpha,n}$. Therefore, we may use (14.9) or, equivalently, the algorithm presented above, in order to obtain derivatives of the form $dL(\theta)/d\theta$. The sample performance functions we will consider in our analysis are all over a finite horizon. Letting $C(X(t,\theta))$ be a bounded cost associated with operating the system at state $X(t,\theta)$, we define (see also [19])

$$L_T(\theta) = \int_0^T C(X(t,\theta))dt \tag{14.12}$$

$$L_M(\theta) = \int_0^{T_M} C(X(t,\theta))dt \tag{14.13}$$

$$L_{\alpha,M}(\theta) = \int_0^{T_{\alpha,M}} C(X(t,\theta))dt. \tag{14.14}$$

Here, $L_T(\theta)$ measures the total cost over an interval of time $[0,T]$ for some given finite T. On the other hand, $L_M(\theta)$ is the total cost measured over exactly M event occurrences, while $L_{\alpha,M}(\theta)$ is the corresponding cost over exactly M occurrences of some event type α. These functions cover many (but not all) useful performance measures encountered in practice. In queueing systems, for instance,

$X(t, \theta)$ is usually the queue length, and by setting $C(X(t, \theta)) = X(t, \theta)$ in (14.12), we can estimate the *mean queue length* over $[0, T]$ as $L_T(\theta)/T$. The *mean waiting time* over M customers can also be obtained as follows. Observe that the system time of the kth customer is given by $S_k(\theta) = T_{d,k}(\theta) - T_{a,k}(\theta)$, where d is a departure event and a is an arrival event. Thus, setting $C(X(t, \theta)) = 1$ in (14.14) yields $S_k(\theta) = L_{d,k}(\theta) - L_{a,k}(\theta)$. The mean waiting time over M customers is then given by the sum of M such differences divided by M.

Under the non-interruption condition for all events in our models, and assumptions $(\textbf{A1})$ and $(\textbf{A2})$, it is not difficult to see that the derivatives of $L_T(\theta)$, $L_M(\theta)$, and $L_{\alpha, M}(\theta)$ exist (with probability 1) and are given by

$$\frac{dL_T}{d\theta} = \sum_{k=1}^{N(T)} \frac{dT_k}{d\theta}[C(X_{k-1}) - C(X_k)], \qquad (14.15)$$

$$\frac{dL_M}{d\theta} = \sum_{k=0}^{M-1} C(X_k)\left[\frac{dT_{k+1}}{d\theta} - \frac{dT_k}{d\theta}\right], \qquad (14.16)$$

$$\frac{dL_{\alpha, M}}{d\theta} = \sum_{k=0}^{N(T_{\alpha, M})-1} C(X_k)\left[\frac{dT_{k+1}}{d\theta} - \frac{dT_k}{d\theta}\right], \qquad (14.17)$$

where $N(T)$ counts the total number of events observed in $[0, T]$. Formal derivations of these expressions are given by Glasserman [19]. The crucial observation is that in DEDS's the state remains unchanged at X_k in any interval $(T_k, T_{k+1}]$. This allows us to rewrite the integrals (14.12)-(14.14) in simpler summation forms. As an example, (14.13) is rewritten as

$$L_M = \sum_{k=0}^{M-1} C(X_k)[T_{k+1} - T_k].$$

In this form, it is not difficult to see that differentiation with respect to θ yields (14.16). Detailed derivations can be found in [19].

We can now see how (14.9) may be used in conjunction with (14.15)-(14.17) to obtain sample derivatives: (14.9) allows us to evaluate the event time derivatives in the expressions above, and hence transform perturbations in event times into perturbations in sample performance functions.

Example 14.2. *IPA for the GI/G/1 queueing system (continued).* To illustrate (14.15)-(14.17), let us return to the departure time perturbations of the form $dT_{d,k}/d\theta$, which were evaluated earlier for a GI/G/1 system, where θ is the mean of the service time distribution $G_d(x; \theta)$. Let us see how to use this information to evaluate the derivative of the mean waiting time over M customers (where M is fixed). As pointed out above, setting $C(X(t, \theta)) = 1$ in (14.14), we get $S_k(\theta) = T_{d,k}(\theta) - T_{a,k}(\theta)$. Thus, the derivative $dS_k/d\theta$ can be obtained directly from the event time perturbations evaluated earlier using (14.9). In fact, since $dT_{a,k}/d\theta = 0$ for all arrivals, we have

$$dS_k/d\theta = dT_{d,k}/d\theta,$$

where $dT_{d,k}/d\theta$ is given by the value of the accumulator Δ_d in the algorithm provided above when the kth departure time occurs. Recall (14.10) where $\Delta_d := \Delta_d + dV_d/d\theta$ when d occurs and does not end a busy period. In other words, after i customer departures within a busy period, we get

$$\frac{dS_i}{d\theta} = \sum_{j=1}^{i} \frac{dV_{d,j}}{d\theta}.$$

Suppose M happens to be the number of customers in a busy period. Then, since the mean waiting time over a busy period consisting of M customers is

$$L_M(\theta) = \frac{1}{M} \sum_{i=1}^{M} S_i,$$

its derivative is obtained by combining the last two equations:

$$\frac{dL_M}{d\theta} = \frac{1}{M} \sum_{i=1}^{M} \sum_{j=1}^{i} \frac{dV_{d,j}}{d\theta}.$$

In addition, recall (14.11), where $\Delta_d := \Delta_a + dV_d/d\theta = dV_d/d\theta$ when event a occurs initiating a new busy period. This allows us to evaluate the sample derivative, $dL_M/d\theta$, over a fixed number M, of customers observed over several busy periods. Let n_b be the index of the last customer in the bth busy period, and $b = 1, 2, \ldots$, with $n_0 = 0$. In general, the Mth customer falls into some busy period whose index we denote by B. To keep notation simple, we agree that in the last, generally incomplete, busy period, n_B is the index that corresponds to the Mth customer. Then, the derivative of the mean waiting time over M customers, $dL_M/d\theta$, is given by

$$\frac{dL_M}{d\theta} = \frac{1}{M} \sum_{b=1}^{B} \sum_{i=n_{b-1}+1}^{n_b} \sum_{j=n_{b-1}+1}^{i} \frac{dV_{d,j}}{d\theta}. \qquad (14.18)$$

This expression is of special importance in the development of PA, as it represents the first formal result derived from first principles (see also [11], [35]).

14.3.4 Unbiasedness and the commuting condition

As mentioned in the conclusion of Subsection 14.3.2, IPA consists of two components. The first is the derivation of a sample function derivative $dL(\theta, \xi)/d\theta$ from the information observed on a nominal sample path under θ alone, as discussed in the last subsection. We now address the second component of IPA: Assuming we have obtained an expression for $dL(\theta, \xi)/d\theta$, under what conditions can we guarantee that it is an unbiased estimate of the performance measure derivative $dJ/d\theta$? In other words, when is the interchange of expectation and differentiation valid in (14.5)? Before getting into details, one might immediately suspect that this interchange may be prohibited when $L(\theta, \xi)$ exhibits discontinuities in θ. Such discontinuities may arise when a change in θ causes various event order changes. However, as will be discussed next, some event order changes may in fact occur without violating the continuity of $L(\theta, \xi)$. Intuitively, we can characterize the effect of a perturbation $\Delta\theta$ on a nominal sample path under θ as discussed below.

No matter how small we make $\Delta\theta$, the nominal and the perturbed sample paths, (θ, ξ) and $(\theta + \Delta\theta, \xi)$, respectively, must sooner or later differ. Equivalently, the ensemble of sample paths $(\theta + \Delta\theta, \xi)$ will differ in at least one member from the ensemble, (θ, ξ), provided these ensembles are large enough. If we limit ourselves to finite horizon problems, the frequency of occurrences of such a difference is of order $\Delta\theta$. Furthermore, if the sample path difference is also small and of order $\Delta\theta$, then the average net effect of the difference in the nominal and the perturbed performance will be of order $(\Delta\theta)^2$, which is negligible for the first derivative calculation. In other words, if the differences caused by $\Delta\theta$ between the nominal and the perturbed sample paths are small and only temporary, then IPA will give an unbiased estimate of $dJ/d\theta$.

A precise characterization of this condition was provided by Glasserman [19] and is known as the *commuting condition*:

(C) Let $x, y, z_1 \in X$ and $\alpha, \beta \in \Gamma(x)$ such that $p(z_1; x, \alpha) \cdot p(y; z_1, \beta) > 0$. Then, there exists $z_2 \in X$, such that: $p(z_2; x, \beta) = p(y; z_1, \beta)$ and $p(y; z_2, \alpha) = p(z_1; x, \alpha)$.

Moreover, for any $x, z_1, z_2 \in X$ such that $p(z_1; x, \alpha) = p(z_2; x, \alpha) > 0$, we have: $z_1 = z_2$.

In other words, (C) requires that if a sequence of events $\{\alpha, \beta\}$ takes state x to state y, then the sequence $\{\beta, \alpha\}$ must also take x to y. Moreover, this must happen in such a way that every transition triggered by α or β in this process occurs with the same probability. The last part of (C) also requires that if an event α takes place at state x, the next state is unique, unless the transition probabilities to distinct states z_1, z_2 are not equal. The commuting condition is particularly simple to visualize through the diagram of Figure 2.

Condition (C) is easy to verify by inspection of the state transition diagram of the system of interest, as the following example illustrates.

Example 14.3. *IPA for the GI/G/1 queueing system (continued).* Figure 3 depicts the transition diagram of this system. Every state other than $x = 0$ has a feasible event set $\Gamma(x) = \{a, d\}$, and $p(x+1; x, a) = p(x-1; x, d) = 1$. It is immediately obvious from the diagram that the event sequence $\{a, d\}$ applied to any $x > 0$ results in the state sequence $\{x + 1, x\}$, whereas the event sequence $\{d, a\}$ results in the state sequence $\{x - 1, x\}$. Condition (C) is satisfied, since the final state x is always the same.

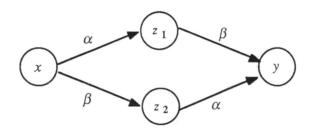

Figure 2. The commuting condition (C).

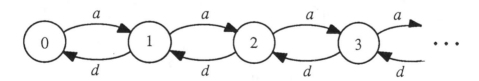

Figure 3. State transition diagram for the GI/G/1 queueing system.

Another way of making the same observation is illustrated in Figure 4. In the nominal sample path, the sequence of the d and a events results in the state sequence $\{1, 0, 1\}$. When the service times are perturbed as shown, the order of these two events changes and the resulting state sequence becomes $\{1, 2, 1\}$, which still satisfies condition (C).

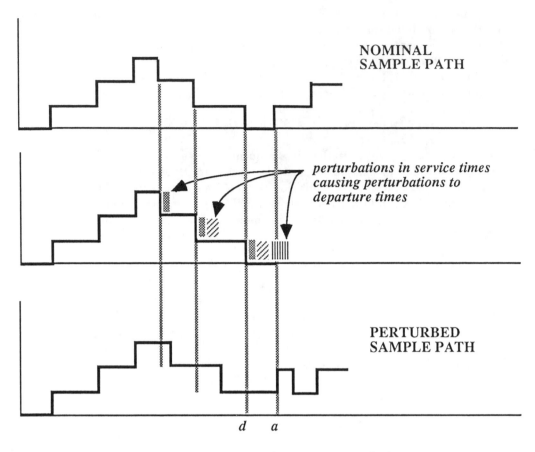

Figure 4. Nominal and perturbed sample paths of the GI/G/1 queueing system.

The commuting condition plays a crucial role in IPA because it guarantees that sample functions $L(\theta)$ of the form (14.12)-(14.14) are continuous in θ. This is stated as a theorem below (for details see Glasserman [19]).

Theorem 14.4. *Under assumptions* ($A1$)-($A2$) *and condition* (C), *the sample functions* $L_T(\theta)$, $L_M(\theta)$, *and* $L_{\alpha,M}(\theta)$ *(for finite* $T_{\alpha,M}$) *defined in* (14.12)-(14.14) *are* (*almost surely*) *continuous in* θ.

A simple illustration of this theorem is shown in Figure 5. The top part is the sample path of a GI/G/1 queueing system. Suppose that the parameter θ affects departure times only. Thus, as θ varies, the d event shown in this sample path approaches the a event shown. As it does so, the total area under the sample path changes smoothly with $\Delta \to 0$, and the striped rectangle gradually shrinks to zero. At the point where $\Delta = 0$ and the a and d events are about to change order, the total area is A as shown. Next, looking at the bottom part, we see that as $\Delta \to 0$, the striped area is gradually being added to the total area under the sample path, and when the a and d events are about to change order, the total area is once again A. Thinking of the area as the sample function of interest $L(\theta)$, we see that $L(\theta)$ is continuous in θ, *despite the event order change*. The crucial observation is that the striped area in the top part becomes zero *precisely at the instant of the event order change*. The role of the commuting condition here is in ensuring that this event order change does not affect the future state evolution of the system.

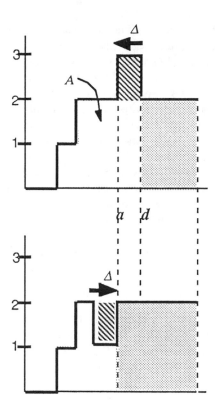

Figure 5. Illustrating the continuity of a sample function $L(\theta)$
for the GI/G/1 queueing system.

The fact that IPA derivate estimators are unbiased is formally stated in the following theorem (for details see Glasserman [19]).

Theorem 14.5. *Suppose assumptions (**A1**)-(**A2**) hold and condition (**C**) is satisfied. In addition, assume that $|dV_{\alpha,k}/d\theta| \leq B \cdot (V_{\alpha,k}+1)$ for some $B > 0$. Then, if $E[sup_\theta N(T)^2] < \infty$, for $L_T(\theta)$ defined in (14.12),*

$$\frac{d}{d\theta}E[L_T(\theta)] = E\left[\frac{dL_T(\theta)}{d\theta}\right]. \tag{14.19}$$

If $E[sup_\theta T_M] < \infty$, then for $L_M(\theta)$ defined in (14.13),

$$\frac{d}{d\theta}E[L_M(\theta)] = E\left[\frac{dL_M(\theta)}{d\theta}\right]. \tag{14.20}$$

If $E[sup_\theta(T_{\alpha,M})^2] < \infty$, and $E[sup_\theta(N(T_{\alpha,M}))^2] < \infty$, for $L_{\alpha,M}(\theta)$ defined in (14.14),

$$\frac{d}{d\theta}E[L_{\alpha,M}(\theta)] = E\left[\frac{dL_{\alpha,M}(\theta)}{d\theta}\right]. \tag{14.21}$$

The commuting condition can be extended (see [21]) and stated as a *monotonicity condition*. This requires that if the nominal and perturbed event sequences differ only in the order and not in the total number of each event type present, then the system from which these event sequences are generated must be in the same state (or states that have the same feasible event set). This is also known as the *permutability* condition.

There are several useful classes of queueing systems for which simple IPA algorithms can be used to obtain unbiased derivative estimators for sample performance functions of the form (14.12)-(14.14). These include the following:

A. *Simple queueing networks:* A simple network consists entirely of FCFS, infinite buffer, single-server nodes and a single class of customers routed through the network with a state-independent Markovian routing mechanism. A GI/G/1 queueing system and networks with general service and arrival distributions that satisfy the conditions of Theorem 14.5 are examples of simple queueing networks.

B. *Simple networks with multiclass customers,* in which every node visited by more than one class of customers is directly fed by only a single source. In such networks, a customer from some class arriving at a node cannot change order with any other customer from a different class.

C. *Networks with blocking,* in which every node with a finite buffer is directly fed by a single source. In such a network, no server can directly block two or more servers simultaneously. A cyclic queueing system with finite buffers is an example of such a network.

D. *Some networks with state-dependent routing mechanisms* satisfying the following: For any (arrival or departure) event α, let $\{X_1(\alpha),...,X_N(\alpha)\}$ be a partition of the state space such that x and x' are in the same subset $X_i(\alpha)$ if and only if the transition probabilities for the customer moving upon the occurrence of α are the same in x and x'. Then, the interchangeability is valid if the event α is the only event that can trigger a state transition between two states that belong to two different subsets $X_i(\alpha)$ and $X_j(\alpha)$, $i \neq j$.

The main virtue of IPA is its extreme computational simplicity. If IPA is used in conjunction with a simulation model, only minimal changes need to be made to enable it (see, for example, Appendix E and Ch. 3 in [35]). Moreover, increasing the dimension of a parameter vector θ adds very little to the computational burden. Numerical stability is also a virtue, since no divisions by $\Delta\theta$ are ever involved. Under certain assumptions, one can also prove that a derivative estimate obtained through IPA is the minimum variance estimate, exploiting the fact that it is entirely based on common random numbers. The low variance property of IPA estimators has also been corroborated by experimental observations.

The nonintrusive nature of IPA algorithms can also be exploited to provide derivative estimates for on-line control and optimization of queueing systems. In other words, IPA algorithms do not interfere with the normal operation of a system in real time; it simply processes data that are anyway available, such as time stamps on packets in a communication network or detecting the beginning or end of an idle period at some server. Examples where IPA is used in conjunction with optimization algorithms for queueing systems may be found in [8], [14], and [35, pp. 285-286].

14.3.5 Consistency of IPA estimators

Returning to the sample performance measures defined in (14.12)-(14.14), note that they all apply to a finite horizon, specified either through a given time instant or through a given number of events. Thus, strictly speaking, we should always explicitly write $L_N(\theta)$ when referring to a sample function based on N events.

Adopting the notation $L'_N(\theta) = dL_N/d\theta$, then $L'_N(\theta)$ is a *strongly consistent* estimator of $dJ/d\theta$ if, with probability 1,

$$\lim_{N \to \infty} L'_N(\theta) = \frac{dJ}{d\theta}, \tag{14.22}$$

where $J(\theta)$ is some performance measure of the system in the steady state, such as the stationary mean waiting time of customers in a GI/G/1 queueing system or its stationary mean queue length. It is reasonable to expect that if IPA provides an unbiased estimator for any finite N and if $L_N(\theta)$ is itself a strongly consistent estimator of $J(\theta)$, then (14.22) is also satisfied. The first formal proof of strong consistency for an IPA estimator is due to Suri and Zazanis [45] for the case of derivatives of the steady-state mean waiting time of an M/G/1 queue, in which case $dJ/d\theta$ can be analytically evaluated. In general, however, such analytical expressions are not available. Nevertheless, establishing strong consistency under certain technical conditions is possible. For details, the reader is referred to [20] and [48].

14.4 EXTENSIONS OF IPA

It is, of course, easy to make up examples for which IPA will lead to biased estimates of a performance gradient, i.e., when the interchangeability of expectation and differentiation in (14.5) is invalid. This situation typically arises when the perturbed parameters are routing probabilities or when a multiclass queueing system is considered. In such cases, *infinitesimal* parameter perturbations result in *finite* performance differences due to event order changes that cannot be ignored by virtue of the commuting (or monotonicity) condition. In other words, such perturbations in θ lead to discontinuities in the sample function $L(\theta)$. Moreover, there are many examples where discontinuities arise because a performance measure is inherently discrete; for example, if $L(\theta)$ is the number of customers served during a busy period. In such cases, direct computation of the sample derivative $dL(\theta, \xi)/d\theta$ often yields a value of zero. Yet, for any ξ, there is a value of $\Delta\theta$ that will cause a finite discontinuity in $L(\theta, \xi)$; and for any $\Delta\theta$, there exists a.s. some ξ that gives rise to a finite discontinuity in $L(\theta, \xi)$. Nevertheless, averaging these discontinuous $L(\theta, \xi)$ in ξ, gives a smooth $J(\theta) = E[L(\theta, \xi)]$ that has a well-behaved nonzero slope (see [35], p. 80). Such problems are said to be *non-IPA-applicable*, since IPA gives rise to a biased estimate of $dJ/d\theta$. To overcome this difficulty, extensions of IPA evolve in two major directions, which we will briefly describe in this section.

The first and most popular direction is based on what might be called *probability-based extensions*. These are based on the idea of "smoothing": Infrequent occurrences of finite perturbations are statistically equivalent to frequent occurrences of infinitesimal perturbations. Let us illustrate this idea through a specific example which we can use in many guises. Consider a stream of Poisson events with rate λ and note that it can be modified to represent another Poisson stream with rate $\lambda - \Delta\lambda$ in the following two equivalent ways:

(*i*) Stretch the time axis of the Poisson stream by the factor $\lambda/(\lambda - \Delta\lambda)$. This corresponds to perturbing every event time by an infinitesimal amount.

(*ii*) Delete each event with probability $\Delta\lambda/\lambda$. This corresponds to occasionally perturbing interarrival times by a finite number.

The simplest example where this idea can be applied is in the case where the parameter of interest is some routing probability. Consider the case of routing customers to one or two exponential servers with different mean service times s_1 and s_2, and let θ be the probability that a customer is routed to the first server. Let X be the actual service time assigned to a customer, which can be specified as follows:

$$X = \begin{cases} -s_1\ln(V) & \text{if } 0 \leq U \leq \theta \\ -s_2\ln(V) & \text{if } 0 < U \leq 1, \end{cases} \tag{14.23}$$

where U and V are independent random variables uniformly distributed over $[0, 1]$. Notice that under the condition $0 \leq U \leq \theta$ (respectively $\theta < U \leq 1$), U/θ (respectively $(1-U)/(1-\theta)$) is also a random variable uniformly distributed over $[0, 1]$. Therefore, instead of using (14.23) to determine the service time X, we can use a single random variable U and set

$$X = \begin{cases} -s_1\ln\left(\dfrac{U}{\theta}\right) & \text{if } 0 \leq U \leq \theta \\ -s_2\ln\left(\dfrac{1-U}{1-\theta}\right) & \text{if } \theta < U \leq 1. \end{cases} \tag{14.24}$$

It can be seen that if (14.23) is used, a finite discontinuous change can occur in X when θ is infinitesimally perturbed and U happens to be very close to θ. In other words, we can have $U > \theta$, but $U < \theta + \Delta\theta$ causes a jump in $X(\theta)$, since a customer originally routed to server 2 would now be routed to server 1. On the other hand, if we use (14.24) instead, the service time of *every* customer changes when θ is changed. The important advantage of using (14.24) is that the service time changes of customers who switch between servers, because of the perturbation in θ are no longer finite. These changes are now infinitesimal, since switching takes place only at values of U (not V) that are close to θ, i.e., when X is close to $\ln(1) = 0$. It is clear that in this case the service time changes are of order $\Delta\theta$. Thus, when switching takes place, the discontinuities will be of the same order. On the other hand, the probability that a customer in the nominal path is to be switched is also of order $\Delta\theta$. Therefore, the expected change in performance that is caused by switched customers is of order $(\Delta\theta)^2$, and this can be ignored in calculating the first derivative. This is equivalent to saying that event order changes may be ignored and that IPA will, therefore, provide unbiased estimates for performance derivatives, with respect to the routing probability θ. Another version of this idea is to convert the routing probability perturbation into an equivalent arrival rate perturbation to the two servers (see [35], Ch. 5). It is worth emphasizing that the basic idea of smoothing by *converting finite but infrequent perturbations into frequent but infinitesimal perturbations or vice versa*, is present in both of the above cases.

The basic idea illustrated in (14.23)-(14.24) can be used to develop estimators for various problems involving discontinuities in parameter θ, which are no more complex than simple IPA-based ones. For example, a multiclass queueing system with r arrival streams (one for each class) and admission control applied to each stream was analyzed by Gong et al. in [26] to obtain unbiased derivative estimators of various performance measures with respect to the r arrival rate parameters of this model.

The same idea has also been used by Brémaud and Vazquez-Abad [3] in the development of the so-called *Rare* PA (or RPA) estimators. The term "rare" is motivated by the observation that when comparing a Poisson stream of rate λ with another one of rate $\lambda - \Delta\lambda$, the latter is rarely different from the former over a finite interval. One can then view events that are present in the "λ-stream" but not in the "$(\lambda - \Delta\lambda)$-stream" as *marked* ones; conversely, from the point of view of the $(\lambda - \Delta\lambda)$-stream, these events are *phantoms*. When such a stream represents a process of customers arriving at a queueing system, one can then obtain unbiased derivative estimates of various performance measures with respect to λ by evaluating the effect of a marked (respectively phantom) customer on an observed sample path. Using this approach, Gong and Schulzrinne [27] have developed derivative estimators of performance measures with respect to routing probabilities under rather general conditions. This idea, however, can also be extended to various discrete resources in a queueing system, which one can similarly treat as "marked" (those that would be absent if a parameter were perturbed) or "phantom" (those that would become available if a parameter were perturbed). Cassandras and Julka [12] have studied a class of scheduling problems through this approach by treating "time slots" to be allocated to different customer classes as the resources that can be "marked" or "phantomized".

In order to generalize the idea of smoothing, Gong and Ho [25] developed an approach leading to the so-called *Smoothed* PA (SPA). Returning to the interchangeability issue in (14.5), we first decompose the expectation over the sample derivative into two parts: a conditional expectation and an expectation over the conditioning variables, i.e.

$$\frac{d}{d\theta}E[L(\theta,\xi)] = \frac{d}{d\theta}E[E[L(\theta,\xi) \mid Z]] = E\left[\frac{dE[L(\theta,\xi) \mid Z]}{d\theta}\right]. \qquad (14.25)$$

We can then expect $E[L(\theta,\xi) \mid Z]$ to be smoother than $L(\theta,\xi)$; hence, the interchange of E and $d/d\theta$ is valid. Note that even though $L(\theta,\xi)$ may be discontinuous, "sufficient averaging" can render $E[L(\theta,\xi)]$ differentiable. In fact, the usual finite-difference approach of computing sensitivities via

$$\frac{dL}{d\theta} = \lim_{\substack{n\to\infty \\ \Delta\theta\to 0}} \frac{1}{n\Delta\theta}\left[\sum_{i=1}^{n} L(\theta + \Delta\theta, \xi_i) - \sum_{i=1}^{n} L(\theta, \xi_i)\right]$$

is a simple statement of the above observation. Now, the trick in (14.25) is to average just enough to avoid discontinuities, but not too much so as to require the generation of an entire additional sample path, as in the "brute-force" finite-difference approach. Between the extremes of differentiating the expectation and taking the expectation of the derivative, there exists a whole spectrum of partial expectation and smoothing possibilities. In fact, Glasserman and Gong [22] have shown explicitly that, for a large class of problems with inherent discontinuities in performance, SPA can yield unbiased derivative estimates by converting the differentiation of $L(\theta,\xi)$ to a differentiation of the probability of occurrence of such discontinuities via appropriate conditioning. Further extensions of this idea are given in [16]. In addition, SPA can be used to extend PA techniques to the estimation of *second-order derivatives*, where event order changes, even if those satisfying the commuting condition, generally cause effects which cannot be ignored if unbiased estimates are to be obtained. Bao and Cassandras [1] have recently used SPA to derive estimators for the second derivatives of throughput, and mean delays with respect to mean service times in closed serial (tandem) and simple queueing networks (refer back to Subsection 14.3.4 for the definition of a simple network).

Taking a closer look at (14.25), an obvious issue that arises is that of deciding what to use for the conditioning variable "Z". In some cases, the choice of Z is rather obvious and the resulting SPA estimators are only marginally more complicated than simple IPA ones, as in the problems studied by Wardi et al. in [49, 50]. More generally, however, both earlier and recent developments in [3, 4, 15, 44] suggest a natural decomposition and conditioning of the difference $L(\theta + \Delta\theta, \xi) - L(\theta, \xi)$ due to the fact that each event in the nominal path has a positive probability of being deleted due to a perturbation in θ (recall again the example given earlier regarding the two equivalent ways of generating a Poisson stream with rate $\lambda - \Delta\lambda$ from one with rate λ). Specifically,

$$\frac{dJ}{d\theta} = \lim_{\Delta\theta \to 0} \frac{1}{\Delta\theta} E \left[L(\theta + \Delta\theta, \xi) - L(\theta, \xi) \right]$$

$$= \sum_{\text{all } i} \lim_{\Delta\theta \to 0} \frac{1}{\Delta\theta} E[L_{-i}(\theta, \Delta\theta, \xi) - L(\theta, \xi)], \qquad (14.26)$$

where $L_{-i}(\theta, \Delta\theta, \xi)$ is the sample performance with the ith event deleted, as a result of $\Delta\theta$, and the summation is over all possible i that could be deleted due to this perturbation. Next, we evaluate the conditional expectation above as follows:

$$\sum_{\text{all } i} \lim_{\Delta\theta \to 0} \frac{1}{\Delta\theta} E[L_{-i}(\theta, \Delta\theta, \xi) - L(\theta, \xi)]$$

$$= \sum_{\text{all } i} \lim_{\Delta\theta \to 0} \frac{1}{\Delta\theta} [L_{-i}(\theta, \Delta\theta, \xi) - L(\theta, \xi)] \frac{dP_{-i}(\theta, \Delta\theta)}{d\theta} \Delta\theta$$

$$= \sum_{\text{all } i} [L_{-i}(\theta, \Delta\theta, \xi) - L(\theta, \xi)] \frac{dP_{-i}(\theta, \Delta\theta)}{d\theta}, \qquad (14.27)$$

where $[dP_{-i}(\theta, \Delta\theta)/d\theta]\Delta\theta$ is the probability that the ith event will be deleted due to the perturbation $\Delta\theta$. Note, once again, that we avoid the problem of dividing through by $\Delta\theta$ by being able to differentiate the probability $P_{-i}(\theta, \Delta\theta)$. Moreover, $P_{-i}(\theta, \Delta\theta)$ is usually directly related to the given elementary random variable distributions. One can view (14.27) as a form of an "improved" likelihood ratio method, where the usual difficulty of the variance increase due to differentiating the sample path distribution is avoided, and the term $L_{-i}(\theta, \Delta\theta, \xi)$ is used as a control variate to further minimize the variance [44]. We should point out, however, that (14.27) merely represents a form of "intelligently" generating two sample paths, one for $L(\theta, \xi)$ and the other for $L_{-i}(\theta, \Delta\theta, \xi)$, by using as much information as possible from the nominal path. In general, the term $L_{-i}(\theta, \Delta\theta, \xi) - L(\theta, \xi)$ may or may not be easy to compute. In the latter case, one can seek to exploit structural properties of the underlying system in order to maximize the efficiency of such computations. For example, the presence of regenerative cycles in the system can help in the sense of reducing the summation over i in (14.27) by restricting attention to a single such cycle. Similarly, some Markovian structures can often simplify computations involved. Finally, alternatives to the difference $L_{-i}(\theta, \Delta\theta, \xi) - L(\theta, \xi)$, such as $L_{+i}(\theta, \Delta\theta, \xi) - L(\theta, \xi)$ and $L_{+i}(\theta, \Delta\theta, \xi) - L_{-i}(\theta, \Delta\theta, \xi)$ (with the obvious notational interpretations), provide additional computational flexibility (see [44]). More recently, Dai and Ho [15] and Brémaud and Gong [4] have shown other alternatives, such as perturbing the consequence of an event (the next state), instead of deleting the event.

The second direction in extending IPA is *calculus-based*. To elaborate, we start by considering an ensemble of sample paths and event sequences of the form $\sigma = \{e_1, e_2, \ldots, e_i, \ldots\}$, where e_i is the ith event of a sample path. Letting $X(t; \theta, \xi)$ denote the state sequence, we can write

$$J(\theta) = \sum_{\sigma} E[L(X(t; \theta, \xi)) \mid \sigma]$$

and

$$\frac{dJ(\theta)}{d\theta} = \sum_{\sigma} \frac{dE[L(X(t;\theta,\xi)) \mid \sigma]}{d\theta}, \tag{14.28}$$

where the conditional expectation is taken over all sample paths that are "deterministically similar" to the event sequence σ, in the sense that all event sequences are identical (event times will differ, however). The boundary of the integration (conditional expectation) is a hyper cube $R(\sigma)$ in the multidimensional event sequence space with edges defined by the upper and lower limit on the timing of the events before they change order with their neighboring events. Thus,

$$\frac{dE[L(X(t;\theta,\xi)) \mid \sigma]}{d\theta} = \frac{d}{d\theta} \int_{R(\sigma)} L(X(t;\theta,\xi)) dF_{X \mid \sigma},$$

and by elementary calculus,

$$\frac{d}{d\theta} \int_{R(\sigma)} L(X(t;\theta,\xi)) dF_{X \mid \sigma} = \int_{R(\sigma)} \frac{dL(X(t;\theta,\xi))}{d\theta} dF_{X \mid \sigma} + \frac{dR}{d\theta}. \tag{14.29}$$

The first term on the right-hand side of (14.29) is simply the usual IPA term, which can be calculated by simple IPA rules. The interchange of integration and differentiation is valid by the definition of deterministic similarity above. The second term is the correction term that must be added. Gaivoronski et al. [18] provide many examples showing the validity of (14.29) and the solution of otherwise non-IPA-applicable problems. The term $dR/d\theta$ can be thought of as a bias correction term and it may or may not be easy to calculate. However, (14.29) offers a valuable insight; for example, consider the Lindley equation for the system time T_n in a GI/G/1 queueing system, where n indexes arriving customers. Letting A_n and S_n denote the nth interarrival and service times, respectively, we have

$$T_{n+1} = \begin{cases} T_n + S_{n+1} - A_{n+1} & \text{if } T_n - A_{n+1} \geq 0 \\ S_{n+1} & \text{otherwise} \end{cases}$$

and, whenever IPA can be used as an unbiased derivative estimator of the mean system time,

$$\frac{dT_{n+1}}{d\theta} = \begin{cases} \dfrac{dT_n}{d\theta} + \dfrac{dS_{n+1}}{d\theta} & \text{if } T_n - A_{n+1} \geq 0 \\ \dfrac{dS_{n+1}}{d\theta} & \text{otherwise,} \end{cases} \tag{30}$$

where θ is the mean service time. However, in order to prove (14.30) we must essentially prove that perturbations in the boundary condition $T_n - A_{n+1} = 0$ will not adversely affect (i.e., to first order) the validity of (14.30). Returning to our discussion of the commuting condition and Figure 4, this fact is a consequence of Theorem 14.4 in Subsection 14.3.4.

14.5 FINITE PERTURBATION ANALYSIS (FPA) AND THE SAMPLE PATH CONSTRUCTABILITY PROBLEM

The PA techniques discussed in previous sections are aimed directly at obtaining *gradient* estimates from observed sample path information. However, it is fruitful to view PA as a mind set, which, in more general terms, is concerned with the efficient exploration of a performance response surface $J(\theta)$ through multiple sample paths generated under different values of θ. Ideally, we would like to be able to generate all such sample paths using information extracted from a single sample path under a given value of θ.

Finite PA (FPA) takes the viewpoint that simulation is a mapping of a system parameter vector θ and a sequence ξ of i.i.d. random numbers to the sample performance function $L(\theta, \xi)$. Then, $L(\theta + \Delta\theta, \xi)$ is merely a slightly different mapping. In particular, in view of the framework presented in Section 14.2, we note that the clock structure, which is based on ξ, need not be duplicated. In principle, generating a sample path $(\theta + \Delta\theta, \xi)$ involves selecting and assembling different pieces of the same clock structure according to the state transition mechanism defined for the system (see [32, 33]).

One can go further, however, by attempting to extend this idea to *discrete* system parameters, which may affect not only the event lifetime distributions in a model, but also the actual state transition structure itself. Examples include queue capacities, customer population sizes in closed network models, batch sizes in bulk service models, and threshold parameters in various routing or scheduling policies encountered in practice. This motivation has led to the *Augmented System Analysis* (ASA) approach, first proposed by Cassandras and Strickland [6] for Markov chains, and later extended to include at most one non-Markovian event process [7].

This discussion leads to a general problem referred to as "sample path constructability". Returning to the framework of Section 14.2, let $(E, X, \Gamma, f, x_0, G)$ be a DEDS and $\Theta = \{\theta_1, \ldots, \theta_m\}$ be a finite discrete parameter set. Suppose that some value of the parameter (for simplicity, scalar) $\theta = \theta_1$ is fixed, and let (θ_1, ξ) be a sample path under θ_1. In particular, this sample path gives rise to a timed event sequence $\{e_k^1, t_k^1\}, k = 1, 2, \ldots$. Then, assuming that all events e_k^1 and event times t_k^1 for $k = 1, 2, \ldots$ are directly observable, the problem is to construct a sample path $\{e_k^j, t_k^j\}, k = 1, 2, \ldots$ for any θ_j, $j = 2, \ldots, m$, as shown in Figure 6. We refer to this as the constructability problem (see [7, 11, 13]). Ideally, we would like this construction to take place on line, i.e., as the observed sample path evolves. Moreover, we would like the construction of all $m - 1$ sample paths for $j = 2, \ldots, m$ to be carried out in parallel.

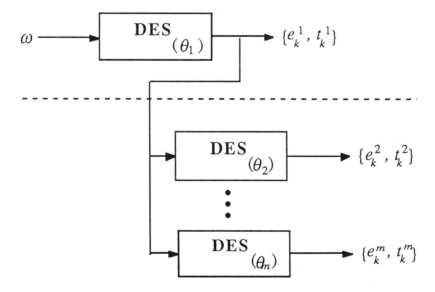

Figure 6. The sample path constructability problem.

A constructability condition for the problem above was given by Cassandras and Strickland [7]. Details on this condition and approaches for solving the constructability problem form the basis of a new area that has emerged in recent years from the FPA branch of PA (see Ch. 9 of [11]). The main idea is that generation of sample paths of *structurally similar but parametrically different* DEDS's can be efficiently carried out *in parallel* by minimizing duplications such as the repeated generation of variates for each sample path. In the case of simulation, massively parallel SIMD (Single-Instruction-Multiple-Data) computers are ideally suited to carry out such parallel simulations; in Figure 6, one can visualize m parallel processors, one for each sample path construction under a different value of θ. Since experiments are distributed (or parallelized), this approach avoids some serious synchronization problems encountered when one must parallelize the simulation procedure itself [17, 41]. Moreover, one can expect that scalability (in the number of sample paths to be constructed) should approach the theoretical maximum. In the case of on-line applications, it is still possible to use the framework of Figure 1 and develop optimization algorithms over discrete parameter sets, as recently shown in [13].

Although the constructability condition mentioned above is generally constraining, it is still possible to accomplish the task of parallel sample path generation. This comes at some cost, and the resulting tradeoffs have opened up some challenging new research directions well beyond the scope of this chapter. To date, there are two main approaches that have been proposed for this purpose: one is the ASA approach mentioned earlier, and the other is the *Standard Clock* (SC) approach, proposed by Vakili [47]. The latter is a radical departure from traditional simulation methodologies; for details, the reader is referred to [23, 9, 38] and Ch. 9 of [11].

Finally, it is worth mentioning that these recent developments have also fostered efforts to tackle large-scale, often combinatorially complex, optimization problems. We limit ourselves here to the mention of an idea proposed by Ho et al. [37], based on performing *ordinal* optimization before *cardinal* optimization. The main virtue of ordinal optimization is as follows: The relative order of system performance values as a function of parameter values is insensitive to errors in performance estimation. Given this fact, instead of trying to successively improve upon a system design sequentially (e.g., by the traditional way of hill climbing/steepest descent), and expending a large effort to insure accurate estimation of these intermediate results (which will be eventually discarded), an alternative is to simultaneously and approximately evaluate many designs in concert with simultaneous simulation experiments as discussed above (see also Figure 6).

14.6 CONCLUSIONS

The area of PA can be viewed narrowly as a methodology for estimating performance gradients with respect to parameters of discrete event systems from information extracted from a single observed sample path. More broadly, however, PA may be viewed as a mind set leading to a spectrum of new techniques for estimating global performance response surfaces for both continuous and discrete parameters. In the former case, IPA is a technique for evaluating derivatives of sample functions, based on observed sample path data, and using them as estimates of performance derivatives. The classes of systems and performance measures for which

IPA provides unbiased estimates were discussed in Subsection 14.3.4. Beyond these classes, it is still possible to use PA techniques at the expense of more sample path data and more computation. In recent years, a whole range of such techniques has emerged; for the most up-to-date taxonomy of these techniques, see [34] and p. 714 of [11].

From a broader perspective, PA seeks to generate multiple sample paths of a system under different (discrete or continuous) parameter settings from a single observed sample path. From this perspective, a whole range of new conceptual and analytical problems in simulation and modeling is awaiting exploration and solution. This promises to be an exciting prospect when one takes into account the dynamics of DEDS's, the impact of new hardware technology, and a new mind set.

BIBLIOGRAPHY

[1]   Bao, G. and Cassandras, C.G., First and second derivative estimators for closed Jackson-like queueing networks using perturbation analysis techniques, *Proc. of 32nd IEEE Conf. Decision and Control* (1993), 698-703.

[2]   Bello, M., *The Estimation of Delay Gradient for Purpose of Routing in Data Communication Networks*, S.M. Thesis, Elect. Eng. Dept., M.I.T. 1977.

[3]   Brémaud, P. and Vazquez-Abad, F.J., On the pathwise computation of derivatives with respect to the rate of a point process: the phantom RPA method, *Queueing Sys.* **10** (1992), 249-270.

[4]   Brémaud, P. and Gong, W.B., Derivatives of likelihood ratios and smoothed perturbation analysis for the routing problem, *ACM Trans. on Modeling and Simulation* (1991), (to appear).

[5]   Cao, X.R., Convergence of parameter sensitivity estimates in a stochastic experiment, *IEEE Trans. Autom. Contr.* **AC-30** (1985), 834-843.

[6]   Cassandras, C.G. and Strickland, S.G., On-line sensitivity analysis of Markov chains, *IEEE Trans. on Autom. Control* **AC-34**:1 (1989), 76-86.

[7]   Cassandras, C.G. and Strickland, S.G., Observable augmented systems for sensitivity analysis of Markov and semi-Markov processes, *IEEE Trans. on Autom. Control* **AC-34**:10 (1989), 1026-1037.

[8]   Cassandras, C.G., Abidi, M.V. and Towsley, D., Distributed routing with on-line marginal delay estimation, *IEEE Trans. on Commun.* **COM-38**:3 (1990), 348-359.

[9]   Cassandras, C.G., Lee, J.I. and Ho, Y.C., Efficient parametric analysis of performance measures for communications networks, *IEEE J. on Selected Areas in Commun.* **8**:9 (1990), 1709-1722.

[10]   Cassandras, C.G., Gong, W.-B. and Lee, J.I., On the robustness of perturbation analysis estimators for queueing systems with unknown distributions, *J. Optim. Theory Appl.* **30**:3 (1991), 491-519.

[11]   Cassandras, C.G., *Discrete Event Systems: Modeling and Performance Analysis*, Irwin Publ. 1993.

[12]   Cassandras, C.G. and Julka, V., Scheduling policies using marked/phantom slot algorithms, *Queueing Systems: Theory and Appl.* (1993), (submitted).

[13]   Cassandras, C.G. and Pan, J., Parallel sample path generation for discrete event systems and the traffic smoothing problem, *J. Disc. Event Dyn. Sys.* (1994), (submitted).

[14]   Chong, E. and Ramadge, P., Convergence of recursive optimization algorithms using infinitesimal perturbation analysis estimates, *J. of Disc. Event Dyn. Systems* 1:4 (1992), 339-372.

[15]   Dai, L.-Y. and Ho, Y.C., Structural infinitesimal perturbation analysis (SIPA) for derivative estimation in discrete event dynamic systems, *IEEE Trans. Autom. Contr.* (1994), (to appear).

[16]   Fu, M. and Hu, J.Q., Extensions and generalizations of smoothed perturbation analysis in a generalized semi-Markov process framework, *IEEE Trans. on Autom. Control* **AC-37**:10 (1992), 1483-1500.

[17]   Fujimoto, R.M., Parallel discrete event simulation, *Comm. of ACM* **33**:10 (1990), 31-53.

[18]   Gaivoronski, A., Shi, L. and Sreenivas, R., Augmented infinitesimal perturbation analysis: An alternate explanation, *J. Disc. Event Dyn. Sys.* 2:2 (1992), 99-120.

[19]   Glasserman, P., *Gradient Estimation via Perturbation Analysis*, Kluwer, New York 1990.

[20]   Glasserman, P., Hu, J.Q. and Strickland, S.G., Strong consistency of steady state derivative estimations, *Prob. Eng. Info. Sci.* **5** (1991), 391-413.

[21]   Glasserman, P. and Yao, D., Algebraic structure of some stochastic discrete event systems with applications, *J. Disc. Event Dyn. Sys.* 1:1 (1991), 7-36.

[22]   Glasserman, P. and Gong, W.-B., Smoothed perturbation analysis for a class of discrete event systems, *IEEE Trans. on Autom. Control* **AC-35**:11 (1991), 1218-1230.

[23]   Glasserman, P. and Vakili, P., Comparing Markov chains simulated in parallel, *Prob. in the Engin. and Info. Sciences* (1994), (to appear).

[24]   Glynn, P., Likelihood ratio gradient estimation: An overview, *Proc. of the 1987 Winter Simulation Conf.* (1987), 366-375.

[25]   Gong, W.-B. and Ho, Y.C., Smoothed perturbation analysis of discrete event dynamic systems, *IEEE Trans. Autom. Contr.* **AC-32**:10 (1987), 858-866.

[26]   Gong, W.-B., Cassandras, C.G. and Pan, J., Perturbation analysis of a multiclass queueing system with admission control, *IEEE Trans. Autom. Contr.* **AC-36**:6 (1991), 707-723.

[27]   Gong, W.-B. and Schulzrinne, H., Application of smoothed perturbation analysis to probabilistic routing, *Math. Comp. Simulation* **34** (1992), 467-485.

[28]   Ho, Y.C., Eyler, A. and Chien, T.T., A gradient technique for general buffer storage design in a serial production line, *Int. J. Prod. Research* **17**:6 (1979), 467-485.

[29]   Ho, Y.C. and Cassandras, C.G., A new approach to the analysis of discrete event dynamic systems, *Automatica* **19**:2 (1983), 149-167.

[30]   Ho, Y.C., Cao, X.R. and Cassandras, C.G., Infinitesimal and finite perturbation analysis of discrete event dynamic systems, *Automatica* **19**:4 (1983), 439-445.

[31]   Ho, Y.C. and Cao, X.R., Perturbation analysis and optimization of queueing networks, *J. Optim. Theory and Appl.* **40**:4 (1983), 559-582.

[32]   Ho, Y.C. and Li, S., Extensions of the perturbation analysis techniques for discrete event dynamic systems, *IEEE Tran. Autom. Cont.* **AC-33**:5 (1988), 427-438.

[33]   Ho, Y.C., Li, S. and Vakili, P., On the efficient generation of discrete event

sample paths under different parameter values, *Math. and Comp. in Simulation* **30** (1988), 347-370.

[34] Ho, Y.C. and Strickland, S.C., A taxonomy of PA techniques, In: *Introduction to Discrete Event Dynamic Systems*, IEEE Press 1991.

[35] Ho, Y.C. and Cao, X.R., *Perturbation Analysis of Discrete Event Dynamic Systems*, Kluwer Academic Publ. 1991.

[36] Ho, Y.C., Perturbation analysis: Concepts and algorithms, *Proc. of 1992 Winter Simulation Conference.*

[37] Ho, Y.C., Sreenivas, R. and Vakili, P., Ordinal optimization of discrete event dynamic systems, *J. Disc. Event Dyn. Sys.* **2**:2 (1992), 61-88.

[38] Ho, Y.C., Cassandras, C.G. and Makhlouf, M., Parallel simulation of real-time systems via the standard clock approach, *J. Math. Comp. Simulation* **35** (1993), 33-41.

[39] Law, A.M. and Kelton, W.E., *Simulation Modeling and Analysis*, McGraw-Hill, New York 1991.

[40] Mohanty, B.P. and Cassandras, C.G., The effect of model uncertainty on some optimal routing problems, *J. of Optim. Theory and Appl.* **77**:2 (1993), 257-290.

[41] Rego, V.J. and Sunderam, V.S., Experiments in concurrent stochastic simulation: The EcliPSE paradigm, *J. Paral. Dist. Comput.* **14** (19920, 66-84.

[42] Reimann, M.I. and Weiss, A., Sensitivity analysis via likelihood ratio, *Proc. of the 1986 Winter Simulation Conf.* (1986), 285-289.

[43] Rubinstein, R., *Monte Carlo Optimization, Simulation and Sensitivity of Queueing Networks*, Wiley, New York 1986.

[44] Shi, L., Discontinuous perturbation analysis, *IEEE Trans. Autom. Contr.* (1992), (submitted).

[45] Suri, R. and Zazanis, M., Perturbation analysis gives strongly consistent sensitivity estimates for the M/G/1 queue, *Mgmt. Science* **34**:1 (1988), 39-64.

[46] Suri, R. and Leung, Y.T., Single run optimization of discrete event simulations - An empirical study using the M/M/1 queue, *IIE Trans.* **21** (1989), 35-49.

[47] Vakili, P., A standard clock technique for efficient simulation, *Opn. Res. Letters* **10** (1991), 445-452.

[48] Wardi, Y. and Hu, J.Q., Strong consistency of infinitesimal perturbation analysis for tandem queueing networks, *J. Disc. Event Dyn. Sys.* **1**:1 (1991), 37-59.

[49] Wardi, Y., Gong, W.-B., Cassandras, C.G. and Kallmes, M.H., Smoothed perturbation analysis for a class of piecewise constant sample performance functions, *J. Disc. Event Dyn. Sys.* **1**:4 (1992), 393-414.

[50] Wardi, Y., Kallmes, M.H., Cassandras, C.G. and Gong, W.-B., Perturbation analysis algorithms for estimating the derivatives of occupancy-related functions in serial queueing networks, *Ann. Opns. Res.* **39** (1992), 269-293.

[51] Woodside, C.M., Response time sensitivity measurement for computer systems and general closed queueing networks, *J. Perf. Eval.* **4** (1984), 199-210.

Chapter 15

Polynomial time algorithms for estimation of rare events in queueing models[1]

Vladimir Kriman and Reuven Y. Rubinstein

ABSTRACT This chapter presents a framework for analyzing time complexity of rare events estimators for queueing models. In particular, it deals with polynomial and exponential time *switching regenerative* (SR) estimators for the steady state probabilities of excessive backlog in the GI/GI/1 queue and some of its extensions. The SR estimators are based on *large deviation theory* and *exponential change of measure*, which is parametrized by a scalar t. We show how to find the optimal value w of parameter t, which leads to the *optimal exponential change of measure* (OECM), and we find conditions under which the OECM generates *polynomial time* estimators. Finally, we investigate the "robustness" of the proposed SR estimators, in the sense that we find how much one can perturb the optimal value w in the OECM such that the SR estimator still leads to a dramatic variance reduction and is still useful in practice. Our numerical results suggest that if optimal parameter value w is perturbed up to 20%, we only lose 2-3 orders of magnitude of variance reduction compared to the orders of tenth under the optimal value w.

CONTENTS

15.1 INTRODUCTION

Estimation of rare events is important for many modern applied systems, in particular, for asynchronous transfer mode multiplexers in broadband integrated switch-

---

[1]Work supported by the Technion V.P.R. Fundent charitable trust-nonmilitary research fund and the Center for Economic Research at Tilburg University.

---

ing digital network (see [7]). It is well known that under the original probability measures, (crude Monte Carlo) estimation of rare events is very time consuming and, therefore, extremely costly. Instead, a method based on changing the underlying distribution, called *importance sampling* (IS), to speed up the simulation is typically used [29]. In the past decade, IS has been applied to a variety of problems arising in the analysis of rare events in queueing systems (see e.g., [1, 2-14, 16-18, 19, 20, 26, 27, 30-36, and 38]). The main idea of the IS approach is to make the new probability measure. Then, in order to obtain an unbiased estimator of the desired rare event, the simulated events are weighted by using the likelihood ratio.

Note that most papers on rare events utilize large deviations (LD) theory and exponential change of measure (ECM) as the main mathematical tools (see Section 15.9, Appendix A for the definition of ECM). LD techniques give the optimal change of measure (in the sense of minimal variance) *within the class of all feasible ECM*. We call it the *optimal exponential change of measure* (OECM) (see e.g. [36] and [11]). In this chapter we raise the following questions: *What will be the computational cost under this 'optimal' distribution? Does expanding the class of admissible simulation distributions produce any more asymptotically efficient solutions?* (See, e.g. [30]). To answer these questions we propose a framework for simulation speedup analysis, which is based on the theory of complexity. In particular, we deal with polynomial and exponential time of the so-called *switching regenerative* (SR) estimators for evaluating the steady state probabilities of excessive backlog in the GI/GI/1 queue and some of its extensions. The idea is to use initially, in each regenerative cycle, a more congested simulation distribution that will lead to quick occurrence of rare events, and then change the simulation distribution, so that the system will be driven back to the regeneration state. Empirical studies of this technique are given, for example, in [12, 20, 26, 29]. For parallel work on complexity of rare events estimators, see Asmussen and Rubinstein [6].

Identification of the OECM for complex queueing models reduces, at best, to a difficult numerical problem (see, e.g. [27] and [34]). Hence, typically one obtains a computational error (exceptions are some cases where we have explicit solutions; see, e.g. [16] and [17]). In particular, we consider the following problem: *Given that the simulation distribution generates an exponential time IS algorithm, how high will the computational cost T be?* We show that $T = \exp(1/\epsilon z + o(1/\epsilon))$, where $\epsilon \equiv 1/x$ is a small parameter, while we obtain an analytical expression for the exponential rate z for the switching regenerative estimator. Note that our results agree with [35], in which a rather general problem of estimating rare event is considered.

While we deal here with the steady state behavior, we would like to mention a related transient analysis by Sadowsky [30]. He analyzes transient behavior in the GI/GI/m queue and shows that within the class of all possible importance sampling distributions, the OECM has the following *strong asymptotic optimality property*: as the backlog $z \to \infty$, the computational cost (simulation time required to obtain an estimator of prescribed accuracy) T_x grows *less than exponentially fast* (we show that, in fact, it is *polynomial*) in x, and all other GI/GI/m simulation distributions incur a computational cost that grows at a *strictly positive exponential rate*! There are a few parallel works (see, e.g., [9] and [23]), but they do not directly deal with queueing systems.

The rest of the chapter is organized as follows. In Section 15.2 we present a framework for complexity analysis of Monte Carlo estimators. Sections 3 and 4 deal with switching regenerative estimators and their properties. Section 15.5 investigates the "robustness" of the SR estimator, in the sense that we find how much one can perturb the optimal value in the OECM such that the SR estimator still leads to dramatic variance reduction and might be useful in practice. Section 15.6 presents a number of extensions. In particular, it considers deterministic service times, batch arrivals, and stationary Markovian interarrival times. We also show that similar complexity results hold for sensitivities (gradients) of probabilities of rare events with respect to parameters of the interarrival and service time distributions. In Section 7 we give supporting numerical examples, while in Section 15.8 we give a brief discussion for future research and concluding remarks. The definition of ECM and proofs of the main results are included in the Appendix.

15.2 FRAMEWORK FOR COMPLEXITY ANALYSIS VIA IMPORTANCE SAMPLING

Consider the expected performance

$$\alpha = \mathbf{P}(\mathbf{Y} \in A) = \int_{-\infty}^{\infty} I_A(\mathbf{y})f(\mathbf{y})dy = \mathbb{E}_f[I_A(\mathbf{Y})], \tag{15.1}$$

where the subscript f means that the expectation is taken with respect to the probability density $f(\cdot)$, and $I_A(\mathbf{y})$ is the indicator function of the set A, i.e., $I_A(\mathbf{y}) = 1$ if $\mathbf{y} \in A$ and $I_A(\mathbf{y}) = 0$ otherwise.

Let $g(\mathbf{y})$ be a probability density function. Assume that $g(\mathbf{y})$ dominates $f(\mathbf{y})$ in the absolutely continuous sense, that is, $\text{supp}\{f(\mathbf{y})\} \subset \text{supp}\{g(\mathbf{y})\}$. Using density $g(\mathbf{y})$ we can represent α in (15.1) as

$$\alpha = \int I_A\left(\mathbf{y}\frac{f(\mathbf{y})}{g(\mathbf{y})}g(\mathbf{y})\right)dy = \mathbb{E}_g[I_A(\mathbf{Y})W(\mathbf{Y})], \tag{15.2}$$

where $W(\mathbf{y}) = f(\mathbf{y})/g(\mathbf{y})$ is called the *likelihood ratio* (LR).

An unbiased estimator of α is given by

$$\bar{\alpha}_N(g) = \frac{1}{N}\sum_{n=1}^{N} I_n W_n(\mathbf{Y}_n), \tag{15.3}$$

that is, α can be estimated by taking a random sample $\mathbf{Y}_1, \ldots, \mathbf{Y}_N$ from a density $g(\cdot)$, and then the output $(I_n \equiv I_A(\mathbf{Y}_n))$ is made unbiased if we multiply it by the likelihood ratio $W_n = W_n(\mathbf{Y}_n)$. Sampling from a different density is called a *change of measure*; the density g is called the *importance sampling density* (or if $g(\mathbf{y}) = dG(\mathbf{y})/dy$, G is called the *importance sampling distribution*); and finally, $\bar{\alpha}_N(g)$ is called the *IS estimator* (and the associated Monte Carlo algorithm is called the *IS algorithm*). In the particular case where there is no change of measure ($g = f$), we have $W = 1$, and the IS estimator reduces to the so-called *crude Monte Carlo (CMC) estimator*:

$$\bar{\alpha}_N(f) = \frac{1}{N}\sum_{n=1}^{N} I_n, $$

where $\mathbf{Y}_1, \ldots, \mathbf{Y}_N$ is a random sample from density $f(\cdot)$.

Since essentially any dominating density g can be used for sampling, a natural problem is to find the optimal one, i.e., the density that minimizes the variance of $\bar{\alpha}_N(g)$:

$$\min_g \text{Var}(\bar{\alpha}_N(g)). \tag{15.4}$$

Selecting $g(\mathbf{y} \equiv g^*(\mathbf{y}) = f(\mathbf{y})/\alpha$ for $\mathbf{y} \in A$ and $g^*(\mathbf{y}) = 0$ otherwise, results in $Z_n = I_n f(\mathbf{Y}_n)/g^*(\mathbf{Y}_n) = \alpha$, with probability one. Since the variance of a constant is zero, $g^*(\mathbf{y})$ is the optimal change of measure (and one sample from g^* gives exactly α). The optimal change of measure thus has the interpretation of being simply the original distribution conditioned on the rare event having occurred. Unfortunately, there are several difficulties with this optimal density g^*. First, it depends explicitly on α, the unknown quantity that we are trying to estimate. If, in fact, α were known, there would be no need to run the simulation experiment at all. Second, even if α or its approximation were known, it might be impractical to sample efficiently from g^*, since typically it cannot be specified in a closed form.

Assume below that $g(\mathbf{y}) = f(\mathbf{y}, \mathbf{v}_0)$, that is, $g(\mathbf{y})$ comes from the same parametric family of distributions as $f(\mathbf{y}) = f(\mathbf{y}, \mathbf{v})$. The parameter vector $\mathbf{v}_0$ is called the *reference* parameter. In this case, the likelihood ratio W in (15.2) reduces to

$$W(\mathbf{y}, \mathbf{v}_0) = \frac{f(\mathbf{y}, \mathbf{v})}{f(\mathbf{y}, \mathbf{v}_0)},$$

and instead of the problem (15.4), we can consider the *simpler* problem

$$\min_{\mathbf{v}_0} \mathrm{Var}_{\mathbf{v}_0}\{I_A(\mathbf{Y})W(\mathbf{Y}, \mathbf{v}_0)\}, \qquad (15.5)$$

which is the same as

$$\min_{\mathbf{v}_0} \mathcal{L}(\mathbf{v}_0) = \min_{\mathbf{v}_0} \mathbb{E}_{\mathbf{v}_0}\{I_A(\mathbf{Y})W(\mathbf{Y}_i, \mathbf{v}_0)^2\}. \qquad (15.6)$$

The optimal solution of this problem, say $\mathbf{v}_0^*$, is typically not available analytically. We can, however, estimate $\mathbf{v}_0^*$ by minimizing the so-called *stochastic counterpart* of (15.6), that is,

$$\min_{\mathbf{v}_0} \bar{\mathcal{L}}(\mathbf{v}_0) = \min_{\mathbf{v}_0}\left[N^{-1} \sum_{i=1}^{N}\{I_A(\mathbf{Y}_i)W(\mathbf{Y}_i, \mathbf{v}_0)^2\}\right], \qquad (15.7)$$

and then take its optimal solution, say $\bar{\mathbf{v}}_{0N}^*$, as an estimator of $\mathbf{v}_0^*$. This yields the following algorithm.

Algorithm 15.1.
1. Estimate the optimal vector of reference parameters $\mathbf{v}_0^*$ of problem (15.5) from the solution of the stochastic counterpart (15.7).
2. Estimate the rare probability α by using estimator (15.3) with $g(\mathbf{y}) = g(\mathbf{y}, \mathbf{v}_0^*)$.

The solution of the stochastic counterpart (15.7), however, might be time consuming, especially when the system is complex and the dimensionality of the reference parameter vector $\mathbf{v}_0$ is high. In some cases, large deviations theory (see e.g., [16-18], [27]) provides simple algorithms for identification of the asymptotically optimal density $g(\mathbf{y}, \mathbf{v}_0^*)$ within the parametric family of the exponential changes of measure (see Section 15.4 for some details). Its application, however, is limited. For example, Frater et al. [17] noted that for Jackson-type open networks the identification of the optimal exponential change of measure reduces to the solution of a complex minimax problem.

We now introduce a framework for *complexity analysis*, which is related to the basic concepts of complexity theory (see e.g., [37]).

Let $\ell(x) \equiv \mathbb{E}\varphi(L, x) > 0$, where L is a sample performance, x is a deterministic parameter, and φ is a given function, say $\varphi(L, x) = I_{\{x, \infty\}}(L)$. Let $\bar{\ell}_N$ be an IS estimator of $\ell(x)$.

Definition 15.2. We say that $\bar{\ell}_N$ is an (ϵ, δ)-*accurate estimator* of $\ell(x)$ $(0 < \epsilon, \delta < 1)$ if

$$\mathbf{P}(\,|\,\overline{\ell}_N - \ell(x)\,| < \epsilon\ell(x)) > 1 - \delta. \tag{15.8}$$

For example, a (0.05, 0.10)-accurate estimator ensures that the relative error does not exceed 5% with probability more than 90%.

Consider the *squared coefficient of variation* (SCV) of $\overline{\ell}_N$, that is,

$$\kappa^2(x) = \frac{N\,\mathrm{Var}(\overline{\ell}_N)}{\ell^2(x)}.$$

By the Central Limit Theorem (CLT) we have that $N \approx \gamma\kappa^2(x)$, where $\gamma = \Phi^{-1}(1 - \delta/2)^2\epsilon^{-2}$.

Definition 15.3. An IS estimator is called (ϵ, δ)-*polynomial*, if (15.8) is guaranteed by a sample size (computational cost) $N = O(p(x))$ for some polynomial function $p(\cdot)$. Any IS estimator whose computational cost $N \equiv N(x)$ cannot be bounded by a polynomial function is called an *exponential time* estimator.

It follows that the estimator $\overline{\ell}_N$ will need a polynomial time if the SCV $\kappa^2(x)$ is bounded in x by a polynomial function $p(x)$.

For better insight into polynomial and exponential time IS estimators, consider the following simple example.

Example 15.4. Suppose we are interested in estimating $\ell = P(Y > x)$, where the random variable Y has an exponential distribution with rate v, i.e., $Y \sim f(y) = v\exp(-vy)$.

Taking into account that in this case $\ell = e^{-vx}$, it is readily seen that the SCV for the CMC estimator is $\kappa^2(x) \approx e^{vx}$, provided x is large. Hence, the CMC estimator is exponential in x.

Let $g(y) = v_0\exp(-v_0 y)$ be the IS density, and assume that we want to choose v_0 so as to minimize the variance of the LR estimator $\overline{\ell}_N(v_0)$.

The second moment of the random variable (rv) IW is

$$\mathbb{E}_g\{IW\}^* = \mathbfcal{L}(v_0) = \frac{v^2}{v_0}\int_{y = x}^{\infty} e^{(v_0 - 2v)y}dy = \frac{v^2 e^{-(2v - v_0)x}}{v_0(2v - v_0)}, \tag{15.9}$$

which is infinite for $v_0 \geq 2v$.

The optimal value of the reference parameter $v_0^* = v_0^*(x)$, which minimizes $\mathbfcal{L}(v_0)$, is $v_0^*(x) = v + x^{-1} - (v^2 + x^{-2})^{1/2}$. Suppose $x^{-1} \ll v$, hence, $P(Y > x)$ is small, say less than 10^{-6}. In this case we have

$$v_0^*(x) \approx x^{-1} \tag{15.10}$$

and $\mathbb{E}_g[I_{(x,\infty)}(Y)] \approx e^{-1}$.

Consider the relative efficiency $\epsilon(v_0, x)$ defined as $\epsilon(v_0, x) = \dfrac{\mathrm{Var}\{\overline{\ell}_N(v_0)\}}{\mathrm{Var}\{\overline{\ell}_N\}}$.
For $v_0 = v_0^*$ we obtain

$$\epsilon^* = \epsilon(v_0^*, x) \approx 0.5xve^{1 - vx}.$$

If we take, for example, $v = 1$ and $x = 12$, then $\mathbb{E}I_{[12,\infty)}(L) = P(L > 12) = e^{-12} \approx 10^{-6}, v_0^*(x) \approx 1/12$ and $\epsilon^* \approx 10^{-4}$. Thus, using the optimal value $v_0^* \approx 1/12$ we obtain a dramatic variance reduction of order 10^4.

Consider now the SCV for the LR estimator $\overline{\ell}_N(v_0)$. It is not difficult to see that

$$\kappa^2(v_0, x) = \frac{v^2 e^{v_0 x}}{v_0(2v - v_0)} - 1.$$

For $v_0^* = x^{-1}$ it reduces to

$$\kappa^2(v_0^*, x) \approx 0.5xve. \tag{15.11}$$

That is, for large x, the SCV of the CMC estimator and the optimal LR estimator increase in x exponentially and linearly, respectively ($\kappa^2(x) \approx e^{vx}$ and $\kappa^2(v_0^*, x) \approx 0.5xve$, respectively). In other words, the CMC and IS estimators can be viewed as *exponential* and *polynomial* time. The required sample size (computational costs) N is $N = O(e^x)$ and $N = O(x)$, respectively.

If we choose $v_0 = kv_0^*$ instead of the optimal value v_0^*, then for large x we obtain $\kappa^2(v_0, x) = 0.5k^{-1}xve^k$. For example, if $k = 2$ (that is, $v_0 = 2v_0^* \approx 2/x$) we obtain $\kappa^2(v_0, x) \approx 0.25xve^2$. In this case the relative efficiency equals $\epsilon(v_0, x) \approx 0.5ee^*$, so perturbing v_0^* ($k = 2$) by 100% increases the variance approximately only $0.5e$ times.

Table 1 displays the relative efficiency $\epsilon(v_0, x)$ as a function of v_0 for $v = 1$ and $x = 20$.

Table 1. The relative efficiency $\epsilon(v_0, x)$ of the IS estimator as a function of v_0, given $v = 1$ and $x = 20$.

v_0	0.6	0.4	0.2	0.1	0.075	0.05*	0.025
$\epsilon(v_0, x)$	$4.0 \cdot 10^{-4}$	$9.6 \cdot 10^{-6}$	$3.1 \cdot 10^{-7}$	$7.8 \cdot 10^{-8}$	$6.2 \cdot 10^{-8}$	$5.5 \cdot 10^{-8*}$	$6.7 \cdot 10^{-8}$

15.3 STANDARD AND SWITCHING REGENERATIVE ESTIMATORS FOR RARE EVENTS

This section deals with the so-called *switching regenerative* (SR) estimators for the evaluation of p_x, the steady state probability of excessive backlog, in a stable GI/GI/1 queue. Before introducing the SR estimators (§15.3.2), we need to cite some material on standard regenerative likelihood ratio estimators or, simply, *standard regenerative* estimators.

15.3.1 Standard regenerative estimators

The steady state probability of excessive backlog can be written as

$$p_x = \mathbf{P}_x(Q_n > x) = \frac{\mathbb{E}_G \sum_{n=1}^{\tau} I_{(x,\infty)}(Q_n)W_n}{\mathbb{E}_G \sum_{n=1}^{\tau} W_n}, \qquad (15.12)$$

where Q_n is the number of customers in the queue just before the nth customer arrives at the system, I_A is the indicator function of the event A, τ is the length of a regenerative cycle, and $W_n (n = 1, \ldots, \tau)$ is the *likelihood ratio process* defined as

$$W_n = \prod_{k=1}^{n} \frac{F_A(dA_k)}{G_A(dA_k)} \prod_{k=1}^{D(n)} \frac{F_B(dB_k)}{G_B(dB_k)}, \qquad (15.13)$$

where $\{A_k\}$ and $\{B_k\}$ are the sequences of interarrival and service times with distributions $F_A(\cdot)$ and $F_B(\cdot)$, respectively; $G(\cdot)$ dominates the distribution $F(\cdot)$ in the absolutely continuous sense (see e.g., [29]); G is called the *importance sampling* (IS) distribution, or sometimes the *dominating* distribution (see [5]); and $D(n)$ is the number of customers being served just before the nth customer arrives at the queue ($D(n) < n$). Note that in order for the GI/G/1 queue to remain stable under the probability distribution $G = (G_A, G_B)$, the inequality

$\mathbb{E}_{G_A}A_k < \mathbb{E}_{G_B}B_k$ must hold. An unbiased estimator of p_x is

$$\bar{p}_{x,N} = \frac{\sum_{i=1}^{N}\sum_{n=1}^{\tau_i}I_{\{Q_{ni}>x\}}W_{ni}}{\sum_{i=1}^{N}\sum_{n=1}^{\tau_i}W_{ni}}, \tag{15.14}$$

where τ_i is the length of the ith regenerative cycle, and N is the number of generated regenerative cycles.

Under the assumption that G_A and G_B come from a parametric family of distributions as $F_A = F_A(\mathbf{y},\mathbf{v}_1)$ and $F_B = F_B(\mathbf{y},\mathbf{v}_2)$ do, Asmussen, Rubinstein, and Wang [5] explicitly calculated the variances of the estimator $\bar{p}_{x,N}$ for the steady state waiting time in the M/M/1 queue. They showed that in order to obtain variance reduction with $\bar{p}_{x,N}$ relative to the crude Monte Carlo (CMC) method, one has to choose G_A and G_B such that the associated traffic intensity ρ_0 (under (G_A,G_B)) is moderately larger than the original traffic intensity ρ (under (F_A,F_B)). Asmussen and Rubinstein [6] proved that the LR estimator $\bar{p}_{x,N}$ is of *exponential time*. Notice that for regenerative estimators we assume that the computational cost T is not $N = N(x)$ (see Definition 15.3), but $T \equiv N\mathbb{E}_\tau$, where N is the number of required cycles and τ is the cycle length. In this way, we measure the computational cost in terms of *time per customer rather than time per cycle*.

15.3.2 Switching regenerative (SR) estimators

A natural generalization of the LR estimator $\bar{p}_{x,N}$ in (15.14) uses *dynamic* IS distributions G_k at each step k of the cycle instead of a fixed G. In other words, let $G_{A,k}(\cdot)$ and $G_{B,k}(\cdot)$ $(k = 1,\ldots,\tau)$ be the IS distributions used for the kth interarrival and kth service times, respectively. That is, assume that system behavior during a cycle of length τ is driven by the set of distribution functions $\widetilde{\pi} = [(G_{A,1},G_{B,1}),\ldots,(G_{A,\tau},G_{B,\tau})]$, which we call the *set of IS policies*. Assume next that a policy $\widetilde{\pi}$ of choosing the probability measures during a cycle is completely defined by the system's evolution up to the last arrival prior to this moment. Then one can see that the realizations of the policies will be identical, provided the sample paths

$$[(Q_1,\widetilde{B}_1),\ldots,(Q_\tau,\widetilde{B}_\tau)], \tag{15.15}$$

are the same, where $\widetilde{B}_i$ is the remaining service time for the customer served at the moment the ith customer arrives at the system during the cycle. Notice that we might also allow dependence of the current service and interarrival time distributions on the index of the last arriving customer; in particular, notice its relationship with a finite family of stopping times $\Xi = \{\xi_r\}$, where

$$\xi_k = \min[t, 0 \le t \le \tau \mid Q_t = k]. \tag{15.16}$$

(The above means that our policies $\widetilde{\pi}$ cover both the deterministic and random policies introduced in [29].) With this in hand, we can extend W_n in (15.13) to

$$W_n \equiv \prod_{k=1}^{n}\frac{F_A(d,A_k)}{G_{A,k}(dA_k)}\prod_{k=1}^{D(n)}\frac{F_B(dB_k)}{G_{B,k}(dB_k)}, \tag{15.17}$$

where G is redefined as

$$G \equiv \widetilde{\pi}. \tag{15.18}$$

The algorithm for estimating p_x according to (15.14), with

$$W_{ni} = \prod_{k=1}^{n} \frac{F_A(dA_{ki})}{G_{A,k,i}(dA_{ki})} \prod_{k=1}^{D(n)} \frac{F_B(dB_k)}{G_{B,k,i}(dB_{ki})}, \tag{15.19}$$

and τ_i being the length of ith cycle, $1 \le i \le N$, can be written as follows.

Algorithm 15.5

1. Specify a policy $\tilde{\pi}$ by choosing a particular set of distributions $(G_{A,1}, G_{B,1})$, $\ldots, (G_{A,\tau}, G_{B,\tau})$ at each cycle, and generate N corresponding regenerative cycles.

2. Let
$$\tilde{\pi}_i = [(G_{A,1,i}, G_{B,1,i}), \ldots, (G_{A,\tau_i,i}, G_{B,\tau_i,i})]$$

   be a realization of the policy $\tilde{\pi}$ at the ith cycle. Then, compute X_i and Y_i as follows:
$$X_i = \sum_{n=1}^{\tau_i} I_{\{Q_{ni} > x\}} W_{ni} \text{ and } Y_i = \sum_{n=1}^{\tau_i} W_{ni},$$

   where W_{ni} was given in (15.19).

3. Calculate the point estimator $\bar{p}_{x,N}$ (see (15.14)) as $\bar{p}_{x,N} = \dfrac{\bar{X}}{\bar{Y}}$ and a $100(1-\delta)\%$ confidence interval for p_x:
$$\tilde{I} = [\bar{p}_{x,N} \pm \frac{z_\delta s}{\bar{Y} N^{1/2}}],$$

   where $z_\delta = \Phi^{-1}(1 - \delta/2)$, $\bar{X}(1/N) \sum_{i=1}^{N} X_i$, $\bar{Y} = (1/N) \sum_{i=1}^{N} Y_i$,
$$s^2 = s_{11} - 2\bar{p} s_{12} + \bar{p}^2 s_{22}, \quad s_{11} = \frac{1}{N-1} \sum_{i=1}^{N} (X_i - \bar{X})^2,$$

$$s_{22} = \frac{1}{N-1} \sum_{i=1}^{N} (Y_i - \bar{Y})^2, \text{and } s_{12} = \frac{1}{N-1} \sum_{i=1}^{N} (X_i - \bar{X})(Y_i - \bar{Y}).$$

We shall call the estimator (15.14), (15.19) the *switching regenerative* (SR) estimator.

15.4 PROPERTIES OF THE SWITCHING REGENERATIVE ESTIMATOR

In this section we characterize a subclass $\Pi^{**} \subset \Pi$ of IS policies, which gives rise to the polynomial time SR estimators of the type (15.14), (15.19) (see Theorem 15.9). We start with the following definition.

Definition 15.6. We say that a policy $\tilde{\pi}$ belongs to the class Π if the following conditions are satisfied:

1. For all $\tilde{\pi} = [(G_{A,k}, G_{B,k}), \ k = 1, \ldots, \tau] \in \Pi$, we have that $G_{A,k} \in \mathcal{G}_A$ and $G_{B,k} \in \mathcal{G}_B$, where $\mathcal{G}_A$ and $\mathcal{G}_B$ are sets of the dominating distribution functions for F_A and F_B, respectively; $0 < \mathbb{E}_{G_{A,k}} A^2 < \infty$, $0 < \mathbb{E}_{G_{B,k}} B^2 < \infty$, $k = 1, \ldots, \tau$.

2. $G_{B,k}(\cdot) \equiv G_{B,j}(\cdot)$ if $R(k) = R(j)$; that is, we allow a change of probability distribution only at the arrival epochs k, $1 \le k \le \tau$. (Here $R(k)$ is the number of the last arriving customer just before the kth customer starts service.) Moreover, we assume that the distribution between the kth and $(k+1)$th arrivals, $G_{A,k}$ and $G_{B,j}$, $j = D(k)+1, \ldots, D(k+1)$, are completely defined by k and $\Xi = \{\xi_r, r = 1, \ldots, r_0\}$, where $\{\xi_k\}$ is the finite set of stopping times. (See (15.15) and (15.16) for the definition of $\tilde{B}_k$ and ξ_k, respectively.)

3. $\mathbb{E}_{G_{A,k}} A_k - \mathbb{E}_{G_{B,j}} B_j < 0$, $D(k) < j < D(k+1)$, $1 \le k \le \xi_1 \equiv \xi$; that is, the simulating system is *unstable* until the occurrence of the first overflow, and

$\mathbf{P}_{\underset{\tilde{\pi}}{}}(\xi < \infty) = 1$ for all x.

4. $\mathbb{E}_{G_{A,k}}A_k - \mathbb{E}_{G_{B,j}}B_j > 0$, $D(k) < j < D(k+1)$, $k \in [\xi, ..., \tau] \backslash I_3$, where $I_3 \subset [\xi, ..., \tau]$ is a finite set; that is, after the first overflow untill the end of the cycle, the system becomes *stable* and $\mathbf{P}\{\tau - \xi < \infty; \xi < \tau\} = 1$ and $\lim_{x \to \infty} P(Q_t = \psi x) = 0$, for $1 < \psi < \infty$.

5. All parameters of $G_{A,k}$ and $G_{B,k}$ are independent of x.

Definition 15.7.   Let $\pi^* = [(G_{A,1}^*, G_{B,1}^*), ..., (G_{A,\tau}^*, G_{B,\tau}^*)] \in \Pi$ be a policy with

$$G_{A,k}^* = \begin{cases} F^{(w)}, & 1 \le k < \xi, \\ F_A, & \xi \le k \le \tau, \end{cases} \tag{15.20}$$

and

$$G_{B,k}^* = \begin{cases} F^{(-w)}, & 1 \le R(k) < \xi, \\ F_B, & \xi \le R(k) \le \tau, \end{cases} \tag{15.21}$$

respectively, where $F_A^{(w)}$ and $F_B^{(-w)}$ are the *optimal exponential change of the measures* (OECM) for the interarrival and service time distributions, defined as

$$F_A^{(w)}(dx) = \exp(-wx - \Lambda_A(w))F_A(dx) \tag{15.22}$$

and

$$F_B^{(-w)}(dx) = \exp(wx - \Lambda_B(-w))F_B(dx), \tag{15.23}$$

respectively; in (15.22) $\Lambda_X(w) = \log(\mathbb{E}e^{-wX})$ denotes the cumulant function, w is a unique positive solution of the equation

$$\Lambda(w) \equiv \Lambda_A(w) + \Lambda_B(-w) = 0$$

(for more details see Section 15.9, Appendix A), and as before, $R(k)$ is the number of the customers arriving at the system just before the kth customer starts its service (clearly, $R(k) > \xi$ implies $k > \xi - x$).

As examples of $F_X^{(t)}$ for $t \in \{t > 0, \Lambda_X(x) < \infty\}$, let

(a) $Y \sim \exp(v)$; then $F^{(t)} = \exp(v + t)$
(b) $Y \sim \text{Gamma}(\lambda, \beta)$; then $F^{(t)} = \text{Gamma}(\lambda + t, \beta)$
(c) $Y \sim N(\mu, \sigma^2)$; then $F^{(t)} = N(\mu - 2\sigma^2 t, \sigma^2)$
(d) $Y \sim \text{Geometric}(p)$; then $F^{(t)} = \text{Geometric}(1 - (1 - p)e^{-t})$.

Note that under policy π^*, the LR process W_{ni} in (15.19) reduces (see also formula (4.5.6) of Rubinstein and Shapiro [29]) to

$$W_{ni} = \begin{cases} \prod_{k=1}^{n} \dfrac{F_A(d_{A_{ki}})}{G_{A,k}^*(d_{A_{ki}})} \prod_{k=1}^{D(n)} \dfrac{F_B(d_{B_k})}{G_{B,k}^*(d_{B_{ki}})}, & 1 \le n \le \xi \\ \\ \prod_{k=1}^{\xi} \dfrac{F_A(d_{A_{ki}})}{G_{A,k}^*(d_{A_{ki}})} \prod_{k=1}^{D(\xi)} \dfrac{F_B(d_{B_k})}{G_{B,k}^*(d_{B_{ki}})}, & \xi < n \le \tau. \end{cases} \tag{15.24}$$

Formulas (15.20), (15.21), (15.24) mean that from the beginning of the cycle until the first overflow we use the OECM pair $(F_A^{(w)}, F_B^{(-w)})$, and then accomplish the cycle by switching to the original distributions.

Definition 15.8.   Let Π^* be a class of policies $\tilde{\tilde{\pi}} \in \Pi$ such that

$$G_{A,k} = F_A^{(w)}, \quad k \in [1, ..., \xi_x] \backslash I_1, \quad \text{(a.s.)}$$

and

$$G_{B,k} = F_B^{(-w)}, R(k) \in [1, \ldots, \xi_x] \backslash I_2, \text{ (a.s.)}$$

where $I_1 \subset [1, \ldots, \xi_x]$ and $I_2 \subset [1, \ldots, D(\xi_x)]$ are sets with $o(x)$ elements.

Theorem 15.9. *Assume the following conditions hold:*

Condition A1. $0 < \mathbb{E}B < \mathbb{E}A < \infty$ *and* $P(B - A > 0) > 0$;

Condition A2. *There exists a scalar w satisfying*

$$\Lambda(w) \equiv \Lambda_A(w) + \Lambda_B(-w) = 0, \tag{15.25}$$

such that $\Lambda_A(w) < \infty$, $\Lambda_B(-w) < \infty$, and $0 < \Lambda'(w) < \infty$ (see also Appendix A, Section 15.9). Then,

$$\pi^* \in \Pi^{**}, \tag{15.26}$$

$$\Pi^{**} \subset \Pi^*, \tag{15.27}$$

where $\Pi^{**} \subset \Pi$ is a subclass of Π generating the polynomial time IS estimators.

Proof. The proof of Theorem 15.9 is given in Appendix B (Section 15.9).

Expressions (15.26) and (15.27) of Theorem 15.9 can be interpreted as follows. Expression (15.26) states that the estimator (15.14), (15.24) based on policy π^* is polynomial, while formula (15.27) states that for Algorithm 15.5 to be polynomial it is necessary that $\tilde{\pi} \in \Pi^*$, that is, it is necessary to use the OECM until the first overflow of ξ excluding, perhaps, $o(x)$ steps.

It is our belief that Theorem 15.9 is the first result characterizing the complexity of switching regenerative estimators and specifying a subclass $\Pi^{**} \in \Pi$ that generates polynomial time estimators.

We shall call the parameter w in (15.20), (15.21), (15.22), (15.23) satisfying (15.25) the *optimal parameter* (OP) in the OECMs $F_A^{(w)}$ and $F_B^{(-w)}$, respectively.

Assume that the original interarrival and service time pdf's in the GI/G/1 queue come from the exponential family

$$f(S, \mathbf{v}) = a(\mathbf{v}) \exp(b(\mathbf{v})c(S))d(S), \tag{15.28}$$

where $c(S)$ is bounded by a polynomial function and $a(\mathbf{v})$ and $b(\mathbf{v})$ are twice continuously differentiable functions in parameter vector $\mathbf{v}$. Taking into account (15.22), (15.23), and the fact that the OP w is a scalar, it follows that the OECM ($F_A^{(w)} = F_A^{(w)}(\mathbf{v}_{01}^*)$, $F_B^{(-2)} = F_B^{(w)}(\mathbf{v}_{02}^*)$) coincides with the *original distributions* ($F_A(\mathbf{v}_1)$, $F_B(\mathbf{v}_2)$), up to a *single* parameter; that is, *only a single parameter*, in the vector $\mathbf{v}_{01}^*$ of the OECM $F_A^{(w)}(\mathbf{v}_{01}^*)$, differs from the vector $\mathbf{v}_1$ of the original interarrival time distribution $F_A(\mathbf{v}_1)$, and similarly for the service time distribution.

Table 2 presents the optimal parameter w and the optimal reference parameter vector $\mathbf{v}_0^* = (\mathbf{v}_{01}^*, \mathbf{v}_{02}^*)$ for several commonly used exponential families.

Table 2. The optimal parameter w and the optimal reference parameter vector $\mathbf{v}_0^* = (\mathbf{v}_{01}^*, \mathbf{v}_{02}^*)$ for several commonly used exponential families.

F_A, F_B	$\mathbf{v}_1, \mathbf{v}_2$	$\mathbf{v}_{01}^*, \mathbf{v}_{02}^*$	w
$\exp(\cdot)$	λ, μ	μ, λ	$\mu - \lambda$
Gamma$(\cdot)$	$(\lambda, \beta), (\mu, \gamma)$	$(\lambda + 2, \beta), (\mu - w, \gamma)$	$(w > 0: (\lambda + w)^\beta (\mu - w)^\gamma = \lambda^\beta \mu^\gamma$
Poisson$(\cdot)$	λ, μ	μ, λ	$\log(\lambda\mu)$

15.5 ROBUSTNESS OF SWITCHING ESTIMATORS

In this section we address the following robustness question: *How much can one perturb the optimal parameter w in the OECM such that the SR estimators* (15.14), (15.24) *still obtain low variance?* To answer this question, we argue as follows. We first replace $F_A^{(w)}$ and $F_B^{(-w)}$ in (15.22), (15.23) by G_A' and G_B', respectively, assuming that $(G_A', G_B') \neq (F_A^{(w)}, F_B^{(-w)})$.

Definition 15.10. Let Π' be a subclass of the IS policies,

$$\widetilde{\pi} = [(G_{A,1}, G_{B,1}), \ldots, (G_{A,\tau}, G_{B,\tau})] \in \Pi',$$

where

$$G_{A,k} = \begin{cases} G_A', & 1 \leq k < \xi_x, \\ F_A, & \xi_x \leq k \leq \tau, \end{cases}$$

and

$$G_{B,k} = \begin{cases} G_B', & 1 \leq R(k) < \xi_x, \\ F_B, & \xi_x \leq R(k) \leq \tau. \end{cases}$$

It will follow from Theorem 15.11 below that in this case the estimator (15.14), (15.24) has exponential time complexity. Clearly, when $G_A' = F_A^{(w)}$ and $G_B' = F_B^{(-w)}$, we have $\widetilde{\pi} \equiv \pi^*$ and, thus, a polynomial time estimator (15.14), (15.24) with the exponential rate z of the computational cost $T_{\widetilde{\pi}}$ ($T_{\widetilde{\pi}} = \exp(z_{\widetilde{\pi}} x + o(x))$) equal to 0.

Next we derive explicit expressions for the exponential rate z of the computational cost $T_{\widetilde{\pi}}$ under $(G_A', G_B') \neq (F_A^{(w)}, F_B^{(-w)})$. This z will characterize the robustness of our IS estimator. In addition, this exponential rate z might give useful insight into time complexity of IS estimators for more general queueing systems. We define the following new (but not probabilistic!) finite measures:

$$K_A(dA) \equiv K_A(G_A)(dA) \equiv \frac{F_A}{G_A}(dA) F_A(dA) \tag{15.29}$$

and

$$K_B(dB) \equiv K_B(G_B)(dB) \equiv \frac{F_B}{G_B}(dB) F_B(dB). \tag{15.30}$$

We also define

$$\Lambda_{K_A}(u) = \log \mathrm{II}_{K_A} e^{-uA_k} \equiv \log \int_0^\infty e^{-uA_k} K_A(dA_k)$$

and

$$\Lambda_{K_B}(u) = \log \mathrm{II}_{K_B} e^{-uB_k} \equiv \log \int_0^\infty e^{-uB_k} K_A(dB_k).$$

Note that the symbol II is used here instead of the conventional symbol $\mathbb{E}$ (for expectation); the functions $\Lambda_{K_A}(u)$ and $\Lambda_{K_B}(u)$ have a meaning similar to the cumulants Λ_A and Λ_B, respectively.

We finally define $v \equiv \sup[u \mid \Lambda_K(u) \leq 0]$, where $\Lambda_K(u) = \Lambda_{K_A}(u) = \Lambda_{K_B}(-u)$, and we assume that the following regularity condition holds.

Condition A3. There exists a value u such that $\Lambda_K(u) \leq 0$.

Note that since $\Lambda(u) \to \infty$ as $u \to \infty$, condition **A3** directly implies that there exists a unique value v such that

$$\Lambda_K(v) = 0 \text{ and } 0 < \Lambda'(v) < \infty. \tag{15.31}$$

Theorem 15.11. *Assume that conditions* **A1-A3** *hold. Then for any policy* $\widetilde{\pi} \in \Pi'$, *the exponential rate* z *equals*

$$z_{\widetilde{\pi}} \equiv \lim_{x \to \infty} x^{-1} \log T_{\widetilde{\pi}} = \Lambda_{K_A}(v) - 2\Lambda_A(w). \tag{15.32}$$

Proof. The proof of Theorem 15.11 is given in Appendix B (Section 15.9). □

Corollary 15.12. *Assume that conditions* **A1-A3** *hold. Let* $\widetilde{\pi}_+ = \widetilde{\pi}(G'_A, G'_B) \in \Pi'$, *with* $G'_A = F_A^{(t)}$ *and* $G'_B = F_B^{(-t)}$. *Then,*

$$z_{\widetilde{\pi}_t} = \Lambda_A(t) + \Lambda_A(v - t) - 2\Lambda_A(w). \tag{5.5}$$

Proof. Corollary 15.12 follows direction from (15.32) by substituting $G'_A = F_A^{(t)}$ and $G'_B = F_B^{(-t)}$. □

Using (15.33), we obtain various efficiency characteristics of the SR estimator (15.14), (15.24), namely the computational cost $T_{\widetilde{\pi}_t}$, the exponential rate $z_{\widetilde{\pi}}$, the optimal speedup factor $S_{opt} = T_F/T_{\pi^*}(z_{\pi^*} = 0)$, and the speedup factor $S_t = T_F/T_{\widetilde{\pi}_t}$.

Table 3 presents (analytic) efficiency values of the SR estimator (15.14), (15.24) under both the optimal parameter w and the perturbed parameter t, as functions of the traffic intensity ρ, for the probabilities of excessive backlog $p_x = 10^{-15}$ in the M/M/1 queue with the service rate $\mu = 1$. In particular (under the OECM ($F_A^{(w)}$, $F_B^{(-w)}$), $w = \mu - \lambda$)), it represents S_{opt}, called the *optimal speedup factor*, and (under the ECM ($F_A^{(t)}, F_B^{(-t)}$) it represents the S_t and $z_{\widetilde{\pi}_t}$, called the *speedup factor* $S_t = T_F/T_{\widetilde{\pi}_t}$ and the *exponential rate*. Notice that both S_t and $z_{\widetilde{\pi}_t}$ were calculated for t being *less* than optimal value w by 20%. Notice also that the values x in the last row of the table correspond to $p_x = 10^{-15}$.

It follows from the table that the OECM leads to dramatic variance reduction (11-13 orders of speedup out of 15 orders), and that perturbing w by 20% ($t < w$), we *lose only 2-3 orders of speedup.*

Table 3. The efficiencies S_{opt}, S_t and $z_{\widetilde{\pi}_t}$ of the SR estimator

(15.14), (15.24) as functions of the traffic intensity ρ for the probability of excessive backlog $p_x = 10^{-15}$ in the M/M/1 queue with $\mu = 1$.

ρ	0.2	0.4	0.6	0.8
w	0.8	0.6	0.4	0.2
S_{opt}	10^{13}	10^{12}	10^{12}	10^{11}
t	0.64	0.48	0.32	0.16
S_t	10^{11}	10^{10}	10^9	10^8
$z_{\widetilde{\pi}_t}$	$1.0 \cdot 10^{-1}$	$5.6 \cdot 10^{-2}$	$3.1 \cdot 10^{-2}$	$1.3 \cdot 10^{-2}$
x	21	36	62	107

15.6 EXTENSIONS

Here we present some extensions of the results of Sections 15.4 and 15.5. In particular, we consider estimation of:

(*i*)   rare events in the GI/D/1 queue (§15.6.1),

(*ii*)   rare events in the GI/G/1 queue with dependent arrivals (§15.6.2),

(*iii*) rare events in the GI/G/1 queue with batch arrivals (§15.6.3),

(*iv*)   sensitivities (derivatives) of the probability of the excessive backlog $p_x(\mathbf{v})$, with respect to the parameter vector $\mathbf{v}$ of the interarrival time distributions (§15.6.4).

15.6.1 The GI/D/1 queue

By definition, this queue has deterministic constant service times, independent and identically distributed interarrivals with the pdf $f_A(A) \equiv dF_A(A)/dA$. Presented in [18] is a heuristic result on quick simulation of rare events in the M/D/1 queue and its extensions to batch arrivals with exponential distribution. The problem here is that the deterministic service time excludes the possibility of introducing a corresponding change of measure with the concomitant LR process W_n in the standard way. To circumvent this problem, we use the so-called "*push out*" (PO) technique introduced in [28]. It is shown in [28] how to incorporate the PO technique for the GI/D/1 queue in the standard regenerative setting, in order to reduce the original problem to an equivalent associated one. In short, one merely needs to "push out" the deterministic service time parameter, say d, to the interarrival time random variable $Y \sim f_A(y)$, then introduce a new auxiliary random variable $\widetilde{Y} = A - d$ distributed according to $\widetilde{f}_A(y+d)$ and a new auxiliary sample performance $\widetilde{Q}_n(\widetilde{\mathbf{Y}}_n)$, $\widetilde{\mathbf{Y}}_n = (\widetilde{Y}_1,...,\widetilde{Y}_n)$, instead of f_A and $Q_n(\mathbf{A}_n)$, $\mathbf{A}_n = (A_1,...,A_n)$, respectively, and incorporate them into the following asymptotical consistent regenerative estimator of p_x:

$$\overline{p}_{x,N} = \frac{\sum_{i=1}^{N}\sum_{n=1}^{\tau} I_{\{\widetilde{Q}_{ni} > x\}} W_{ni}}{\sum_{i=1}^{N}\sum_{n=1}^{\tau_i} W_{ni}}, \tag{15.34}$$

where (similar to (15.14)) τ_i is the length of the ith regenerative cycle; N is the number of regenerative cycles; $\widetilde{\mathbf{Y}}_{ni} = (\widetilde{Y}_{1i},...,\widetilde{Y}_{ni})$, $n = 1,...,\tau_i$, $i = 1,...,N$; $\widetilde{Q}_{ni}(\widetilde{\mathbf{Y}}_{ni}) = Q_{ni}(\mathbf{Y}_{ni})$;

$$W_{ni} = W_{ni}(\widetilde{\mathbf{Y}}_{ni}) = \prod_{k=1}^{n} \frac{\widetilde{f}(\widetilde{Y}_{ki})}{\widetilde{g}(\widetilde{Y}_{ki})};$$

and $\widetilde{\mathbf{Y}}_{ni}$ is a random sample from $\widetilde{g}$, which is a dominating pdf for $\widetilde{f}$.

In order to adopt the "push out" method, we need to replace the OECM in the SR estimator of the type (15.14), (15.24) by a new OECM defined below, with all other data remaining the same. (Similar results can be obtained for the D/GI/1 queue.) It is not difficult to see that under conditions **A1-A2** we have

$$w = w_1 \equiv \sup[\theta \geq 0 \mid \Lambda_{\widetilde{Y}}(\theta) < 0],$$

where $\Lambda_{\widetilde{Y}}(\theta) = \mathbb{E}e^{-\theta\widetilde{Y}}$ with $\widetilde{Y} \sim \widetilde{f}(\widetilde{Y}) = f_A(y+d)$. Let $k(y) \equiv \dfrac{\widetilde{f}(y)^2}{\widetilde{g}(y)}$, where $\widetilde{g}(\cdot)$ is a dominating pdf for $\widetilde{f}(\cdot)$, and

$$\Lambda_k(u) \equiv \mathrm{II}_k e^{-uy} \equiv \log \int_0^{\infty} e^{-uy} k(y) dy.$$

Then, the following proposition follows.

Proposition 15.13. *Consider an IS estimator based on the policy $\widetilde{\pi} \in \Pi'$. Assume that conditions **A1**, **A2** hold, and there exists a value u such that $\Lambda(u) < 0$. Let v_1 be the solution of the equation $\Lambda_k(u) = 0$, $0 < \Lambda_k'(u) < \infty$. Then, the exponential part $z_{\widetilde{\pi}}$ of the computational cost $T_{\widetilde{\pi}}$ is given by*

$$z_{\widetilde{\pi}} = d(2w_1 - v_1).$$

Proof. The proof of this and the following two propositions (namely Propositions 15.6.2 and 15.6.3) are omitted because they are similar to the proof of Theorem 16.5.1. (Their proofs are available from the authors.)

Notice that for the IS pdf $\widetilde{g}(y) = g^{(w_1)}(y + d)$, we obtain $z_{\widetilde{\pi}} = 0$ and hence a polynomial time estimator (15.34).

15.6.2 Rare events with dependent arrivals

First, we consider the case with interarrival times driven by stationary uniform recurrent Markov chains (see [9]), which admit a strong Perron-Frobenius theory (see, e.g. [8]). Note that the class of such distributions covers the TES (Transform-Expand-Sample) processes (see, e.g. [25] and Chapter 10 in this book), which by itself captures a wide variety of sample path behaviors with autocorrelation input sequences; TES processes are used for modeling video and data sources.

We now introduce the following notations for the cumulant function and ECM for the interarrival distribution given by stationary uniform recurrent Markov chains:

- $\Lambda_A(\theta) = \log(\lambda_A(\theta))$, where $\lambda_A(\theta)$ is a unique simple real eigenvalue of the kernel $\exp(-\theta A_k) F_A(db, a)$ with the following properties: $\lambda_A(\theta)$ is analytic and strictly convex, and the related eigenfunction $\phi(a; \theta)$ is bounded.
- The interarrival time distribution equals

$$F_A^{(\theta)}(a, db) = e^{-\theta b - \Lambda_A(\theta)} \frac{\phi(b; \theta)}{\phi(a; \theta)} F_A(a, db). \tag{15.35}$$

Consider now a single-server queue with iid service times, but with interarrival time distributions given by a uniform recurrent stationary (time-homogeneous) Markov chain. We call such a queue the GM/G/1 queue. In this case we define a *cycle* as a piece of trajectory starting the system at a steady state regime, provided the system is empty, and terminating the cycle on the first occasion when the system is empty again.

It is important to note that in this case we need to simulate (in parallel) two samples paths (trajectories): one with and one without change of measure (CMC). When the second process runs long enough to be treated as in the steady state, we use the instants when the system is empty to start a cycle of the first process by an IS simulation. Note that unlike the standard regenerative cycles, these cycles are not independent. However, the ratio estimator remains asymptotically unbiased, provided the cycles are initialized under a steady state distribution (see [20]). The associated IS algorithm is the same as Algorithm 15.5.

Proposition 15.14. *Consider a GM/M/1 queue. Let*

$$K_A(a, db) \equiv \frac{F_A(a, db)}{G_A(a, db)} F_A(a, db),$$

where $G_A(\cdot, \cdot)$ dominates the distribution $F_A(\cdot, \cdot)$ and $\Lambda_{K_A}(u) \equiv \log(\lambda_{K_A}(u))$.

Furthermore, let $\lambda_{K_A}(u)$ be defined as a unique simple real eigenvalue of the kernel $\exp(\theta A_k) K_A(dy,x)$ having the following properties: $\lambda_{K_A}(u)$ is analytic and strictly convex and the related eigenfunction $\phi_{K_A}(x;\theta)$ is bounded. Finally, assume that there exists a unique value u such that $\Lambda_K'(u) < 0$. Then, under conditions A1-A2, the estimator (15.14), (15.24), based on the IS policy $\widetilde{\pi} \in \Pi'$ with the simulation distributions $G_A(a,db)$ (stationary Markov chain) and $G_B(b)$, possesses the following exponential rate of the computational cost:

$$z_{\widetilde{\pi}} = \Lambda_{K_A}(v_2) - 2\Lambda_A(w_2), \tag{15.36}$$

where v_2 and w_2 are the solutions of the equations $\Lambda_K(v_2) = 0$, $0 < \Lambda_K'(v_2) < \infty$, and $\Lambda(w_2) = 0$, $0 < \Lambda'(w_2) < \infty$, respectively.

For the particular case $G_A = F_A^{(w_2)}$ and $G_B = F_B^{(-w_2)}$, we have $z_{\widetilde{\pi}} = 0$ and, hence, a polynomial time IS estimator.

Example 15.15. Consider the case where the interarrival process $\{A_k\}$ is represented by an AR(1) process $A_{k+1} = rA_k + (1-r)X, 0 \le r \le 1$, $X \sim \exp(\lambda)$. For this case the real eigenvalue and the associated eigenvector are given by

$$\lambda_A(\theta) = \frac{\lambda}{\lambda + \theta} \quad \text{and} \quad \phi(b;\theta) = e^{\frac{b\lambda r}{1-r}}.$$

Substituting these expressions into (15.35), we see that the ECM (in particular, the OECM) for this case is the same as for the independent interarrivals with distribution $\exp(\lambda)$. In other words, the family of the exponential changes of measure for this case is the family of independent exponential distributions.

15.6.3 Rare events in a GI/G/1 queue with batch arrivals

Let T be a random variable denoting the size of an arriving batch (i.e., T is the number of customers in a batch). Assume that T is distributed according to a discrete distribution, denoted by $F_T(\cdot)$. We call such a queue the GB/GI/1 queue. Let us make the following assumptions.

Condition B1. (analogous to A1) $0 < \mathbb{E}B/\mathbb{E}T < \mathbb{E}A < \infty$ and $\mathbb{P}(BT - A) > 0) > 0$; $\Lambda_T(\nu) \equiv \log \mathbb{E}[\exp(-\nu T)] < \infty$ for all ν.

Condition B2. There exists w_3 such that $\Lambda(w_3) = 0$ and $0 < \Lambda'(w_3) < \infty$, where we define $\Lambda(\theta) = \Lambda_A(\theta) + \Lambda_T(-\Lambda_B(-\theta))$.

Condition B3. Let $K_T(n) = F_T(n)^2/G_T(n)$, $\Lambda_{K_T}(u) = \log \mathbb{I}_{K_T}\exp(-sT) = \log \sum_{n=0}^{\infty} \exp(-sn)K_T(n)$ and $\Lambda_K(\theta) = \Lambda_{K_A}(u) + \Lambda_{K_T}(-\Lambda_{K_B}(-u))$. Then there exists a value v_3 such that $\Lambda_K(v_3) = 0$ and $0 < \Lambda_K'(v_3) < \infty$.

Proposition 15.16. *Under conditions B1-B3, the exponential part of the computational cost in the GB/GI/1 queue is given by*

$$z_{\widetilde{\pi}} = \Lambda_{K_A}(v_3) - 2\Lambda_A(w_3).$$

For the particular case with $G_A = F_A^{(w_3)}$, $G_B = F_B^{(-w_3)}$, and $G_N = F_T^{(\nu)}$, where $\nu = -\Lambda_B^{(-w_3)}$, we have $z_{\widetilde{\pi}} = 0$ and hence a polynomial time IS estimator.

Example 15.17. Consider the M/M/1 queue with batch arrivals and assume that the batch size has a Poisson distribution; that is, $A \sim \exp(\lambda)$, $B \sim \exp(\mu)$, and $T \sim \text{Poisson}(\gamma)$. In this case, the ECM (with the parameter θ) is given by $G_A = F_A^{(\theta)} = \exp(\lambda + \theta)$, $G_B = F_B^{(-\theta)} = \exp(\mu - \theta)$, and $G_T = F_T^{(\nu)} = \text{Poisson}(\gamma(e^{-\nu}-1))$. With $\nu = -\Lambda_B(-\theta)$, the distribution $F_T^{(\nu)}$ reduces to Poisson $(\gamma(\mu/(\mu-\theta)-1))$. By Proposition 15.16, we have a polynomial time IS estimator for $\theta = w_3$, where w_3 satisfies $\dfrac{\lambda}{\lambda + w_3} e^{\gamma(\frac{\mu}{\mu - w_3} - 1)} = 1$.

15.6.4 Sensitivity analysis

Consider now the sensitivity (gradient) of p_x with respect to the parameter vector $\mathbf{v}$ of the interarrival time pdf $f_A(x, \mathbf{v})$. Assume that $f_A(x, \mathbf{v})$ belongs to the exponential family (15.28). Differentiating $\bar{p}_x$ in (15.14) with respect to $\mathbf{v}$, we obtain the following asymptotically consistent estimator of ∇p_x [29]:

$$\nabla \mathbf{v} p_x \equiv \bar{p}_x^{(1)} - \bar{p}_x \bar{p}_x^{(2)}, \tag{15.37}$$

where

$$\bar{p}_x = \frac{\sum_{i=1}^{N} \sum_{t=1}^{\tau_i} I_{ti} W_{ti}}{\sum_{i=1}^{N} \sum_{t=1}^{\tau} W_{ti}}, \quad \bar{p}_x = \frac{\sum_{i=1}^{N} \sum_{t=1}^{\tau_i} I_{ti} S_{ti} W_{ti}}{\sum_{i=1}^{N} \sum_{t=1}^{\tau} W_{ti}},$$

$$\bar{p}_x^{(2)} = \frac{\sum_{i=1}^{N} \sum_{t=1}^{\tau_i} S_{ti} W_{ti}}{\sum_{i=1}^{N} \sum_{t=1}^{\tau} W_{ti}}, \quad I_{ti} = I_{\{Q_{ti} > x\}},$$

and (see (15.28))

$$S_{ti} = t \frac{\nabla_{\mathbf{v}} a(\mathbf{v})}{a(\mathbf{v})} + \nabla \mathbf{v} b(\mathbf{v}) \sum_{k=1}^{t} c(A_{ki}). \tag{15.38}$$

Note that S_t is called the *score function process* (see e.g., [29]).

    Proposition 15.18. *Under the conditions of Theorem* 15.11, *the computational cost* $T_{\widetilde{\pi}}^{(1)} = \exp(z_{\widetilde{\pi}}^{(1)}(x) + o(x))$ *of the estimator* (15.37) *coincides with* $T_{\widetilde{\pi}} = \exp(z_{\widetilde{\pi}}(x) + o(x))$ *of the SR estimator* (15.14), (15.24).

    Proof. In Appendix B (Section 15.9) we present a sketch of the proof of Proposition 15.18 which is similar to that of Theorem 15.11.   □

    It follows from Proposition 15.18 that for efficient simultaneous estimators of p_x and ∇p_x, we can use the same change of measures. In particular, π^* results in a polynomial time IS estimator.

15.7 NUMERICAL EXAMPLES

Here we present numerical results on the efficiency of the SR estimator (15.14), (15.24) for the M/M/1 queue (with the service rate $\mu = 1$) and some of its extensions. In all cases we simulated 10^6 customers and estimated the probability p_x of excessive backlog x. We ran the SR estimator with $\widetilde{\pi} \in \Pi'$ (see Definition 15.10) for the following two cases: with the optimal parameter w, corresponding to the distributions $(G'_A, G'_B) = (F_A^{(w)}, F_B^{(-w)})$ and with the perturbed values of w, denoted by t and corresponding to $(G'_A, G'_B) = (F_A^{(t)}, F_B^{(-t)})$. In all our tables the values in the first column marked with $*$ corresponds to w.

    Table 4 presents the computational cost (simulation time) $\bar{T}_{\widetilde{\pi}}$ required to obtain a $(0.5, 0.5)$-accurate estimators (see Definition 15.2); the ratio $R_t = \bar{T}_{\widetilde{\pi}_t}/\bar{T}_{\pi^*} = S_{opt}/S_t$, called the *loss factor*, and the exponential rate z, (approximation of $\exp(zx)$), as functions of t (and the relative perturbation $\delta = (t - w)/w$, while estimating the probability $p_x = 5.34 \cdot 10^{-10}$ ($x = 40$) in the M/M/1 queue with $\rho = 0.6$.

Table 4. The efficiency of the SR estimator (15.14), (15.24) as a function of the perturbed parameter t for the M/M/1 queue with $\rho = 0.6$, $\mu = 1$, and $p_x = 5.34 \cdot 10^{-10}$ $(x = 40)$.

t	δ	$\bar{T}_{\widetilde{\pi}}$	R_t	$\exp(zx)$	z
0.30	-0.25	$1.20 \cdot 10^6$	4.5	6.1	$4.53 \cdot 10^{-2}$
0.35	-0.125	$5.87 \cdot 10^5$	2.2	1.7	$1.33 \cdot 10^{-2}$
0.40*	0.00	$2.76 \cdot 10^5$	1.0	1.0	0.0
0.43	0.075	$4.27 \cdot 10^5$	1.6	1.3	$1.3 \cdot 10^{-2}$
0.46	0.15	$1.52 \cdot 10^6$	5.5	5.2	$4.13 \cdot 10^{-2}$

Tables 5, 6, and 7 present data similar to those of Table 4 for the M/D/1 queue with $\rho = 0.3$, $d = 1$, and $p_x = 1.04 \cdot 10^{-9}$ $(x = 20)$; the MM/M/1 queue with $\rho = 0.5$, $\mu = 1$, $p_x = 5.34 \cdot 10^{-10}$ $(x = 40)$, and the interarrival sequence being the AR(1) sequence, $A_{k+1} = \theta A_k + (1 - \theta)X$, $X \sim \exp(\lambda)$, $\theta = 0.3$, and the derivative $\nabla_\lambda p_x$ with respect to the arrival rate λ in the M/M/1 queue with $\rho = 0.3$, $\mu = 1$, $p_x = 1.44 \cdot d^{-16}$ $(x = 30)$, $\nabla_\lambda p_x = 1.42 \cdot 10^{-14}$, respectively.

Table 5. The efficiency of the SR estimator (15.14), (15.24) for the M/D/1 queue with $\rho = 0.3$, $d = 1$, and $p_x = 1.04 \cdot 10^{-9}$ $(x = 20)$.

t	δ	$\bar{T}_{\widetilde{\pi}}$	R_t	$\exp(zx)$	z
1.3	-0.368	$1.11 \cdot 10^7$	39.42	27.1	$1.65 \cdot 10^{-1}$
1.6	-0.22	$1.55 \cdot 10^5$	5.48	3.46	$6.20 \cdot 10^{-2}$
2.0645*	0.0	$2.84 \cdot 10^4$	1.0	1.0	0.0
2.4	0.19	$7.51 \cdot 10^4$	2.65	2.27	$4.10 \cdot 10^{-2}$
2.7	0.31	$1.06 \cdot 10^7$	37.43	48.4	$1.94 \cdot 10^{-2}$

Table 6. The efficiency of the SR estimator (15.14), (15.24) for the MM/M/1 queue with $\rho = 0.5$, $\mu = 1$, $\theta = 0.3$, and $p_x = 5.34 \cdot 10^{-10}$ $(x = 40)$.

t	δ	$\bar{T}_{\widetilde{\pi}}$	R_t	$\exp(zx)$	z
0.40	-0.2	$1.08 \cdot 10^6$	4.2	3.51	$1.22 \cdot 10^{-2}$
0.45	-0.1	$5.13 \cdot 10^5$	2.0	1.44	$4.19 \cdot 10^{-2}$
0.50*	0.0	$2.55 \cdot 10^5$	1.0	1.0	0.0
0.54	0.08	$5.43 \cdot 10^5$	2.1	1.45	$1.23 \cdot 10^{-2}$
0.58	0.16	$1.91 \cdot 10^6$	7.5	10.3	$7.78 \cdot 10^{-2}$

Table 7. The efficiency of the SR estimator (15.37), (15.24) for the
M/M/1 queue with $\rho = 0.3$, $\mu = 1$, $p_x = 1.44 \cdot 10^{-6}$ $(x = 30)$,
and $\nabla_\lambda p_x = 1.42 \cdot 10^{-4}$.

t	δ	$\bar{T}_{\tilde{\pi}}$	R_t	$\exp(zx)$	z
0.50	-0.285	$1.03 \cdot 10^7$	40.7	53.9	$1.33 \cdot 10^{-2}$
0.60	-0.143	$8.75 \cdot 10^5$	3.75	3.53	$4.21 \cdot 10^{-2}$
0.70*	0.0	$2.53 \cdot 10^5$	1.0	1.0	0.0
0.75	0.07	$3.68 \cdot 10^5$	1.45	1.9	$2.14 \cdot 10^{-2}$
0.80	0.143	$3.17 \cdot 10^7$	124.9	171.2	$1.71 \cdot 10^{-1}$

It follows from these tables that our simulation results are in agreement with theory. In particular, the SR estimator (15.14), (15.24) is robust with respect to small and moderate perturbations in w, in the sense that for the relative perturbation $|\cdot| < 0.2$ we still have a dramatic variance reduction. Similar results were obtained for the GI/G/1 queue with different interarrival and service distributions.

Our extensive numerical results also suggest that the optimal parameter w, once found, can be used for estimating *simultaneously all probabilities* p_x (for different values x) of the order 10^{-2} or less.

15.8 FURTHER RESEARCH AND CONCLUDING REMARKS

As for further research, we give some guidance on how to approximate the optimal parameter w (for the above GI/G/1 queue and its extensions) without resorting to the solution of the equation of type

$$\Lambda(t) \equiv \Lambda_A(t) + \Lambda_B(-t) = 0. \tag{15.39}$$

To do so, consider minimization of the variance of the switching regenerative estimator (15.14), (15.24), but with the OP w in (15.24) replaced by t, that is, consider the following problem:

$$\min_t \mathrm{Var}\{\bar{p}_{x, N}(t)\}. \tag{15.40}$$

It readily follows from the definition of $T_{\tilde{\pi}_t}$ that asymptotically in N the optimal solution of (15.40) coincides with the optimal solution w of the problem

$$\min_t T_{\tilde{\pi}_t}, \tag{15.41}$$

where $\tilde{\pi}_t = \tilde{\pi}(G'_A, G'_B) \in \Pi'$ and $G'_A = F_A^{(t)}$, $G'_B = F_B^{(-t)}$. One of our main goals will be to show that the SR estimator (15.14), (15.24) still remains polynomial if we replace the OECM pair $(F_A^{(w)}(y_1, \mathbf{v}_1), F_B^{(-w)}(y_2, \mathbf{v}_2))$ in W_{ni} (see (15.24)) by the pair $(G_A^*, G_B^*) = (F_A(y_1, \mathbf{v}_{01}^*), F_B(y_2, \mathbf{v}_{02}^*))$, where the pair $(\mathbf{v}_{01}^*, \mathbf{v}_{02}^*)$ represents the optimal solution of problem (15.40), with t replaced by the vector $(\mathbf{v}_{01}, \mathbf{v}_{02})$.

In practice, however, the exact optimal solution $(\mathbf{v}_{01}^*, \mathbf{v}_{02}^*)$ of the problem

$$\min_{(\mathbf{v}_{01}, \mathbf{v}_{02})} \tilde{\mathrm{Var}}\{\bar{p}_{x, N}(\mathbf{v}_{01}, \mathbf{v}_{02})\} \tag{15.42}$$

is not available, so it must be estimated by simulation, that is, it is the solution of the stochastic counterpart (see e.g., [29]):

$$\min_{\mathbf{v}_{01}, \mathbf{v}_{02})} \widetilde{\text{Var}}\{\bar{p}_{x,N}(\mathbf{v}_{01}, \mathbf{v}_{02})\}. \tag{15.43}$$

Here, $\widetilde{\text{Var}}(\cdot)$ is the sample variance of $\text{Var}(\cdot)$.

Problem (15.43) can be readily extended to more general queueing models, such as

$$\min_{\mathbf{v}_0}\{\widetilde{\text{Var}}\bar{p}_{x,N}(\mathbf{v}_0)\}.$$

Let $\widehat{\mathbf{v}}_{0N}^*$ be the optimal solution of this problem. One of our further main goals is to show that the SR estimators of type (15.14), (15.24), with the estimated optimal value $\widehat{\mathbf{v}}_{0N}^*$ in the LR $W_{ni}(\mathbf{y}, \widehat{\mathbf{v}}_{0N}^*)$ will still possess reasonably low variance. Such optimism relies on:

(a) the robustness and low variance of the SR estimator (15.14), (15.24) with the OP w;

(b) the convergence of $\widehat{\mathbf{v}}_{0N}^*$ to $\mathbf{v}_0^*$ (see [29]), from which it follows that one can always take a sample N (perhaps, large enough) such that

$$\| \widehat{\mathbf{v}}_{0N}^* - \mathbf{v}_0^* \| / \| \mathbf{v}_0^* \| \leq 0.2$$

holds with a high probability. Note that 0.2 is associated with the robustness of SR estimator (15.14), (15.24) in the sense that for relative perturbations $|\delta| = |t - w|/w \leq 0.2$ it still has manageable variance.

Concluding remarks

In this chapter we presented a framework for complexity analysis of rare event estimators. In particular, we defined polynomial and exponential time SR estimators for the evaluation of the steady state probabilities of excessive backlog p_x in the GI/GI/1 queue and some of its extensions. The proposed SR estimators are based on large deviation theory and the use of exponential change of measure, parametrized by scalar t. We showed how to find the optimal value w of parameter t, which results in the optimal exponential change of measures (OECM) and found conditions (see Theorem 15.9) under which the SR estimator (15.14), (15.24) is polynomial. We investigated the robustness of the proposed SR estimators, in the sense that we found how much one can perturb the optimal value w while the SR estimator still results in dramatic variance reduction and remains useful in practice. In particular, our numerical results suggest that if the optimal parameter w is perturbed up to 20%, we only lose 2-3 orders of magnitude of variance reduction, compared with 10 orders under the optimal value w.

15.9 APPENDICES

Appendix A: Optimal exponential change of measure

Large deviation theory allows us to identify the exponential part of statistics of rare events, namely

$$\lim_{x \to \infty} \frac{1}{x}\log \alpha_x \sim I$$

or

$$\alpha_x \approx \exp(-xI) \quad (\text{UTLE}),$$

where UTLE is the acronym for "up to logarithmic equivalence".

We shall use LD results that deal with identifying the OECM. We now explain this concept referring to the GI/GI/1 queue case.

Let X be a random variable (rv) with distribution $F_X(\cdot)$. Let

$$\Lambda_X(\theta) = \log(\mathbb{E}e^{-\theta X}), \quad \theta \in R^1,$$

denote the cumulant function, and define

$$\mathfrak{D}_X \equiv \{\theta \in R^1 \mid \Lambda_X(\theta) < \infty\}.$$

For any $\theta \in \mathfrak{D}_X$, define the exponential change of measure as follows:

$$F_X^{(\theta)}(dx) = \exp(-\theta x - \Lambda_X(\theta))F_X(dx).$$

The GI/GI/1 queue is determined by iid interarrival and service times sequences $\{A_k\}$ and $\{B_k\}$. Define

$$\Lambda(\theta) = \Lambda_A(\theta) + \Lambda_B(-\theta)$$

and

$$w = \sup[\theta \in \mathfrak{D}_A \cap \mathfrak{D}_B \mid \Lambda(\theta) \le 0].$$

The classical LD result states that if $0 < w < \infty$, $\Lambda(w) = 0$, and $\Lambda'(w) < \infty$, then

$$P_\pi(Q_n \ge x) \approx \exp(x\Lambda(w)) \quad (\text{UTLE}).$$

Moreover,

$$\theta = w, \tag{15.44}$$

specified by

$$F_A^{(\theta)}(dx) = \exp(-\theta x - \Lambda_A(\theta))F_A(dx), \tag{15.45}$$

and

$$F_B^{(\theta)}(dx) = \exp(\theta x - \Lambda_B(-\theta))F_B(dx) \tag{15.46}$$

specifies the optimal (in terms of minimal variances of the IS estimators) exponential change of measure within the class of all feasible $(\theta \in \mathfrak{D}_A \cap \mathfrak{D}_B)$ exponential changes of measure (called the optimal exponential change of measure (OECM)).

Recall that OECM leads to an unstable queueing process (see, e.g. [17] for some discussions). For example, for the M/M/1 queue with traffic intensity $\rho < 1$ the OECM corresponds to the M/M/1 queue with traffic intensity $\rho_0 \equiv \lambda_0/\mu_0 = \rho^{-1} = \mu/\lambda > 1$.

Appendix B: Proof of main results

Proof of Theorem 15.9 (sketch). Part A. Using the well known-formula for the variance of ratio estimators (see e.g., [29]), we have (after simple transformations) that the SCV $\kappa(x)$ of the SR estimator (15.14), (15.24) is given by

$$\kappa^2(x) = \frac{\mathbb{E}_{\widetilde\pi} X^2}{(\mathbb{E}_{\widetilde\pi} X)^2} + \frac{\mathbb{E}_{\widetilde\pi} Y^2}{(\mathbb{E}_{\widetilde\pi} Y)^2} - 2\frac{\mathbb{E}_{\widetilde\pi} XY}{\mathbb{E}_{\widetilde\pi} X \mathbb{E}_{\widetilde\pi} Y} = \kappa_1 + \kappa_2 + \kappa_3, \tag{15.47}$$

where $X \equiv \sum_{t=1}^\tau I_t W_t$, $Y \equiv \sum_{t=1}^\tau W_t$, $I_t \equiv I_{[x,\infty)}(Q_t)$,

$$W_t = \frac{\prod_{k=1}^{D(t)} F_B(dB_k)\prod_{k=1}^t F_A(dA_k)}{\prod_{k=1}^{D(t)} G_{B,k}(dB_k)\prod_{k=1}^t G_{A,k}(dA_k)},$$

and $\prod_{k=i}^{j} C_k \equiv 1$ and $\sum_{k=i}^{j} C_k \equiv 0$ for any sequence $\{C_k\}$ if $j < i$. Using the IS policy $\tilde{\pi} = \pi^*$ (see Definition 15.7), we obtain

$$W_t = \exp(x\Lambda_A(w) + wZ_t), \ t = \xi,$$

$$= W_\xi, \quad \xi < t < \tau, \tag{15.48}$$

where

$$Z_t \equiv \sum_{k=1}^{t} A_k - \sum_{k=1}^{t-x} B_k. \tag{15.49}$$

Now consider each term of the right-hand side of (15.47) separately:

$$\mathbb{E}_{\pi^*} X^2 = \mathbb{E}_{\pi^*} [(\sum_{t=\xi}^{\tau} I_t W_t)^2; \xi \leq \tau]$$

$$= \mathbb{E}_{\pi^*}[W_\xi^2 \sum_{t=\xi}^{\tau} I_t; \xi \leq \tau] + 2\mathbb{E}_{\pi^*}[\sum_{t=\xi}^{\tau} I_t W_t \sum_{s=t+1}^{\tau} I_s W_s; \xi \leq \tau]$$

$$= \mathbb{E}_{\pi^*}[e^{2Z_\xi} \xi^w h_x; \xi \leq \tau] e^{2\Lambda_A(w)x}, \tag{15.50}$$

where $h_x = \mathbb{E}[\sum_{t=\xi}^{\tau} I_t(1 + 2\sum_{s=t+1}^{\tau} I_s) \mid \mathcal{F}_\xi]$, with $\mathcal{F}_\xi = \sigma(\widetilde{B}_\xi)$. Hence we can write $h_x \equiv h_x(\widetilde{B}_\xi)$.

Lemma 15.19. *Under assumptions A1, A2 we have*

1. *The pointwise monotone limit $h_x(\cdot) \to h(\cdot)$, where $h(\cdot)$ is a measurable bounded function.*

2.

$$\lim_{x\to\infty} \mathbb{E}_w[e^{2Z_\xi} \xi^w h(\widetilde{B}_\xi); \xi \leq \tau] = O(1)\lim_{x\to\infty} \mathbb{E}[e^{2Z_\xi} \xi^w; \xi \leq \tau].$$

Proof. The proof of this lemma is, in fact, a specification of the proofs of Lemma 5.2 and Lemma 5.4 in [33]. $\qquad\qquad\square$

Write now,

$$\xi = \min[1 \leq t \leq \tau \mid Z_t < 0].$$

By renewal theory (see, e.g., [3] or [15]) Z_ξ has a limiting distribution

$$\lim_{x\to\infty} \mathbb{E}_w[e^{2Z_\xi} \xi^w; \xi \leq \tau] = \psi,$$

where ψ is a constant $0 < \psi < \infty$. Thus, $\mathbb{E}_{\pi^*} X^2 = O(1)\exp(2\Lambda_A(w)x)$. Noting that $(\mathbb{E}X)^2 = (\mathbb{E}\tau P_\pi(Q \geq x))^2 = O(1)\exp(2\Lambda_A(w)x)$ (see e.g., [3]), we obtain that κ_1 is bounded by a constant function in x.

Consider κ_3. We have (see [5])

$$\mathbb{E}Y^2 = \mathbb{E}_{\pi^*}[(\sum_{s=t}^{\tau} W_t)^2]$$

$$= \mathbb{E}_{\pi^*}[(\sum_{s=t}^{\tau} W_t^2)] + 2\mathbb{E}_{\pi^*}[\sum_{t=1}^{\tau} W_t \sum_{s=t+1}^{\tau} W_s]$$

$$= \mathbb{E}_{\pi^*}[(\sum_{s=t}^{\tau} W_t^2)(1 + 2\mathbb{E}[\tau - t \mid \mathcal{F}_t])]. \tag{15.51}$$

Since $t \geq D(t)$, $S_t \equiv \sum_{k=1}^{t} A_k - \sum_{k=1}^{D(t)} B_k \leq 0$, $1 \leq t \leq \tau$, and $\Lambda_A(w) < 0$, we obtain $W_t \equiv \exp((t - D(t))\Lambda_A(w) + wS_t) \leq 1$, $1 \leq t \leq \xi$, and thus,

$$\mathbb{E}_{\pi^*}Y^2 \leq \mathbb{E}_{\pi^*}[\sum_{t=1}^{\tau} W_t] + 2\mathbb{E}_{\pi^*}[\sum_{t=1}^{\tau} W_t \mathbb{E}[\tau - s \mid \mathcal{F}_t]]$$
$$= \mathbb{E}\tau + 2\mathbb{E}[\sum_{t=1}^{\tau} \mathbb{E}[\tau - t \mid \mathcal{F}_t]]$$
$$\leq \mathbb{E}\tau + 2(\mathbb{E}\tau)^2 = O(1).$$

Then, $\mathbb{E}_{\pi^*}Y = \mathbb{E}\tau$; hence, $\kappa_2(x) = O(1)$.

Noting next that $N < \gamma(\kappa_1(x) + \kappa_2(x))$, we have $N = O(1)$. (Recall that N is the number of cycles required to obtain an estimator of (ϵ, δ)-accuracy and $\gamma = \Phi^{-1}(1 - \delta/2)^2 \epsilon^{-2}$.) It is not difficult to show that

$$\lim_{x \to \infty} P_{\pi^*}(\xi \leq \psi x; \xi < \tau) = 1, \tag{15.52}$$

where $\psi < \infty$. Indeed, note that the waiting time of the $(\xi - 1)$th customer,

$$L_\xi = \sum_{k=1}^{\xi-1} (B_k - A_k) < \sum_{k=\xi-x}^{\xi-1} B_k,$$

provided $\xi < \tau$. This implies

$$\sum_{k=1}^{\xi-1} (B'_k - A_k) < \sum_{k=\xi-x}^{\xi-1} B'_k,$$

where

$$B'_k = B_k, \quad 1 \leq k \leq \xi - x$$

$$= S_k, \quad k > \xi - x,$$

with $S_k \sim G_B$. It is clear that $\xi = \xi(x) > x$. Hence, by the strong law of large numbers (SLLN), we obtain

$$\xi^{-1} \sum_{k=1}^{\xi} (B'_k - A_k) \to \mathbb{E}_{G_B} B - \mathbb{E}_{G_A} A \equiv \delta_1 > 0$$

(a.s.) as $x \to \infty$. Besides, by the SLLN, we get

$$x^{-1} \sum_{k=\xi-x}^{\xi-1} B'_k \to \mathbb{E}_{G_B} B \equiv \delta_2$$

(a.s.) as $x \to \infty$. Hence, $\lim_{x \to \infty} P_{\pi^*}(\xi \delta_1 < x\delta_2; \xi < \tau) = 0$ for $0 < \delta_2/\delta_1 < \infty$. Moreover, using the CLT, we can obtain by routine evaluation $P_{\pi^*}(\xi/x = \tilde{\psi}_1; \xi < \tau) = O(\exp(-r_1 \tilde{\psi}_1 x))$ for $\delta_2/\delta_1 < \tilde{\psi}_1 < \infty$, $r < 0$. Therefore,

$$\mathbb{E}_{\pi^*}[\xi; \xi < \tau] \leq \psi_1 x.$$

Similarly,

$$\mathbb{E}_{\pi^*}[\tau - \xi; \tau > \xi] \leq \psi_2 x, \tag{15.53}$$

where ψ_1 and ψ_2 are some finite constants. We also note that

$$P_{\pi^*}(\tau = \psi x; \tau < \xi) = P_{\pi^*}(\sum_{k=1}^{\psi x} (B_k - A_k) < 0; \tau < \xi).$$

But $(\psi x)^{-1} \sum_{k=1}^{\psi x} (B_k - A_k) \to \mathbb{E}_{G_B} B_k - \mathbb{E}_{G_A} A_k > 0$ (a.s.) as $x \to \infty$. Hence by the CLT,

$$P_{\pi^*}(\tau = \psi x; \tau < x) = O(\exp(-r_2\psi x)),$$

with $0 < \psi < \infty$, $0 < r_2 < \infty$ for large x, and therefore $\mathbb{E}[\tau; \tau < \xi] < \psi_3 x$, $\psi_3 < \infty$. Thus, we have

$$\mathbb{E}_{\pi^*}\tau = \mathbb{E}_{\pi^*}[\xi; \tau > \xi] + \mathbb{E}_{\pi^*}[\tau - \xi; \tau > \xi] + \mathbb{E}_{\pi^*}[\tau; \tau \leq \xi].$$

Hence $\mathbb{E}_{\pi^*}\tau \leq (\psi_1 + \psi_2 + \psi_3)x$.

Finally, $T_{\pi^*} = N\mathbb{E}_{\pi^*}\tau = O(x)$. $\square$

Part B: Note that, according to (15.47), (15.50) and (15.51), we have

$$\frac{\mathbb{E}_{\widetilde{\pi}}[X^2]}{\mathbb{E}_{\widetilde{\pi}}[Y^2]} > \frac{\mathbb{E}_{\widetilde{\pi}}[\sum_{t=1}^{\tau} I_t W_t^2]}{\mathbb{E}_{\widetilde{\pi}}[\sum_{t=1}^{\tau} W_t^2(1 + 2\mathbb{E}[\tau - t \mid \mathcal{F}_t]]}.$$

Then, for $\widetilde{\pi} \in \Pi$ and $H_t \equiv \mathbb{E}[\tau - t \mid \mathcal{F}_t]$, we have $H_t = O(Q_t)$, given $\mathcal{F}_t = (Q_t, \widetilde{B}_t)$, with $Q_t \to \infty$ and $\widetilde{B}_t < \infty$ (similar to (15.52)). Taking the latter into account, according to the definition of $\widetilde{\pi} \in \Pi$, we have

$$P_{\widetilde{\pi}}(Q_t = \widetilde{\psi}_1 x) = O(\exp(-r\widetilde{\psi}_1 x))$$

for $\widetilde{\psi}_1 > 1$, $0 < r < \infty$, and

$$\mathbb{E}_{\widetilde{\pi}}[W_t^2 H_t] < \psi x \mathbb{E}_{\widetilde{\pi}}[W_t^2],$$

where $\psi < \infty$.

Thus,

$$\frac{\mathbb{E}_{\widetilde{\pi}}[X^2]}{\mathbb{E}_{\widetilde{\pi}}[Y^2]} > (1 + 2\psi x)^{-1} \frac{\mathbb{E}_{\widetilde{\pi}}[\sum_{t=1}^{\tau} I_t W_t^2]}{\mathbb{E}_{\widetilde{\pi}}[\sum_{t=1}^{\tau} W_t^2]}.$$

Now we introduce the following notation (similar to (15.29)-(15.30)):

$$K_{A,k}(dA) = \frac{F_{A,k}(dA)}{G_{A,k}(dA)} F_{A,k}(dA),$$

$$K_{B,k}(dB) = \frac{F_{B,k}(dB)}{G_{B,k}(dB)} F_{B,k}(dB),$$

and we let $\widetilde{\pi}_K$ and $\Pi_{\widetilde{\pi}_K}$ with K_A and K_B be defined in the same way as $\widetilde{\pi}$ and $\mathbb{E}_{\widetilde{\pi}}$ with G_A and G_B, respectively. Now we can write

$$\frac{\mathbb{E}_{\widetilde{\pi}}[X^2]}{\mathbb{E}_{\widetilde{\pi}}[Y^2]} > (1 + 2\psi x)^{-1} \frac{\Pi_{\widetilde{\pi}_K}[\sum_{t=1}^{\tau} I_t]}{\Pi_{\widetilde{\pi}_K}[\sum_{t=1}^{\tau} 1]}.$$

Assume without loss of generality that $\Pi_{\widetilde{\pi}_K}\tau < \infty$. Using the basic result for regenerative processes (see, e.g., [3], (5.1.1)), we obtain

$$\frac{\mathbb{E}_{\widetilde{\pi}}[X^2]}{\mathbb{E}_{\widetilde{\pi}}[Y^2]} > 91 + 2\psi x)^{-1}\Pi_{\pi_K}I_t,$$

where the subscript π_K indicates that the "expectation" Π_{π_K} is taken in the steady state mode, provided the behavior of the queue is "driven" by the IS policy $\widetilde{\pi}_K$. Now we use the fact that $I_t = I_{[-\infty, 0)}(\max_{s, 1 \leq s \leq t} \widetilde{Q}_s)$, where $\widetilde{Q}_{t+1} =$

$\tilde{Q}_t + 1 - D_t$, and D_t is the number of customers departing from the queue between the tth and $(t+1)$th arrivals, given it is not empty during this period. (It is clear that $Q_{t+1} = \max[Q_t + 1 - D_t, 0]$, i.e., $\tilde{Q}_t$ is an unreflected modification of Q_t.) Denote $\tilde{\xi} \equiv \min[t \mid \tilde{Q}_t > x]$. We obtain (see, e.g., [15]) that $\mathrm{II}_{\pi_K} I_t = \mathrm{II}_{\tilde{\pi}_K}[\tilde{\xi} < \infty]$. Thus, we have

$$\frac{\mathbb{E}_{\tilde{\pi}}[X^2]}{\mathbb{E}_{\tilde{\pi}}[Y_2]} > (1 + 2\psi x)^{-1} \mathrm{II}_{\tilde{\pi}_K}[\tilde{\xi} < \infty].$$

Parallel to the proof of Lemma 3 in [23], we obtain

$$\mathrm{II}_{\tilde{\pi}_K}[I_{[0,\infty)}(\tilde{\xi})] > \mathrm{II}_{\tilde{\pi}_{K_u}}[I_{[0,\infty)}(\tilde{\xi})\exp(P_u)],$$

where

$$P_u = \sum_{t=1}^{\tilde{\xi}}(\Lambda_{K_A, t}(u) + \sum_{t=1}^{\tilde{\xi}-x}\Lambda_{K_B, t}(-u)); \qquad (15.54)$$

the subscript $\tilde{\pi}_{K_u}$ indicates that we use $K_A^{(u)}$ and $K_B^{(-u)}$ instead of K_A and K_B. Once can choose u such that

$$\lim_{x \to \infty} \mathrm{II}_{\tilde{\pi}_{K_u}}[I_{[0,\infty)}(\tilde{\xi})] > \psi_1 > 0$$

(see, e.g., [24], page 227 or [23], Lemma 3).

By Jensen's inequality, $\Lambda_{K_A, t}(2u) \geq 2\Lambda_A(w)$ and $\Lambda_{K_B, t}(-2u) \geq 2\Lambda_B(-w)$ with equalities if and only if $G_{A,t} = F^{(w)}$ and $G_{B,t} = F^{(-w)}$, respectively. Let $\tilde{\pi} \notin \mathrm{II}^*$; then,

$$\lim_{x \to \infty}\frac{P_u - 2\Lambda_A(w)}{x} = \tilde{\psi} > 0.$$

Thus,

$$\frac{\kappa_1}{\kappa_2} > \frac{O(1)}{1 + 2\psi x}\frac{\mathbb{E}_{\tilde{\pi}}X^2}{\mathbb{E}_{\tilde{\pi}}Y^2\exp(2\Lambda_A(w)x)} = \frac{O(1)}{1 + 2\psi x}\exp(zx), \qquad (15.55)$$

where $z > 0$. Finally, note that by Cauchy's inequality,

$$T_{\tilde{\pi}} \geq N \geq \gamma(\sqrt{\kappa_1} - \sqrt{\kappa_2})^2.$$

We have proved that $\tilde{\pi} \notin \mathrm{II}^*$ implies $\tilde{\pi} \notin \mathrm{II}^{**}$, which directly implies statement B of the theorem. □

Proof of Theorem 15.11 (sketch). Parallel to the proof of Theorem 15.9, Part B for $\tilde{\pi} \in \mathrm{II}'$, and using that $P_v = \Lambda_{K_A}(v)x$ in (15.53), we obtain that

$$\frac{\kappa_1}{\kappa_2} > \frac{O(1)}{1 + 2\psi x}\exp(x(\Lambda_{K,A}(v) - 2\Lambda_A(w)));$$

that is, $z_{\tilde{\pi}} > \Lambda_{K,A}(v) - 2\Lambda_A(w) \geq 0$. (For $z_{\tilde{\pi}} > 0$, we have by (15.54) that κ_1 is the dominating term in (15.47).) Then, taking into account that $\tilde{B}_\xi$ has a limiting distribution (by renewal theory) and $h(\tilde{B}) = \mathbb{E}(\sum_{t=\xi}^{\tau}I_t(1 + 2\sum_{s=t+1}^{\tau}I_s) \mid \tilde{B}_\xi]$ is a bounded function (by Lemma 15.8.1), we have

$$\mathbb{E}_{\tilde{\pi}}[h(\tilde{B}_\xi) \mid \mathcal{F}_\xi] < \psi < \infty \text{ (a.s.)}.$$

Thus we obtain an upper bound

$$\kappa_1 < \exp(-2\Lambda_A(w)x)\mathbb{E}_{\widetilde{\pi}}[W_\xi^2 h(\widetilde{B}_\xi)] < \psi \exp(-2\Lambda_A(w)x)\mathbb{E}_{\widetilde{\pi}}[W_\xi^2].$$

Then, under **A1-A3** we have by Theorem 3.1 [30],

$$\mathbb{E}_{\widetilde{\pi}}[W_\xi^2] = \widetilde{\psi}\exp(x\Lambda_{K,A}(v)),$$

where $0 < \widetilde{\psi} < \infty$. Finally, we obtain $z_{\widetilde{\pi}} = \Lambda_{K,A}(v) - 2\Lambda_A(w)$.   □

We omit the proofs of Propositions 15.6.1-15.6.3, since they differ from the proof of Theorems 15.4.1 and 15.5.1 only in their routine calculations.

Proof of Proposition 15.18 (sketch). We can write $\nabla p_x \equiv \nabla \mathbf{v} p_x$ as follows (see [29]):

$$\nabla p_x = \frac{\mathbb{E}\sum_{t=1}^{\tau} I_t S_t}{\mathbb{E}\tau} - \frac{\mathbb{E}\sum_{t=1}^{\tau} I_t}{\mathbb{E}\tau}\frac{\mathbb{E}\sum_{t=1}^{\tau} S_t}{\mathbb{E}\tau}$$

$$\equiv p_x^{(1)} - p_x p^{(2)}.$$

Let $T_{\widetilde{\pi}}^{(1)}$, $T_{\widetilde{\pi}}$, $T_{\widetilde{\pi}}^1$, $T_{\widetilde{\pi}}^2$, and $T_{\widetilde{\pi}}^{0,2}$ be the simulation costs required to obtain IS estimators of prescribed accuracy for ∇p_x, p_x, $p_x^{(1)}$, $p^{(2)}$, and $p_x p_x^{(2)}$, respectively. It is clear that

$$\max[T_{\widetilde{\pi}_1}, T_{\widetilde{\pi}_2}^{0,2}] < T_{\widetilde{\pi}_i}^{0,2} < \psi_1 T_{\widetilde{\pi}_1} + \psi_2 T_F^{0,2}, \quad i = 1,2.$$

Here, $\widetilde{\pi}_i$, $i = 1,2$ are certain IS policies from Π, ψ_i, $i = 1,2$ are some finite constants, and $T_F^{0,2} = T_{\widetilde{\pi}_2}^{0,2}$ for the CMC estimator. Taking into account that $T_F^{0,2}$ does not depend on x, we obtain

$$T_{\widetilde{\pi}} < T_{\widetilde{\pi}}^{0,2} = O(T_{\widetilde{\pi}})$$

for large x. This implies that

$$T_{\widetilde{\pi}} < T_{\widetilde{\pi}}^{(1)} < \psi_3 T_{\widetilde{\pi}}^1 + \psi_4 T_{\widetilde{\pi}},$$

when $\psi_i < \infty$, $i = 3,4$. It remains to prove that the simulation cost $T_{\widetilde{\pi}}^1 = O(T_{\widetilde{\pi}})$. Similar to (15.47), we have

$$T_{\widetilde{\pi}}^1 = \mathbb{E}_{\widetilde{\pi}}\tau\gamma\kappa(x) = \frac{\mathbb{E}_{\widetilde{\pi}} X_1^2}{(\mathbb{E}_{\widetilde{\pi}} X_1)^2} + \frac{\mathbb{E}_{\widetilde{\pi}} Y^2}{(\mathbb{E}_{\widetilde{\pi}} Y)^2} - \frac{2\mathbb{E}_{\widetilde{\pi}} X_1 Y}{\mathbb{E}_{\widetilde{\pi}} X_1 \mathbb{E}_{\widetilde{\pi}} Y}$$

$$= \kappa_{1,1} + \kappa_{2,1} + \kappa_3, \tag{15.56}$$

where $X_1 \equiv \sum_{t=1}^{\tau} I_t S_t W_t$, and $Y \equiv \sum_{t=1}^{\tau} W_t$. We will show that $\kappa_{1,1} = O(\kappa_1)$ (see (15.47)), which implies by the proof of Theorem 15.9 (see in particular (15.54)) that $T_{\widetilde{\pi}}^1 = O(T_{\widetilde{\pi}})$.

From the proof of Theorem 3.1 [30] we have that $\lim_{x\to\infty} P_{\widetilde{\pi}}(\xi \leq \tau < \psi_2 x) = 1$. Then by the definitions of $f_A(\cdot)$ and S_t we have

$$S_t = \psi_3 t + \psi_4 \sum_{k=1}^{t} c(A_k)$$

and $\mathbb{E}c(A_k)^2 < \infty$. Hence,

$$\lim_{x\to\infty} P_{\widetilde{\pi}}(|S_t| < \psi_6 x) = 1$$

for $t \leq \tau$. Also, by the CLT we obtain, using routine evaluation, that

$$\lim_{x\to\infty} P_{\widetilde{\pi}}(|S_t| > \psi_7 x) = O(\exp(-r\psi_7 x)),$$

where $r > 0$. Therefore,

$$\mathbb{E}_{\widetilde{\pi}}(X_1^2) = \mathbb{E}_{\widetilde{\pi}}((\sum_{t=1}^{\tau} I_t W_t S_t)^2) = O(x^2)\mathbb{E}(X^2).$$

We also have

$$\mathbb{E}_{\widetilde{\pi}} X_1 \mathbb{E}_{\widetilde{\pi}}((\sum_{t=1}^{\tau} I_t W_t S_t)) = (\nabla p_x - p_x p^{(2)})\mathbb{E}\tau = O(x p_x) = O(x)\mathbb{E}X.$$

Hence, we get

$$\kappa_{1,1} = O(\kappa_1).$$

The statement of the proposition directly follows from the last equation. $\square$

ACKNOWLEDGEMENT

We would like to thank Adam Schwartz at Technion and Jack Kleijnen at Tilburg University for several valuable suggestions on the earlier draft of this work.

BIBLIOGRAPHY

[1]   Anantharam, V., Heidelberger, P. and Tsoucas, P., Analysis of rare events in continuous time Markov chains via time reversal and fluid approximation, *Research Report*, IBM Research Division 1990.

[2]   Asmussen, S., Conjugate processes and the simulation of ruin problems, *Stoch. Proc. Appl.* **20** (1985), 213-229.

[3]   Asmussen, S., *Applied Probability and Queues*, John Wiley & Sons, New York 1987.

[4]   Asmussen, S., Risk theory in a Markovian environment, *Scand. Actuarial J.* (1989), 69-100.

[5]   Asmussen, S., Rubinstein, R.Y., and Wang, C.L., Regenerative rare events simulation via likelihood ratios, *J. Appl. Prob.* **31**:3 (1994), 797-815.

[6]   Asmussen, S. and Rubinstein, R.Y., Complexity properties of steady state rare events simulation in queueing models, in: *Advances in Queueing; Theory, Methods, and Open Problems*, (429-461) ed. by J.H. Dshalalow, CRC Press, Boca Raton 1995.

[7]   Bolotin, V.A., Kappel, J.G. and Kuehn, P.J., Teletraffic analysis of ATM systems, *IEEE J. Select. Areas Commun.* **9** (1991), 281-283.

[8]   Bucklew, J.A., *Large Deviation Techniques in Decision, Simulation, and Estimation*, Wiley, New York 1990.

[9]   Bucklew, J.A., Ney, P. and Sadowsky, J.S., Monte Carlo simulation and large deviations theory for uniformly recurrent Markov chains, *J. Appl. Prob.* **27** (1990), 44-59.

[10]   Chang, C.S., Heidelberger, P., Juneja, S. and Shahabuddin, P., Effective bandwidth and fast simulation of ATM intree networks, *Research Report*, IBM Research Division, T.J. Watson Research Center, New York 1992.

[11]   Cottrel, M., Fort, J.C. and Malgoures, G., Large deviations and rare events in the study of stochastic algorithms, *IEEE Trans. Auto. Control* **AC-28** (1983), 907-918.

[12]   Devetsikiotis, M. and Townsend, K.R., On the efficient simulation of large communication networks using importance sampling, In: *Proc. of IEEE Globecom '92*, IEEE Computer Society 1992.

[13] Devetsikiotis, M. and Townsend, K.R., A dynamic importance sampling methodology for the efficient estimation of rare events probabilities in regenerative simulations of queueing systems, In: *Proc. of IEEE Globecom '92* (1992), 1290-1297.

[14] Devetsikiotis, M. and Townsend, K.R., Statistical optimization of dynamic importance sampling parameters for efficient simulation of communication networks (1993), preprint.

[15] Feller, W., *An Introduction to Probability Theory and its Application*, Vol. 2, John Wiley & Sons, New York 1966.

[16] Frater, M.R., Lennon, T.M. and Anderson, B.D.O., Optimally efficient estimation of the statistics of rare events in queueing networks, *IEEE Trans. Auto. Control* **AC-36** (1991), 1395-1405.

[17] Frater, M.R. and Anderson, B.D.O., Fast estimation of the statistics of excessive backlogs in tandem networks of queues, *Austral. Telecomm. Res.* **23** (1989), 49-55.

[18] Frater, M.R., Walrand, J., and Anderson, B.D.O., Optimality efficient estimation of the buffer overlow in queues with deterministic service times, *Austral. Telecomm. Res.* **24** (1990), 1-8.

[19] Glynn, P.W. and Iglehart, D.L., Importance sampling for stochastic simulations, *Mgmt. Sci.* **35**:11 (1989), 1367-1392.

[20] Heidelberger, P., Fast simulation of rare events in queueing and reliability models, *IBM Research Report* **RC 19028**, Yorktown Heights, New York 1993. Preliminary version published in *Perf. Eval. of Comp. and Commun. Sys.*, Springer Lecture Notes in Computer Science **729**, 165-202.

[21] Kingman, J.F.C., On the algebra of queues, *Ann. Math. Statis.* **3** (1966), 285-326.

[22] Kriman, V., Sensitivity analysis of GI/GI/m/B queues with respect to buffer size by the score function method, *Stoch. Models* **11**:1 (1995), 171-194.

[23] Lehtonen, T. and Nyrhinen, H., Simulating level crossing probabilities by importance sampling, *Adv. Appl. Prob.* **Dec.**, (1992).

[24] Martin-Lof, A., Entropy, a useful concept in risk theory, *Scand. Actuarial J.* (1986), 223-235.

[25] Melamed, B., TES: A class of method for generating autocorrelation uniform variates, *ORSA J. on Computing* **3** (1991), 317-329.

[26] Nicola, V.F., Shahabuddin, P., Heidelberger, P., and Glynn, P.W., Fast simulation of steady-state availability in non-Markovian highly dependable systems, In: *Proc. of the Twentieth Intern. Symp. on Fault-Tolerant Comp.*, IEEE Computer Society Press (1993), 491-498.

[27] Parekh, S. and Walrand, J., A quick simulation method for excessive backlogs in networks of queues, *IEEE Trans. Automat. Contr.* (1989), 54-66.

[28] Rubinstein, R.Y., The 'push-out' method for sensitivity analysis of discrete event systems, *Ann. Op. Res.* **39** (1992), 229-250.

[29] Rubinstein, R.Y. and Shapiro, A., *Discrete Event Systems: Sensitivity Analysis and Stochastic Optimization via the Score Function Method*, Wiley, New York 1993.

[30] Sadowsky, J.S., Large deviations theory and efficient simulation of excessive backlogs in a GI/GI/m queue, *IEEE Trans. Automat. Contr.* **AC-36** (1991), 1383-1394.

[31] Sadowsky, J.S. and Bucklew, J.A., On large deviations theory and asymptotically efficient Monte Carlo estimation, *IEEE Trans. Inform. Theory* **IT-36**

(1990), 579-588.

[32]   Sadowsky, J.S. and Szpankowski, W., The probability of large queue length
        and waiting times in a heterogeneous multiserver queue, Part I: Tight
        limits, Manuscript, School of Elect. Eng. and Dept. of Comp. Sci., Purdue
        Univ., West Lafayette, IN 47907 USA (1992).

[33]   Sadowsky, J.S. and Szpankowski, W., The probability of large queue length
        and waiting times in a heterogeneous multiserver queue, Part II: Positive re-
        currence and logarithmic limit, Manuscript, School of Elect. Eng. and Dept.
        of Comp. Sci., Purdue Univ., West Lafayette, IN 47907 USA (1992).

[34]   Sadowsky, J., On the optimality and stability of exponential twisting in
        Monte Carlo simulation, *IEEE Trans. Inform. Theory* **IT-39** (1993), 119-
        128.

[35]   Sadowsky, J., Monte Carlo estimation of large deviation probabilities,
        Manuscript, School of Elect. Eng. and Dept. of Comp. Sci., Purdue Univ.,
        West Lafayette, IN 47907 USA (1994).

[36]   Shwartz, A. and Weiss, A., Large deviation for performance analysis:
        queues, communication and computers, Manuscript, Dept. of Elect. Eng.,
        Technion-IIT, Haifa 32000, Israel (1994).

[37]   Stockmeyer, L.J., Computation complexity, *Handbooks in OR & MS* **3**
        (1992).

[38]   Tsoucas, P., Rare events in series of queues, *J. of Appl. Prob.* (1992), 168-
        175.

Chapter 16

Parametric estimation of tail probabilities for the single-server queue[1]

Peter W. Glynn and Marcelo Torres

ABSTRACT In this chapter, we consider the question of how long the arrival process to the single-server queue needs to be observed in order to accurately estimate the long-run fraction of time that the workload exceeds y. We assume that the arrival process can be modeled parametrically. In such a parametric context, our results suggest that one typically needs to observe the arrival process over a time horizon that is large relative to y^2. This conclusion appears to hold regardless of whether the arrival process model exhibits complex dependencies or not.

CONTENTS

16.1 INTRODUCTION

Consider a packet-switched communications network, in which admission control of additional sources to the network is determined by constraints on the long-run fraction of time, which network switches expect to see packet buffer content that exceeds predetermined levels. This call acceptance criteria is natural when end-to-end network delay and packet loss probabilities are of principle importance in measuring network performance. In such a setting, one expects that the admissions control policy will need to either implicitly or explicitly estimate the long-run fraction of time that the packet buffer content is greater than a given level y. Unless a great deal is known about the source traffic to the network, such estimates will need to be based on real-time estimates of traffic.

In this chapter, we consider a simplified version of the above problem, in the hope that it offers some general insight into the issues likely to be encountered in developing such network estimators. We simplify the problem in three fundamental ways. Firstly, rather than dealing with a network, we consider only the traffic congestion as modeled by a single station with a constant rate server. Secondly, we approximate the behavior of the finite buffer workload process with that of the

---

[1]This research was supported by the Army Research Office under contract No. DAAL03-91-G-0319.

---

infinite capacity analog. For instance, when considering the long-run fraction of packets lost due to buffer overflow, we replace the finite-buffer loss probability by the corresponding probability that the infinite-capacity workload process has a steady state workload greater than the finite system's buffer capacity; this probability is commonly used as a surrogate to the loss probability desired and is generally a good approximation when the buffer is (very) large relative to the utilization. Finally we assume, in this paper that the arrival process to the queue can be modeled parametrically.

In particular, Section 16.2 is concerned with estimation issues related to the tail of the steady-state workload distribution of the M/M/1 queue. The uncertainty in the source traffic to the queue is captured as uncertainty in the underlying Poisson arrival rate to the system. Thus, Section 16.2 discusses the question of how to construct estimators for the tail probabilities in such a queue, when the arrival rate is unknown and needs to be estimated from real-time traffic. In Section 16.3, we consider the workload tail estimation problem again, except that now the source traffic is modeled as a Markov-modulated fluid, with a parametrically determined generator. This extension is of some interest, in view of the fact that such Markov-modulated source models are widely used to characterize the complex dependencies exhibited by packet arrivals in modern communication networks; see for example [1, 2, 3, 14, 19]. Section 16.4 considers how our theory is affected by the traffic intensity. Finally, Section 16.5 describes open research problems.

The most important single qualitative conclusion reached in our analysis is that the amount of real-time traffic that one must observe in order to accurately estimate the steady-state probability that workload exceeds y must be large relative to y^2, when y is out in the extreme right tail. This behavior is exhibited in both our M/M/1 analysis and Markov-modulated analysis. While the amount of data that one needs to collect might appear to be large, it turns out that much more data needs to be collected if the source traffic is modeled non-parametrically. In particular, the amount of data collected must increase much faster than y^2 in such problems; see Glynn and Torres [10] for details. Thus, assuming parametric modeling of the source traffic is appropriate to the application considered, it can provide better performance predictions to the admission control entity during the real-time operation of a network.

We note that some of the theory developed in this paper also relates to the question of how "continuous" a queue is, as a function of the arrival process (see Theorem 16.3, for example). Thus, some of our results can be viewed as being in the same spirit as work previously published by, for example, Stoyan [20] and Borovkov [7].

16.2 PARAMETRIC ESTIMATION OF TAIL PROBABILITIES FOR THE M/M/1 QUEUE

In this section, we will offer an analysis of the basic estimation issues in the context of the simplest possible (interesting) single-server queueing model, namely the M/M/1 queue. To illustrate our basic theory, we choose to focus on the workload process $W = (W(t): t \geq 0)$ associated with the single-server queue.

To be more specific, let $N = (N(t): t \geq 0)$ be a Poisson process running at rate λ^* and let $V = (V_n: n \geq 1)$ be an independent sequence of i.i.d. exponential random variables having mean $1/\mu^*$. We can interpret $N(t)$ as the total number of customers to arrive to the queue in $[0, t]$, whereas V_n can be viewed as the processing time of the nth customer. Assuming the queue starts off with no work present, it follows that

$$W(t) = S(t) - \min_{0 \leq u \leq t} S(u),$$

where

$$S(t) = \sum_{i=1}^{N(t)} V_i - t.$$

If $\rho \equiv \lambda^* / \mu^*$, it is well known (see, for example, Asmussen [4], pg. 96) that

$$W(t) \Rightarrow W(\infty)$$

as $t \to \infty$, where

$$P(W(\infty) \geq y) = \rho \exp\left(-(\mu^* - \lambda^*)y\right) \tag{16.1}$$

for $y \geq 0$. Thus, if λ^* and μ^* are known, (16.1) offers a complete solution as to how to compute the steady-state probability that the workload in the M/M/1 queue exceeds y.

However, in many real-world settings, one may have imperfect knowledge of the statistical characteristics of the source that is generating arrivals to the system. In the M/M/1 queue context, this might translate into λ^* and/or μ^* being unknown *a priori*. To fix our ideas, let us suppose that only λ^* is unknown and that we wish to predict the long-run fraction of time that the workload will exceed y. By observing the arrival process N over $[0, t]$, we can estimate the arrival rate λ^* via the empirically observed rate

$$\lambda(t) \equiv \frac{1}{t} N(t).$$

($\lambda(t)$ is, in fact, the maximum likelihood estimator of λ^*.) Based on (16.1), it therefore appears reasonable to compute our prediction for the long-run fraction of time that the workload will exceed y via

$$\alpha(t; y) \equiv g(\lambda(t), y),$$

where

$$g(\lambda, y) = \frac{\lambda}{\mu^*} \exp\left(-(\mu^* - \lambda)y\right).$$

The asymptotic theory associated with maximum likelihood estimation suggests that our estimator $\alpha(t; y)$ for $\alpha(y) \overset{\triangle}{=} P(W(\infty) \geq y)$ is, in some sense, asymptotically optimal; see for example, Ibragimov and Has'minskii [12].

A natural question that arises here is the extent to which uncertainty in our estimate of λ^* propagates into uncertainty in our prediction of $\alpha(y)$. But note that

$$\alpha(t; y) - \alpha(y) = g(\lambda(t), y) - g(\lambda^*, y)$$

$$\approx g'(\lambda^*, y)(\lambda(t) - \lambda^*)$$

$$= (1 + \lambda^* y) \exp\left(-(\mu^* - \lambda^*)y\right)/\mu^* \cdot (\lambda(t) - \lambda^*).$$

Since $\lambda(t) - \lambda^*$ is asymptotically normal, this suggests the following result.

Proposition 16.1. *If $\rho < 1$, then*

$$t^{1/2}(\alpha(t; y) - P(W(\infty) \geq y)) \Rightarrow \sigma(y) N(0, 1),$$

where

$$\sigma^2(y) = \rho(1 + \lambda^* y)^2 \exp\left(-2(\mu^* - \lambda^*)y\right)/\mu^*.$$

A rigorous proof is straightforward; see, for example, Serfling [18].

Suppose now that we are particularly interested in computing the fraction of time that the steady-state workload exceeds y, where y is large. More specifically, suppose that we wish to offer such a prediction that is valid to a high level of relative accuracy. Thus, we are concerned with the magnitude of $\alpha(t;y)/P(W(\infty) \geq y)$ when y is large.

Our next result addresses this issue.

Proposition 16.2. *Suppose $\rho < 1$.*

If $t_y/y^2 \to +\infty$, then

$$\frac{\sqrt{t_y}}{y}\left(\frac{\alpha(t_y;y)}{\alpha(y)} - 1\right) \Rightarrow \sqrt{\lambda^*}N(0,1) \tag{16.2}$$

as $y \to \infty$.
If $t_y/y^2 \to \gamma > 0$, then

$$\frac{\alpha(t_y;y)}{\alpha(y)} \Rightarrow \exp\left(\sqrt{\frac{\lambda^*}{\gamma}}N(0,1)\right) \tag{16.3}$$

as $y \to \infty$.

Proof. Note that

$$\alpha(t;y)/\alpha(y) = g(\lambda(t),y)/g(\lambda^*,y) \tag{16.4}$$

$$= \frac{\lambda(t)}{\lambda^*}\exp\left((\lambda(t) - \lambda^*)y\right). \tag{16.5}$$

Assuming $t_y/y^2 \to +\infty$, we obtain

$$g(\lambda(t_y),y)/\alpha(y) = \frac{\lambda(t_y)}{\lambda^*}\left(1 + \exp\left(\xi(t_y)\right)\right)(\lambda(t_y) - \lambda^*)y$$

$$= (1 + O_p(t_y^{-1/2}))(1 + \exp\left(\xi(t_y)\right))(\lambda(t_y) - \lambda^*)y,$$

where $\xi(t_y) \Rightarrow 0$ as $y \to \infty$. Hence,

$$\frac{\sqrt{t_y}}{y}\left(\frac{\alpha(t_y,y)}{\alpha(y)} - 1\right) = O_p(y^{-1}) + \exp\left(\xi(t_y)\right)(1 + O_p(t_y^{-1/2}))\sqrt{t_y}(\lambda(t_y) - \lambda^*).$$

Since $t^{1/2}(\lambda(t) - \lambda^*) \Rightarrow \sqrt{\lambda^*}N(0,1)$ as $t \to \infty$, (16.2) follows. For (16.3), observe that if $t_y/y^2 \to \gamma > 0$, then

$$(\lambda(t_y) - \lambda^*)y \Rightarrow \sqrt{\frac{\lambda^*}{\gamma}}N(0,1)$$

as $y \to \infty$. The result is then immediate from (16.5).   □

Proposition 16.2 makes clear that in order to predict the workload tail probability when y is large, the arrival process N must be observed for a period of time t_y that is large relative to y^2. Without observing N over such a time scale, the tail probability cannot be reasonably estimated to a given degree of relative accuracy. Thus, even under the stringent M/M/1 assumptions made here, the amount of time over which the arrival process must be observed to make such workload predictions increases rapidly as a function of y (with the critical time scale being of order y^2.)

16.3 PARAMETRIC ESTIMATION OF TAIL PROBABILITIES FOR MARKOV MODULATED QUEUES

We study here the issue of the degree to which the results of Section 16.2 generalize to Markov-modulated queues. Such queues are widely viewed as offering the modeling flexibility required to cope with the complexity of traffic sources typical of modern communications networks.

To describe such a queue, let $X = (X(t): t \geq 0)$ be a continuous-time Markov chain living on a finite state space S. The infinitesimal generator of X is assumed to belong to the set $\{\bar{A}(\theta): a < \theta < b\}$; we denote the "true" value of $\theta \in (a, b)$ as θ^*. We require $\bar{A}(\theta^*)$ to be irreducible; we further assume that $\bar{A}(\cdot)$ is continuously differentiable.

For $f: S \to (0, \infty)$, let

$$\Lambda(t) = \int_0^t f(X(s)) ds; \tag{16.6}$$

we interpret $\Lambda(t)$ as the total work to arrive to the queue in $[0, t]$. As (16.6) suggests, we view the work as arriving in "fluid" form. Set

$$S(t) = \int_0^t f(X(s)) ds - t.$$

Then, assuming that the queue starts off with no work present, the workload process $W = (W(t): t \geq 0)$ is given by

$$W(t) = S(t) - \min_{0 \leq u \leq t} S(u).$$

Let $\bar{P}_\theta(\cdot)$ be the probability measure on the path-space of X, under which X has generator $\bar{A}(\theta)$ and initial distribution $\pi(\theta)$, where $\pi(\theta)$ is the (unique) stationary distribution associated with $\bar{A}(\theta)$. Then,

$$\bar{P}_\theta(S(t) - \min_{0 \leq u \leq t} S(u) \in \cdot) = \bar{P}_\theta(\max_{0 \leq u \leq t} (\int_u^t [f(X(s)) - 1] ds) \in \cdot) \tag{16.7}$$

$$= \bar{P}_\theta(\max_{0 \leq u \leq t} (\int_0^u [f(X(t-s)) - 1] ds) \in \cdot) \tag{16.8}$$

$$= P_\theta(\max_{0 \leq u \leq t} (\int_0^u [f(X(s)) - 1] ds) \in \cdot) \tag{16.9}$$

where $P_\theta(\cdot)$ is the probability measure on the path-space of X, under which X has initial distribution $\pi(\theta)$ and generator $A(\theta) = (A(\theta, x, y): x, y \in S)$, where $A(\theta, x, y) = \bar{A}(\theta, y, x) \pi(\theta, y) / \pi(\theta, x)$.   (Thus, P_θ describes the time-reversed dynamics of X under $\bar{P}_\theta(\cdot)$).

Let $M = (M(t): t \geq 0)$ be the maximum process defined via

$$M(t) \equiv \max_{0 \leq u \leq t} S(u).$$

Relations (16.7)-(16.9) assert that

$$\bar{P}_\theta(W(t) \in \cdot) = P_\theta(M(t) \in \cdot).$$

Since M is nondecreasing, it follows that $M(t) \uparrow M(\infty)$ a.s., and consequently,

$$\bar{P}_\theta(W(t) \in \cdot) \Rightarrow P_\theta(M(\infty) \in \cdot)$$

as $t \to \infty$. It follows that if one knows θ^*, one can predict the long-run fraction of time that the workload exceeds y via the equation

$$\alpha(y) \equiv P_{\theta^*}(M(\infty) \geq y).$$

Our interest here is, however, in the case where the arrival source is imperfectly known. In particular, suppose that we have only an estimator $\theta(t)$ for θ^* available, rather than θ^* itself. (If X itself were observed under $\bar{P}_{\theta^*}$, one can apply standard maximum likelihood estimation methods to construct such an estimator $\theta(t)$; see, for example, Billingsley [6]. If f is not one-to-one and only $(f(X(s)): 0 \leq s \leq t)$ is observed, then more complicated estimators of θ^* would need to be considered, as described in Bickel and Ritov [5] and Ryden [17]). Our prediction for the probability that the workload exceeds y in steady-state is then

$$\alpha(t; y) = g(\theta(t), y),$$

where

$$g(\theta, y) = P_\theta(M(\infty) \geq y). \tag{16.10}$$

In order that $M(\infty)$ have a proper distribution (and hence $W(\infty)$) under the true parameter θ^*, we need to assume that the long-run rate at which work enters the system is less than one. Specifically, we shall assume that $\rho(\theta^*) < 1$, where

$$\rho(\theta) \triangleq \sum_{x \in S} \pi(\theta, x) f(x).$$

Set $T(y) = \inf\{t \geq 0 : S(t) \geq y\}$. From (16.10), it is evident that we can express $g(\theta, y)$ as a "level-crossing" probability, namely

$$g(\theta, y) = P_\theta(T(y) < \infty).$$

In order to proceed further, we shall now invoke a "change-of-measure" argument. Let D be the diagonal matrix defined by $D = \text{diag}(f(x) : x \in S)$, and let

$$A(\theta, \gamma) = A(\theta) + \gamma(D - I)$$

for $\gamma \in \mathfrak{R}$. Because $A(\theta, \gamma)$ is an irreducible M-matrix, it follows that there exists a strictly positive column eigenvector $h(\theta, \gamma)$ and a real eigenvalue $r(\theta, \gamma)$ such that

$$A(\theta, \gamma) h(\theta, \gamma) = r(\theta, \gamma) h(\theta, \gamma).$$

Using a similar argument to that presented on page 126 of Bucklew [9] for discrete-time Markov chains, we can prove that $r(\theta, \cdot)$ is convex with $r(\theta, \gamma) \to +\infty$ as $\gamma \to +\infty$. Furthermore, $\frac{\partial}{\partial \gamma} r(\theta, \gamma)|_{\gamma=0} < 0$ for θ in a neighborhood of θ^*. Hence, it is clear that there exists a positive root $\gamma(\theta)$ to the equation

$$r(\theta, \gamma) = 0.$$

Set

$$B(\theta, x, y) = A(\theta, \gamma(\theta), x, y) \frac{h(\theta, \gamma(\theta), y)}{h(\theta, \gamma(\theta), x)}$$

and note that $B(\theta) = (B(\theta, x, y) : x, y \in S)$ is the generator of an irreducible continuous-time Markov chain on S. Furthermore,

$$A(\theta, x, y) = \frac{h(\theta, \gamma(\theta), x)}{h(\theta, \gamma(\theta), y)} K(\theta, x, y),$$

where $K(\theta) = B(\theta) - \gamma(\theta)(D - I)$. It is trivial to verify that

$$(\exp(A(\theta)t))(x,y) = \frac{h(\theta,\gamma(\theta),x)}{h(\theta,\gamma(\theta),y)}(\exp(K(\theta)t))(x,y). \tag{16.11}$$

Set $u_\theta(t,x,y) = (\exp(K(\theta)t))(x,y)$ and note that $u_\theta(t) = (u_\theta(t,x,y): x,y \in S)$ solves the linear system of ordinary differential equations defined by

$$u_\theta'(t) = (B(\theta) - \gamma(\theta)(D-I))u_\theta(t) \tag{16.12}$$

$$u_\theta(0) = I.$$

Let $\underline{E}_\theta(\cdot)$ and $\underline{P}_\theta(\cdot)$ be the expectation operator and probability distribution on the path-space of X associated with the generator $B(\theta)$ and initial distribution $\pi(\theta)$. It is well known that the solution to (16.12) can be expressed probabilistically as

$$u_\theta(t,x,y) = \underline{E}_\theta[\exp(-\gamma(\theta)S(t))I(X(t)=y) \mid X(0)=x]. \tag{16.13}$$

Combining (16.11) and (16.13), we conclude that

$$P_\theta(X(t)=y \mid X(0)=x) = \underline{E}_\theta[\exp(-\gamma(\theta)S(t))I(X(t)=y)$$

$$\cdot \frac{h(\theta,\gamma(\theta),X(0))}{h(\theta,\gamma(\theta),X(t))} \mid X(0)=x]. \tag{16.14}$$

Letting $E_\theta(\cdot)$ be the expectation operator associated with $P_\theta(\cdot)$, it is straightforward to extend (16.14) to the identity

$$E_\theta\zeta = \underline{E}_\theta\zeta \exp(-\gamma(\theta)S(t))\frac{h(\theta,\gamma(\theta),X(0))}{h(\theta,\gamma(\theta),X(t))}, \tag{16.15}$$

for any non-negative r.v. ζ that is measurable with respect to $\sigma(X(s): 0 \le s \le t)$. Since $T(y)$ is a stopping time, (16.15) in turn easily implies the identity

$$P_\theta(T(y) < \infty) = \underline{E}_\theta \exp(-\gamma(\theta)S(T(y)))\frac{h(\theta,\gamma(\theta),X(0))}{h(\theta,\gamma(\theta),X(T(y)))}I(T(y)<\infty)$$

$$= \exp(-\gamma(\theta)y)\underline{E}_\theta\frac{h(\theta,\gamma(\theta),X(0))}{h(\theta,\gamma(\theta),X(T(y)))}I(T(y)<\infty).$$

Let $\nu(\theta) = (\nu(\theta,x): x \in S)$ be the (unique) stationary distribution associated with $B(\theta)$. Again, by following an argument similar to that given in Bucklew [9] for discrete-time chains, one can show that

$$\frac{\partial}{\partial\gamma}r(\theta,\gamma) \mid_{\gamma=\gamma(\theta)} = \sum_{x \in S}(f(x)-x)\nu(\theta,x).$$

Since $\frac{\partial}{\partial\gamma}r(\theta,\gamma)$ must clearly be positive at the root $\gamma(\theta)$, it follows that $S(\cdot)$ must have positive drift under $\underline{P}_\theta$ for θ in a neighborhood of θ^*. Consequently, we may simplify (16.16) to the identity

$$P_\theta(T(y) < \infty) = \exp(-\gamma(\theta)y)\underline{E}_\theta\frac{h(\theta,\gamma(\theta),X(0))}{h(\theta,\gamma(\theta),X(T(y)))}.$$

In other words,

$$g(\theta,y) = \exp(-\gamma(\theta)y)\underline{E}_\theta\frac{h(\theta,\gamma(\theta),X(0))}{h(\theta,\gamma(\theta),X(T(y)))}. \tag{16.16}$$

Our analysis of Section 16.2 suggests that the derivative $g'(\theta^*,y)$ plays an important role in the estimation theory associated with predicting $\alpha(y)$. Our next results proves that $g(\cdot,y)$ is smooth. It can be viewed as a strengthening of the various "continuity" results available for the single-server queue (see, for example, page 194 of Asmussen [4]) to "differentiability". Set $\Gamma(\theta) = \{(x,y): \bar{a}(\theta, x, y) \neq 0\}$ and $h(\theta, x) = h(\theta, \gamma(\theta), x)$. Also, let $J(t)$ denote the number of jumps of X over $[0, t]$, and let $Y = (Y_n : n \geq 0)$ denote the embedded discrete-time Markov chain associated with X.

Theorem 16.3. *Assume that $\rho(\theta^*) < 1$ and that $\bar{A}(\theta^*)$ is irreducible. If $\bar{A}(\cdot)$ is continuous differentiable in a neighborhood of θ^* and $\Gamma(\theta) = \Gamma(\theta^*)$ in a neighborhood of θ^*, then $g(\cdot, y)$ is differentiable at θ^* for each $y \geq 0$. Furthermore,*

$$g'(\theta^*, y) = \exp(-\gamma(\theta^*)y) \cdot E_{\theta^*} \frac{h(\theta^*, X(0))}{h(\theta^*, X(T(y)))}$$

$$\cdot \left[\frac{\pi'(\theta^*, X(0))}{\pi(\theta^*, X(0))} + \sum_{i=0}^{J(T(y))-1} \frac{A'(\theta^*, Y_i, Y_{i+1})}{A(\theta^*, Y_i, Y_{i+1})} + \int_0^{T(y)} A'(\theta^*, X(s), Xs)) ds \right]. \quad (16.17)$$

Proof. Our "absolute continuity" condition involving the set $\Gamma(\cdot)$ implies the existence of a likelihood ratio process $(L(\theta, t): t \geq 0)$ such that

$$g(\theta, y) = \exp(-\gamma(\theta)y) E_\theta \frac{h(\theta, X(0))}{h(\theta, X(T(y)))} L(\theta, T(y));$$

see Brémaud [8]. Setting $\lambda(\theta, x) = -B(\theta, x, x), L(\theta, t)$ is given by

$$L(\theta, t) = \frac{\pi(\theta, X(0))}{\pi(\theta^*, X(0))} \prod_{i=0}^{J(t)-1} \frac{B(\theta, Y_i, Y_{i+1})}{B(\theta^*, Y_i, Y_{i+1})}$$

$$\cdot \exp\left(-\int_0^t [\lambda(\theta, X(s)) - \lambda(\theta^*, X(s))] ds\right).$$

But

$$B(\theta, Y_i, Y_{i+1}) = A(\theta, Y_i, Y_{i+1}) \frac{h(\theta, Y_{i+1})}{h(\theta, Y_i)}$$

and

$$\int_0^t \lambda(\theta, X(s)) ds = -\left(\int_0^t A(\theta, X(s), X(s)) ds + \gamma(\theta) S(t)\right).$$

Thus,

$$g(\theta, y) = \exp(-\gamma(\theta^*)y) E_\theta \beta(\theta),$$

where

$$\beta(\theta) \triangleq \exp((\gamma(\theta^*) - \gamma(\theta))y) \frac{h(\theta, X(0))}{h(\theta, X(T(y)))} L(\theta, T(y)) \quad (16.18)$$

$$= \frac{\pi(\theta, X(0))}{\pi(\theta^*, X(0))} \frac{h(\theta^*, X(0))}{h(\theta^*, X(T(y)))} \prod_{i=0}^{J(t)-1} \frac{A(\theta, Y_i, Y_{i+1})}{A(\theta^*, Y_i, Y_{i+1})}$$

$$\cdot \exp\left(\int_0^{T(y)} [A(\theta, X(s), X(s)) - A(\theta^*, X(s), X(s))] ds\right). \quad (16.19)$$

Consequently, assuming that we can interchange the derivative and expectation operators, it follows that $g'(\theta, y) = \exp(-\gamma(\theta^*)y) E_{\theta^*} \beta'(\theta)$, where

$$\frac{\beta'(\theta)}{\beta(\theta)} = \frac{d}{d\theta} \log \beta(\theta) \tag{16.20}$$

$$= \frac{\pi'(\theta, X(0))}{\pi(\theta, X(0))} + \sum_{i=0}^{J(T(y))-1} \frac{A'(\theta, Y_i, Y_{i+1})}{A(\theta, Y_i, Y_{i+1})} + \int_0^{T(y)} A'(\theta, X(s), X(s)) ds. \tag{16.21}$$

Since $\beta(\theta^*) = h(\theta^*, X(0))/h(\theta^*, X(T(y)))$, this leads immediately to (16.17). So, it remains to justify the interchange of derivative and expectation.

We need to show that the difference quotient

$$D(h) = \frac{\beta(\theta + h) - \beta(\theta)}{h}$$

can be dominated (uniformly in h small) by a $\underline{P}_{\theta^*}$ integrable r.v. Note that $r(\cdot)$ is continuous [13] and $r(\theta, \cdot)$ is convex. Hence, $\gamma(\cdot)$ is continuous in a neighborhood of θ^*, as is $h(\cdot, x)$ for $x \in S$. It is then straightforward to show that for each $\epsilon > 0$, there exist (deterministic) positive constants h_0 and c such that

$$\sup_{|h| < h_0} |D(h)| \le \sup_{|h| < h_0} |\beta'(h)| \tag{16.22}$$

$$\le c \exp(\epsilon T(y) + \epsilon J(T(y))). \tag{16.23}$$

To show that the right-hand side of (16.23) is integrable, we use a martingale argument. Recall that

$$A(\theta^*, \gamma(\theta^*)/2) h(\theta^*, \gamma(\theta^*)/2) = r(\theta^*, \gamma(\theta^*)/2) h(\theta^*, \gamma(\theta^*)/2).$$

In other words,

$$[B(\theta^*) - (\gamma(\theta^*)/2)(D - I)]k = \eta k,$$

where $k(x) = h(\theta^*, \gamma(\theta^*)/2, x)/h(\theta^*, \gamma(\theta^*), x)$ and $\eta = r(\theta^*, \gamma(\theta^*)/2)$. Let $C = (C(x,y): x,y \in S)$ be defined by

$$C(x,y) = B(\theta^*, x, y) = [\gamma(\theta^*)(f(x)-1)/2 + \eta]\delta_{xy},$$

and note that $(C(x,y)k(y)/k(x): x,y \in S)$ is the generator of a continuous-time Markov chain in S. Hence,

$$\sum_y \exp(Ct)(x,y) \cdot \frac{k(y)}{k(x)} = 1.$$

As in the derivation leading to (16.13), this can be written probabilistically as

$$\underline{E}_{\theta^*}[\exp(-\gamma(\theta^*)S(t)/2 - \eta t)\frac{k(X(t))}{k(X(0))}] = 1.$$

It follows easily that

$$\exp(-\gamma(\theta^*)S(t)/2 - \eta t)\frac{k(X(t))}{k(X(0))}$$

is a $\underline{P}_{\theta}^*$-martingale. The optional sampling theorem then yields

$$\underline{E}_{\theta^*}M(t \wedge T(y)) = 1$$

for $t \ge 0$. But $S(t \wedge T(y)) \le y$ and $\gamma(\theta^*) > 0$, so evidently,

$$M(t \wedge T(y)) \ge \exp(-\gamma(\theta^*)y/2) \inf_{x,z}\frac{k(x)}{k(z)} \cdot \exp(-\eta(t \wedge T(y))).$$

Hence,

$$\underline{E}_{\theta^*}\exp\left(-\eta(t\wedge T(y))\right)\le\exp\left(\gamma(\theta^*)y/2\right)\sup_{x,z}\frac{k(x)}{k(z)}.$$

Since $r(\theta^*,0)=r(\theta^*,\gamma(\theta^*))=0$ and $r(\theta^*,\cdot)$ is convex, we can infer that $\eta<0$. The monotone convergence theorem then establishes the finiteness of $\underline{E}_{\theta^*}\exp\left(-\eta T(y)\right)$, yielding the existence of a positive ϵ for which $\underline{E}_{\theta^*}\exp\left(\epsilon T(y)\right)$ is finite.

To handle $\exp\left(\epsilon J(T(y))\right)$, note that for $\nu>0$,

$$\underline{E}_{\theta^*}\exp\left(\nu J(T(y))\right)=\sum_{n=0}^{\infty}\underline{E}_{\theta^*}\exp\left(\nu J(T(y))\right)I(n\le T(y)<n+1) \qquad (16.24)$$

$$\le\sum_{n=0}^{\infty}\underline{E}_{\theta^*}\exp\left(\nu J(N+1)\right)I(T(y)\ge n) \qquad (16.25)$$

$$\le\sum_{n=0}^{\infty}\underline{E}_{\theta^*}^{1/2}\exp\left(2\nu J(n+1)\right)\underline{P}_{\theta^*}^{1/2}(T(y)\ge n). \qquad (16.26)$$

But $J(\cdot)$ can be stochastically dominated by a Poisson process having rate $\widetilde{\lambda}\triangleq\max\left(\lambda(\theta^*,x):x\in S\right)$, and so

$$\underline{E}_{\theta^*}^{1/2}\exp\left(2\nu J(n+1)\right)\le\exp\left(\widetilde{\lambda}\,(n+1)(e^{2\nu}-1)/2\right).$$

The finiteness of $\underline{E}_{\theta^*}\exp\left(\epsilon T(y)\right)$ then guarantees the finiteness of $\underline{E}_{\theta^*}\exp\left(\nu J(T(y))\right)$ for some positive ν. The Cauchy-Schwartz inequality then yields (16.23) for some suited chosen positive ϵ. $\qquad\Box$

Theorem 16.3 proves that $g(\cdot,y)$ is differentiable for each $y\ge0$, and offers a representation for the derivative that is surprisingly simple. In particular, note that the function f plays no explicit role in the expression for $g'(\theta^*,y)$ (although it does serve to implicitly define the expectation $\underline{E}_{\theta^*}(\cdot)$ under which the expectation is computed).

Assume now that θ^* is unknown and that it must be statistically estimated from data collected over the time interval $[0,t]$. Let $\theta(t)$ be the corresponding estimator for θ^* Most well-behaved estimators obey a central limit theorem (CLT) of the form

$$t^{1/2}(\theta(t)-\theta^*)\Rightarrow\delta N(0,1) \qquad (16.27)$$

as $t\to\infty$, for some suitably defined constant $\delta>0$; see, for example, the extensive central limit theory available for maximum likelihood estimators in [12].

Corollary 16.4. *Assume $(\theta(t):t\ge0)$ obeys (27). Then, under the conditions of Theorem 16.3, it follows that*

$$t^{1/2}(\alpha(t;y)-\alpha(y))\Rightarrow g'(\theta^*,y)\delta N(0,1)$$

as $t\to\infty$.

Corollary 16.4 is the Markov-modulated analog of Proposition 16.1.

We turn now to the analog of Proposition 16.2, developed earlier in the M/M/1 context. Equation (16.16) suggests that the r.v. $X(T(y))$ plays an important role in determining the behavior of $g(\theta,y)$. With this in mind, note that $S(t)\to+\infty$ P_θ a.s. for all θ in a neighborhood of θ^* and hence $T(y)<\infty$ P_θ a.s. for all $y\ge0$ under such values of θ. Set $\underline{X}(y)=X(T(y))$ for $y\ge0$. Then, $\underline{X}=(\underline{X}(y):y\ge0)$ is a well-defined stochastic process under P_θ for θ in a neighborhood of θ^*. Furthermore, it turns out that $\underline{X}$ is itself a time-homogeneous continuous-time Markov chain with a single closed communicating class $\underline{S}\subset S$; [2, 3, 16].

Let $\underline{\pi}(\theta) = (\pi(\theta, x): x \in S)$ be the (unique) stationary distribution of $\underline{X}$ under $\underline{P}_\theta$. Since $\underline{X}(y)$ is asymptotically independent of $\underline{X}(0)$, this suggests the following approximation for $g(\theta, y)$, derived from (16.16):

$$g(\theta, y) \approx \exp\left(-\gamma(\theta)y\right) \cdot \sum_{x \in S} \pi(\theta, x) h(\theta, \gamma(\theta), x) \cdot \underline{\pi}(\theta, z) / h(\theta, \gamma(\theta), z)$$

$$\stackrel{\Delta}{=} \exp\left(-\gamma(\theta)y\right) \cdot v(\theta).$$

From the above approximations, the conclusions of Theorem 16.5 are intuitively clear.

Theorem 16.5. *Assume the condition of Theorem 16.3 and suppose that $(\theta(t): t \geq 0)$ satisfies (27). If $t_y/y^2 \to +\infty$ and $y/\log t_y \to +\infty$, then*

$$\frac{\sqrt{t_y}}{y}\left(\frac{\alpha(t_y; y)}{\alpha(y)} - 1\right) \Rightarrow \delta\frac{v'(\theta^*)}{v(\theta^*)} N(0, 1)$$

as $y \to \infty$. If $t_y/y^2 \to \gamma > 0$, then

$$\frac{\alpha(t_y; y)}{\alpha(y)} \Rightarrow \exp\left(\delta\gamma'(\theta^*)\gamma^{-1/2}N(0, 1)\right)$$

as $y \to \infty$.

Proof. Our starting point is (16.16). Let $v(\theta, y) = \underline{E}_\theta h(\theta, \gamma(\theta), X(0))/h(\theta, \gamma(\theta), X(T(y)))$. We need to show that replacing $v(\theta, y)$ by $v(\theta)$ holds uniformly in a neighborhood of θ^* as $y \to +\infty$. To this end, we write

$$v(\theta, y) = \sum_{x, z} \pi(\theta, x) h(\theta, \gamma(\theta), x) \underline{P}_\theta(\underline{X}(y) = z \mid X(0) = x) \cdot h(\theta, \gamma(\theta), z)^{-1}.$$

We will show that for ϵ sufficiently small, there exists $a > 0$ such that

$$\sup_{|\theta - \theta^*| < \epsilon} |\underline{P}_\theta(\underline{X}(y) = z \mid \underline{X}(0) = x) - \underline{\pi}(\theta, z)| = O(e^{-ay}) \tag{16.28}$$

as $y \to \infty$. The finiteness of the state space then guarantees that

$$\sup_{|\theta - \theta^*| < \epsilon} |v(\theta, y) - v(\theta)| = O(e^{-by}) \tag{16.29}$$

as $y \to \infty$, for some suitably chosen $b > 0$. To verify (16.28), we invoke a "uniform coupling" argument. Note that

$$\underline{P}_\theta(\underline{X}(y) = z \mid \underline{X}(0) = x) = \underline{P}_\theta(X(T(y)) = z \mid X(0) = x).$$

The same argument as that applied in the proof of Theorem 16.3 establishes that the right-hand side is differentiable at θ^* for each $y \geq 0$. Now,

$$\inf_{\substack{x \in S \\ z \in \underline{S}}} \underline{P}_\theta^*(X(T(1)) = z \mid X(0) = x) > 0$$

follows as a consequence of the fact that $\underline{X}$ is a continuous-time Markov chain on finite state space. Because the transition probabilities are differentiable at θ^*, there therefore exists $\epsilon > 0$ such that

$$\inf_{\substack{x \in S \\ z \in \underline{S} \\ |\theta - \theta^*| < \epsilon}} \underline{P}_\theta(\underline{X}(1) = z \mid \underline{X}(0) = x) \stackrel{\Delta}{=} m > 0.$$

A standard coupling argument (see, for example, Lindvall [15]) then proves that

$$\inf_{\substack{x \in S \\ z \in S \\ |\theta - \theta^*| < \epsilon}} |\, \underline{P}_\theta(\underline{X}(y) = z \mid \underline{X}(0) = x) - \underline{\pi}\,(\theta, z)\,| \le (1 - c)^{\lfloor y \rfloor},$$

yielding (16.28). In view of (16.16) and (16.29), we can write

$$\frac{\alpha(t_y; y)}{\alpha(y)} = \exp\big(-(\gamma(\theta(t)) - \gamma(\theta^*))y\big) \cdot \left\{ \frac{v(\theta(t))}{v(\theta^*)} + O(e^{-by}) \right\}.$$

The conclusions of Theorem 16.5 then follow from precisely the same style of argument as that used for Proposition 16.2, provided only that we show $v(\cdot)$ to be differentiable at θ^*. (Note that the additional hypothesis that $y/\log t_y \to +\infty$ serves only to guarantee that $O(e^{-by})$ is of (much) smaller magnitude than $t_y^{-1/2}$, the magnitude of $\theta(t_y) - \theta^*$, thereby permitting us to discard $O(e^{-by})$ in our asymptotic analysis.)

To prove that $v(\cdot)$ is differentiable at θ^*, recall our earlier remark that the matrix of transition probabilities

$$(\underline{P}_\theta(\underline{X}(y) = z \mid \underline{X}(0) = x); x, z \in \underline{S})$$

is differentiable at θ^*. The stationary distribution $\underline{\pi}\,(\theta)$ can be viewed as the (unique) stationary distribution of the above transition matrix. Since the transition matrix is differentiable at θ^*, we may conclude that $\underline{\pi}\,(\cdot)$ is differentiable at θ^*; see Glynn [10]. It then trivially follows that $v(\cdot)$ is differentiable at θ^*.   □

Theorem 16.5 reinforces the qualitative insight gained from Section 16.2. If the arrival process is modeled parametrically, then it must be observed over a time scale t_y, which is large relative to y^2, in order that (relatively) accurate predictions of the tail workload probability $P_{\theta^*}(W(\infty) > y)$ be available.

16.4 FURTHER ASYMPTOTIC ANALYSIS

In Sections 2 and 3, we considered the question of how much data needs to be collected in order to parametrically estimate tail and workload probabilities in two related single-server models, as a function of the parameter y. In this section, we briefly consider the issue of how much data needs to be collected, as a function of the traffic intensity of the queue. We restrict our attention to the M/M/1 queue, because its parametrization lends itself naturally to such analysis (whereas, in the Markov-modulated setting, there is no obvious connection between θ and the traffic intensity of the queue).

The starting point is relation (16.5) of Section 16.2. Note that $\alpha(t; y)/\alpha(y)$ is independent of μ^*, so that the relative error associated with estimating $\alpha(y)$ is independent of μ^*. In particular, the relative error is (perhaps surprisingly) independent of how close μ^* is to λ^*, and thereby, independent of the traffic intensity. Of course, the magnitude of the workloads y that are of interest depend heavily on the traffic intensity ρ.

For example, as $\rho \to 1$ in "heavy traffic", it is clear that y must be of order $(1 - \rho)^{-1}$, in order that $P_{\theta^*}(W(\infty) \ge y)$ be held constant. The analysis of Section 16.16.2 then suggests that the time horizon t_y over which the arrival process should be observed ought to be large relative to $(1 - \rho)^{-2}$. Since $(1 - \rho)^{-2}$ is the time-scaling over which $W(\cdot)$ exhibits significant (relative) fluctuations, this indicates that the non-parametric estimator

$$\alpha_N(t;y) \triangleq \frac{1}{t} \int_0^t I(W(s) \geq y)ds$$

should also provide good estimation behavior over time horizons that are large relative to $(1-\rho)^{-2}$. In other words, parametric modeling of the arrival process does not have a dramatic effect on the ability to estimate tail workload probabilities when the value y being considered is in "the middle" of the distribution. Its primary impact is in estimation of the extreme tail.

16.5 OPEN RESEARCH PROBLEMS

A number of different research problems are suggested by the line of inquiry pursued in this chapter.

Given that this chapter focuses on parametric modeling of queues, an important question relates to the types of parametric models that lend themselves best to specific application contexts. A particularly intriguing variant on this issue is the question of how to effectively model complex dependent arrival processes via stochastic processes that are parametrized by low-dimensional vectors. Ideally, the parametrized class of processes chosen should give rise to a class of queues that is numerically tractable, and for which the parameter estimation can be implemented in a satisfactory fashion. Furthermore, one prefers, everything else equal, a parametrization in which each of the parameters entering the model has some physically intuitive meaning.

Given our emphasis in this chapter on parametric queueing models, it is natural to ask about how the results developed here would be affected by a non-parametric formulation. In particular, if one is unwilling to assume that the arrival process is well-described via some (finite-dimensional) parametric model, how long must one observe the arrival process in order to accurately estimate tail probabilities for (very) large buffer levels?

Turning now to the theory developed in this chapter, it is worth noting that computing the quantity $g(\theta(t), y)$ (our estimator for $\alpha(y)$), given $\theta(t)$, may be itself a highly non-trivial activity. For example, there is no closed-form for this quantity in the Markov-modulated context. Instead, one needs to compute $g(\theta(t), y)$ either via some conventional numerical solver, or via a Monte Carlo simulation of some kind. Given the real-time applications that motivate this class of problem, the issue of how to rapidly compute $g(\theta(t), y)$ could potentially be of importance. In addition, if confidence intervals for $\alpha(y)$ are desired, rapid computation of $g'(\theta(t), y)$ (see Proposition 16.1 and Theorem 16.3) plays an important role. Such computational issues should ideally be considered in concert with the statistical questions that were central in this chapter.

BIBLIOGRAPHY

[1]   Anick, D., Mitra, D. and Sondhi, M.M., Stochastic theory of a data handling system with multiple sources, *Bell Sys. Tech. J.* **61** (1982), 1871-1894.

[2]   Asmussen, S., Stationary distributions for fluid flow models with or without

Brownian noise, *Stochastic Models* **11**:1 (1995), 21-49.

[3]   Asmussen, S., Busy period analysis, rare events and transient behavior in fluid flow models, *J. of Appl. Math. and Stoch. Anal.* **7**:3 (1994), 269-299.

[4]   Asmussen, S., *Applied Probability and Queues*, Wiley, New York 1987.

[5]   Bickel, P.J. and Ritov, Y., Inference in hidden Markov models I. Local asymptotic normality in the stationary case, *Bernoulli* (1996), to appear.

[6]   Billingsley, P., *Statistical Inference for Markov Processes*, University of Chicago Press, Chicago 1961.

[7]   Borovkov, A.A., *Stochastic Processes in Queueing Theory*, Springer-Verlag, New York 1976.

[8]   Brémaud, P., *Point Processes and Queues: Martingale Dynamics*, Springer-Verlag, New York 1981.

[9]   Bucklew, J.A., *Large Deviations Techniques in Decision, Simulation, and Estimation*, Wiley, New York 1990.

[10] Glynn, P.W., Sensitivity analysis for the stationary probabilities of a Markov chain, In: *Proc. of the 4th Army Conference on Applied Math and Comp.* (1986), 917-932.

[11] Glynn, P.W. and Torres, M., Nonparametric estimation of tail probabilities for the single-server queue, Forthcoming technical report, Dept. of Operations Res., Stanford University, Stanford, CA 1996.

[12] Ibragimov, I.A. and Has'minskii, R.Z., *Statistical Estimation: Asymptotic Theory*, Springer-Verlag, New York 1981.

[13] Kato, T., *Perturbation Theory for Linear Operators*, Wiley, New York 1976.

[14] Kesidis, G., Walrand, J. and Chang, C.S., Effective bandwidths for multiclass Markov fluids and other ATM sources, *IEEE/ACM Trans. on Networking* **1**:4 (1993), 424-428.

[15] Lindvall, T., *Lectures on the Coupling Method*, Wiley, New York 1982.

[16] Rogers, C.L.G., Fluid models in queueing theory and Wiener-Hopf factorization of Markov chains, submitted.

[17] Ryden, T., Estimates for hidden Markov models, *Ann. Stat.* **22**:4 (1994), 1884-1945.

[18] Serfling, R.J., *Approximation Theorems of Mathematical Statistics*, Wiley, New York 1980.

[19] Stern, T. and Elwalid, E., Analysis of separable Markov-modulated rate models for information-handling systems, *Adv. Appl. Prob.* **23** (1991), 105-139.

[20] Stoyan, D., *Comparison Methods for Queues and Other Stochastic Models*, Wiley, New York 1983.

Index